PROGRESS IN BRAIN RESEARCH

VOLUME 174

MIND AND MOTION: THE BIDIRECTIONAL LINK BETWEEN THOUGHT AND ACTION

Other volumes in PROGRESS IN BRAIN RESEARCH

Volume 137: Spinal Cord Trauma: Regeneration, Neural Repair and Functional Recovery, by L. McKerracher, G. Doucet and S. Rossignol (Eds.) – 2002, ISBN 0-444-50817-1.

Volume 138: Plasticity in the Adult Brain: From Genes to Neurotherapy, by M.A. Hofman, G.J. Boer, A.J.G.D. Holtmaat, E.J.W. Van Someren, J. Verhaagen and D.F. Swaab (Eds.) – 2002, ISBN 0-444-50981-X.

Volume 139: Vasopressin and Oxytocin: From Genes to Clinical Applications, by D. Poulain, S. Oliet and D. Theodosis (Eds.) – 2002, ISBN 0-444-50982-8.

Volume 140: The Brain's Eye, by J. Hyönä, D.P. Munoz, W. Heide and R. Radach (Eds.) – 2002, ISBN 0-444-51097-4.

Volume 141: Gonadotropin-Releasing Hormone: Molecules and Receptors, by I.S. Parhar (Ed.) – 2002, ISBN 0-444-50979-8.

Volume 142: Neural Control of Space Coding, and Action Production, by C. Prablanc, D. Pélisson and Y. Rossetti (Eds.) – 2003, ISBN 0-444-509771.

Volume 143: Brain Mechanisms for the Integration of Posture and Movement, by S. Mori, D.G. Stuart and M. Wiesendanger (Eds.) – 2004, ISBN 0-444-513892.

Volume 144: The Roots of Visual Awareness, by C.A. Heywood, A.D. Milner and C. Blakemore (Eds.) – 2004, ISBN 0-444-50978-X.

Volume 145: Acetylcholine in the Cerebral Cortex, by L. Descarries, K. Krnjević and M. Steriade (Eds.) – 2004, ISBN 0-444-51125-3.

Volume 146: NGF and Related Molecules in Health and Disease, by L. Aloe and L. Calzà (Eds.) – 2004, ISBN 0-444-51472-4.

Volume 147: Development, Dynamics and Pathology of Neuronal Networks: From Molecules to Functional Circuits, by J. Van Pelt, M. Kamermans, C.N. Levelt, A. Van Ooyen, G.J.A. Ramakers and P.R. Roelfsema (Eds.) – 2005, ISBN 0-444-51663-8.

Volume 148: Creating Coordination in the Cerebellum, by C.I. De Zeeuw and F. Cicirata (Eds.) – 2005, ISBN 0-444-51754-5.

Volume 149: Cortical Function: A View from the Thalamus, by V.A. Casagrande, R.W. Guillery and S.M. Sherman (Eds.) – 2005, ISBN 0-444-51679-4.

Volume 150: The Boundaries of Consciousness: Neurobiology and Neuropathology, by Steven Laureys (Ed.) – 2005, ISBN 0-444-51851-7.

Volume 151: Neuroanatomy of the Oculomotor System, by J.A. Büttner-Ennever (Ed.) – 2006, ISBN 0-444-51696-4.

Volume 152: Autonomic Dysfunction after Spinal Cord Injury, by L.C. Weaver and C. Polosa (Eds.) – 2006, ISBN 0-444-51925-4.

Volume 153: Hypothalamic Integration of Energy Metabolism, by A. Kalsbeek, E. Fliers, M.A. Hofman, D.F. Swaab, E.J.W. Van Someren and R.M. Buijs (Eds.) – 2006, ISBN 978-0-444-52261-0.

Volume 154: Visual Perception, Part 1, Fundamentals of Vision: Low and Mid-Level Processes in Perception, by S. Martinez-Conde, S.L. Macknik, L.M. Martinez, J.M. Alonso and P.U. Tse (Eds.) – 2006, ISBN 978-0-444-52966-4.

Volume 155: Visual Perception, Part 2, Fundamentals of Awareness, Multi-Sensory Integration and High-Order Perception, by S. Martinez-Conde, S.L. Macknik, L.M. Martinez, J.M. Alonso and P.U. Tse (Eds.) – 2006, ISBN 978-0-444-51927-6.

Volume 156: Understanding Emotions, by S. Anders, G. Ende, M. Junghofer, J. Kissler and D. Wildgruber (Eds.) – 2006, ISBN 978-0-444-52182-8.

Volume 157: Reprogramming of the Brain, by A.R. Møller (Ed.) – 2006, ISBN 978-0-444-51602-2.

Volume 158: Functional Genomics and Proteomics in the Clinical Neurosciences, by S.E. Hemby and S. Bahn (Eds.) – 2006, ISBN 978-0-444-51853-8.

Volume 159: Event-Related Dynamics of Brain Oscillations, by C. Neuper and W. Klimesch (Eds.) – 2006, ISBN 978-0-444-52183-5.

Volume 160: GABA and the Basal Ganglia: From Molecules to Systems, by J.M. Tepper, E.D. Abercrombie and J.P. Bolam (Eds.) – 2007, ISBN 978-0-444-52184-2.

Volume 161: Neurotrauma: New Insights into Pathology and Treatment, by J.T. Weber and A.I.R. Maas (Eds.) – 2007, ISBN 978-0-444-53017-2.

Volume 162: Neurobiology of Hyperthermia, by H.S. Sharma (Ed.) – 2007, ISBN 978-0-444-51926-9.

Volume 163: The Dentate Gyrus: A Comprehensive Guide to Structure, Function, and Clinical Implications, by H.E. Scharfman (Ed.) – 2007, ISBN 978-0-444-53015-8.

Volume 164: From Action to Cognition, by C. von Hofsten and K. Rosander (Eds.) – 2007, ISBN 978-0-444-53016-5.

Volume 165: Computational Neuroscience: Theoretical Insights into Brain Function, by P. Cisek, T. Drew and J.F. Kalaska (Eds.) – 2007, ISBN 978-0-444-52823-0.

Volume 166: Tinnitus: Pathophysiology and Treatment, by B. Langguth, G. Hajak, T. Kleinjung, A. Cacace and A.R. Møller (Eds.) – 2007, ISBN 978-0-444-53167-4.

Volume 167: Stress Hormones and Post Traumatic Stress Disorder: Basic Studies and Clinical Perspectives, by E.R. de Kloet, M.S. Oitzl and E. Vermetten (Eds.) – 2008, ISBN 978-0-444-53140-7.

Volume 168: Models of Brain and Mind: Physical, Computational and Psychological Approaches, by R. Banerjee and B.K. Chakrabarti (Eds.) – 2008, ISBN 978-0-444-53050-9.

Volume 169: Essence of Memory, by W.S. Sossin, J.-C. Lacaille, V.F. Castellucci and S. Belleville (Eds.) – 2008, ISBN 978-0-444-53164-3.

Volume 170: Advances in Vasopressin and Oxytocin – From Genes to Behaviour to Disease, by I.D. Neumann and R. Landgraf (Eds.) – 2008, ISBN 978-0-444-53201-5.

Volume 171: Using Eye Movements as an Experimental Probe of Brain Function — A Symposium in Honor of Jean Büttner-Ennever, by Christopher Kennard and R. John Leigh (Eds.) – 2008, ISBN 978-0-444-53163-6.

Volume 172: Serotonin–Dopamine Interaction: Experimental Evidence and Therapeutic Relevance, by Giuseppe Di Giovanni, Vincenzo Di Matteo and Ennio Esposito (Eds.) – 2008, ISBN 978-0-444-53235-0.

Volume 173: Glaucoma: An Open Window to Neurodegeneration and Neuroprotection, by Carlo Nucci, Neville N. Osborne, Giacinto Bagetta and Luciano Cerulli (Eds.) – 2008, ISBN 978-0-444-53256-5.

PROGRESS IN BRAIN RESEARCH

VOLUME 174

MIND AND MOTION: THE BIDIRECTIONAL LINK BETWEEN THOUGHT AND ACTION

EDITED BY

MARKUS RAAB

Institute of Psychology, German Sport University Cologne, Cologne, Germany

JOSEPH G. JOHNSON

Department of Psychology, Miami University, Oxford, Ohio, USA

HAUKE R. HEEKEREN

Neurocognition of Decision Making Group, Max Planck Institute for Human Development, Berlin, Germany

ELSEVIER

AMSTERDAM – BOSTON – HEIDELBERG – LONDON – NEW YORK – OXFORD
PARIS – SAN DIEGO – SAN FRANCISCO – SINGAPORE – SYDNEY – TOKYO

Elsevier
360 Park Avenue South, New York, NY 10010-1710
Linacre House, Jordan Hill, Oxford OX2 8DP, UK
Radarweg 29, PO Box 211, 1000 AE Amsterdam, The Netherlands

First edition 2009

British Library Cataloguing in Publication Data
A catalogue record for this book is available from the British Library

Library of Congress Cataloging-in-Publication Data
A catalog record for this book is available from the Library of Congress

ISBN: 978-0-444-53356-2 (this volume)
ISSN: 0079-6123 (Series)

For information on all Elsevier publications
visit our website at elsevierdirect.com

Printed and bound in Hungary

09 10 11 12 13   10 9 8 7 6 5 4 3 2 1

List of Contributors

O.H. Azar, Department of Business Administration, Guildford Glazer School of Business and Management, Ben-Gurion University of the Negev, Beer Sheva, Israel
M. Bar-Eli, Department of Business Administration, Guildford Glazer School of Business and Management, Ben-Gurion University of the Negev, Beer Sheva, Israel
S.L. Beilock, Department of Psychology, University of Chicago, Chicago, IL, USA
R. Brand, Department of Sport and Exercise Psychology, University of Potsdam, Potsdam, Germany
L. Calmeiro, School of Social and Health Sciences, University of Abertay, Dundee, United Kingdom
H. Colonius, Department of Psychology, Oldenburg University, Oldenburg, Germany
J.A. Conlin, Center for Adaptive Behavior and Cognition, Max Planck Institute for Human Development, Berlin, Germany
L. Damisch, Department of Psychology, University of Cologne, Cologne, Germany
R.F. de Oliveira, Institute of Psychology, German Sport University Cologne, Cologne, Germany
D.A. DeCaro, Department of Psychology, Miami University, Oxford, OH, USA
A. Diederich, School of Humanities and Social Sciences, Jacobs University, Bremen, Germany
R.C. Eklund, Department of Educational Psychology and Learning Systems, Florida State University, Tallahassee, FL, USA
N. Green, Neurocognition of Decision Making Group, Max Planck Institute for Human Development, Berlin, Germany
B.D. Hatfield, Department of Kinesiology, University of Maryland, College Park, MD, USA
H.R. Heekeren, Neurocognition of Decision Making Group, Max Planck Institute for Human Development, Berlin, Germany; Max Planck Institute for Human Cognitive and Brain Sciences, Leipzig, Germany
T. Heinen, Institute of Psychology, German Sport University Cologne, Cologne, Germany
E.-J. Hossner, Department of Sport Science, Liverpool Hope University, Liverpool, United Kingdom
J.G. Johnson, Department of Psychology, Miami University, Oxford, OH, USA
W.M. Land, Department of Educational Psychology and Learning Systems, Florida State University, Tallahassee, FL, USA
Y. Lurie, Department of Management, Guildford Glazer School of Business and Management, Ben-Gurion University of the Negev, Beer Sheva, Israel
J. Munzert, Institute for Sports Science, Justus-Liebig University, Giessen, Germany
T. Mussweiler, Department of Psychology, University of Cologne, Cologne, Germany
A. Nieuwenhuys, Research Institute MOVE, Faculty of Human Movement Sciences, VU University Amsterdam, Amsterdam, The Netherlands
D. O'Hare, Department of Psychology, University of Otago, Dunedin, New Zealand
R.R.D. Oudejans, Research Institute MOVE, Faculty of Human Movement Sciences, VU University Amsterdam, Amsterdam, The Netherlands
H. Plessner, Institute of Psychology 1, University of Leipzig, Leipzig, Germany
M. Raab, Institute of Psychology, German Sport University Cologne, Cologne, Germany
S. Razon, Department of Educational Psychology and Learning Systems, Florida State University, Tallahassee, FL, USA

H. Ritter, Neuroinformatics Group and Center of Excellence "Cognitive Interaction Technology" (CITEC) and Research Institute for Cognition and Robotics (CoR-Lab), Bielefeld University, Bielefeld, Germany

T. Schack, Neurocognition and Action Research Group and Center of Excellence "Cognitive Interaction Technology" (CITEC) and Research Institute for Cognition and Robotics (CoR-Lab), Bielefeld University, Bielefeld, Germany

G. Schweizer, Department of Sport and Exercise Psychology, University of Potsdam, Potsdam, Germany

G. Tenenbaum, Department of Educational Psychology and Learning Systems, Florida State University, Tallahassee, FL, USA

J. Trommershäuser, Department of Psychology, Giessen University, Giessen, Germany

J.N. Vickers, Faculty of Kinesiology, University of Calgary, Calgary, AB, Canada

K.G. Volz, Max Planck Institute for Neurological Research, Cologne, Germany

D.Y. von Cramon, Max Planck Institute for Neurological Research, Cologne, Germany; Max Planck Institute for Human Cognitive and Brain Sciences, Leipzig, Germany

M. Weigelt, Neurocognition and Action Research Group and Center of Excellence "Cognitive Interaction Technology" (CITEC), Bielefeld University, Bielefeld, Germany

N. Wenderoth, Motor Control Laboratory, Department of Biomedical Kinesiology, KU Leuven, Belgium

A.M. Williams, Research Institute for Sport and Exercise Sciences, Liverpool John Moores University, Liverpool, United Kingdom

U. Wolfensteller, Max Planck Institute for Human Cognitive and Brain Sciences, Leipzig, Germany; Section of Systems Neuroscience, Department of Psychiatry and Psychotherapy, Medical Faculty Carl Gustav Carus, Technical University of Dresden, Dresden, Germany

K. Zentgraf, Institute for Sports Science, Justus-Liebig University, Giessen, Germany

Preface

This volume focuses on the topic of mind and motion, the bidirectional link between thought and action, a study that by nature is multidisciplinary. Funded by the Center for Interdisciplinary Research (ZiF) at Bielefeld University and the Alexander von Humboldt Foundation, in May 2008 a workshop in Bielefeld, Germany brought together an international group of scholars from a variety of disciplines such as movement science, neuroscience, neuroinformatics, psychology, robotics, and sport science.

The idea behind bringing this group together is that the bidirectional link between thought and action is a topic of study that by nature is multidisciplinary. One coherent interest of these experts is achieving a joint understanding of decision making that has heretofore been advanced mainly in isolation within the distinct research fields. Despite the contributions from such diverse viewpoints, the locus of human decision making has resided almost exclusively in the cognitive realm of the mind. Our work expands this notion to include the important dimension of action (motion), thereby expanding, refining, and generally improving our knowledge of decision making. Rather than viewing observable actions as merely the necessary extension of latent cognitive processes, we envision a system that appreciates the close, bidirectional interaction of mind and motion in forming a coherent system. The focus on the bidirectional link adds to the topic of embodied cognition recently presented in volume 164 of *Progress in Brain Research* "From action to cognition" (Von Hofsten and Rossander, 2007) and the *Attention and Performance volume* XXII on the nature of "Sensorimotor foundations of higher cognition" (Haggard et al., 2008).

The goals of the symposium held in May 2008 were to (a) explore the range of implications this view has on existing research, (b) develop novel, testable hypotheses for future research, and (c) adapt existing research methods in both movement science and decision research to provide the best tools for empirical study.

The workshop addressed questions that are central for the understanding of the bidirectional link between thought and action: how do we anticipate the consequences of choices and how is the brain able to represent these options and the potential consequences? How are different options evaluated and how is a preferred option implemented? These questions were discussed in three interdisciplinary groups of experts providing a unique environment in which seeking answers to these questions also generated stimulating input for further research.

We hope that this book will increase the interaction among the research fields represented and start a dialogue that we are confident will be enduring and productive to a broad audience.

This volume is structured into three sections corresponding to the three guiding questions. In each section the associated question will be answered via individual chapters as well as an interdisciplinary group report of all authors led by a group reporter. This volume provides an extensive body of knowledge and an attempt to go beyond individual chapters by integrating group reports as a runway into the future.

Markus Raab
Joseph G. Johnson
Hauke R. Heekeren

Acknowledgment

The ZiF-conference "*Mind and motion: the bidirectional link between thought and action*" was sponsored by the Center of Interdisciplinary Research, Bielefeld, Germany and the Alexander von Humboldt Foundation (Transcoop 2-Deu/119109).

Contents

See Color Plate Section at the end of this book

SECTION I

How are Choice Options and Their Consequences Perceived and Represented?

M. Raab et al. (Eds.)
Progress in Brain Research, Vol. 174
ISSN 0079-6123

CHAPTER 1

Grounding cognition in action: expertise, comprehension, and judgment☆

Sian L. Beilock*

Department of Psychology, University of Chicago, Chicago, IL, USA

Abstract: Recent work demonstrating that both the observation and planning of actions share common neural substrates suggests that merely thinking about action may call upon motor-based neural processes. As a result, higher-level cognitive processes not directly involved with motor production, such as language comprehension or the preference judgments one makes for objects and items in their environment, may be rooted in the sensorimotor systems. In this chapter we not only explore the links between cognition and action, but ask how such cognition–action links may differ as a function of one's experience performing and seeing actions related to the language one hears or the items one is making judgments about. Together, the work presented here suggests that a complete understanding of high-level performance not only requires consideration of how cognition drives action, but vice versa — a bidirectional link between cognition and action.

Keywords: embodied cognition; expertise; judgment; comprehension

Introduction: an embodied perspective

I teach a class at the University of Chicago entitled "Cognitive psychology" and because of this I have the opportunity to examine a majority of the cognitive psychology textbooks that are available for use. Although these text books are rather variable in terms of the topics they cover, one consistency jumps out. Out of the 20 or so cognitive psychology textbooks I own, only one of them has a chapter devoted to motor learning and control. And, it is not just that these books do not devote an entire chapter to the motor system, many of these textbooks do not even mention the word motor at all. It is as if the field of cognitive psychology has — in some sense — proclaimed "Who needs the body in the study of cognition?"

The above point regarding the absence of motor-related topics in cognitive psychology is not limited to my own observations, but has been made by prominent psychologists interested in motor control issues as well (e.g., Rosenbaum, 2005). This lack of motor representation in cognitive psychology is perhaps not surprising given that traditional views of cognitive psychology characterize the mind as an abstract information processor largely divorced from the body and the environment. However, more recent theories of embodied cognition suggest that our ability to represent objects, events, and even abstract concepts (e.g., metaphor) is subserved by the sensorimotor systems we rely on to navigate throughout the world (e.g., Glenberg, 1997;

☆This research was supported by NSF Grant BCS-0601148 to Sian L. Beilock.

*Corresponding author.
Tel.: +773-834-3713;
E-mail: beilock@uchicago.edu

DOI: 10.1016/S0079-6123(09)01301-6

Barsalou, 1999; Zwaan, 1999; Wilson, 2002; Gallese and Lakoff, 2005). This embodied viewpoint has roots in ecological psychology's refutation of a distinction between perception and action (Gibson, 1979/1986) and finds support across multiple levels of psychological inquiry.

For example, reading action words associated with the leg and arm (e.g., "kick," "pick") activates brain areas implicated in the movements of these body parts (Hauk et al., 2004; Tettamanti et al., 2005). And, Longcamp et al. (2003, 2005) have found that presenting single English letters to experienced English speakers activated premotor areas involved in writing (i.e., Exner's area) — even though there is no intention to actually write in the perceiver. Moreover, sensibility judgments of sentences such as "Can you squeeze a tomato?" are facilitated when individuals are primed with a sentence-associated hand shape (a clenched hand) relative to an inconsistent hand shape (a pointed finger; Klatzky et al., 1989). Thus, rather than our representations of the objects and events we hear, see, or read about being limited to amodal or propositional code arbitrarily related to the concepts they represent (Pylyshyn, 1986), our representations appear to be grounded in action. In other words, our knowledge is embodied in the sense that it consists of sensorimotor information about potential interactions the objects or events we encounter may allow (Wilson, 2002).

Despite recent interest in this embodied cognition viewpoint, less work has focused on how individual differences in visual and/or motor experience (e.g., motor skill expertise) shape embodied knowledge representations. Moreover, little work has examined how embodied representations might affect the explicit choices and judgments individuals make in situations when there is no intention to act. Using both behavioral and neuroimaging techniques, my colleagues and I are currently exploring (a) how motor and visual experience in a particular domain changes language comprehension in that domain — an activity previously thought to be largely amodal in nature (Holt and Beilock, 2006; Beilock et al., 2008) — and (b) how sensorimotor experience gives rise to the automatic simulation of action possibilities associated with the stimuli one encounters which, in turn, can influence preferences and judgments for objects and events (see Beilock and Holt, 2007; Yang et al., submitted). Together, our work suggests that cognition is deeply rooted in action — irrespective of one's intention to act — and that experience operating in particular environments directly affects knowledge representations, preference judgments, and even memory. Below I describe, in more detail, some of the research we have been conducting in an attempt to support this experience-driven embodied viewpoint. Specifically, I describe work examining how activities as diverse as language comprehension and likeability judgments for the objects and items individuals encounter may be driven, at least in part, by the motor system.

Language comprehension

Sport is unlike most human activities, inspiring those who play as well as those who merely watch. For those who aspire to achieve elite performance levels, intensive practice is a must. However, whether athletic experience carries implications beyond the playing field (i.e., beyond action perception and production) is less well understood. Importantly, this issue has been an increasing topic of interest as researchers interested in how one goes about judging the performance of themselves versus others (see Chapter 2: On the relativity of athletic performance: a comparison perspective on performance judgments in sports) and researchers interested in how high-pressure situations might impact the realization of the action possibilities afforded by one's environment (see Chapter 4: Perceiving and moving in sports and other high-pressure contexts) begin to explore the bidirectional link between cognition and action.

Across diverse research areas, investigators of human performance are trying to understand the implications that athletic experience may carry for performance "beyond the playing field" by building on the idea that cognition not only drives action, but the other way around as well. Our work has been motivated by the assumption that there is a bidirectional link between cognition and action and we have been especially interested in looking for connections between an individual's action experiences (in terms of both watching others and performing themselves) and their ability to perform activities that, at the outset, appear to be largely amodal and abstract in nature — that is, not heavily dependent on specific sensorimotor systems. One activity that we have been especially interested in is language comprehension.

In a recent study, Beilock et al. (2008) showed that sports experience changes the neural basis of language comprehension — even when there is no intention to act based on the language one hears. As it turns out, people with different motor skill experiences not only rely on different cognitive and neural operations in overt action execution (Beilock and Carr, 2001), but also in the comprehension of action-related language. To demonstrate this, my colleagues and I asked professional and intercollegiate ice-hockey players ($n = 12$), ice-hockey fans with no ice-hockey playing experience but a significant amount of ice-hockey watching experience ($n = 8$), and hockey novices with no ice-hockey playing *or* watching experience ($n = 9$) to passively listen to sentences depicting ice-hockey action scenarios (e.g., "The hockey player finished the stride") or everyday actions scenarios (e.g., "The individual pushed the bell") during functional neuroimaging (functional magnetic resonance imaging, fMRI). Everyone then performed a comprehension task outside the fMRI scanner that gauged their understanding of the sentences they heard inside the scanner.

In this comprehension task (see Fig. 1), participants listened to the sentences describing ice-hockey actions and everyday actions that they had heard in the scanner. Following each sentence, participants were presented with a picture of a target individual who was performing an action that either matched or mismatched the action implied in the sentence. The participant's task was to judge as quickly as possible whether the target individual was mentioned in the sentence (on some trials, pictures of individuals not mentioned in the sentence were presented and the action was not part of the directed decision). If individuals comprehend the perceptual and action qualities described in the sentences they hear (Zwaan et al., 2002; Holt and Beilock, 2006), then responses should be facilitated for pictured individuals whose actions match those implied in the sentence relative to pictured individuals whose actions do not match those implied in the sentence. We termed this index of comprehension the *action-match effect*.

All participants, regardless of hockey experience were able to comprehend the everyday action scenarios (i.e., they all showed a significant action-match effect). This is not surprising given that everyone should have experience viewing and performing everyday actions. However, participants with more hockey experience (i.e., players and fans) were better hockey language comprehenders (i.e., they showed a larger action-match effect; for confirmatory results see Holt and Beilock, 2006). More interestingly, this experience-driven hockey comprehension effect was fully mediated by increased neural activity in the left dorsal premotor cortex [Talairach center-of-gravity = $(\pm 45,9,41)$] that occurred while subjects initially listened to the ice-hockey scenarios during fMRI. Put another way, effective auditory comprehension of action-based language was accounted for by experience-dependent activation of the left dorsal premotor cortex, a region thought to support the selection of well-learned action plans and procedures — often in response to learned symbolic associations (Grafton et al., 1998; Wise and Murray, 2000; Schluter et al., 2001; Toni et al., 2002; Rushworth et al., 2003; O'Shea et al., 2007).

Interestingly, facilitated comprehension of hockey action sentences was not limited to our participants with significant hockey motor experience (i.e., hockey players): ice-hockey *fans* also showed increased comprehension of hockey language scenarios. This comprehension effect in fans was accompanied by activation in the left dorsal premotor cortex while listening to the hockey action scenarios that was significantly above baseline. Ice-hockey players showed a similar pattern of neural activation whereas ice-hockey novices did not. Moreover, increased dorsal premotor activation during hockey sentence comprehension was seen bilaterally for the

Hockey Action Sentence	Picture	
(A) The hockey player finished the stride.	(A)	
(B) The hockey player finished the shot.	(B)	

Everyday Action Sentence	Picture	
(A) The individual pushed the bell.	(A)	
(B) The individual pushed the cart.	(B)	

Fig. 1. Examples of the post-scan comprehension task stimuli. Picture A serves as a "match" for Sentence A and a "mismatch" for Sentence B. Picture B serves as a "match" for Sentence B and a "mismatch" for Sentence A. Adapted from Beilock et al. (2008).

fans but not players. This bilateral premotor activation may be indicative of more effortful action selection in fans versus players, which would be generally consistent with the overall longer response times seen in the fans versus players for hockey sentence comprehension. Nonetheless, it is important to point out that only activation in the left dorsal premotor cortex — and not bilateral premotor activation — mediated the impact of hockey experience on language comprehension. The fact that the left (but not right) premotor cortex was the only significant mediator of our experience-language comprehension relation is in line with the finding that the left premotor cortex plays a dominant role in higher-level action selection, regardless of the side of the body involved (Schluter et al., 2001; Haaland et al., 2004; Grafton and Hamilton, 2007).

The relation between hockey experience and hockey sentence comprehension was also fully mediated by bilateral dorsal primary sensory-motor activity [Talairach center-of-gravity = (±22,–21,53)] while listening to hockey action sentences. Just as with the left premotor region, when activity in these bilateral regions was used to predict hockey language comprehension along with hockey experience, the relation between hockey experience and comprehension was rendered nonsignificant. In this case, however, only a strong *negative* relation between bilateral primary sensory-motor activity and hockey comprehension remained. Primary sensory-motor regions are thought to be heavily involved in instantiating the specific step-by-step movements needed to carry out a novel task (Rizzolatti and Luppino, 2001; Grafton and Hamilton, 2007). Less hockey experience was associated with increased activity in bilateral primary sensory-motor regions during hockey language listening and decreased comprehension. Those without the ability to associate plans for action with linguistic cues — most likely because such actions are not a dominant part of their motor skill repertoires (novices) — instead show increased activity (vs. more experienced counterparts) in neural areas known to be involved in the instantiation of simple movements. Such activation actually hurts comprehension, possibly because it does not embody the higher-level action plans effective comprehension relies on.

In summary, the data presented above show that sports experience enhances the understanding of sports-related language even when there is no intention to act because of the recruitment of neural areas normally involved in action planning and execution (areas outside the purview of traditional language processing). Substantial prior experience viewing or performing ice-hockey actions enhances hockey language comprehension, likely by enabling individuals to associate linguistically described action scenarios with motor plans for execution. This, in turn, gives individuals the type of robust and multimodal representation that is the hallmark of optimal language comprehension.

Explicit judgments

The above findings suggest that we represent language, at least in part, via covert sensorimotor simulation of how we might execute a described behavior and that this representation may be fundamentally different depending on one's experience viewing and performing the actions described in the language in question. But, do these action-cognition links extend beyond understanding? For example, might the covert simulation of action carry implications beyond comprehension, influencing — for example — individuals' explicit judgments about the stimuli they encounter. We have explored this idea in terms of both explicit memory judgments and explicit preferences individuals have for the objects and events they encounter.

Memory judgments

Fluency, or the ease with which an item is processed, is thought to lead individuals to have a subjective feeling of remembering that can often serve as a useful heuristic in recognition. However, fluency does not always result in accurate memory judgments because it can arise independently of whether one has actually seen the item they are judging before. For example, a word such as "test" presented in a semantically predictive sentence (e.g., "the anxious student took a *test*") is more likely to be recognized as old (i.e., having been seen or studied previously) than when this same word is presented in a nonpredictive sentence (e.g., "later in the afternoon he/she took a *test*"). This is because semantic expectancy increases the *conceptual* fluency of this word (Whittlesea, 1993). Similarly, manipulating an item's visual clarity alters its *perceptual* fluency. The easier an item is to visually process, the more likely individuals will say that they have seen it before (Whittlesea et al., 1990).

In a recent series of studies, we (Yang et al., submitted) asked whether memory errors might arise from a source rather different from the semantic or visual context of a given stimulus. Specifically, we examined whether fluency effects might be tied to motor plans that are automatically activated in association with the items individuals encounter — even in situations where there is no actual intention to act.

As mentioned above, recent behavioral and neurophysiological work suggest that we represent our surroundings, at least in part, via covert motor simulation of how we might execute an observed behavior or act on the objects around us. We reasoned that if individuals covertly simulate actions associated with the items they observe, and if this gives rise to information about ease or fluency of action, then this may in turn impact memory judgments for these items. In other words, fluency might not only arise from the conceptual context in which an item is viewed or the visual qualities of the item itself, but also from the automatic activation of motor plans for action associated with the stimuli one is making memory judgments about.

To test this idea, we turned to the domain of typing. Recent work has shown that an integral part of letter processing — at least for experienced typists — is the motor simulation of typing the letters themselves. Specifically, in a Stroop-like task, Rieger (2004, 2007) found that typing experts' manual responses were faster when the finger used to indicate the color of a letter the typists were presented with was congruent with the finger typically used to type the letter using standard touch-typing conventions. Such work suggests that when typing experts perceive letters, they automatically activate motor plans for typing them. If this motor simulation carries information about how easy it would be to produce such letters, then individuals' propensity to recognize letters as previously studied items in a memory judgment task might be a function of how easy it would be to actually type the letters. This would be the case despite the fact that the individuals have no intention to type.

Skilled and novice typists studied a list of letter dyads and then took a subsequent recognition memory test. To manipulate the typing ease of the dyads, we varied whether the presented letter dyads would be typed, using standard touch-typing methods, with the same finger (e.g., FV) or different fingers (e.g., FK). The interval between typing consecutive letters using the same finger is longer than the interval between typing consecutive letters with different fingers (Viviani and Laissard, 1996). This is because typing is a parallel process in which consecutive letters are programmed simultaneously and a given finger can only be in one place at a time. Thus, letter dyads typed by the same finger should cause more motor interference or be harder to type than dyads typed with different fingers because, the latter case can be planned and performed more so in parallel than the former (Rumelhart and McClelland, 1982; Rumelhart and Norman, 1982).

If dyad recognition memory judgments are driven by motor fluency, individuals' propensity to recognize a dyad as old should be higher for those dyads that would be easier (i.e., different-finger dyads) versus harder (i.e., same-finger dyads) to type. However, this should only hold for skilled typists who have extensive typing experience and have thus formed consistent mappings between specific letters and the fingers used to type them. This is exactly what was found. Skilled typists made more *false recognition errors* (i.e., indicting they had studied a dyad when in fact they had not) for same-finger dyads in comparison to different-finger dyads.

Preference judgments

The above findings suggest that explicit memory judgments can be influenced by the motor associations individuals have with the items being judged. But, can we take this a step further? If (a) individuals mentally simulate acting on the objects they perceive in their environment, and (b) if this mental simulation of action differs as a function of skill level, and (c) if people prefer to act in ways that create less motor interference (i.e., are more fluent), then individuals should report higher ratings of *liking* for objects that are easier versus harder to act on — even though these individuals have no intention to act. In other words, the fluency associated with covertly simulating typing letter dyads may not only impact one's memory judgments, but one's preferences for one dyad over another.

To explore this idea, we (Beilock and Holt, 2007) again turned to the domain of typing. Skilled and novice typists were simultaneously presented with two separate letter dyads on a screen and asked to indicate the dyad they preferred. Specifically, participants were informed that they would see two letter dyads and that they should verbally indicate which of the two dyads they preferred, using their first impressions of the letters while avoiding choosing dyads based on their associations with any initials or abbreviations. As in the memory judgment study, the dyads fell into one of two categories: dyads that would be typed with the same finger using standard typing methods (e.g., FV) or dyads that would be typed with different fingers (e.g., FJ). Each dyad pair presented to participants always involved one dyad from each category — a paradigm first introduced by Van den Bergh et al. (1990). As mentioned above, because typing is thought to involve the overlap of successive key strokes (Rumelhart and Norman, 1982), typing two letters with the same finger should result in more motor interference than typing two letters with different fingers, as the former case requires that the same digit essentially be in two places at once (or in very close succession).

Results demonstrated that skilled typists preferred dyads typed with different fingers (i.e., dyads that would be easier to type) significantly more than chance (note that chance is 0.5 because half of the letter dyads that were judged involved the same finger and half different fingers). Novices showed no preference (see Fig. 2,

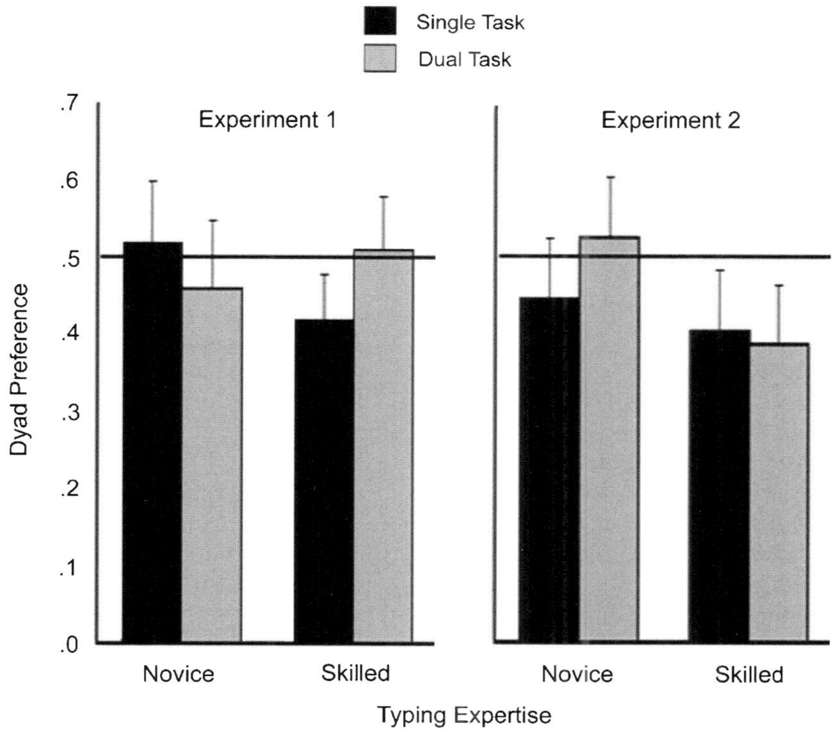

Fig. 2. Letter dyad preferences in the single-task and dual-task blocks for novice and skilled typists in Experiment 1 and Experiment 2. Dyad preference: To create a dependent variable indicating preferences for same-finger or different-finger dyads, a score was calculated such that every time a same-finger dyad was preferred, a 1 was assigned to that trial. A different-finger dyad preference received a 0. These values were summed and divided by the total number of trials in each block. Thus, a score of 0.5 indicates no preference, less than 0.5 indicates a preference for different-finger (easier to type) dyads, and greater than 0.5 indicates a preference for same-finger (harder to type) dyads. Error bars represent 95% confidence intervals. Adapted from Beilock and Holt (2007).

experiment 1, black bars). Importantly, participants were recruited for a study examining "cognitive task performance" to minimize associations with the study and typing. It was only when the study was completed that individuals were categorized as skilled or novice typists. As a result, participants were unaware of the link between the study and typing, and when asked, could not explicate how the letter dyads typed with the same versus different finger differed.

Why might skilled typists show the letter dyad preference that novices do not? If typing experience results in the association between specific letters and the motor programs used to type them, and perceiving letters automatically activates these motor plans (Rieger, 2004; see also Prinz's common coding theory, 1997), then such covert simulation of typing should provide information about the relative interference involved in acting on the letters one is presented with. And, if individuals prefer to act in ways that reduce interference, one should prefer letter dyads that, if acted on, would result in the least amount of motor interference.

To explicitly test these claims, while making their preference judgments on some trials, participants held a finger press pattern in memory that involved the same fingers that would be used to type the presented dyads. If holding this pattern utilizes motor system resources that could otherwise be used to inform typists' preference judgments, such preferences should disappear — exactly what occurred (see Fig. 2, experiment 1, gray bars). A second experiment showed that this motor interference was specific to the digits actually involved in typing the dyads. When expert typists held a motor pattern in memory involving fingers *not* used to type the dyads, the preference remained (see Fig. 2, experiment 2, gray bars — note black bars replicate experiment 1's single-task effect). Thus, covert mental simulation of acting on the information one is presented with not only impacts preference judgments, but this influence is limited to information motorically resonant with the specific effectors involved in the simulated action.

Conclusions

In conclusion, the embodied viewpoint finds support across multiple levels of psychological inquiry from behavioral studies of memory to neuroimaging studies of language comprehension. Although such work supports the notion that the cognitive and neural systems that subserve action are also engaged during cognitive tasks that, on the surface, involve no intention to action, less attention has been focused on how individuals' motor skill experiences modulate the content of their embodied knowledge representations. Our work demonstrates that one's experience on the playing field, ice rink, or even computer keyboard fundamentally changes the extent to which (and how) cognition is grounded in action.

References

Barsalou, L. W. (1999). Perceptual symbol systems. *Behavioral and Brain Sciences*, *22*, 577–660.

Beilock, S. L., & Carr, T. H. (2001). On the fragility of skilled performance: what governs choking under pressure? *Journal of Experimental Psychology — General*, *130*, 701–725.

Beilock, S. L., & Holt, L. E. (2007). Embodied preference judgments: can likeability be driven by the motor system? *Psychological Science*, *18*, 51–57.

Beilock, S. L., Lyons, I. M., Mattarella-Micke, A., Nusbaum, H. C., & Small, S. L. (2008). Sports experience changes the neural processing of action language. *Proceedings of the National Academy of Sciences of the United States of America*, *105*, 13269–13273.

Gallese, V., & Lakoff, G. (2005). The brain's concepts: the role of the sensory-motor system in Reason and Language. *Cognitive Neuropsychology*, *22*, 455–479.

Gibson, J. J. (1979/1986). *The ecological approach to visual perception*. Hillsdale, NJ: Lawrence Erlbaum Associates.

Glenberg, A. M. (1997). What memory is for. *Behavioral and Brain Sciences*, *20*, 1–55.

Grafton, S. T., & Hamilton, A. F. (2007). Evidence for a distributed hierarchy of action representation in the brain. *Human Movement Science*, *26*, 590–616.

Grafton, S. T., Fagg, A. H., & Arbib, M. A. (1998). Dorsal premotor cortex and conditional movement selection: a PET functional mapping study. *Journal of Neurophysiology*, *79*, 1092–1097.

Haaland, K. Y., Elsinger, C. L., Mayer, A. R., Durgerian, S., & Rao, S. M. (2004). Motor sequence complexity and performing hand produce differential patterns of hemispheric lateralization. *Journal of Cognitive Neuroscience*, *16*, 621–636.

Hauk, O., Johnsrude, I., & Pulvermuller, F. (2004). Somatotopic representation of action words in human motor and premotor cortex. *Neuron*, *41*, 301–307.
Holt, L. E., & Beilock, S. L. (2006). Expertise and its embodiment: examining the impact of sensorimotor skill expertise on the representation of action-related text. *Psychonomic Bulletin & Review*, *13*, 694–701.
Klatzky, R. L., Pellegrino, J. W., McCloskey, B. P., & Doherty, S. (1989). Can you squeeze a tomato? The role of motor representations in semantic sensibility judgments. *Journal of Memory and Language*, *28*, 56–77.
Longcamp, M., Anton, J. L., Roth, M., & Velay, J. L. (2003). Visual presentation of single letters activates a premotor area involved in writing. *Neuroimage*, *19*, 1492–1500.
Longcamp, M., Anton, J. L., Roth, M., & Velay, J. L. (2005). Premotor activations in response to visually presented single letters depend on the hand used to write: a study on left-handers. *Neuropsychologia*, *43*, 1801–1809.
O'Shea, J., Sebastian, C., Boorman, E. D., Johansen-Berg, H., & Rushworth, M. F. (2007). Functional specificity of human premotor-motor cortical interactions during action selection. *European Journal of Neuroscience*, *26*, 2085–2095.
Prinz, W. (1997). Perception and action planning. *European Journal of Cognitive Psychology*, *9*, 129–154.
Pylyshyn, Z. W. (1986). *Computational cognition: toward a foundation for cognitive science*. Cambridge, MA: MIT Press.
Rieger, M. (2004). Automatic keypress activation in skilled typing. *Journal of Experimental Psychology — Human Perception and Performance*, *30*, 555–565.
Rieger, M. (2007). Letters as visual action-effects in skilled typing. *Acta Psychologica*, *126*, 138–153.
Rizzolatti, G., & Luppino, G. (2001). The cortical motor system. *Neuron*, *31*, 889–901.
Rosenbaum, D. A. (2005). The Cinderella of psychology: the neglect of motor control in the science of mental life and behavior. *American Psychologist*, *60*, 308–317.
Rumelhart, D. E., & McClelland, J. L. (1982). An interactive activation model of context effects in letter perception. 2. The contextual enhancement effect and some tests and extensions of the model. *Psychological Review*, *89*, 60–94.
Rumelhart, D. E., & Norman, D. A. (1982). Simulating a skilled typist — a study of skilled cognitive-motor performance. *Cognitive Science*, *6*, 1–36.
Rushworth, M. F., Johansen-Berg, H., Gobel, S. M., & Devlin, J. T. (2003). The left parietal and premotor cortices: motor attention and selection. *Neuroimage*, *20*, 89–100.
Schluter, N D., Krams, M., Rushworth, M. F., & Passingham, R. E. (2001). Cerebral dominance for action in the human brain: the selection of actions. *Neuropsychologia*, *39*, 105–113.
Tettamanti, M., Buccino, G., Saccuman, M. C., Gallese, V., Danna, M., Scifo, P., Fazio, F., Rizzolatti, G., Cappa, S. F., & Perani, D. (2005). Listening to action-related sentences activations fronto-parietal motor circuits. *Journal of Cognitive Neuroscience*, *17*, 273–281.
Toni, I., Shah, N. J., Fink, G. R., Thoenissen, D., Passingham, R. E., & Zilles, K. (2002). Multiple movement representations in the human brain: an event-related fMRI study. *Journal of Cognitive Neuroscience*, *14*, 769–784.
Van den Bergh, O., Vrana, S., & Eelen, P. (1990). Letters from the heart — affective categorization of letter combinations in typists and nontypists. *Journal of Experimental Psychology — Learning Memory and Cognition*, *16*, 1153–1161.
Viviani, P., & Laissard, G. (1996). Motor templates in typing. *Journal of Experimental Psychology — Human Perception and Performance*, *22*, 417–445.
Whittlesea, B. W. A. (1993). Illusions of familiarity. *Journal of Experimental Psychology — Learning Memory and Cognition*, *19*, 1235–1253.
Whittlesea, B. W. A., Jacoby, L. L., & Girard, K. (1990). Illusions of immediate memory — evidence of an attributional basis for feelings of familiarity and perceptual quality. *Journal of Memory and Language*, *29*, 716–732.
Wilson, M. (2002). Six views of embodied cognition. *Psychonomic Bulletin & Review*, *9*, 625–636.
Wise, S. P., & Murray, E. A. (2000). Arbitrary associations between antecedents and actions. *Trends in Neuroscience*, *23*, 271–276.
Yang, S., Gallo, D. A., & Beilock, S. L. (submitted). Embodied memory judgments: a case of motor fluency.
Zwaan, R. A. (1999). Embodied cognition, perceptual symbols, and situation models. *Discourse Processes*, *28*, 81–88.
Zwaan, R. A., Stanfield, R. A., & Yaxley, R. H. (2002). Language comprehenders mentally represent the shape of objects. *Psychological Science*, *13*, 168–171.

M. Raab et al. (Eds.)
Progress in Brain Research, Vol. 174
ISSN 0079-6123

CHAPTER 2

On the relativity of athletic performance: a comparison perspective on performance judgments in sports

Lysann Damisch* and Thomas Mussweiler

Department of Psychology, University of Cologne, Cologne, Germany

Abstract: Identifying the best performance at a specific point in time constitutes the central purpose of all competitive sports. In many sports, this decision is based on subjective performance judgments. In the present chapter, we consider how these judgments are formed by emphasizing the relative nature of person judgments and athletic performance judgments in particular. Specifically, we argue that athletic judgments are influenced by comparisons with prior performances thereby leading to similar judgmental consequences as have been observed in other comparison domains (e.g., social comparison). Particularly, we demonstrate that performance judgments are assimilated to judgments of the preceding performance if judges focus on similarities between the two. If judges focus on differences, however, contrast ensues. Strategies for preventing or correcting this judgmental bias are discussed.

Keywords: comparison; performance judgments; sport; similarity; dissimilarity; assimilation; contrast

The relativity of athletic performances

Athletic performances are inherently relative in nature. Because athletic achievements are obtained in competition with others, it is not those who perform well that find their way onto Olympic pedestals and into halls of fame, but those who perform better than others. Eric Moussambani, for example, was one of the best swimmers of Equatorial Guinea, which was mainly because most Guineans do not know how to swim at all. Moussambani had learned to swim in small pools and rivers and 8 months later, due to a wild card from the International Olympic Committee (IOC), he even made it to the Olympics 2000 in Sydney, Australia. In the men's 100-m freestyle race, Moussambani performed a personal best even though he had never before raced in a 50-m pool nor had he managed to swim the whole distance of 100 m without a break. However, despite this personal record, in relation to the performance of all other participants, Moussambani's performance appeared rather weak. The time he needed to swim the 100-m distance was more than twice the time of his faster competitors and remains the slowest time an Olympic 100-m freestyle race was ever finished.

The essential relativity of athletic performances, however, is not only apparent for the performing athletes, as is apparent in this introductory example. It equally holds for the judges who evaluate these performances. In fact, the

*Corresponding author.
Tel.: +49-221-470-1525; Fax: 49-221-470-5105;
E-mail: lysann.damisch@uni-koeln.de

DOI: 10.1016/S0079-6123(09)01302-8

assessment of athletic performances by subjective evaluations is part of many sports such as gymnastics, springboard diving, or figure skating (Landers, 1970; Stefani, 1998). In many ways, such performance judgments are choice situations. Given a number of different performances, it is the judges' task to decide, which one should be chosen as the best performance. From this perspective, the relative nature of performance judgments becomes apparent.

Given the important standing of athletic performance judgments, it seems essential to know how these judgments are formed. How is the psychological context in which athletic performances occur incorporated into the final judgment? In which way can these contextual stimuli influence how a performance is perceived and evaluated? What are the psychological mechanisms that underlie the contextual effects on performance judgments?

Influences of athletic performance judgments

In general, judgments in sport follow the same principles as social judgments in other domains (Plessner, 2005; Plessner and Haar, 2006). In particular, if judgments have to be made under suboptimal conditions, which is often true for judgments in sports (Salmela, 1978a, b), they are influenced by normatively irrelevant factors (Nisbett and Ross, 1980; Kahneman et al., 1982). Some of these factors are rather specific for the sports context. Thus, performance judgments can be influenced by feedback about other judges' scores, leading to conformity (Scheer et al., 1983; Boen et al., 2008), order-related expectancies (Plessner, 1999), and effects of prior processing (Ste-Marie and Lee, 1991; Ste-Marie et al., 2001). Other factors influencing performance judgments in sports, however, are rather general factors that influence person judgments universally. One such influencing factor is comparison processes.

Self–other comparisons

Generally, person judgments are fundamentally comparative in nature. To judge another person as athletic, for example, already implies that this person is more athletic than others and is thus comparative in nature (Huttenlocher and Higgins, 1971). Consistent with this general notion, abundant research attests that when people judge themselves or another person, they rely on comparisons with accessible standards. In fact, the tendency to engage in comparison information processing is remarkably robust and occurs even if judges are not explicitly asked for. When making judgments about themselves, for example, people spontaneously compare themselves to another person (Festinger, 1954; Mussweiler and Rüter, 2003). Similar, when making judgments about another person, people spontaneously compare that person to themselves (Dunning and Hayes, 1996). Recent evidence suggests that this tendency to make spontaneous comparisons when processing information about a given target goes so far that even stimuli that are not consciously perceived because they are presented subliminally are compared to a pertinent standard (Dehaene et al., 1998). The same is true if not the target but the standards are presented so briefly that judges are not consciously aware of them (Mussweiler et al., 2004a; Mussweiler and Englich, 2005). The urge to engage in comparative processing is so strong that even standards that are normatively irrelevant are used for comparisons with a target (Gilbert et al., 1995). This whole body of evidence suggests that comparisons are naturally and spontaneously engaged whenever context information is processed. From that notion, one could derive a certain likelihood that comparisons also are involved when judging performances in the field of sport.

Assimilation and contrast

Less clarity, however, exists about the direction of this comparative influence. At first sight, the evidence that has accumulated in social comparison research (Buunk et al., 1990; Brown et al., 1992; Brewer and Weber, 1994; Lockwood and Kunda, 1997) suggests that comparative processing can lead to opposing judgmental influences, which are hard to predict. Sometimes, comparisons produce contrast effects in target evaluation. In a study by Morse and Gergen (1970), for example, participants judged themselves to be less competent in the context of a competent person than in comparison with an incompetent standard. In another study, the evaluation of one's own attractiveness was more favorable after viewing another unattractive person than after viewing an attractive person (Brown et al., 1992). At other times, however, comparisons yield the opposite outcome by producing assimilation effects in target evaluation. For example, the self may be judged to be more competent after a comparison with a competent rather than an incompetent other (Brewer and Weber, 1994; Pelham and Wachsmuth, 1995). Similarly, participants' evaluations of their athletic abilities are higher in the context of an athletic rather than an unathletic standard (Mussweiler and Strack, 2000). In fact, it had turned out that whether assimilation or contrast is the result of comparison depends on a host of moderators such as the extremity of the standard (Herr et al., 1983; Herr, 1986), the ambiguity of the target (Herr et al., 1983; Pelham and Wachsmuth, 1995), the category membership (Brewer and Weber, 1994; Mussweiler and Bodenhausen, 2002), or the psychological closeness between target and standard (Tesser et al., 1988; Brown et al., 1992; Lockwood and Kunda, 1997). How can these diverging findings be explained?

The selective accessibility model (SAM) of comparison processes

One integration of these diverse findings is suggested by the SAM of comparison mechanisms and consequences (Mussweiler, 2003). From this theoretical perspective, judges engage in one of two alternative comparison processes. The main moderating factor that determines which of the two processes is engaged is the perceived similarity between the standard and the target. Depending on this overall perceived similarity between a given target and a particular standard either assimilative or contrastive comparison consequences will occur. Applying previous literature on hypothesis testing to the current phenomenon of comparative judgments, the once selected hypothesis concerning the perceived degree of similarity between a target and a standard will be tested by focusing on information that is consistent with the first general impression (Snyder and Swann, 1978; Trope and Bassok, 1982; Klayman and Ha, 1987). If judges perceive a target and a standard as rather similar, they engage in the process of similarity testing by selectively searching for information that indicates that target and standard are indeed similar. Thus, target knowledge, which is consistent with the standard, becomes accessible. If judges perceive a target and a standard as rather dissimilar, they engage in the process of dissimilarity testing by selectively searching for information to indicate that the target and standard are different. Thus, target knowledge, which is inconsistent with the standard, becomes accessible. Finally, this temporarily highly accessible information, which had just been gathered during the process of hypothesis testing, is used to make the target judgment. Consequently, similarity testing moves the target judgment closer to the standard (assimilation), whereas dissimilarity testing moves the target judgment away from the standard (contrast) (for a more detailed description of the model see Mussweiler, 2003).

Empirical evidence in support of the SAM

In the realm of social comparison research, manifold empirical support for the basic assumptions of the SAM has been gathered. Thus, it has been shown that perceived similarity between the standard and the self influences the judgment of the self (Mussweiler, 2001; Mussweiler and Bodenhausen, 2002). Mussweiler (2001), for example, used a procedural priming task to manipulate whether participants primarily focus on similarities or on dissimilarities experimentally during a comparison. The result of a subsequent social comparison task showed that participants assimilated their self-judgments to a given standard if they were made to focus on similarities. Participants contrasted their self-judgments away from the comparison standard, however, if they had been manipulated to focus on dissimilarities.

Thus, an induced focus on similarities versus differences determines whether assimilative versus contrastive comparisons result. In turn, assimilative versus contrastive comparisons are accompanied by a generalized focus on similarities versus differences. More specifically, after assimilating the evaluation of their athletic abilities to those of a standard (Bill Clinton), participants focused on similarities in a subsequent unrelated picture comparison task. After contrasting the judgments about their sportsmanship from those of a standard (Pope John Paul), however, participants focused on differences in a picture comparison (Mussweiler et al., 2004b). Taken together with a large body of additional evidence (for a recent review see Mussweiler, 2007), these social comparison findings suggest that a focus on similarities versus differences critically determines whether a comparison leads to assimilation or contrast.

Further evidence for the underlying mechanisms of comparative processing in judgment tasks stems from research examining what specific target knowledge is activated in the case of assimilation versus contrast to context stimuli. Using a specific variant of a lexical decision task (Dijksterhuis et al., 1998), these studies demonstrated that standard-consistent knowledge is more accessible after comparisons with assimilative judgmental outcomes, whereas standard-inconsistent knowledge is more accessible after comparisons with contrastive judgmental consequences (Mussweiler and Strack, 2000; Mussweiler and Bodenhausen, 2002; Smeesters and Mandel, 2006). In a recent study by Förster et al. (2008), it was even demonstrated that the accessibility of standard-inconsistent knowledge, for example, statistically mediated the contrastive self-evaluations of participants' athletic abilities in comparison to those of a pertinent standard.

In summary, the described body of evidence is consistent with the notion that assimilation is closely associated with a focus on similarities and accessibility of standard-consistent knowledge, whereas contrast is closely associated with a focus on differences and accessibility of standard-inconsistent knowledge. The selective accessibility mechanisms of similarity and dissimilarity testing may thus well be the psychological mechanisms that underlie assimilation and contrast effects in a variety of domains, such as performance judgments in sports.

Other–other comparison

The presented findings demonstrate that the selective accessibility mechanisms of similarity and dissimilarity testing contribute to comparison consequences in the context of the self. If judges compare their own characteristics and performances to others, then they appear to engage in one of two alternative comparison processes: similarity or dissimilarity testing. However, previous empirical research has no answer to the question whether the same mechanisms apply to comparisons that do not involve the self. On the one hand, one could argue that social comparisons are unique in that they involve the self as unique psychological entity and are likely to be shaped by motivational processes. This would suggest that self–other comparisons differ from other–other comparisons. On the other hand, however, all person comparisons may share a common psychological architecture in that they all involve a psychological process in which the characteristics of a target person are related to those of a standard. In this respect, it remains unclear whether mechanisms of selective accessibility also play out in those types of person–person comparisons that are also part of athletic performance judgments or other types of sequential judgments such as in oral exams, job interviews, or casting shows. More specifically, the question remains whether judges of athletic performances indeed engage in the process of similarity or dissimilarity testing and thus, yield assimilation or contrast effects. If this were the case, then a comparison perspective on performance judgments would hold a number of interesting implications. First, if a performance judgment depends on a comparison with a previous performance, one could expect that sequential performance judgments of any athletic competition should be related with each other. Would such a relation between consecutively presented performances indeed be attained when analyzing the Olympic performance judgments in gymnastics, for example? Second, if a performance judgment depends on a comparison with a previous performance, one could wonder which direction the comparison process takes and thus, which consequences it yields. Would judges rather assimilate their judgments to an earlier performance or contrast them away from the previous performance? Third, if a performance judgment depends on a comparison with a previous performance, the question arises, which characteristics decide whether the judge engages into a similarity testing comparison or a dissimilarity testing comparison process. Would a simple factor such as the nationality of sequentially performing athletes be enough to have experienced judges engage in the opposing processes of similarity versus dissimilarity testing and thus, produce the opposing outcomes of assimilation or contrast? Our recent research (Damisch et al., 2006) suggests several answers to these questions.

The inter-relation of Olympic gymnastic judgments

Are performance judgments in sport indeed influenced by preceding performances? How could such an influence be obtained? Our recent research suggests a simple analysis of real life date and reveals evidence, which supports the claim. If judges use the performance of an athlete as a comparison standard in order to judge a subsequent performance, we figured that sequential judgments should be correlated with each other instead of being independent.

To test this assumption, we analyzed the performance judgments of the gymnastics competition of the 2004 Olympic Games in Athens, Greece. In sum, this included 1373 individual judgments from different competitive events (preliminary rounds, team finals, individual all-around finals, apparatus finals) for female and male gymnasts. Again, if these judgments, made by highly qualified and trained judges at the most important competition — the Olympic Games — were not influenced by prior performance judgments, no correlation should occur. However, performance judgments of consecutively starting gymnasts should be correlated, if the judgment for a particular performance is influenced by comparing it to a preceding performance. Moreover, given the specific situational conditions in gymnastics competitions, we particularly expected judges to focus on similarities. Consecutively starting gymnasts at a competition share a host of features such as their gender, their approximate age group, parts of the presented routine, and their general level of gymnastic abilities. Here, a natural focus on similarities and thus an assimilative influence of prior gymnasts on performance judgments seemed to be especially likely. Ultimately, if the judgment for a specific athlete is built on the information gathered during testing the hypothesis that this performance was similar to the preceding performance, assimilation, and thus, a positive correlation should be apparent: the better the preceding athlete's performance, the better the target athlete's performance should be judged. This is exactly what we found. Computing several Pearson product–moment correlations between the performance judgments for all gymnasts and the scores of their respective predecessors, we found a significant positive correlation. This correlation was equally high, whether it was controlled for the starting order of the gymnasts or not. Furthermore, we found that the influence of the performance just prior to the target performance was the strongest, whereas this influence decreased the farther in time (two performances prior to the target performance; three performances prior to the target performance) the prior judgment was made to the target judgment. Together, these findings demonstrate that the likelihood for Olympic gymnasts to receive high scores for their exercises is higher if the preceding gymnast presented a good rather than a flawed performance. Putting this positive correlation into other words, one can say that performance judgments indeed are influenced by the context of prior performances.

Assimilation and contrast after procedural priming of comparison focus

The first piece of our evidence suggests that comparison processes appear to shape athletic performance judgments. Which direction does this influence take and what are the factors that decide about the final outcome? A selective accessibility direction suggests that the most crucial factor influencing the direction of the process is the comparison focus judges adopt during the comparison process. Applied to athletic performance judgments, one can expect that judges who focus on similarities between consecutively starting athletes yield assimilative judgments, whereas judges who focus on dissimilarities between sequentially presented performances yield contrast effects. This perspective was supported by findings of our recent line of research. Specifically, we tested our assumption in an experimental study. To do so, participants' foci on similarities versus dissimilarities were directly manipulated before they were asked to judge the performance of two consecutively presented athletes. Specifically, we used a procedural priming task (for an overview see Smith, 1994), in which participants' tendency to rely on a particular processing style is strengthened by engaging them in this type of processing in a preceding unrelated task (Mussweiler, 2001). More specifically, we had participants work on a picture comparison task, thereby either focusing on similarities or differences between the pictures. Participants were given sketches of two scenes and were asked to either list all the similarities or all the differences they could find for the two pictures. Doing so activates participants' focus on either of these two alternative processing styles, which is then thought to carry over to a subsequent task. Thus, participants who focused on similarities between the two pictures were expected to focus also on similarities between the consecutively presented athletes in the subsequent performance judgment task that is otherwise unrelated to the picture comparison task. For the judgment task, participants were presented with two athletic performances in the event of ski jumping. Performance judgments were assessed by having participants, who were students with no experience on this task, estimate the achieved length for each of the presented ski jumps. Whereas the objective performance of the former athlete was manipulated (high vs. low), the length of the latter ski jump was kept constant on a moderate level. Consistent with a selective accessibility perspective on comparison consequences, performance judgments of the target athlete critically depended on whether participants were induced to focus on similarities or differences (see Fig. 1). Judges who were primed to focus on similarities and thus, to engage in similarity testing assimilated their performance judgments toward the standard. These judges estimated the length of the target ski jump to be higher if it was presented after the prior high performance rather than the prior low performance. Judges who were primed to focus on differences and thus, to engage in dissimilarity testing, however, contrasted their performance judgments away from the standard. These judges estimated the length of the target ski jump to be lower after a prior high performance rather than a prior low performance. This pattern of means produced a significant interaction effect, while none of the main effects reached the level of significance.

Assimilation and contrast in experts' gymnastic performance judgments

So far, we have learned that athletic performance judgments seem to be influenced by comparisons with prior performances and that these influences can take the form of assimilation or contrast effects depending on whether judges are induced to focus on similarities or dissimilarities in an unrelated picture task. Would this influence on assimilative versus contrastive judgmental biases also extend to naturally occurring factors that facilitate similarity versus dissimilarity testing? Moreover, would performance judgments be influenced by prior performances, even if the judges were very familiar with that task? Again, our recent research suggests affirming both questions. Specifically, in this line of our research, we conducted a study in the field of gymnastics, in which solely certified and highly qualified, male, German gymnastic judges were used as participants. It was their task to evaluate the performance of two sequentially presented gymnastic routines thereby applying the official rules concerning the evaluation of male gymnasts' routines

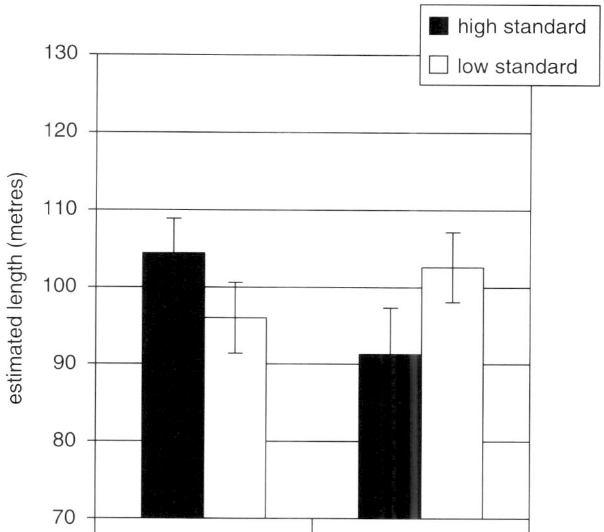

	similarity focus	dissimilarity focus
■ high standard	104.4	91.3
□ low standard	96	102.6

Fig. 1. Means and standard errors for performance judgments of target ski jump (length in meters) by standard performance (high vs. low) and induced focus (similarity vs. dissimilarity).

prescribed by the code of points (Fédération Internationale de Gymnastique, 2001). Whereas the objective performance of the former gymnast was manipulated (high vs. low), the performance of the latter athlete was kept constant on a moderate level. To vary the perceived similarity between the presented gymnasts, we experimentally manipulated their alleged nationalities. By giving the judges the information about the nationalities of the gymnasts, while other characteristics were not mentioned or were controlled, this feature of the gymnasts was considered to be a very salient attribute that would be taken into account to assess the general degree of similarity between the gymnasts. We expected the two presented gymnasts to be perceived as similar if both of them apparently had the same nationality (Australian). In this case, we predicted, judges would engage in similarity testing, hence leading to assimilation effects. If the gymnasts differed in terms of their nationalities (first Canadian, second Australian), however, dissimilarity testing would be more likely, and thus, contrast effects would occur. As depicted in Fig. 2, the results are in line with our assumptions. While performance judgments for the standard gymnast were not affected by the gymnast's ostensible nationality, performance judgments for the target gymnast critically depended on whether the two gymnasts were presented as similar to each other or dissimilar from each other. Judges, who got the information that both gymnasts had the same nationality and thus engaged in similarity testing, assimilated their performance judgment toward the standard. These judges rewarded the target gymnastic performance with a higher score if it was presented after the prior high performance rather than the prior low performance. Judges, who were informed that the gymnasts had different nationalities and thus engaged in dissimilarity testing, however, contrasted their performance judgments away from the standard. These judges evaluated the target routine with a lower score if it was presented after the prior high performance rather than the prior low performance. This pattern of means produced a significant interaction effect,

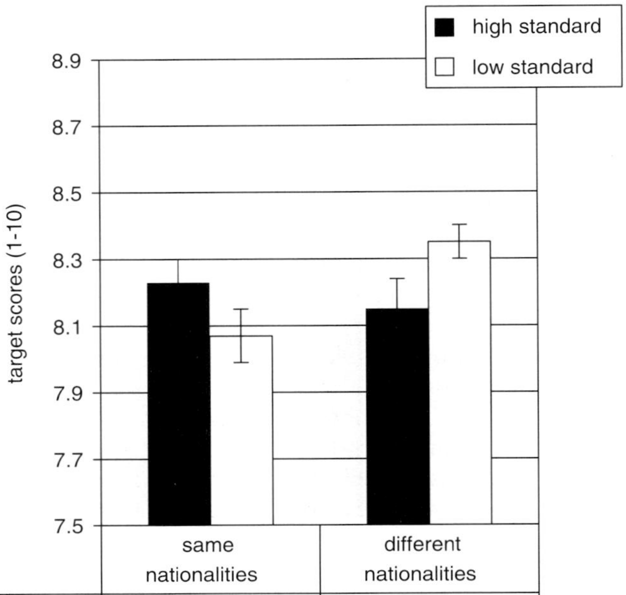

	same nationalities	different nationalities
■ high standard	8.23	8.15
□ low standard	8.07	8.35

Fig. 2. Mean performance judgments and standard errors for target gymnastic routine by standard performance (high vs. low) and alleged nationalities (same vs. different).

while none of the main effects reached the level of significance.

These findings clearly suggest that judgments of athletic performances are influenced by previously encountered achievements. Specifically, our findings demonstrate that a particular performance judgment does not only depend on the quality of the prior performance but also whether the sequentially judged athletes are perceived as similar or dissimilar from each other.

Advantages and disadvantages of comparative performance judgments

Summing up the findings of our research, we find that performance judgments in sports are influenced by previous performances of other athletes. Moreover, we demonstrate that the judgmental consequences and their moderating factor are similar to those that are typical of comparison processes. As is true for comparison consequences, it appears to be the degree of perceived similarity between two consecutively judged targets that determines whether assimilation or contrast is the judgmental outcome. In this respect, the same basic principles that determine comparison consequences also appear to apply to sequential performance judgments. This concordance suggests that comparison processes contribute to the obtained effects.

Does such a comparative evaluation result in homogenous and fair judgments for all athletes or do the assimilative and contrastive influences on the final score mark a judgmental bias, which asks for correction? On the one hand, we would assume, there is nothing wrong with comparative judgments within the realm of sports. In fact, the relativity of performances was the starting point of the present chapter. As the initial example demonstrated, Eric Moussambani performed the best swimming race of his life. However, when compared with all other participants of the swimming event, it becomes clear that his performance was not good enough to get even near an Olympic medal. On the other hand, the findings of our research go above this relative classification of performances. In the swimming example, Moussambani's performance was assessed via an objective measure. No matter whether all other participants were faster or slower than he was, the time he needed to finish the race would always stay the same. As the presented findings indicate, this is different for sports such as gymnastics, for example. Here, it is not just the ranking of performances relative to other athletes in order to choose the best performance, but rather, it is the initial assessment of the gymnasts' performance themselves that is influenced in assimilative or contrastive directions. Thus, one and the same performance of a gymnast can be evaluated differently depending on which comparison this judgment is based. This difference, which constitutes about two tenths of a point in the described gymnastics study, might have a profound effect on the final outcome. Looking at the results of the Olympics in 2004, for example, it becomes apparent that for each of the six apparatuses, on which men perform, the score of the winner and the score of the gymnast ending at the sixth position did not differ by more than two tenths of a point. From this perspective, under some conditions the assimilative and contrastive influences might result in advantages for some athletes, however in disadvantages for other athletes. In this light, it seems necessary to intervene and look for ways to prevent or correct for these biases.

Prevention and correction of comparative biases in athletic performance judgments

One way to correct for the comparative bias in athletic performance evaluation simply seems to prevent these influences from happening. Abundant literature in the field of judgment and decision making suggests that judges are particularly likely to base their judgments on normatively irrelevant information under suboptimal conditions and reduced cognitive capacity (e.g., Nisbett and Ross, 1980). At the same time, it is known that suboptimal conditions such as time pressure or extremely demanding evaluation rules are no exceptions in the task of athletic performance evaluation (Salmela, 1978a; Plessner and Raab, 1999). From this perspective, any intervention that helps the judges to safe cognitive capacity might prevent them from building their performance judgments on irrelevant influences such as the performance of a previous athlete. Thus, for example, more intense training (Ericsson and Lehmann, 1996) as well as a higher specification of the athletic judges (Salmela, 1978a), the adaptation of evaluation rules on psychological boundaries (Plessner and Raab, 1999), and the routine use of technical equipment such as video analyses (Wilson, 1976; Puhl, 1980) might be suggestions to avoid the comparative bias.

However, it has been claimed that the comparison process itself constitutes a heuristic form of information processing (Mussweiler and Epstude, 2009). The frequent use of comparisons caused by the relative nature of all human judgments might lead to a general leaning toward comparative processing. This reasoning is supported by recent findings demonstrating that comparisons often happen in a relatively spontaneous, effortless, and unintentional manner rather than being deliberately and strategically engaged (Gilbert et al., 1995; Mussweiler et al., 2004a; Mussweiler and Englich, 2005). Also in athletic performance evaluation, judges seem to engage in comparisons with previous performances in a rather automatic and effortless way as in neither our experiments nor in real athletic competitions, judges are explicitly urged to engage into a comparison between the athletes. Thus, rather than trying to prevent this automatic tendency of informational processing, it might be a more efficient approach to correct these judgmental biases. However, former research on correction processes exhibited a rather pessimistic perspective concerning people's general ability to correct for contextual influences (Wetzel et al., 1981; Wegner et al., 1985). Wilson and Brekke (1994), for example, doubted that people are able to correct judgmental biases because of their generally weak introspective abilities (Nisbett and Wilson, 1977). In contrast, more recent findings demonstrate that sometimes people do correct their judgments for influences of contextual factors (Schwarz et al., 1991; Martin and Achee, 1992; Petty and Wegener, 1993). This becomes especially likely if judges know about the source, direction, and size of the bias (Stapel et al., 1998). Thereby, this knowledge can either be retrieved from people's naïve theories (Strack, 1992; Strack and Hannover, 1996) or from explicitly given information (Petty and Wegener, 1993; Wegener and Petty, 1995). Applying these findings to the assimilative and contrastive influences on performance judgments in sports, we suggest that educating athletic judges on the contribution of comparative processes in performance evaluation might be a reasonable approach. That is, judges should not just get an unspecific warning of potential contextual influences (Stapel et al., 1998) but rather get detailed information concerning the assumed comparison process, the moderating factor of perceived similarity and dissimilarity between consecutively starting athletes, and the impact of the resulting assimilative and contrastive judgment biases. These factors could be included as part of athletic judges' training. Furthermore, to keep these influences salient, they could be explicitly pointed out in a briefing of the judges right before an athletic event, for example.

In conclusion, we have demonstrated that athletic performance judgments are not independent from one another. Rather, a preceding performance judgment clearly influences the actual target judgment in predictable ways. Specifically, the direction of the influence depends on the perceived similarity between the sequentially presented athletes. If a judge perceives an athlete as generally similar to the prior athlete, an assimilation effect will be the consequence. If consecutively presented athletes are perceived as generally dissimilar, however, contrast effects in the target performance judgment will ensue. Under specific conditions, this judgmental bias clearly results in an advantage for some athletes and a disadvantage for others. In this light, we pointed out strategies to both prevent these biases from happening in the first place as well as motivate judges to correct these biasing influences afterward. In this respect, we hope that our research contributes to a better understanding of athletic performance judgments and might help judges to make the right choice in deciding who will make their way onto the Olympic pedestal.

References

Boen, F., van Hoye, K., Auweele, Y. V., Feys, J., & Smits, T. (2008). Open feedback in gymnastic judging causes conformity bias based on informational influencing. *Journal of Sports Sciences*, *26*, 621–628.

Brewer, M. B., & Weber, J. G. (1994). Self-evaluation effects of interpersonal versus intergroup social comparison. *Journal of Personality and Social Psychology*, *66*, 268–275.

Brown, J. D., Novick, N. J., Lord, K. A., & Richards, J. M. (1992). When Gulliver travels: social context, psychological closeness, and selfappraisals. *Journal of Personality and Social Psychology*, *62*, 717–727.

Buunk, B. P., Collins, R. L., Taylor, S. E., VanYperen, N. W., & Dakof, G. A. (1990). The affective consequences of social comparison: either direction has its ups and downs. *Journal of Personality and Social Psychology*, *59*, 1238–1249.

Damisch, L., Mussweiler, T., & Plessner, H. (2006). Olympic medals as fruits of comparison? Assimilation and contrast in sequential performance judgments. *Journal of Experimental Psychology: Applied*, *12*, 166–178.

Dehaene, S., Naccache, L., Le Clec'H, G., Koechlin, E., Mueller, M., Dehaene-Lambertz, G., van de Moortele, P.-F., & Le Bilhan, D. (1998). Imaging unconscious semantic priming. *Nature*, *395*, 597–600.

Dijksterhuis, A., Spears, R., Postmes, T., Stapel, D. A., Koomen, W., van Knippenberg, A., & Scheepers, D. (1998). Seeing one thing and doing another: contrast effects in automatic behavior. *Journal of Personality and Social Psychology*, *75*, 862–871.

Dunning, D., & Hayes, A. F. (1996). Evidence of egocentric comparison in social judgment. *Journal of Personality and Social Psychology*, *71*, 213–229.

Ericsson, K. A., & Lehmann, A. C. (1996). Expert and exceptional performance: evidence of maximal adaptations to task constraints. *Annual Review of Psychology*, *47*, 273–305.

Fédération Internationale de Gymnastique. (2001). *Code depointage — Gymnastique artistique masculine [Code of points — male gymnastics]*. Switzerland: Moutier.

Festinger, L. (1954). A theory of social comparison processes. *Human Relations*, *7*, 117–140.

Förster, J., Liberman, N., & Kuschel, S. (2008). The effect of global versus local processing styles on assimilation and contrast in social judgments. *Journal of Personality and Social Psychology*, *94*, 579–599.

Gilbert, D. T., Giesler, R. B., & Morris, K. A. (1995). When comparisons arise. *Journal of Personality and Social Psychology, 69*, 227–236.

Herr, P. M. (1986). Consequences of priming: judgment and behavior. *Journal of Personality and Social Psychology, 51*, 1106–1115.

Herr, P. M., Sherman, S. J., & Fazio, R. H. (1983). On the consequences of priming: assimilation and contrast effects. *Journal of Experimental Social Psychology, 19*, 323–340.

Huttenlocher, J., & Higgins, E. T. (1971). Adjectives, comparatives, and syllogisms. *Psychological Review, 78*, 487–504.

Kahneman, D., Slovic, P., & Tversky, A. (1982). *Judgment under uncertainty: heuristics and biases*. Cambridge: Cambridge University Press.

Klayman, J., & Ha, Y.-W. (1987). Confirmation, disconfirmation, and information in hypotheses testing. *Psychological Review, 94*, 211–228.

Landers, D. M. (1970). A review of research on gymnastic judging. *Journal of Health, Physical Education and Recreation, 41*, 85–88.

Lockwood, P., & Kunda, Z. (1997). Superstars and me: predicting the impact of role models on the self. *Journal of Personality and Social Psychology, 73*, 91–103.

Martin, L. L., & Achee, J. W. (1992). Beyond accessibility: the role of processing objectives in judgment. In L. L. Martin & A. Tesser (Eds.), *The construction of social judgments* (pp. 159–216). Hillsdale, NJ: Lawrence Erlbaum Associates.

Morse, S., & Gergen, K. J. (1970). Social comparison, self-consistency, and the concept of self. *Journal of Personality and Social Psychology, 16*, 148–156.

Mussweiler, T. (2001). Focus of comparison as a determinant of assimilation versus contrast in social comparison. *Personality and Social Psychology Bulletin, 27*, 38–47.

Mussweiler, T. (2003). Comparison processes in social judgment: mechanisms and consequences. *Psychological Review, 110*, 472–489.

Mussweiler, T. (2007). Assimilation and contrast as comparison effects: a selective accessibility model. In D. A. Stapel & J. Suls (Eds.), *Assimilation and contrast in social psychology* (pp. 165–185). New York: Psychology Press.

Mussweiler, T., & Bodenhausen, G. (2002). I know you are but what am I? Self-evaluative consequences of judging ingroup and outgroup members. *Journal of Personality and Social Psychology, 82*, 19–32.

Mussweiler, T., & Englich, B. (2005). Subliminal anchoring: judgmental consequences and underlying mechanisms. *Organizational Behavior and Human Decision Processes, 98*, 133–143.

Mussweiler, T., & Epstude, K. (2009). Relatively fast! Efficiency advantages of comparative thinking. *Journal of Experimental Psychology: General, 138*, 1–21.

Mussweiler, T., & Strack, F. (2000). The "relative self": informational and judgmental consequences of comparative self-evaluation. *Journal of Personality and Social Psychology, 79*, 23–38.

Mussweiler, T., & Rüter, K. (2003). What friends are for! The use of routine standards in social comparison. *Journal of Personality and Social Psychology, 85*, 467–481.

Mussweiler, T., Rüter, K., & Epstude, K. (2004a). The man who wasn't there: subliminal social comparison standards influence self-evaluation. *Journal of Experimental Social Psychology, 40*, 689–696.

Mussweiler, T., Rüter, K., & Epstude, K. (2004b). The ups and downs of social comparison: mechanisms of assimilation and contrast. *Journal of Personality and Social Psychology, 87*, 832–844.

Nisbett, R., & Ross, L. (1980). *Human inference: strategies and shortcomings of social judgment*. Englewood Cliffs, NJ: Prentice Hall.

Nisbett, R. E., & Wilson, T. D. (1977). Telling more than we can know: verbal reports on mental processes. *Psychological Review, 84*, 231–259.

Pelham, B. W., & Wachsmuth, J. O. (1995). The waxing and waning of the social self: assimilation and contrast in social comparison. *Journal of Personality and Social Psychology, 69*, 825–838.

Petty, R. E., & Wegener, D. T. (1993). Flexible correction processes in social judgment: correcting for context-induces contrast. *Journal of Experimental Social Psychology, 29*, 137–165.

Plessner, H. (1999). Expectation biases in gymnastics judging. *Journal of Sport and Exercise Psychology, 21*, 131–144.

Plessner, H. (2005). Positive and negative effects of prior knowledge on referee decisions in sports. In T. Betsch & S. Haberstroh (Eds.), *The routines of decision making* (pp. 311–324). Hillsdale, NJ: Lawrence Erlbaum Associates.

Plessner, H., & Haar, T. (2006). Sports performance judgments from a social cognitive perspective. *Psychology of Sport and Exercise, 7*, 555–575.

Plessner, H., & Raab, M. (1999). Kampf- und Schiedsrichterurteile als Produkte sozialer Informationsverarbeitung [Judgments by officials in sports as products of social information processing]. *Psychologie and Sport, 6*, 130–145.

Puhl, J. (1980). Use of video replay in judging gymnastics vaults. *Perceptual and Motor Skills, 51*, 51–54.

Salmela, J. H. (1978a). Gymnastics judging: a complex information processing task, or (who's putting one over on who?) Part 1. *International Gymnast, 20*, 54–56.

Salmela, J. H. (1978b). Gymnastics judging: a complex information processing task, or (who's putting one over on who?) Part 2. *International Gymnast, 20*, 62–63.

Scheer, J. K., Ansorge, C. J., & Howard, J. (1983). Judging bias induced by viewing contrived videotapes: a function of selected psychological variables. *Journal of Sport Psychology, 5*, 427–437.

Schwarz, N., Strack, F., & Mai, H. P. (1991). Assimilation and contrast effects in part-whole question sequences: a conversational-logic analysis. *Public Opinion Quarterly, 55*, 3–23.

Smeesters, D., & Mandel, N. (2006). Positive and negative media effects on the self. *Journal of Consumer Research, 32*, 576–582.

Smith, E. R. (1994). Procedural knowledge and processing strategies in social cognition. In R. S. Wyer & T. K. Srull (Eds.), *Handbook of social cognition* (2nd ed., Vol. 1, pp. 99–152). Hillsdale, NJ: Lawrence Erlbaum Associates.

Snyder, M., & Swann, W. B. (1978). Hypothesis-testing processes in social interaction. *Journal of Personality and Social Psychology*, *36*, 1202–1212.

Stapel, D. A., Martin, L. L., & Schwarz, N. (1998). The smell of bias: what instigates correction processes in social judgments? *Personality and Social Psychology Bulletin*, *24*, 797–806.

Ste-Marie, D. M., & Lee, T. D. (1991). Prior processing effects on gymnastic judging. *Journal of Experimental Psychology – Learning Memory and Cognition*, *17*, 126–136.

Ste-Marie, D. M., Valiquette, S. M., & Taylor, G. (2001). Memory-influenced biases in gymnastic judging occur across different prior processing conditions. *Research Quarterly*, *72*, 420–426.

Stefani, R. (1998). Predicting outcomes. In J. Bennett (Ed.), *Statistics in sport* (pp. 249–275). London: Arnold.

Strack, F. (1992). The different routes to social judgments: experiential versus informational strategies. In L. L. Martin & A. Tesser (Eds.), *The construction of social judgments* (pp. 249–276). Hillsdale, NJ: Lawrence Erlbaum Associates.

Strack, F., & Hannover, B. (1996). Awareness of influence as a precondition for implementing correctional goals. In P. M. Gollwitzer & J. A. Bargh (Eds.), *The psychology of action* (pp. 579–596). New York: Guilford.

Tesser, A., Millar, M., & Moore, J. (1988). Some affective consequences of social comparison and reflection processes: the pain and pleasure of being close. *Journal of Personality and Social Psychology*, *54*, 49–61.

Trope, Y., & Bassok, M. (1982). Confirmatory and diagnostic strategies in social information gathering. *Journal of Personality and Social Psychology*, *43*, 22–34.

Wegener, D. T., & Petty, R. E. (1995). Flexible correction processes in social judgment: the role of naïve theories in corrections for perceived biases. *Journal of Personality and Social Psychology*, *68*, 36–51.

Wegner, D. M., Coulton, G. F., & Wenzlaff, R. (1985). The transparency of denial: briefing in the debriefing paradigm. *Journal of Personality and Social Psychology*, *49*, 338–346.

Wetzel, C. G., Wilson, T. D., & Kort, J. (1981). The halo effect revisited: forewarned is not forearmed. *Journal of Experimental Social Psychology*, *17*, 427–439.

Wilson, T. D., & Brekke, N. (1994). Mental contamination and mental correction: unwanted influences on judgments and evaluations. *Psychological Bulletin*, *116*, 117–142.

Wilson, V. E. (1976). Objectivity, validity, and reliability of gymnastic judging. *Research Quarterly*, *47*, 169–174.

M. Raab et al. (Eds.)
Progress in Brain Research, Vol. 174
ISSN 0079-6123

CHAPTER 3

A cognitive movement scientist's view on the link between thought and action: insights from the "Badische Zimmer" metaphor

Ernst-Joachim Hossner*

Department of Sport Science, Liverpool Hope University, Liverpool, United Kingdom

Abstract: The problem of a bidirectional link between thought and action is approached from the perspective of cognitive movement science. The metaphor of the "Badische Zimmer" — an adaptation of Searle's Chinese room metaphor — is used to illustrate shortcomings in the classical conception of linear information processing and to introduce some features which current theories of movement control and learning should embrace. On this basis, the case is made for a return to an ideomotor view of motor control and learning based on effect prediction (E′) as a function of the situational context (S′) and one's own motor responses (R′). The relevance of the derived concept of sensorimotor chains linking elementary S′R′E′ units in the course of motor learning is finally discussed with respect to potential implications for an integrative theory of perception, action, and decision making.

Keywords: human movement; cognition; motor behavior; decision making; ideomotor principle; serial chaining; time perception; Chinese room

Introduction: classical views on motor control and a need for revision

As these lines were written by a movement scientist whose main focus is on functional aspects of motor behavior, this paper is rooted in classical as well as current ideas on cognitive behavioral control. However, these roots will be developed further with respect to the question to what extent valuable contributions can be made to an integrative view of perception, action, and decision making.

Traditionally, the potential of movement science with respect to contributing to integrative views was limited, as classical cognitive models of motor behavior (e.g., see Schmidt, 1982) postulated a linear processing of information. In this theoretical approach, stimulus identification is defined as the first stage of information processing, followed by response selection as the second stage and response programming and execution as the third and final stage. In linear models, motor actions are assumed to be mere products of decision making and perception, without any backwards effects, that is, from response programming to stimulus identification or response selection. This strict succession from perception to action seems to fit laboratory settings preferred by experimental psychologists perfectly well, as in

*Corresponding author.
Tel.: +41-31-631-5101; Fax: +41-31-631-4631;
E-mail: ernst.hossner@ispw.unibe.ch

DOI: 10.1016/S0079-6123(09)01303-X

such settings participants typically have to respond to signals given by the experimenter in a predefined way. In real-life situations, however, things are often much less structured. In particular, movements quite frequently serve the main purpose of gathering more information about the situation at hand. As a consequence, the functional order is reversed, and succession is no longer from perceiving via deciding to moving, but instead from moving via perceiving to deciding. For this reason, when searching for bidirectional links between thought and action, the fundamental reciprocal character of movement and perception in the context of decision making has necessarily to be taken into account.

There are two ways of dealing with this critique on classical models of motor behavior. The first option is to dispense with the cognitive approach of information processing in its entirety and to consider an alternative theoretical framework that incorporates close links between perception and action. This would more or less mandatorily lead to the adoption of an ecological realistic standpoint (Gibson, 1979/1986) focusing on direct perception–action couplings. The — less radical — second option is to stick to a cognitive theoretical framework and to investigate whether the problem of bidirectional links between perception and action could also be solved *within* the framework. The second option will be pursued here.

For this purpose, the major problem has to be identified first, before possible solutions can be discussed. For this first step of problem identification, it seems worth noting that in classical cognitive models of motor control the information processing from stage to stage is simply implemented as arrows between boxes (for a critique on this sort of linear "boxology" see Neisser, 1976). By means of introspection, this problem does not seem to be a problem at all as it comes to us as indubitable truth that it is we, as individuals, who decide to move on the basis of our own perception, and that our subsequent movement directly results from this decision. The fact that the process of translating perceptions into movements or of movements into perceptions actually *is* a severe theoretical problem has most notably been substantiated by the argument of incommensurability introduced by Wolfgang Prinz (1987).

In what follows, the argument's core will be illustrated on the basis of the "Badische Zimmer" metaphor (Hossner, 2004). Furthermore, it will be shown that the resulting revival of the old idea of ideomotor control leads to some interesting consequences on the temporal organization of motor behavior. Finally, I will investigate to what extent these consequences provide some hints for the future development of integrative theories on perception, action, and decision making.

From China to Heidelberg: a hermetically locked room, internal models, and the ideomotor principle

In his famous Chinese room metaphor, John Searle (1980) describes the situation of an English-speaking man (actually himself) who is in a hermetically locked room. He has no knowledge of Chinese, yet the only input he gets from outside from time to time are questions written on paper in Chinese characters. As his task is to generate appropriate answers to these questions as an output — again by solely using Chinese characters — he has to rely on a rule ledger by looking up the Chinese input characters and writing down the Chinese output characters as described by the rule in the ledger. By this means, from an outer perspective it seems as if the Chinese room system (i.e., the man inside plus his ledger) is able to respond in a meaningful way although the Englishman inside — and this is the crucial point — has not developed any understanding for the whole process at all. In Searle's argument, this lack of understanding serves as a demonstration that algorithmic systems like computers may generate perfect responses to sophisticated questions; however, they will never evolve an understanding of the information they have processed successfully. The argument provides some obvious consequences for theories on intentionality and consciousness.

From a motor control perspective, high-level processes like intentionality and consciousness are of less interest when compared to low-level processes like movement coordination. However, the Chinese room metaphor also relates to those low-level processes, as the basic problem of motor control is to transform inputs (sensations) into outputs (movements). From a motor learning perspective, the main question then would be, how is the person inside the room able to add new rules to his ledger? In order to have a closer look at this question, let us employ the "Badische Zimmer" metaphor (Hossner, 2004).

In the "Badische Zimmer" metaphor, Searle's original configuration is changed such that Latin letters replace Chinese characters and a Chinese person is put into the room instead of an Englishman. The resulting situation is illustrated in Fig. 1. As can be seen, the Chinese person receives sensory input signals about the current situation in the external environment (S) in the transduced form to internal signals (S′). Furthermore, the Chinese person is able to produce internal signals for motor responses (R′) that in turn lead to bodily movements in the external environment (R). Some rules for connecting S′ to R′ can be found pre-existing in the rule ledger. For example, the connection between situation S_x' (standing for a ball rapidly approaching the face) and the appropriate response R_x1' (standing for closing the eyes) may be "hard-wired" as a result of evolution. For other situations, new rules are to be generated by the Chinese him/herself, for example, the rule for connecting the response R_y' (standing for hitting a ball with a tennis racket) to a given situation S_y' (standing for characteristics of the approaching ball and the current position of the own body).

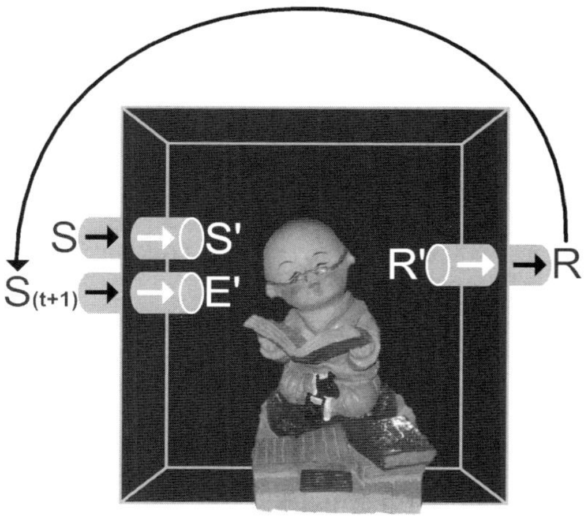

Fig. 1. The "Badische Zimmer" metaphor (based on the Chinese room metaphor introduced by Searle, 1980). The Chinese person inside a hermetically locked room generates new behavioral rules by trying out response signals R′ that in a perceived situational context S′ reliably lead to anticipatable effects E′ by changing the external situation from S to S(t+1) as a consequence of the movement R. The acquired S′R′E′ triplets can then be used in order to actively produce intended effects by inverting the (S′,R′)→E′ predictor into a (S′,E′)→R′ controller.

At this juncture, the reader may like to take some time to figure out how the appropriate response signal can be found by the Chinese person in this task, bearing in mind that the Chinese person solely has access to the internal variables and not to the external ones. A further given is that the Chinese person should not be conceptualized as a human-like cognitive system in order to avoid problems of infinite regress, that is, he/she is incapable of making intelligent inferences about constellations of variables, of abstract thoughts, of considering his/her own situation, of recognizing events, and even of labeling them by use of language. The only things he/she is capable of are to send out response signals R′, to watch constellations of sensory signals S′ and to make notes in his/her rule ledger on interdependencies between these signals. Furthermore, it should be kept in mind that the Chinese person neither "knows" what movements are caused externally by his/her response signals nor has a clue about the "meaning" of the incoming sensory signals in the external world.

The obvious problem the Chinese person faces in this situation is that the input signals S′ cannot be transformed directly into appropriate output signals R′. Because they are not written in the same "language" there is a severe lack of commensurability — the core of the argument of incommensurability given by Prinz (1987). But how can this problem be solved by the Chinese person locked in the "Badische Zimmer"? The answer to this question — and probably the final result of the reader's considerations — is that the *only* way the Chinese person can act on this situation is to try out response signals R′ in the first place (may be in extrapolation of existing experience) and to check whether these signals result in changes on the input side of the room. Of course, those internal changes from S′ to S′(t+1) are to be ascribed to external changes from S to S(t+1). If those changes can be achieved reliably by activating internal signals R′ they acquire the status of anticipatable sensory consequences of one's own action so that the internal signal S′(t+1) can be interpreted as action effect E′. In the "Badische Zimmer" metaphor, these S′R′E′ relations should be written down by the Chinese person as soon as they are detected.

The derived S′R′E′ triplets can be best understood as rules of an internal *predictor*. Whenever a response signal is sent out in a given situation a certain action effect can be predicted. So if a response signal R′ leads externally to the production of a finger movement this movement in turn reliably results in proprioceptive (feeling the finger moving) as well as exteroceptive consequences (seeing the finger moving). These effects E′ may differ with respect to the situational context – perceived as S′ – so that the same response signal may lead to different sensory consequences, for example, the perception of a letter on the screen if the finger is moved on a computer keyboard, or the perception of a tone when the finger is moved on a piano keyboard. This is the reason why not only response–effect relationships are crucial for generating predictor rules, but S′R′E′ triplets standing for a predictable effect E′ as a function of a response signal R′ and a perceived situation S′.

What should have become clear in this first step is that the Chinese person is not able to ascribe any "meaning" to input or output signals in the first place but that meaning is internally generated by relating situational changes to own responses. As such, this makes the derivation of cognitive rules fundamentally embedded in the interrelated context of perception and action. On the basis of this meaning, the acquired internal predictor can be used in order to allow for intentional behavior. If intentional behavior is defined as producing desired effects under given situational constraints, the internal *controller* that is required to achieve intentional goals is a simple inversion of the S′R′E′ rules available in the predictor: if it is known what action effect is predictable as a function of perceived situation and actual response signals, it is a simple task to identify promising response signals as a function of perceived situation and desired action effects.

It is worth noting that, although in Fig. 1 internal variables S′, R′, and E′ are clearly distinguished from external variables S, R, and S(t+1), the metaphor should not be misread as a strong dualistic approach. Instead, it is clearly recognized that the cognitive system illustrated by the "Badische Zimmer" has to construct the "outer" world solely on the basis of its internal accessible signals and that even the construction of oneself in opposition to the world is part of this internal process. Given that, the distinction between internal and external variables looses numerous explanatory power as *everything* must be located internally – including the internal–external distinction.

Developing this thought further, the metaphor may also be considered valuable when it comes to the *appraisal* of theories on motor control and learning as an external-intelligence criterion can be derived. What becomes clear from the "Badische Zimmer" metaphor is that only "internal" signals can be used by the system for optimization purposes. As a straightforward consequence, one should refrain from including "external" variables in theories on motor behavior as, by this, some "external" intelligence would be added to the cognitive system. However, intelligence has to be developed by the cognitive system itself by solely relying on "internal" accessible variables.

That the external-intelligence criterion is not as trivial as it may seem at first sight can be demonstrated by the famous schema theory of motor learning introduced by Schmidt (1975). The schema theory actually fails the external-intelligence criterion because recognition schemata are assumed to be formed by relating initial conditions and actual movement outcomes to sensory consequences of the movement. In this concept, initial conditions and sensory consequences can be defined independently as internally perceived situation S′ and internally perceived movement effect E′, respectively. Movement outcome, however, must be interpreted as external signal S(t+1), which only can be used by the system in the form of its internal transduction to E′. For this reason, an internal process of optimization based on independent values for sensory consequences and movement outcome is not feasible, as sensory consequences E′ and internally transduced movement outcomes E′ always match perfectly. Of course, from an external third-person perspective, the "actual" movement outcome can easily be distinguished from the sensory consequences that are experienced by the moving person from his/her internal first-person perspective. It does not make any theoretical sense, however, to use this knowledge about the external world — this external intelligence — when theorizing about human motor control. In failing the external-intelligence criterion therefore, the idea of recognition schema — which is central to Schmidt's schema theory — has to be rejected.

In brief, the "Badische Zimmer" metaphor provides a criterion for the appraisal of theories on motor control and learning as well as a demonstration of a viable solution of the problem of incommensurability as stated by Prinz (1987). It should be noted, however, that the solution *itself* is not a consequence of the metaphor, but that the metaphor is an illustration of well-established concepts of current cognitive psychology. In particular: the two-step structure of motor learning — an internal predictor as prerequisite for an internal controller — is the central idea of the internal modeling approach (Wolpert et al., 1995); the assumption of SRE triplets is crucial for the concept of *anticipatory behavioral control* proposed by Hoffmann (1993, 2003); and commensurate codes of sensory and motor signals, as "event codes," form the basis for the *theory of event coding* proposed by Hommel et al. (2001; for the underlying idea of "common coding" see Prinz, 1987). Although these approaches significantly differ in detail and although some specific questions still have to be answered, the overarching message seems to be that in the field of motor control and learning, the time has come for a revitalization of the idea of movement control in terms of perceived action effects — the old idea of the "ideomotor principle," which can be traced back to the *Principles* by William James (1890/1981).

As this turn-back makes cognitive approaches on motor control less cognitive, more integrated views of motor behavior are possible, merging approaches on cognitive control on the one hand with more perception-focused or system-related approaches on the other hand (see Chapter 8: The bidirectional links between decision making, perception, and action). For the purpose of identifying bidirectional links between thought and action in the context of decision making, taking a second step will prove to be helpful. This step relates to motor control in time, bearing in mind the important role of effect anticipation for motor control and learning as has been illustrated in this section.

Moving in time: psychological moments, the psychological present, and streams of action

When relating effect anticipation to perception and decision making, two questions arise: (a) assuming that these processes run in parallel (which seems to be reasonable), are there particular points in time for mutual alignment of the processes? (b) If particular points in time for alignment purposes are to be stated, what is the format of the processes' output that allows comparison? When turning to question (a) first, it makes sense to investigate motor control processes with respect to different time scales. Drawing on Pöppel's (1994) seminal work on temporal mechanisms in perception, these may be labeled as microscopic time scales concerning the elements of an elementary action, as mesoscopic time scales concerning the elementary actions themselves, and as macroscopic time scales concerning the linkage of those elementary actions to streams of action (for an overview of psychological models of time see, for example, Block, 1990).

On the microscopic time scale, it is worthwhile having a closer look onto the variables S′ and E′ of the "Badische Zimmer." Both refer to perceived events, that is, signals referring to the situational context and to the anticipated or desired action effect in the format of sensory consequences, respectively. For this reason, they basically fall into the same class of perceptions: S′ could also be defined as effect E′(t–1) of the last elementary action R′(t–1), and E′ could also be defined as initial situational condition S′(t+1) for the next elementary action R′(t+1). In contrast, these perceptions are *limited* as they deliver nothing more than a partial view of "real" situational events (which, see above, are hard to define beyond our access given by the senses); however, they are *enriched* as we typically perceive situational events by more than one sense and integrate these sensory-specific inputs into a holistic percept. With respect to the room metaphor, this process of integration is crucial in order to decide whether a situation is actually given or not, because a situation has to be given in order to be able to predict an effect of one's own action with a sufficient probability. This in turn not only poses the problem of multimodal sensory integration but also the problem of temporal integration, as the internal sensory inputs are typically not delivered to the Chinese person at the same point in physical time. In particular, for events in very close distance to the perceiver, auditory consequences are available sooner than visual ones due to the higher processing speed of the auditory as compared to the visual system of the human body. For distant events, in contrast, this succession in physical time is inverted due to the much higher physical speed of light compared to sound waves. The last mentioned effect becomes obvious in the case of the perception of a lightning strike, when the visual and auditory consequence of one and the same physical event cannot be integrated anymore into a single percept. As a result, two separated events are perceived, a flash first and a thunder later. With the exception of those extraordinary examples, however, the cognitive system is able to merge different sensory inputs perfectly to generate "psychological moments." This process of merging is crucial in order to identify events and to act on this identification appropriately.

On the mesoscopic time scale, it is important for a motor learner to be able to relate perceived situational conditions S′ and action effects E′ to internal response signals R′. With respect to the room metaphor, this poses a further temporal problem to the Chinese person. As long as the effect E′ arrives in the room as the consequence of the changed situation S(t+1), he/she has to store the sheets with the situation and response signals S′ and R′ for future reference. For this task then, a "short-term memory" is needed, thus allowing for variables to be related to each other that otherwise would not be present in the room at the same physical point in time. Hence, besides the capability for the construction of psychological moments, a system capable for motor learning must also include a "psychological present" for the purpose of continuously adding new moments and subtracting old moments. A fluid stream of time perception may be constructed within the time span of the psychological present, however, the general architecture of movement organization in time has to exhibit a more digital character due to the functional necessity of the assumption of psychological moments within a psychological present.

On the macroscopic time scale, it is less a theoretical necessity but more a reasonable theoretical extension to assume that action planning does not exclusively relate to elementary acts. When climbing stairs, for example, it becomes quite clear that not every single step has to be planned as a separate unit (which might fit the scope of a single S′R′E′ triplet best). Climbing stairs seems, rather, to be planned as a compact chunk of units where the final effect, that is reaching the top of the stairs, takes over control for all the steps leading to this final goal. Again, the problem of serial chaining has already been addressed by James (1890/1981), and further refinements have been made by Greenwald (1970); Figs. 2a and 2b draw on both James and Greenwald to illustrate how this process of chunking can be understood on the basis of an effect-related view of motor control and learning. In Fig. 2a, the situation in an early stage of learning is depicted where each elementary behavioral act has to be controlled separately, as

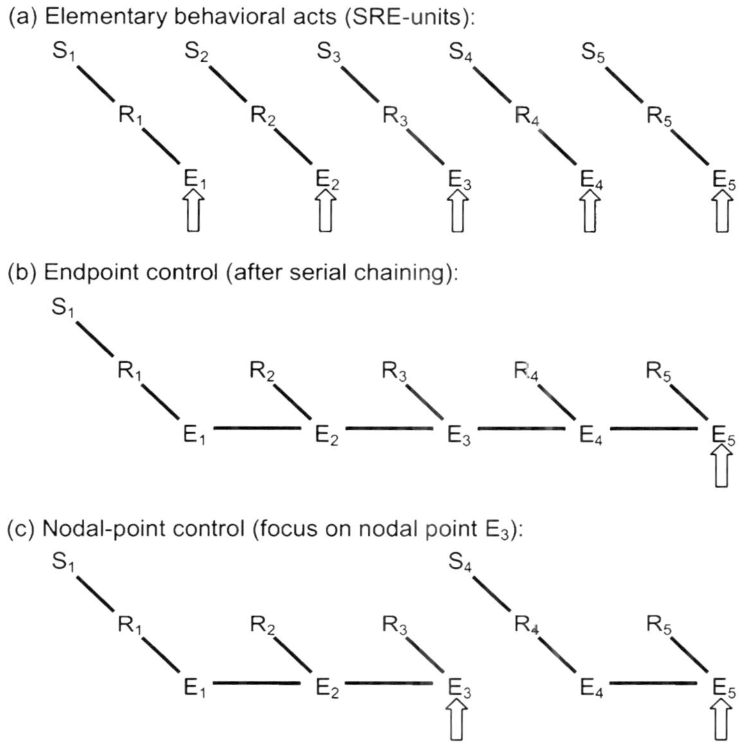

Fig. 2. Elementary behavioral actions in the initial stage of learning (a) and serial chaining after extensive experience to an anticipatory chain of action effects (b). In the expert stage of learning, in a given situational context the intended effect of the chain's endpoint is sufficient in order to start the chain as a behavioral chunk. However, there is still the need to monitor the attainment of subgoals of the chain and the opportunity to split up the chain as a consequence of a mismatch between anticipated and actual perceived effects or as a consequence voluntary pre-planning (c). *Note:* all symbols should be read as S_x', R_x' and E_x'; primes has been dropped for the purpose of clarity.

the achievement of the units' intended effects cannot be predicted to a sufficient degree. However, after extensive experience, as shown in Fig. 2b, the predictability allows for using the effect of a certain S′R′E′ unit as sensory input for the next one. In the expert stage of learning then, the initial conditions of the first unit and the intention of attaining the final unit's effects are sufficient to start the sensorimotor chain as a compact chunk. In this stage, the subunits' effects may still be monitored; the whole action, however, is directed to the endpoint of the chain, so that this type of action control may be best labeled as endpoint control.

For theories of decision making, it seems worth adding the following two specifications to the concept of anticipatory chains of action effects. First, it should be noted that the serial chaining procedure not only results in anticipations of endpoint effects in the more distant future, but also (as elaborated indepth by Hoffmann, 1993) situational constellations are anticipated. These are needed as prerequisites to make a chosen response lead to an intended effect. Given this, the active search for promising situations for putting intentions into practice becomes a crucial part of action control: if there is urgent need for refueling the car, traffic signs indicating a petrol station pop out immediately as soon as they enter sight. Second, it seems worth noting that although in Fig. 2b a succession of effects of equal importance is depicted, the reliability of effect predictions may differ dramatically between the units. An experienced tennis player, for example, would perfectly approach the ball and hit it with the goal of passing the opponent longline; there is much less predictability, however, with respect to the next intentional subgoal as the opponent may or may not succeed in intercepting the ball as a volley. Consequently, the subeffects of a chain should be interconnected as a function of the uncertainty with respect to achieving the next subgoal. The resulting different strengths in connections then will emulate a hierarchy for motor control. There is no need, however, to assume that this hierarchy is implemented as such; the assumption of different strengths of connections between S′R′E′ units is superior with respect to the criterion of parsimony.

In brief, on the basis of the developed theoretical framework, the above posed question (a) on mutual alignment of processes of effect anticipation, perception, and decision making with respect to motor control in time can be answered as follows: experts' motor behavior relies on chunks of elementary actions; these elementary actions can be defined as S′R′E′ units; S′R′E′ units have to be run within the time span of a psychological present; the sensory building blocks of these units correspond to psychological moments. Given that, the moments of effect anticipation under situational constraints seem to be good candidates for the mutual alignment of the processes of perception, action control, and decision making. As a consequence, the follow-up question (b) concerning the format of this alignment, is trivial because: perception corresponds to sensory signals S′; action is assumed to be based on the perceived difference between the current situation S′ and the desired effect E′; and decision-making processes aim at exactly this difference. As *all* these variables are written in the language of perception, the problem of different formats simply disappears.

Final remarks: some speculations on the role of motor control mechanisms for theories on decision making

For scholars focusing on processes of decision making, the message to be derived from what has been discussed from a movement science perspective seems to be encouraging. Not only could a solution of the problem of incommensurability between sensory and motor codes be provided by returning to the idea of the ideomotor principle; beyond that, this solution offers a promising starting point in the search for synergies between approaches that stem from quite diverged traditions — on a molecular level with respect to theories of motor control as well as on a molar level with respect to the integration of theories of action, perception, and decision making.

On a more molecular level, this opportunity for synergies becomes obvious with respect to theories focusing on more top-down processes of cognitive motor control on the one hand and theories focusing on more bottom-up processes of constrained movement coordination on the other hand. Without any doubt, the ideomotor idea of controlling movements in time from one perceived action effect to the next one brings the cognitive-based conceptualization quite close to interacting views of perception and action as specified in the concepts of the perception–action cycle (Neisser, 1976), the "Gestaltkreis" (Gestalt circle, von Weizsäcker, 1973), or even of the ecological view on direct perception (Gibson, 1979/1986). Furthermore, and on a more molar level, it is important to note that the crucial role of intended action effects in ideomotor approaches provides a straightforward link to theories on decision making, as decisions are naturally related to intention and the desired effects of one's own actions. The emphasis of perception as a crucial variable for motor control in turn allows for considerations on bidirectional links between thought and action.

Taken together, these synergies lead directly to an integrative view that is based on the assumption of continuous, circular and parallel processes of perceiving, moving, and decision making, as it is described in more depth in Chapter 8: The bidirectional links between decision making, perception, and action. A specific point that could be added to this model from the special perspective developed throughout this chapter relates to the necessity of discrete points of accentuated check as to whether the current perception fits the current intentional status: if there is a fit, the sensorimotor chain can be processed further; if intended subgoals are not achieved, decision making is demanded.

Although this idea of accentuated checking points is developed here from a more cognitive perspective on motor control, it is worth noting that related concepts can be found in the literature proposed by authors who are deeply rooted in dynamic system theory of movement coordination, in particular Beek's (1989) idea of "anchoring" behavior at meaningful points in time (see also Byblow et al., 1994) and the *referential behavior theory* proposed by Pressing (1998, 1999). This, once again, demonstrates that a converging picture can be possible for the purpose of linking thought and action bidirectionally.

A final comment should be made on the limitations of the overall perspective proposed in this chapter. Although the emerging integrative picture provides promising synergies between current theories of action and perception and although it can been stated on this basis that motor, perceptual, as well as decision-making processes should be thought of as running in parallel, the nature of decision making itself is a question that has been left open. Of course, this question simply cannot be answered from the perspective of a movement scientist; however, some implications are to be identified.

First, and as already stated above, a sufficient condition for interrupting an already running sensorimotor chain is fulfilled, if the monitoring of the chain's subgoals reveals a mismatch between anticipated and actually perceived effects. When failing to take a step when climbing stairs, for example, this mismatch requires new decision-making processes. Second, and as illustrated in Fig. 2c, a splitting up of an already formed chain could also result from voluntary processes of initial action planning. In the expert stage of learning a gymnastic skill, for example, the emphasis of the attentional focus may be placed in advance on a crucial point of the underlying chain — in sport practice often termed "key point" or "nodal point." Third, for the parallel decision-making strand — and definitely beyond the scope of movement science — it might be best to assume a weighting procedure to prioritize options. After this prioritization has been done, an already running sensorimotor chain may serve as a "shield" against other potentially important goals, thereby preventing the actor from a permanent switching between competing goals. In contrast, interrupting an already running chain must always remain an option, as it may either be the case that other goals become more important in the course of pursuing a particular goal or that the environment may change in a nonpredicted manner. In those cases, an entirely new decision has to be made.

Integrating these implications as well as the consequences derived from the perspective of cognitive movement science into a — still to be evolved — overarching theory on the bidirectional link between thought and action seems to be a challenging, but promising endeavor for the future.

References

Beek, P. J. (1989). *Juggling dynamics*. Amsterdam: Vrije Universiteit.

Block, R. A. (1990). Models of psychological time. In R. A. Block (Ed.), *Cognitive models of psychological time* (pp. 1–35). Hillsdale, NJ: Lawrence Erlbaum Associates.

Byblow, W. D., Carson, R. G., & Goodman, D. (1994). Expressions of asymmetries and anchoring in bimanual coordination. *Human Movement Science, 13*, 3–28.

Gibson, J. J. (1979/1986). *The ecological approach to visual perception*. Hillsdale, NJ: Lawrence Erlbaum Associates.

Greenwald, A. G. (1970). Sensory feedback mechanisms in performance control: with special reference to the ideo-motor mechanism. *Psychological Review, 77*, 73–99.

Hoffmann, J. (1993). *Vorhersage und Erkenntnis* [Prediction and reason]. Göttingen, Germany: Hogrefe Verlag.

Hoffmann, J. (2003). Anticipatory behavioral control. In M. Butz, O. Sigaud, & P. Gerard (Eds.), *Anticipatory behavior in adaptive learning systems* (pp. 44–65). Berlin: Springer.

Hommel, B., Müsseler, J., Aschersleben, G., & Prinz, W. (2001). The theory of event coding (TEC): a framework for perception and action. *Behavioral and Brain Sciences, 24*, 869–937.

Hossner, E.-J. (2004). *Bewegende Ereignisse* [Moving events]. Schorndorf, Germany: Hofmann.

James, W. (1890/1981). *The principles of psychology*. Cambridge, MA: Harvard University Press.

Neisser, U. (1976). *Cognition and reality: principles and implications of cognitive psychology*. San Francisco: Freeman.

Pöppel, E. (1994). Temporal mechanisms in perception. *International Review of Neurobiology*, *37*, 185–202.

Pressing, J. (1998). Referential behavior theory. In J. P. Pieck (Ed.), *Motor behavior and human skill* (pp. 357–384). Champaign, IL: Human Kinetics.

Pressing, J. (1999). The referential dynamics of cognition and action. *Psychological Review*, *106*, 714–747.

Prinz, W. (1987). Ideomotor action. In H. Heuer & A. F. Sanders (Eds.), *Perspectives on perception and action* (pp. 47–76). Hillsdale, NJ: Lawrence Erlbaum Associates.

Schmidt, R. A. (1975). A schema theory of discrete motor skill learning. *Psychological Review*, *82*, 225–260.

Schmidt, R. A. (1982). *Motor control and learning. A behavioral emphasis*. Champaign, IL: Human Kinetics.

Searle, J. R. (1980). Minds, brains, and programs. *Behavioral and Brain Sciences*, *3*, 417–457.

von Weizsäcker, V. (1973). *Der Gestaltkreis* [The Gestalt circle]. Stuttgart, Germany: Suhrkamp.

Wolpert, D. M., Ghahramani, Z., & Jordan, M. I. (1995). An internal model for sensorimotor integration. *Science*, *269*, 1880–1882.

M. Raab et al. (Eds.)
Progress in Brain Research, Vol. 174
ISSN 0079-6123

CHAPTER 4

Perceiving and moving in sports and other high-pressure contexts

Raôul R. D. Oudejans* and Arne Nieuwenhuys

Research Institute MOVE, Faculty of Human Movement Sciences, VU University Amsterdam, Amsterdam, The Netherlands

Abstract: From an ecological psychological perspective, the current chapter discusses the influence of anxiety on the perception, selection, and realization of affordances in sports and other high-pressure contexts, particularly police work. Drawing on recent developments in the anxiety-performance literature (i.e., the attentional control theory) the authors' experimental work on the impact of anxiety on perceptual-motor performance is described. Furthermore, several experiments showing positive effects of training with anxiety on performance under pressure are discussed. We argue that through reality-based training with anxiety, processes that underlie performance might be recalibrated to the new task constraints, thereby preventing a degradation of eventual performance in stressful situations.

Keywords: ecological psychology; affordances; attentional control theory; training with anxiety

Introduction

There are two important similarities between professional sports and other high-pressure contexts, such as the work of fire fighters and police officers. First, expert athletes, fire fighters, and police officers often have to perform under conditions where the stakes are high, as a game, a championship, a building, or even several lives are at risk. Performing in those conditions may be accompanied by physical (such as the heat in extinguishing a fire) or mental stress (such as increased arousal or anxiety), putting additional constraints on performance (Cooper et al., 1982; Gould and Udry, 1994; Smith et al., 1997, 2001b; Anderson et al., 2002; Berg et al., 2005). Second, all of these contexts are characterized by several possibilities for and restrictions to action that may or may not be recognized and realized. In soccer, for instance, a player may have several options without the ball (e.g., jump away from an opponent or not, run for the ball, sprint into the open space, or await a better opportunity) while passing, shooting, or dribbling are only possible in ball possession. Similarly, in dealing with a criminal opponent a police officer may shout verbal commands to the opponent, seek cover, or wait for back up, while other options (e.g., shoot at the opponent, or run away) are perhaps not possible. Whether or not certain options are recognized, selected, and realized may determine the eventual performance and the outcome of the event.

The general goal of this chapter is to present our view on the effects of psychological factors such as attention and anxiety on perceiving and

*Corresponding author.
Tel.: +31-20-598-8541; Fax: +31-20-598-8529;
E-mail: r.oudejans@fbw.vu.nl

DOI: 10.1016/S0079-6123(09)01304-1

moving in sports or other high-pressure contexts, particularly police work. One main objective is to unravel the role of anxiety in perceiving and realizing affordances (possibilities for action offered by the situation). As our view is firmly based on Gibson's ecological approach to visual perception (1979/1986), also known as direct perception, we will start with an outline of this theory in the first part of this chapter. Next, we will discuss our research into the effects of anxiety on perception and action, followed by a discussion of our studies on the positive influence of training with anxiety on eventual performance under pressure. The idea behind this line of research is that practicing with anxiety can prevent choking in perceptual-motor performance in experts and beginners alike, as one acclimatizes to the specific processes accompanying anxiety. This holds not just for athletes but also for, for instance, police officers having to fire their weapon in the line of duty. We will end with theoretical implications and concluding remarks.

The ecological approach to perception

Ecological psychology is the approach to the coordination of animal activity that takes as its departure points both animal–environment mutuality and the coupling between perception and action. Animals do not live in a vacuum, but in their environments in which they eat, sleep, mate, move around, etc. In addition, an environment is only an environment when it is occupied by an animal. In other words, animal and environment make an inseparable pair. One cannot explain what an animal is doing without giving due attention to the environmental properties that make the animal's actions possible. Similarly, perception and action are tightly coupled; they imply one another. Not only is perception supposed to enable action (a proposition with which no one would disagree), action is also considered essential for perception (a proposition that is less obvious) as activity creates information (Gibson, 1979/1986).

In the information-processing approach to perception, the stimuli (e.g., retinal images and sound pressure waves) are thought to provide a poor basis for perception. Hence, the act of perception is seen as a recreating process (Wilson, 2002). The observer must elaborate the impoverished input with help of memory and other cognitive processes, that is, processes needed to reconstruct the world, its objects, and events, into meaningful things. The observer has to create his/her own perception of the world "out there" from the meaningless building blocks that enter his/her senses. Despite the impoverished input, perception is recognized as being very rich, elaborate, and accurate. Thus, the richness of the world is considered to exist in perception, but not in the stimuli impinging upon the observer (Michaels and Carello, 1981). This discrepancy has to be solved by the observer with the aid of internal processes and interpretation.

As a result of its emphasis on the way that perceptual information is internally processed, this information-processing approach to perception has initially neglected the importance of the environment for perceiving and acting animals. Theories both in the field of perception and in the field of movement coordination assume in one way or another that it is the animal who gives meaning and structure to perception and to action. A meaningful internal representation of the world out there (a so-called "percept") can only be obtained via a process of association and inference, because the stimulation of receptors only provides an impoverished physical image (the "doctrine of stimulus poverty", Fodor, 1983). Action (meaningful changes of the environment by the animal) too can arise only on the basis of internal representations of the future goal, including the way to achieve this goal.

Gibson proposed that the solution to the problem of how perception is meaningful should not be sought in the animal itself, or in the brain, but in the interaction between animal and environment. Further, he argued that only if the environment is described in animal-relevant terms, and vice versa, can an adequate theory of perception and action arise. Therefore, Gibson (1979/1986) introduced his direct perception account or the theory of information pickup. We now briefly consider three essential concepts in the theory that figure significantly into this chapter. They are *information*, *affordances*, and *active exploration*.

Information

Information in the Gibsonian sense is a concept in which animal and environment clearly come together. Gibson was one of the first to recognize the richness of the information that is available in the flow fields around us. He argued that stimulus information for perception is not poor and meaningless but can specify properties of the environment. Therefore, information in the Gibsonian sense can be conceived as specific to objects, events, and actions. It specifies meaningful relationships between animals and their environments and therefore can guide the behavior of animals. The specificational character of information implies that it is "lawfully related to its source" (Turvey et al., 1981, 1990; Fajen, 2007).

An important aspect of information in the Gibsonian sense is that it points in two directions, to the environment and to the animal. When animals inter*act* with their environments by means of actions, these actions are "neither triggered nor commanded but *controlled*. They are constrained, guided, or steered, and only in this sense are they ruled or governed. In addition, they are controlled not by the brain but by information, that is, by seeing oneself in the world" (Gibson, 1979/1986, p. 225). Thus, information is crucial to the actions of an animal in an environment. By the importance of "seeing oneself in the world", Gibson argues that perception of the environment also involves perception of the self. One cannot perceive something in the environment without at the same time, in one way or another, perceiving oneself. Thus, perception of the self and perception of the environment go together; information about the environment is always information relative to an observer and information for an observer. Therefore, information can be conceived as the bridge between an animal and its environment and it cannot be usefully described without considering both. Given this definition of information, perception of the environment is understood as picking up information specific to the environment. Saying this is easy but how does information become to be specific to its source in the first place?

Optical information is present in the light around us. Ambient light is usually reflected many times and in many directions against the particles in the atmosphere and the surfaces of the environment. It is simply the light that surrounds us. The scatter-reflection of light by surfaces in the environment imparts structure to ambient light. Structured ambient light is called the ambient optic array. Because the optic array is ambient light that is lawfully structured by reflection in the environment, the structure of the ambient optic array can specify surface layout. Hence, the ambient optic array comprises information about the environment and picking up this information is tantamount to perceiving the environment.

To reiterate, information in the Gibsonian sense is the first underpinning of the theory of direct perception. The concept of affordances is the next concept in which animal and environment as well as perception and action come together (Turvey, 1992; Michaels and Beek, 1995).

Affordances

In Gibson's theory of direct perception, perception is seen as an active pickup of meaningful information specifying action possibilities, or affordances. Affordances are the behavioral possibilities of an environmental layout taken with reference to a particular animal. A ball, for instance, affords throwing, hitting, catching, avoiding, or being hit in the head. Water affords drinking, swimming, splashing but it does not offer a surface of support for animals like us, although it may for other organisms. The ground surface affords standing, sitting, lying, running, etc. The important thing is that the description of the environment is in terms of possibilities for behavior. We perceive the environment in terms of possibilities it offers for action, that is, in terms of what it *affords*. The claim is that information specifying these affordances is present in the structure of light, sound, etc. and will be picked up by the active animal. As a result of this pickup the animal will perceive the affordances of the environment; it will perceive what it can do in and with the environment. With the concept of affordances, the perception of meaning is introduced in a very natural way. Perceiving affordances is tantamount to perceiving what the environment means to the observer.

In other words, the concept of an affordance relation is a treatment of meaning; it is intended as a way of describing the surrounding surfaces and substances in animal-*relevant* dimensions, so that an individual animal does not have to *add* meaning to that which is "merely" physical (Shaw et al., 1982, p. 196). Thus, an ecological definition of an environment as a set of affordances "attempts to answer the pragmatic question of what an environment means to an animal. The answer given is that it means what an animal can in principle do or is in practice constrained from doing in that environmental context" (Shaw et al., 1982, p. 196).

Active exploration

Another core idea in which perception and action meet in the theory of direct perception is the idea that perception usually involves active exploration (Gibson, 1979/1986; Gibson, 1988; Turvey et al., 1990; Michaels and Beek, 1995). Perception is no longer conceived as the impingement of stimuli onto a (qua movement) passive observer. Instead, perception entails activity in the sense of moving (e.g., locomoting, hefting, rubbing). Looking around, scanning, and exploring the environment are most common activities of ordinary perception. Thus, an important aspect of the pickup of information is that it is deeply intertwined with movement itself. Not only is perception supposed to enable movement, movement is also considered essential for perception. Endorsing the intimate relation between perceiving and moving has consequences for the study of behavior.

An experimental demonstration of the importance of exploratory activity for the perception of affordances is provided by Mark (1987; Mark et al., 1990). Mark examined the perception of action boundaries of sitting. Observers had to judge the maximal sittable height of a surface they were seeing in front of them. While people appeared very accurate at this, Mark disturbed the functional relation between observers and their environments by having them wear 10 cm high blocks under their feet. Because this changes the action capabilities of the perceiver as well as eye height (which is considered to be involved in the scaling of sittability to the measures of the body), perceivers at first misjudged maximal sittable height in the direction that was expected. However, information-gathering activities, such as locomotion, head turning, and leaning, enabled almost every observer to quickly (within 12 trials) retune judgments of his/her critical action boundary for sitting (Mark, 1987; Mark et al., 1990). In addition, it appeared that not permitting observers to engage in exploratory behavior prevented them from recalibrating their action capabilities to the affordances of the environment (Mark et al., 1990; see also Fajen, 2007). Activity is obviously required to learn about the new relation between observer and environment. However, even in the no-block conditions, exploratory behavior improved performance. Thus, exploratory behavior is crucial to perception. More generally, movement, no matter how small and whether exploratory or performatory, creates information and this creation of information is an essential aspect of perception and action.

Ecological psychology: summary

In sum, the theory of direct perception is about how animals perceive and act in their environments. Contrary to information-processing theories, the claim is that meaningful information is available in the sea of energy around us. This information specifies the affordances of the environment. It specifies what animals can do with and in the environment. Simply picking up this information results in perceiving the affordances. A crucial characteristic of this pickup of information is that it is often an active process in which movement is as important to perception as perception is to movement.

Before we continue our discussion on perceiving and moving in high-pressure contexts, it is important to note that within embodied cognition there are several recent developments within cognitive science and other fields reflecting ideas that are in line with ecological psychology. As such, "there is a growing commitment to the idea that the mind must be understood in the context of its relationship to a physical body that interacts with the world" (Wilson, 2002, p. 625). Within this view, knowledge and perception of the world around us can be considered as based upon neural processes that reflect our previous actions and experiences (e.g., Wilson, 2002; Niedenthal et al., 2005; Proffitt, 2006; Beilock, 2008; Chapter 1: Grounding cognition in action: expertise, comprehension, and judgment; Chapter 17: Perceptual decision making: a bidirectional link between mind and motion). Future work is needed to find out in what regard these new developments add to our understanding of perceptual-motor control in high-pressure contexts.

Perceiving and moving in sport and other high-pressure contexts: effects of anxiety

Regarding sports, Fajen et al. (in press) correctly remark that perceiving and realizing affordances is at the heart of any sport activity. Our claim goes even further, namely that sports exist by virtue of affordances, as the goal of an attacking team or player, for instance, is to create as many possibilities for action (and eventually for scoring) as possible, while the defenders' main aim is to prevent possibilities for action of the attackers. Other sports, such as track and field, consist of comparing absolute action capabilities; who runs a particular distance in the fastest time, who can jump the highest or the furthest, etc. In ball games, of all possible actions in time only a limited number is executed. Somehow there must be a selection both in perception and in the realization of affordances. As for perception, we contend that there is a selection of the information that is picked up. Attention cannot capture all available information at the same time. In selecting affordances, relevant information concerning these affordances should be picked up, while irrelevant information should be ignored. In realizing affordances it is possible that more than one affordance is perceived while only one can be realized. It is impossible to both shoot at goal and pass at the same time. Only the eventual action reveals what the selection has been. Similar examples can be provided for police work, where different situations afford different actions, and where proper perception, selection, and realization of affordances are essential in bringing an event to a proper ending.

As mentioned in the Introduction, an important constraint to perception and action in these contexts is the pressure to perform well, which is often accompanied by anxiety. Our research discussed below aims to understand how anxiety influences perceiving and realizing affordances in such contexts. In the remainder of this section we will describe several studies from our laboratory in which we systematically investigated the influence of anxiety on perceiving and moving. After that we will address the question whether training with anxiety can help in ameliorating performance in stressful situations.

Anxiety and performance

Feelings of performance pressure and, more general, anxiety are likely to arise when a person doubts his/her ability to cope with the demands of a stressful situation (Woodman and Hardy, 2001; Hanton et al., 2004). According to Schwenkmezger and Steffgen (1989, pp. 78–79) "anxiety can be regarded as a broad concept for a number of very complex emotional states and processes that occur as a result of threat. This threat is related to the subjective evaluation of a situation and concerns jeopardy to one's self-esteem during performance or social situations, physical danger, or insecurity and uncertainty." When anxiety and/or the pressure to perform well increase, people tend to perform below expectations given their actual level of skill. In the literature this phenomenon has been termed *choking* (Baumeister, 1984). Choking under pressure is particularly known from sports (e.g., golf putting, Lewis and Linder, 1997; Beilock and Carr, 2001; soccer dribbling, Beilock et al., 2002; and baseball, Gray, 2004), but has been shown to also occur in a variety of other settings, including mathematical problem solving (Beilock et al., 2004) and police work (Murray, 2004).

To explain the mechanisms that are involved in performance decrements under pressure and the influence of anxiety on performance, several theoretical models have been put forward (e.g., Easterbrook, 1959; Wine, 1971; Baumeister, 1984; Sarason, 1988; Eysenck, 1992; Eysenck and Calvo, 1992; Masters, 1992; Eysenck et al., 2007).

Based on our ecological perspective, presented above, we adopt a process-oriented approach to the anxiety–performance relationship, investigating the effects of anxiety on the perception, selection, and realization of affordances. To this end, we will focus primarily on the attentional control theory (Eysenck et al., 2007), an emerging theory that provides a comprehensive account of the mechanisms behind the effects of anxiety on performance and which has recently been developed on the basis of processing efficiency theory (Eysenck and Calvo, 1992). Although the attentional control theory and processing efficiency theory are claimed to have most relevance to cognitive performance, several studies have provided empirical support for the processing efficiency theory with respect to perceptual-motor tasks (e.g., Mullen and Hardy, 2000; Smith et al., 2001a; Williams et al., 2002b; Murray and Janelle, 2003; Mullen et al., 2005; Wilson et al., 2006; Murray and Janelle, 2007; Wilson et al., 2007; Nieuwenhuys et al., 2008). In the next section we will discuss the attentional control theory in more detail and in light of another more prevailing explanation for decrements in perceptual-motor performance under pressure, the explicit monitoring hypothesis (see Beilock and Gray, 2007).

Attentional control theory

In short, the attentional control theory (Eysenck et al., 2007) is based on the assumption that we have two attentional systems: one is goal-directed and based on our intentions, knowledge, and current goals, and the other is stimulus driven, and responds maximally to the most salient or behaviorally relevant cues (Posner and Petersen, 1990; Corbetta and Shulman, 2002). According to Eysenck et al. (2007) anxiety influences performance by reducing the processing and temporary storage capacity of working memory (see Baddeley, 1992), a brain system which is used by the goal-directed attentional system to actively draw attention to a task and to inhibit distraction or interference from task-irrelevant stimuli and/or responses. With anxiety, the balance that normally exists between the goal-directed and stimulus-driven attentional system is disrupted, causing an increase in the influence of the stimulus-driven system and a decrease in the influence of the goal-directed system (e.g., Bishop et al., 2004). Therefore, by reducing *attentional control*, anxiety increases the allocation of attention to threat-related, task-irrelevant stimuli and deciding how to respond, while at the same time reducing on-task attentional focus. It is assumed that threat-related and/or distracting stimuli may exist externally as well as internally. For example, when an athlete focuses his/her attention on offensive crowd noise (an external stimulus) it is likely that relevant information about the task cannot be picked up. In contrast, attention to worry (an internal stimulus) proceeding from, for instance, performance pressure might also lead to missing task-relevant information.

By concentrating on the detection of task-relevant information and the ability to distinguish between task-relevant and task-irrelevant information, the attentional control theory envelops a central tenet of ecological psychology. Furthermore, the theory is generally in line with several earlier models of anxiety and performance (e.g., Wine, 1971; Sarason, 1988; Eysenck, 1992) and more or less incorporates some of the specific predictions of others (Easterbrook, 1959; Baumeister, 1984; Masters, 1992; also see Lewis and Linders, 1997; Beilock and Carr, 2001; Wilson et al., 2007). Going beyond these models, however, the attentional control theory also provides an explanation for why anxiety does not necessarily have to lead to a decrement in performance. According to Eysenck and his colleagues (Eysenck and Calvo, 1992; Eysenck et al., 2007), negative effects of anxiety may be compensated for by self-regulatory processes, involving increased on-task effort and additional activities to improve or maintain performance. As such, anxiety primarily affects the *efficiency* (i.e., process) of task execution while, as a result of the extra effort invested, the *effectiveness* (i.e., outcome) of performance remains unaltered or might even improve.

As an aside, research on the explicit monitoring hypothesis (e.g., Beilock and Carr, 2001; Gray, 2004; Jackson et al., 2006) seems to suggest that in perceptual-motor tasks a conscious step-by-step control of movement execution (skill-focused attention) is responsible for a breakdown of expert performance under pressure, as it disrupts the automatic execution of the task at hand. Thus, according to this line of research, it is conscious control rather than distraction causing these performance decrements. However, in the studies in question attentional focus was always experimentally manipulated, also in the high anxiety conditions. Therefore, it still remains to be seen whether it is skill-focused attention or distraction that (most often) occurs naturally and spontaneously when performers are under pressure. Anecdotal evidence (e.g., Gerrard, 2007), psychological research (Wine, 1971; Sarason, 1988) as well as studies examining athletes' thoughts during actual competition, thus, outside the laboratory setting (e.g., Hatzigeorgiadis and Biddle, 2000, 2001), seem to suggest that rather than skill-focused attention all kinds of distracting thoughts and worries naturally occur when performers are under pressure (see also Wilson and Smith, 2007). Therefore, acknowledging the potentially debilitative effects of explicit monitoring in expert perceptual-motor performance, we feel that in a natural (sporting) environment choking is most often caused by changes in attention to task-irrelevant stimuli rather than to step-by-step skill execution. As such, the attentional control theory (Eysenck et al., 2007) appears to provide a more universal framework in explaining the underlying mechanisms of performing perceptual-motor tasks under pressure (for experts as well as novices). Still, a proper assessment of what happens to attention (most frequently) when performers are put under pressure is needed.

The effects of anxiety on perceiving and moving on a climbing wall

Returning to the predicted changes in attention and information detection due to anxiety, one might contend that anxiety — similar to other state variables such as fatigue or injury (Fajen et al., in press) — may lead to alterations in the perception, selection, and realization of affordances. In fact, recent investigations by Proffitt and his colleagues have shown that differences in individuals' perceptual judgments (e.g., in estimating the geographical slant of a hill or the distance of a certain extent) occur when changes are induced in their physiological or psychological state, for instance, when wearing a heavy backpack, after a tiring run, or when they are afraid of falling (see Proffitt, 2006 for an overview of this line of research). Recently, we conducted a series of experiments in our own laboratory, in which we systematically investigated the influence of anxiety on the perception, selection, and realization of affordances in climbing (Pijpers et al., 2005, 2006; Nieuwenhuys et al., 2008).

Pijpers et al. (2005) asked novices to climb identical traverses low (low-anxiety) and high (high-anxiety) on a climbing wall. As expected, anxiety appeared to affect movement behavior (i.e., the realization of affordances), which was indicated by longer climbing times, longer grasping of holds, and slower movements in the high- compared to the low-anxiety condition.

In a follow-up study, Pijpers et al. (2006) used a similar experimental setup to manipulate anxiety, and reported three experiments on the relation between anxiety and perception. In the first experiment, participants judged their maximal overhead reaching height and executed maximal overhead reaches. As expected, an increase in anxiety (high on the wall) led to decreases in actual as well as perceived maximal reaching height. To test whether these changes in (perceived) action capability were accompanied by changes in the selection of affordances, we conducted a second experiment. In this experiment participants again climbed identical traverses high and low on the wall, which now consisted of an abundance of holds to allow for different action possibilities. As expected, increased anxiety led to increases in climbing times as well as the numbers of holds that were used, implying that participants indeed selected different affordances as a result of the observed changes in perceived reaching ability. However, the actual numbers of holds that were used implied a decrease in reaching distance that was larger than could be expected on the basis of changes in maximal reaching ability alone. To investigate this further, and to gain more insight into the attentional mechanisms that supposedly underlie anxiety-induced changes in perceptual-motor behavior, we conducted a third experiment. In this experiment participants climbed the same traverse, but were now also asked to detect lights that were projected around them on the climbing wall while they were climbing. As fewer lights were reported in the high-anxiety condition we concluded that with increased anxiety, attention was more narrowly focused on information that was relevant for the climbing task, whereas information that was less relevant for the climbing task (i.e., projected lights) was overlooked. In other words, in addition to the observed changes in action capability, anxiety apparently led to changes in information detection (Eysenck et al., 2007; see also Bootsma et al., 1992; Stefanucci et al., 2008; Teachman et al., 2008).

From the perspective of the attentional control theory (Eysenck et al., 2007), one could contend that the longer climbing times and greater numbers and durations of movements that were found, in fact, indirectly indicate a decrease in processing efficiency as anxiety increased. Namely, although performance *effectiveness* (operationalized in this experiment as being able to reach the other side of the climbing wall without slipping or falling down) remained unaltered, as participants still managed to perform the climbing task in the high-anxiety condition, additional resources (e.g., longer climbing times and movement durations) and activities (e.g., more movements) were needed to achieve this (i.e., performance *efficiency* decreased). To test this hypothesis, and to better understand what happens with attention when performers are anxious we recently conducted an experiment in which information detection, and the perception, selection, and realization of affordances were measured all at once by analyzing both movement and gaze behavior in the same setting (Nieuwenhuys et al., 2008).

In this experiment, novices were again asked to climb horizontal traverses low (low-anxiety) and high (high-anxiety) on a climbing wall. Similar to the study by Pijpers et al. (2006) the traverses consisted of an abundance of holds to allow for the selection of different action possibilities. Replicating the findings of Pijpers et al. (2005, 2006), participants showed longer climbing times, grasped holds longer, spent more time moving their hands and feet, and executed more movements in the high-anxiety condition than in the low-anxiety condition. In addition, changes in movement behavior were accompanied by increases in total and average fixation duration and the number of fixations, which were primarily directed at the holds used for climbing. Furthermore, participants appeared to need more time to extract relevant information from the handholds when they were anxious compared with when they were not anxious. These findings provide evidence for a decrease in processing efficiency as anxiety increased (Eysenck and Calvo, 1992; Eysenck et al., 2007). Finally, it also appeared that while the mean distance of hand movements decreased with anxiety, the mean distance over which participants executed their fixations did not. This confirmed that in the high-anxiety condition different action possibilities were selected, as anxious participants apparently preferred handholds that were closer over handholds that were further away, despite the fact that they had visually explored the same handholds as without anxiety.

Conclusions

Based on these experiments (i.e., Pijpers et al., 2005, 2006; Nieuwenhuys et al., 2008), we can conclude that performing with increased anxiety influences perceptual-motor behavior in such a way that it alters (a) perception of action capability, as was indicated by a decrease in perceived maximal reaching height; (b) information detection, as was indicated by detecting fewer lights and increases in the number of (explorative) fixations; (c) perception of affordances, as was indicated by our finding that participants needed more time to recognize task-relevant information as a possibility for action; (d) selection of affordances, as was indicated by increases in the number of holds used for climbing (i.e., preferring handholds that are closer over handholds that are further away); and (e) realization of affordances, as was indicated in our experiments by the increased duration of climbing movements from one hold to another and, consequently, an increase in climbing time.

Training with anxiety

Knowing the influence of anxiety on performance in general and, more specifically, some important perceptual-motor processes underlying it (see above), the question arises whether or not we can do something about it. On the one hand one might think of attempts to reduce feelings of anxiety (see Gould and Udry, 1994 for an overview). On the other hand, however, given the nature of professions such as elite sports and police work, one is always going to be confronted with situations in which the pressures are high. As we have seen, action capabilities (e.g., maximum reaching height) and information detection can change across short time scales as a result of anxiety. Similar to how athletes and other professionals (e.g., fire fighters, police officers, biathlon rifle shooters) learn to deal with the physiologically straining environments in which they have to perform (see Smith et al., 1997, 2001b; Anderson et al., 2002; Vickers and Williams, 2007), we believe that it is possible, through training, to adjust, or — in terms of ecological psychology — recalibrate and reattune to the action constraints that are imposed by experiences of anxiety and performance pressure (Newell, 1986; Davids et al., 2008; Fajen et al., in press; see also Bhalla and Proffitt, 1999; Proffitt, 2006). However, before turning to how we think recalibration and reattunement in response to anxiety might occur, and in what way these processes are facilitated by reality-based practice under pressure, it is important to first provide evidence that training with anxiety is effective.

Baumeister (1984) already suggested that getting used to the pressures accompanying performance situations might be beneficial, as it reduces the chances for choking. In a series of experiments, Baumeister tested the influence of pressure-induced increases in self-consciousness and self-awareness on the performance of individuals who were accustomed to such an internal focus of attention (i.e., who scored high on a dispositional measure of self-consciousness; Fenigstein et al., 1975) and individuals who were not (i.e., who scored low on this dispositional measure of self-consciousness). Consistently, the group that was high on dispositional self-consciousness outperformed the group that was low on dispositional self-consciousness when the incentives to perform well were high. These results led Baumeister (1984) to hypothesize that, either through practice or experience, one might be able to acclimatize to the effects that heightened self-awareness and/or pressure normally have on performance. Lewis and Linder (1997) and Beilock and Carr (2001) provided more direct support for this "acclimatization hypothesis," demonstrating that self-consciousness training reduced choking on a golf-putting task.

Recently, Oudejans and Pijpers conducted a series of experiments with which they provide evidence that reality-based practice under pressure reduces the chances of choking for beginners and experts alike, across a variety of settings (Oudejans, 2008; Oudejans and Pijpers, in press, 2009). In two experiments, Oudejans and Pijpers examined whether training with anxiety can prevent choking in experts performing perceptual-motor tasks. In the first experiment, two teams of expert basketball players practiced free throws over a 5-week period with (experimental team) or without (control team) induced anxiety. Anxiety was induced using various ego-stressor methods such as alleged camera recordings, winning a prize, and expert evaluation, each of which has been applied successfully in earlier research (e.g., Baumeister, 1984; Lewis and Linder, 1997; Beilock et al., 2002). Only after training with anxiety, performance no longer deteriorated during the (mild) anxiety post-test. In the second experiment expert dart players practiced dart throwing from a position high or low on a climbing wall, thus, with or without anxiety. Again, only after training with anxiety, performance was maintained during the anxiety post-test, despite higher levels of anxiety, heart rate, and perceived effort.

In another experiment, Oudejans (2008) examined whether reality-based practice under pressure may help in preventing degradation of handgun shooting performance under pressure for police officers. One group of police officers practiced handgun shooting under pressure evoked by an opponent who fired back using marking (colored soap) cartridges. The control group practiced handgun shooting on standard cardboard targets instead of real opponents. Both groups received three training sessions of 1 hour. While at the outset both groups performed worse in front of an opponent firing back compared to the cardboard targets, after the training sessions, shooting performance of the experimental group no longer deteriorated with an opponent while performance of the control group was equally harmed. These results are similar to those of Oudejans and Pijpers (in press) and indicate that training exercises involving increased pressure can acclimatize shooting performance of ordinary police officers to those situations with elevated pressure they may encounter during their police work.

Apparently training with anxiety is effective in preventing choking under pressure. One might contend, however, that anxiety levels induced during practice approach, but are not as high as, anxiety experienced in the actual events (e.g., handgun shooting in the line of duty, taking a decisive penalty kick during the final of the World Championships, and performing during the finals of the Olympic Games). Therefore, in a final study, Oudejans and Pijpers (2009) investigated whether practicing with *mild* levels of anxiety also prevents choking when performing under *high* levels of anxiety. In this case, as training with anxiety seems to have positive effects both for experts and novices, novices throwing darts from the climbing wall were investigated. Participants practiced dart throwing from low on the wall either with or without experimentally induced mild levels of anxiety (using similar ego-stressor methods as in the first experiment by Oudejans and Pijpers, in press). After practice, participants were tested with (a) low levels of anxiety (low on the wall), (b) mild levels of induced anxiety (low on the wall), and (c) high levels of anxiety (high on the wall). With increasing levels of anxiety during the tests processing efficiency appeared to decrease, as was shown by the increases in perceived effort over the tests. Despite increasing levels of anxiety, and perhaps due to more invested effort, dart performance of the experimental group remained unharmed. Performance of the control group, who had not practiced under mild levels of anxiety, could also be maintained with increased effort under mild anxiety but this was no longer the case when anxiety was further increased during the high-anxiety test. These results suggest that practicing under mild anxiety did help in maintaining performance under high anxiety, a finding with clear practical implications as training with simulated pressure seems sufficient to prevent choking during the actual event when pressures are high.

To conclude, despite variations in context (field or laboratory), task, experience of the participants, as well as in the manipulation and level of anxiety the main findings were replicated in all four experiments. Each time the group that practiced with anxiety managed to maintain performance during the anxiety post-test, while performance of those who had not practiced with anxiety systematically decreased. Therefore, we conclude that practicing under anxiety prevents choking in perceptual-motor performance for experts and novices alike, as one seems to acclimatize to the specific processes accompanying anxiety. We will try to elucidate this in the next section.

Theoretical implications

The question that arises is how positive effects of training with anxiety can be interpreted within the frameworks of the attentional control theory and ecological psychology. According to the attentional control theory, anxiety may not only lead to worry about task performance but also to attempts to reduce or eliminate the negative effects of anxiety on performance by investing additional effort. With additional effort an attempt is made to maintain active attention on the task while inhibiting distraction or interference from task-irrelevant information (Eysenck et al., 2007). As we have shown these attempts become increasingly more successful in the process of acclimatization to anxiety (Oudejans and Pijpers, in press).

From an ecological perspective, as we have already indicated and found, anxiety may change the perception, selection, and realization of affordances. As a consequence of training with anxiety, performers better learn to adjust their actions to the observed changes better. This is in line with the recent ideas by Fajen et al. (in press) on the perception of affordances and control of action in sport. As they explain, following changes in action capabilities due to factors such as fatigue or injury "some form of learning must be involved in the perception of affordances to allow actors to adapt to such changes. Such learning can be thought of in terms of perceptual-motor *(re)calibration*" (p. 25). As a performer's action capabilities change due to constraints such as anxiety, it may be necessary to continually retune and recalibrate to ensure successful control of action (Fajen et al., in press). We propose that in training with anxiety, actions of performers become better calibrated to the new constraints and performers better learn to quickly recalibrate their perceptual-motor control when confronted with changing constraints. Whatever the constraints or context in question, specific activities involving these constraints are obviously required to learn about the new relations between performer and environment. Training with anxiety provides an example of such activities that lead to better calibration and more efficient and effective recalibration processes for performing in situations where the pressure to perform, and consequently anxiety, is increased.

In our view, improved self-regulatory processes (as described in the attentional control theory) and recalibration (as proposed in ecological psychology) may involve similar processes. One might argue that recalibration also manifests itself in extra invested effort and involves attempts to reoptimize detection of task-relevant information (which has changed as well due to changes in action capabilities) while preventing distraction by task-irrelevant information. Based on our results we contend that individuals who have trained with anxiety and are used to recalibrate to their changed action capabilities under such circumstances will invest their extra effort more efficiently and effectively, while individuals who have not trained with anxiety and are not used to recalibrate to their changed action capabilities under such circumstances, may also invest extra effort but not in an effective manner.

Effective acclimatization processes may not be restricted to improved subconscious processes but may also involve the development of conscious strategies leading to, for instance, improved pre-performance routines. Exposure to elevated levels of anxiety during practice may be used to develop strategies to deal with pressure situations and anxiety effectively. For instance, evidence shows that well-learned pre-performance routines for tasks as free-throw, penalty, and rifle shooting (e.g., fixed order of activities, number of dribbles, number of breaths taken) may lead to higher levels of performance (Konttinen and Lyytinen, 1993; Singer, 2000, 2002; Williams et al., 2002a). Eventually, efficient recalibration has the effect that (despite the anxiety experienced) one is better capable of (a) detecting task-relevant information; (b) perceiving and selecting the most optimal affordances, and (c) realizing those affordances with relatively small losses of efficiency.

Concluding remarks

In this chapter we have discussed ecological psychology as a framework in understanding the perception, selection, and realization of action possibilities in sports and other high-pressure contexts, particularly police work. Rather than resorting to internal representations about actions and environment, the ecological approach seeks solutions in the reciprocity of performer and environment and the mutual dependence of perceiving and moving. Representations for action possibilities are present in the interaction between performer and environment, more specifically, in the perceptual information specifying the environment in terms of the performer, hereby inherently carrying meaning.

Anxiety is one of the constraints that are present in several high-pressure contexts. Examples from practice itself as well as several of our studies have shown how anxiety may negatively influence the perception and realization of affordances, in part because of its impact on the detection of task-relevant information. Practicing with anxiety may counter the negative effects of anxiety, because, as we argue, self-regulatory processes improve and perception and action are recalibrated in light of the new constraints and accompanying changes. In learning to perceive, select, and realize affordances in several settings it is important to gain experience in task execution with the same constraints (physical, technical, tactical, as well as psychological) as those encountered in the actual performance environment, be it a decisive event for an athlete or a life-threatening shoot-out for a police officer. It is a challenge for several fields of practice to develop training settings in which relevant constraints are present or at least simulated, hereby approaching the actual performance setting as much as possible.

References

Anderson, G. S., Litzenberger, R., & Plecas, D. (2002). Physical evidence of police officer stress. *Policing: an International Journal of Police Strategies & Management*, *25*, 399–420.

Baddeley, A. (1992). Working memory. *Science*, *255*, 556–559.

Baumeister, R. F. (1984). Choking under pressure: self-consciousness and paradoxical effects of incentives on skillful performance. *Journal of Personality and Social Psychology*, *46*, 610–620.

Beilock, S. L. (2008). Beyond the playing field: sport psychology meets embodied cognition. *International Review of Sport and Exercise Psychology*, *1*, 19–30.

Beilock, S. L., & Carr, T. H. (2001). On the fragility of skilled performance: what governs choking under pressure? *Journal of Experimental Psychology — General*, *130*, 701–725.

Beilock, S. L., Carr, T. H., MacMahon, C., & Starkes, J. L. (2002). When paying attention becomes counterproductive: impact of divided versus skill-focused attention on novice and experienced performance of sensorimotor skills. *Journal of Experimental Psychology — Applied*, *8*, 6–16.

Beilock, S. L., & Gray, R. (2007). Why do athletes choke under pressure. In G. Tenenbaum & R. C. Eklund (Eds.), *Handbook of sport psychology* (3rd ed., pp. 425–444). New York: Wiley.

Beilock, S. L., Kulp, C. A., Holt, L. E., & Carr, T. H. (2004). More on the fragility of performance: choking under pressure in mathematical problem solving. *Journal of Experimental Psychology — General*, *133*, 584–600.

Berg, A. M., Hem, E., Lau, B., Håseth, K., & Ekeberg, Ø. (2005). Stress in the Norwegian police service. *Occupational Medicine*, *55*, 113–120.

Bhalla, M., & Proffitt, D. R. (1999). Visual-motor recalibration in geographical slant perception. *Journal of Experimental Psychology — Human Perception and Performance*, *25*, 1076–1096.

Bishop, S., Duncan, J., Brett., M., & Lawrence, A. D. (2004). Prefrontal cortical function and anxiety: controlling attention to threat-related stimuli. *Nature Neuroscience*, *7*, 184–188.

Bootsma, R. J., Bakker, F. C., Van Snippenberg, F. J., & Tdlohreg, C. W. (1992). The effects of anxiety on perceiving the reachability of passing objects. *Ecological Psychology*, *4*, 1–16.

Cooper, C. L., Davidson, M. J., & Robinson, P. (1982). Stress in the police service. *Journal of Occupational Medicine*, *24*, 30–36.

Corbetta, M., & Shulman, G. L. (2002). Control of goal-directed and stimulus-driven attention in the brain. *Nature reviews. Neuroscience*, *3*, 201–215.

Davids, K., Button, C., & Bennett, S. (2008). *The dynamics of skill acquisition: a constraint led approach*. Champaign, IL: Human Kinetics.

Easterbrook, J. A. (1959). The effect of emotion on the utilization and the organization of behavior. *Psychological Review*, *66*, 183–201.

Eysenck, M. W. (1992). *Anxiety: the cognitive perspective*. Hove, UK: Lawrence Erlbaum Associates.

Eysenck, M. W., & Calvo, M. G. (1992). Anxiety and performance: the processing efficiency theory. *Cognition Emotion*, *6*, 409–434.

Eysenck, M. W., Derakshan, N., Santos, R., & Calvo, M. G. (2007). Anxiety and cognitive performance: attentional control theory. *Emotion*, *7*, 336–353.

Fajen, B. R. (2007). Affordance-based control of visually guided action. *Ecological Psychology*, *19*, 383–410.

Fajen, B. R., Riley, M. A., & Turvey, M. T. (in press). Information, affordances, and the control of action in sport. *International Journal of Sport Psychology*.

Fenigstein, A., Scheier, M. F., & Buss, A. H. (1975). Public and private self-consciousness: assessment and theory. *Journal of Consulting and Clinical Psychology*, *43*, 522–527.

Fodor, J. A. (1983). *The modularity of mind*. Cambridge, MA: MIT Press.

Gerrard, S. (2007). *Gerrard: my autobiography*. London, UK: Bantam Books.

Gibson, E. J. (1988). Exploratory behaviour in the development of perceiving, acting, and the acquiring of knowledge. *Annual Review of Psychology*, *39*, 1–41.

Gibson, J. J. (1979/1986). *The ecological approach to visual perception*. Hillsdale, NJ: Lawrence Erlbaum Associates.

Gould, D., & Udry, E. (1994). Psychological skills for enhancing performance: Arousal regulation strategies. *Medicine and Science in Sports and Exercise*, *26*, 478–485.

Gray, R. (2004). Attending to the execution of a complex sensorimotor skill: expertise differences, choking, and slumps. *Journal of Experimental Psychology — Applied*, *10*, 42–54.

Hanton, S., Mellalieu, S. D., & Hall, R. (2004). Self-confidence and anxiety interpretation: a qualitative investigation. *Psychology of Sport and Exercise*, *5*, 477–495.

Hatzigeorgiadis, A., & Biddle, S. J. H. (2000). Assessing cognitive interference in sport: development of the thought occurrence questionnaire for sport. *Anxiety Stress Coping*, *13*, 65–86.

Hatzigeorgiadis, A., & Biddle, S. J. H. (2001). Athletes' perceptions of how cognitive interference during competition influences concentration and effort. *Anxiety Stress Coping*, *14*, 411–429.

Jackson, R. C., Ashford, K. J., & Norsworthy, G. (2006). Attentional focus, dispositional reinvestment, and skilled motor performance under pressure. *Journal of Sport and Exercise Psychology*, *28*, 49–68.

Konttinen, N., & Lyytinen, H. (1993). Individual variability in brain slow wave profile in skilled sharpshooters during the aiming period in rifle shooting. *Journal of Sport & Exercise Psychology*, *15*, 275–289.

Lewis, B. P., & Linder, D. E. (1997). Thinking about choking? Attentional processes and paradoxical performance. *Personality and Social Psychology Bulletin*, *23*, 937–944.

Mark, L. S. (1987). Eyeheight-scaled information about affordances: a study of sitting and stair climbing. *Journal of Experimental Psychology — Human Perception and Performance*, *13*, 361–370.

Mark, L. S., Balliett, J. A., Craver, K. D., Douglas, S. D., & Fox, T. (1990). What an actor must do in order to perceive the affordance of sitting. *Ecological Psychology*, *2*, 325–366.

Masters, R. S. W. (1992). Knowledge, knerves and know-how: the role of explicit versus implicit knowledge on the breakdown of a complex motor skill under pressure. *British Journal of Psychology*, *83*, 334–358.

Michaels, C. F., & Beek, P. J. (1995). The state of ecological psychology. *Ecological Psychology*, *7*, 259–278.

Michaels, C. F., & Carello, C. (1981). *Direct perception*. Englewood Cliffs, NJ: Prentice-Hall.

Mullen, R., & Hardy, L. (2000). State anxiety and motor performance: testing the conscious processing hypothesis. *Journal of Sports Science*, *18*, 785–799.

Mullen, R., Hardy, L., & Tattersall, A. (2005). The effects of anxiety on motor performance: a test of the conscious processing hypothesis. *Journal of Sport & Exercise Psychology*, *27*, 212–225.

Murray, K. R. (2004). *Training at the speed of life, Vol. 1: the definitive textbook for police and military reality based training*. Gotha, FL: Armiger Publications.

Murray, N. P., & Janelle, C. M. (2003). Anxiety and performance: a visual search examination of the processing efficiency theory. *Journal of Sport & Exercise Psychology*, *25*, 171–187.

Murray, N. P., & Janelle, C. M. (2007). Event-related potential evidence for the processing efficiency theory. *Journal of Sports Science*, *25*, 161–171.

Newell, K. M. (1986). Constraints on the development of coordination. In M. G. Wade & H. T. A. Whiting (Eds.), *Motor development in children: aspects of coordination and control* (pp. 341–360). Dordrecht: Martinus Nijhoff.

Niedenthal, P. M., Barsalou, L. W., Winkielman, P., Krauth-Gruber, S., & Ric, F. (2005). Embodiment in attitudes, social perception, and emotion. *Personality and Social Psychology Review*, *9*, 184–211.

Nieuwenhuys, A., Pijpers, J. R., Oudejans, R. R. D., & Bakker, F. C. (2008). The influence of anxiety on visual attention in climbing. *Journal of Sport & Exercise Psychology*, *30*, 171–185.

Oudejans, R. R. D. (2008). Reality-based practice under pressure improves handgun shooting of police officers. *Ergonomics*, *51*, 261–273.

Oudejans, R.R.D. and Pijpers, J.R. (in press). Training with anxiety has a positive effect on expert perceptual-motor performance under pressure. *The Quarterly Journal of Experimental Psychology*.

Oudejans, R.R.D. and Pijpers, J.R. (2009) *Training with mild levels of anxiety can prevent choking under high levels of anxiety*. Manuscript submitted for publication.

Pijpers, J. R., Oudejans, R. R. D., & Bakker, F. C. (2005). Anxiety-induced changes in movement behavior during the execution of a complex whole-body task. *Quarterly Journal of Experimental Psychology*, *58A*, 421–445.

Pijpers, J. R., Oudejans, R. R. D., Bakker, F. C., & Beek, P. J. (2006). The role of anxiety in perceiving and realizing affordances. *Ecological Psychology*, *18*, 131–161.

Posner, M. I., & Petersen, S. E. (1990). The attention system of the human brain. *Annual Review of Neuroscience*, *13*, 25–42.

Proffitt, D. R. (2006). Embodied perception and the economy of action. *Perspectives on Psychological Science*, *1*, 110–122.

Sarason, I. G. (1988). Anxiety, self-preoccupation and attention. *Anxiety Research*, *1*, 3–7.

Schwenkmezger, P., & Steffgen, G. (1989). Anxiety and motor performance. In B. Kirkcaldy (Ed.), *Normalities and abnormalities in human movement* (pp. 78–99). Basel: Karger.

Shaw, R., Turvey, M. T., & Mace, W. M. (1982). Ecological psychology: the consequence of a commitment to realism. In W. B. Weimer & D. S. Palermo (Eds.), *Cognition and the symbolic processes* (Vol. 2, pp. 159–226). Hilldale, NJ: Lawrence Erlbaum Associates.

Singer, R. N. (2000). Performance and human factors: considerations about cognition and attention for self-paced and externally paced events. *Ergonomics*, *43*, 1661–1680.

Singer, R. N. (2002). Preperformance state, routines, and automaticity: what does it take to realize expertise in self-paced events? *Journal of Sport & Exercise Psychology*, *24*, 359–375.

Smith, D. L., Manning, T. S., & Petruzzello, S. J. (2001b). Effect of strenuous live-fire drills on cardiovascular and psychological responses of recruit firefighters. *Ergonomics*, *44*, 244–254.

Smith, D. L., Petruzzello, S. J., Kramer, J. M., & Misner, J. E. (1997). The effects of different thermal environments on the physiological and psychological responses of firefighters to a training drill. *Ergonomics*, *40*, 500–510.

Smith, N. C., Bellamy, M., Collins, D., & Newell, D. (2001a). A test of processing efficiency theory in a team sport context. *Journal of Sports Science*, *19*, 321–332.

Stefanucci, J. K., Proffitt, D. R., Clore, G. L., & Parekh, N. (2008). Skating down a steeper slope: fear influences the perception of geographical slant. *Perception*, *37*, 321–323.

Teachman, B. A., Stefanucci, J. K., Clerkin, E. M., Cody, M. W., & Proffitt, D. R. (2008). A new mode of fear expression: perceptual bias in height fear. *Emotion*, *8*, 296–301.

Turvey, M. T. (1992). Affordances and prospective control: an outline of the ontology. *Ecological Psychology*, *4*, 173–187.

Turvey, M. T., Carello, C., & Kim, N.-G. (1990). Links between active perception and the control of action. In H. Haken & M. Stadler (Eds.), *Synergetics of cognition* (pp. 269–295). Heidelberg: Springer-Verlag.

Turvey, M. T., Shaw, R. E., Reed, E. S., & Mace, W. M. (1981). Ecological laws of perceiving and acting: in reply to Fodor and Pylyshyn (1981). *Cognition*, *9*, 237–304.

Vickers, J. N., & Williams, A. M. (2007). Performing under pressure: the effects of physiological arousal, cognitive anxiety, and gaze control in biathlon. *Journal of Motor Behavior*, *39*, 381–394.

Williams, A. M., Singer, R. N., & Frehlich, S. G. (2002a). Quiet eye duration, expertise, and task complexity in near and far aiming tasks. *Journal of Motor Behavior*, *34*, 197–207.

Williams, A. M., Vickers, J. N., & Rodrigues, S. (2002b). The effects of anxiety on visual search, movement kinematics, and performance in table tennis: a test of Eysenck and Calvo's processing efficiency theory. *Journal of Sport & Exercise Psychology*, *24*, 438–455.

Wilson, M. (2002). Six views of embodied cognition. *Psychonomic Bulletin & Review*, *9*, 625–636.

Wilson, M., & Smith, N. C. (2007). A test of the predictions of processing efficiency theory during elite team competition using the thought occurrence questionnaire for sport. *International Journal of Sport Psychology*, *38*, 245–262.

Wilson, M., Smith, N. C., Chattington, M., Ford, M., & Marple-Horvat, D. E. (2006). The role of effort in moderating the anxiety-performance relationship: testing the predictions of processing efficiency theory in simulated rally driving. *Journal of Sports Science*, *24*, 1223–1233.

Wilson, M., Smith, N. C., & Holmes, P. S. (2007). The role of effort in influencing the effect of anxiety on performance: testing the conflicting predictions of processing efficiency theory and the conscious processing hypothesis. *British Journal Psychology*, *98*, 411–428.

Wine, J. (1971). Test anxiety and direction of attention. *Psychological Bulletin*, *76*, 92–104.

Woodman, T., & Hardy, L. (2001). Stress and anxiety. In R. N. Singer, H. A. Hausenblas, & C. M. Janelle (Eds.), *Handbook of sport psychology* (2nd ed., pp. 290–318). New York: Wiley.

M. Raab et al. (Eds.)
Progress in Brain Research, Vol. 174
ISSN 0079-6123

CHAPTER 5

How do people perceive and generate options?

Markus Raab*, Rita F. de Oliveira and Thomas Heinen

Institute of Psychology, German Sport University Cologne, Cologne, Germany

Abstract: This chapter examines the question of how people perceive and generate options. Although most current theoretical approaches explain option perception and option generation as requiring extensive and costly cognitive processes, we will argue that people perceive their possibilities for action and generate options without such elaborate means. Borrowing from the contributions of direct perception and bounded rationality, we present and analyze data from three sports tasks: the basketball jump shot, ball allocation in handball, and trampoline somersaulting. Data from kinematics, eye-tracking, and choice behavior experiments are presented and reanalyzed to draw the parallels and distinctions between the two approaches. In conclusion, integrating research on perception, action, and cognition provides a promising description of the bidirectional link between mind and motion.

Keywords: bounded rationality; direct perception; movement control; option generation; gaze behavior; sports

Introduction

For those who watched the gymnastics competitions at the Beijing Olympics, you may still recall the fascinating trampoline moves of gold medalist Lu Chunlong in his triple-piked somersault with a half twist. Although the movements looked fluid and easy, they are in reality quite complex, and it is unclear how such complex actions are regulated. In fact, when a gymnast leaves the trampoline bed, the inertia and angular momentum delimit the possibilities for action, but there is a manifold of options that can be perceived and/or generated that would result in upright landing.

The explanations of how options for action are perceived and generated have typically involved elaborate cognitive processes that are needed to understand information and optimize the generation and choice of options. In what follows, we will argue that people perceive their possibilities for action and generate options without such processes. We will discuss the contributions of direct perception and bounded rationality to our understanding of such extraordinary behavior as the triple piked somersault.

Most current theoretical approaches hold the view that to perceive or generate options people use elaborate cognitive processes. To perceive options within a goal-directed task, one would first need to interpret sensory input by comparing it to previous interpretations stored in memory. These theories contend that people first need these processes to understand their environment and thereafter are capable of perceiving their options. Thus, before choosing an option, one would need to repeat the process of perception to generate a list of possible options. After all the options are

*Corresponding author.
Tel.: +49-221-4982-5491; Fax: +49-221-4982-8320;
E-mail: raab@dshs-koeln.de

DOI: 10.1016/S0079-6123(09)01305-3

generated, people would list the features on which to evaluate their options and then choose the option that ranked highest. This is a description that separates each of the processes and orders them in a logical sequence, but is it plausible?

Many authors have argued that such reliance on elaborate cognitive processes would significantly delay perception and option generation. This is a critical caveat to traditional views because people often act in very dynamic contexts where information is constantly changing. Perceiving one action possibility enhances the generation of other, similar options, and actions themselves create new options on the fly. Thus perception and option generation develop together and interact. Another caveat is that people have limited cognitive resources, which would render the use of extensive cognitive processes ineffective. First, it would be costly to generate all possible options since most would have little applicability at the outset. Second, further cognitive resources would have to be used to evaluate the options against evaluation criteria before choosing one option. The final caveat is that traditional theorizing does not provide the most parsimonious explanations for the mechanisms underlying option perception and generation.

To address these and other limitations, several researchers have departed from the current mainstream theoretical framework to investigate the mechanisms underlying option perception and option generation. To perceive or generate options, people need information about themselves and about their environment. When the requisite information is available, it can be used to guide actions directly. However, when information is scarce or unspecific, people use the available information to generate and choose options. These two settings are best understood by the theories of direct perception and bounded rationality.

Direct perception and bounded rationality

The theory of direct perception has direct realism as a starting point, in that it considers the immediate objects of perception to be events, qualities, and physical objects that exist in the world and can be perceived without cognitive interpretation (Gibson, 1979/1986). Hence, the accuracy of perception is due to the accuracy of the information available to the perceptual systems. For instance, in vision, the light patterns that reach the eye are invariantly related to their sources in the environment (e.g., each object reflects light in a way that is particular to it, therefore that pattern of light constitutes visual information through which the object can be directly perceived (Michaels and Carello, 1981). For perception to be accurate, a person first needs to attend to information sources that are specific to the object (or event or property). This so-called education of attention consists of learning what sources of information can be relied upon for veridical perception. Although there is a multitude of information sources available in the environment, only one or two of those sources will have a direct mapping onto the object of interest (e.g., the light reflected from the object or the shadow it casts). Thus, the education of attention is about establishing the invariant relation between object and perception. Another requirement for perception to be accurate is the calibration of that link between informational variables and perception (Withagen and Michaels, 2005).

Besides information, other hallmarks of direct perception are the mutuality of perception and action and the concept of affordances (see Chapter 4: Perceiving and moving in sports and other high-pressure contexts). In brief, perceptual information can guide actions while, simultaneously, action informs perception in a continuous and dynamic manner. When looking for an object or driving through the city, for instance, a person provokes new perceptions through either exploratory behavior or purposeful actions, and those new perceptions elicit new actions, such as grasping the object or slowing down before a zebra crossing. Affordances are action possibilities that can be best situated in the link between the person and his/her environment (Chemero, 2003). For a driver, a zebra crossing affords cautious behavior, such as slowing down, but for a child the same zebra crossing may afford a new game. One object may offer different possibilities for action to different people.

options is evident. Imagining option possibilities in direct perception equates to using option validity in the bounded rationality approach. Both theories allow for imagining the most relevant option in a given setting instead of all options (cf., Chapter 12: Embodied cognition of movement decisions: a computational modeling approach). We have so far discussed a number of important features concerning option perception and option generation. We have focused on the perception of options that can guide actions, and on the processes through which subsequent options can be generated. The tasks of basketball shooting and option generation in handball are relatively simple and allow straightforward explanations from direct perception and bounded rationality. In tasks of greater complexity it is inherently more difficult to understand how options are perceived and generated.

As mentioned in the Introduction, when a gymnast leaves the trampoline bed, the linear and angular momentum are determined. While airborne, the athlete can only change his/her moment of inertia (e.g., by arm or hip flexion) to accelerate or decelerate somersault rotation, or to initiate or end twisting (Yeadon, 2000). Therefore, a defined time of flight and a defined angular momentum constrain the availability of options, if the athlete is to land on the trampoline in an upright position. This constraint is diametrical in nature because, for instance, a large angular momentum and long flight duration enable options for high-risk maneuvers, (e.g., tucked triple forward somersault with a half-twist) and disable options for low-risk maneuvers (e.g., straight backward somersault). Thus, it is likely that the options for different maneuvers are generated and decided upon before initiating the routine. During their routine, athletes need to pick up information to locate the trampoline bed while airborne, to control both in-flight rotations and landing. It is well established that vision is important in landing somersaults (Rézette and Amblard, 1985; Bardy and Laurent, 1998; Davlin et al., 2001; Hondzinski and Darling, 2001). Overall, participants are better at landing single somersaults when vision is allowed (normal or reduced vision) than when it is prevented (Davlin et al., 2001). Although participants are capable of performing straight double somersaults without vision, they require in-flight instructions to make the necessary adjustments (Yeadon and Mikulcik, 2000). These results call for the measurement of eye movements during trampoline somersaulting to investigate the visual control of somersaulting.

In such an investigation, the gaze behavior of participants was measured during trampoline routines (Heinen, in preparation). Expert and novice gymnasts performed tucked backward somersaults with a constant flight time and two preparatory leaps. Compared to novices, experts showed significantly longer fixations and experts also fixated on the wall and ceiling during takeoff and while airborne. Following up on this result, we examined in-flight fixation locations for different routines. Compared to novices, experts used more fixations on the wall and ceiling in the tucked and straight somersaults. By using the wall or ceiling as a visual anchor point, as it were, athletes may gain information about their rotation (e.g., rotation speed and direction). Besides using this perceptual information to make fine in-flight adjustments to rotation, participants may also control their landing in preparation for the following jump. Figure 2 shows that during a straight single somersault the eye rotates upward (until about 20°) during the stance phase and the athlete fixates on the ceiling at the moment of takeoff from the trampoline. This fixation is maintained during the first part of the flight phase (when changes in moment of inertia about the somersault axis are small in a straight somersault) and its duration is longer compared to a tucked somersault (when changes in moment of inertia are twice as high). Immediately after this fixation, the eye rotates upward again (until about 35°) to anticipate the trampoline bed in preparation for the landing.

A subsequent analysis revealed significant correlations between distinct values of the change of moment of inertia about the somersault axis and parameters of gaze behavior (e.g., fixation duration). For instance, the smaller the change in moment of inertia from takeoff to a fixed airborne body configuration, the longer the fixation duration from takeoff to the first part of the flight

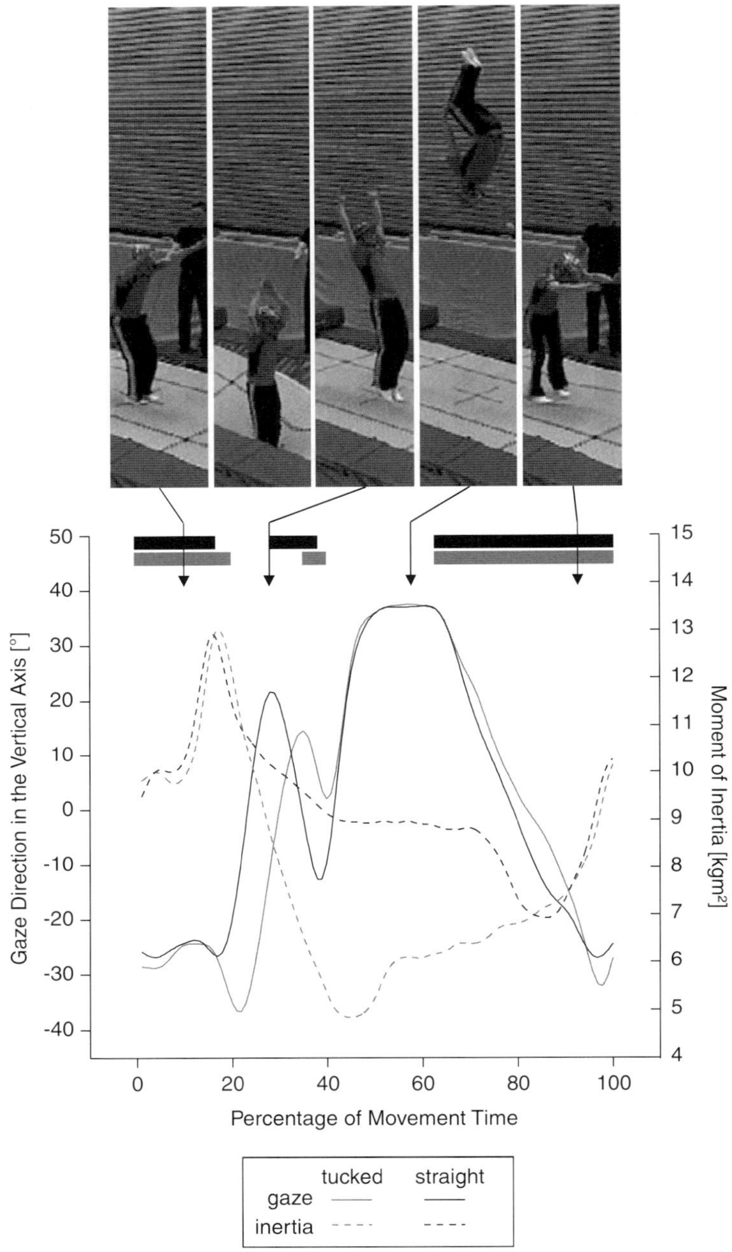

Fig. 2. Gaze direction and moment of inertia during trampoline somersaulting. Top: Different phases of a tucked somersault and their correspondence to movement time (black arrows). The participant is wearing a telemetric eye-tracking system during performance. Bottom: The rotation of the eye (gaze direction) in the vertical direction is indicated by the thick lines. Positive values indicate looking upward and negative values indicate looking below the horizontal. The bars on the top of the graph indicate distinct gaze fixations. The moment of inertia with respect to the somersault axis is indicated by the dashed lines. Data pertaining to the tucked and the straight somersault are indicated, respectively, by the gray and black lines.

phase. Exploratory analyses of the gaze behavior and movement kinematics of more complex skills support the idea that gaze behavior is adapted to movement execution (Heinen, in preparation). Experts performed double somersaults and double somersaults with a full twist (so-called double-doubles) while wearing the eye-tracking device. Verbal reports together with the analysis of gaze behavior and movement kinematics revealed that, when mastered, the double-double is easier to control because athletes can maintain their gaze on the trampoline bed during almost the entire flight phase and control their moment of inertia predominantly on the basis of their current visual information. However, when performing skills with extremely high angular velocities, such as a triple somersault with a half-twist, it may not be possible to maintain gaze on a specific area due to the movement itself. Therefore other sources of information will be more appropriate to generate and perceive options for controlling rotation.

Trampoline jumping is a dynamic whole-body task that is constrained, from the moment of takeoff until landing, by the inertia and angular momentum of the gymnast. While airborne, athletes need to initiate, adjust, or end their rotation. There are many perceptual information options available but only a few relate to the rotation speed and direction and can be used to control rotation. We have argued above that expert trampolinists use visual fixations (visual anchor points) as a means to pick up information about their rotation. The differences between novice and expert gaze strategies indicate that gymnasts need to learn to establish the link between gaze strategy and movement control. Importantly, in-flight adjustments directly influence the control of landing and hence the takeoff parameters of the subsequent jump. In this sense, the information picked up while airborne can also be used to generate options for the subsequent jump. Importantly in this task, the theory of direct perception and the theory of bounded rationality can offer a parsimonious explanation of what perceptual information the trampolinist may perceive and how options may be generated for subsequent movements.

Discussion

Although most current theoretical approaches explain option perception and option generation through extensive and costly cognitive processes, we have argued that people perceive their possibilities for action and generate options without such elaborate means. Borrowing from the contributions of direct perception and bounded rationality, we presented and analyzed data from three sport tasks: the basketball jump shot, ball allocation in handball, and trampoline somersaulting. In so doing we exposed the parallels and distinctions of the two theories. In summary, the two theories are typically applied to different settings of human behavior. Whereas direct perception explains how people perceive and act in information-rich natural environments, bounded rationality aptly explains how people generate options under time or knowledge constraints. However, in these settings, both theories argue that people use a limited number of information sources or cues. Although many information sources exist in natural environments, people tend to pick up and use one or two informational options that are available and specific to the task at hand. Likewise, from all the options that are theoretically available, people generate only one or two and use simple heuristics to choose between them.

We have presented three examples from sports settings to show how the theories of direct perception and bounded rationality can explain various findings. In the basketball jump shooting experiment, the action goal is predetermined by the experimenter but the player needs to pick up the proper information to guide the movement. Players appear to use only one source of information that is sufficient for accurate shooting. In a playmaking experiment in handball, participants were asked to generate several options for action. Although they managed to generate quite a number of them, their first option was consistently the best and subsequent options appeared to be derived from the first one. Finally, in trampoline somersaulting, it appeared that gymnasts used a specific control strategy where gaze fixation was correlated with moment of inertia. By anchoring their gaze on fixed positions, gymnasts may be allowing themselves to pick up information about rotations that they can use for in-flight adjustments and to prepare the somersault landing. Because options for movements are restricted in-flight by inertia and angular momentum, gymnasts should generate movement options before takeoff. Their posture at landing may provide the best cues for generating options for the next trampoline routine.

In this chapter we have focused on the potential benefits of integrating the theories of bounded rationality and direct perception by finding their communalities and determining their different applications. The three examples provide challenges for both theories. The basketball example is about picking up information for action, which has not been researched in bounded rationality but is a hallmark endeavor in direct perception. In contrast, the handball example of option generation is a recurrent research theme in bounded rationality but not in direct perception. Finally, the trampolinist example is particularly challenging, mainly because of the complexity of the task. We argued that by integrating explanations from direct perception and bounded rationality we can best understand such complex movement skills.

It is important to note that our aim was not to integrate the theories but rather to learn from their contributions. The strongest common denominator of direct perception and bounded rationality is the view that people solve their option-perception and option-generation tasks in a very parsimonious fashion. When information is available and specific they use it while neglecting other sources of information. When information is not available or is unspecific they use simple heuristics that result in very fast and satisfactory results. Some may argue that heuristics cannot describe lower cognitive processes such as those involved in perception (e.g., Norman, 2002, including commentaries; Shepard, 2001), and others may argue that direct perception is not able to explain mental imagery or other internal higher cognitive processes (e.g., see Boschker, 2001 for a debate). However, our examples from research on direct perception concerned with mental imagery (Boschker, 2001) or heuristics in the area of perception (Raab and Johnson, 2007) show that these phenomena can be studied by theories outside of their carved territories (see Raab and Gigerenzer, 2005 for further argumentation).

What needs to be done next? First, a more thorough theoretical analysis is warranted. For example, it is necessary to analyze whether the concept of information in the two theories can be brought closer together. This is relevant because it would allow the study of perceptual aspects in bounded rationality and higher-level cognition in direct perception. In so doing the ground could be laid for understanding the interaction between people, objects, and imagination. Second, we should adapt the research paradigms such that they allow research into more complex behaviors. This could include analyses of the structure of the environment and the capabilities and limitations of people in using various sources of information. Our example of trampoline routines demonstrates that this would require some changes in methods but it has the potential to provide valuable insights. Finally, the integration of both approaches could be used in field applications, for instance in perceptual-motor rehabilitation or in sports performance optimization. These are challenging prospects for the study of people's perception and generation of options.

References

Bardy, B. G., & Laurent, M. (1998). How is body orientation controlled during somersaulting. *Journal of Experimental Psychology: Human Perception and Performance*, *24*(3), 963–977.

Boschker, M. S. J. (2001). *Action-based imagery. On the nature of mentally imagined motor actions* (Doctoral dissertation, VU University Amsterdam). Amsterdam: PrintPartners Ipskamp B. V.

Brancazio, P. J. (1981). Physics of basketball. *American Journal of Physics*, *49*, 356–365.

Chemero, A. (2003). An outline of a theory of affordances. *Ecological Psychology*, *15*(2), 181–195.

Davlin, C. D., Sands, W. A., & Shultz, B. B. (2001). Peripheral vision and back tuck somersaults. *Perceptual and Motor Skills*, *93*, 465–471.

de Oliveira, R. F., Huys, R., Oudejans, R. R. D., van de Langenberg, R., & Beek, P. J. (2007). Basketball jump shooting is controlled online by vision. *Experimental Psychology*, *54*(3), 180–186.

de Oliveira, R. F., Oudejans, R. R. D., & Beek, P. J. (2008). Gaze behavior in basketball shooting: further evidence for online visual control. *Research Quarterly for Exercise and Sport*, *78*(4), 399–404.

de Oliveira, R. F., Oudejans, R. R. D., & Beek, P. J. (in press). Experts appear to use angle of elevation information in basketball shooting. *Journal of Experimental Psychology: Human Perception and Performance*.

Gibson, J. J. (1979/1686). *The ecological approach to visual perception*. Hillsdale, NJ: Lawrence Erlbaum Associates.

Goldstein, D. G., & Gigerenzer, G. (2002). Models of ecological rationality: the recognition heuristic. *Psychological Review*, *109*, 75–90.

Heinen, T. (in preparation). What does a trampolinist perceive during a somersault?

Hondzinski, J. M., & Darling, W. G. (2001). Aerial somersault performance under three visual conditions. *Motor Control*, *3*, 281–300.

Johnson, J., & Raab, M. (2003). Take the first: option-generation and resulting choices. *Organizational Behavior and Human Decision Processes*, *91*, 215–229.

Karnath, H.-O., Sievering, D., & Fetter, M. (1994). The interactive contribution of neck muscle proprioception and vestibular stimulation to subjective 'straight ahead' orientation in man. *Experimental Brain Research*, *101*, 140–146.

Klein, G., Wolf, S., Militello, L., & Zsambok, C. (1995). Characteristics of skilled option generation in chess. *Organizational Behavior and Human Decision Processes*, *62*(1), 63–69.

Michaels, C. F., & Carello, C. (1981). *Direct perception*. Englewood Cliffs, NJ: Prentice-Hall.

Norman, J. (2002). Two visual systems and two theories of perception: an attempt to reconcile the constructivist and ecological approaches. *Behavioral and Brain Sciences*, *25*, 73–144.

Raab, M., & Gigerenzer, G. (2005). Intelligence as smart heuristics. In R. J. Sternberg, J. Davidson, & J. Pretz (Eds.), *Cognition and intelligence* (pp. 188–207). Cambridge: Cambridge University Press.

Raab, M., & Johnson, J. (2007). Option-generation and resulting choices. *Journal of Experimental Psychology: Applied*, *13*, 158–170.

Rézette, D., & Amblard, B. (1985). Orientation versus motion visual cues to control sensorimotor skills in some acrobatic leaps. *Human Movement Sciences*, *4*, 297–306.

Selten, R. (2001). What is bounded rationality? In G. Gigerenzer & R. Selten (Eds.), *Bounded rationality: the adaptive toolbox* (pp. 13–36). Cambridge, MA: MIT Press.

Shepard, R. N. (2001). On the possibility of universal mental laws: a reply to my critics. *Behavioral and Brain Sciences*, *24*, 712–748.

Volz, K. G., Schooler, L., Schubotz, R., Raab, M., Gigerenzer, G., & von Crammon, D. Y. (2006). Why you think Milan is larger than Modena: neural correlates of the recognition heuristic. *Journal of Cognitive Neuroscience*, *18*(11), 1924–1936.

Withagen, R., & Michaels, C. F. (2005). The role of feedback information for calibration and attunement in perceiving length by dynamic touch. *Journal of Experimental Psychology: Human Perception and Performance*, *31*(6), 1379–1390.

Yeadon, M. R. (2000). Aerial movement. In V. M. Zatsiorsky (Ed.), *Biomechanics in sport: performance enhancement and injury prevention* (pp. 273–283). Oxford: Blackwell Science.

Yeadon, M. R., & Mikulcik, E. C. (2000). Stability and control of aerial movements. In B. M. Nigg, B. R. MacIntosh, & J. Mester (Eds.), *Biomechanics and biology of movement* (pp. 211–221). Champaign, IL: Human Kinetics.

M. Raab et al. (Eds.)
Progress in Brain Research, Vol. 174
ISSN 0079-6123

CHAPTER 6

How the orbitofrontal cortex contributes to decision making — A view from neuroscience

Kirsten G. Volz[1,*] and D. Yves von Cramon[1,2]

[1]Max Planck Institute for Neurological Research, Cologne, Germany
[2]Max Planck Institute for Human Cognitive and Brain Sciences, Leipzig, Germany

Abstract: In the present contribution, the various functional interpretations concerning the putative function of the orbital prefrontal cortex are reviewed since this region and adjacent areas are considered the neural substrate of social behavior in general, and decision-making behavior in particular. This literature review revealed different but related interpretations as to the function of the orbital prefrontal cortex (including the ventromedial prefrontal cortex (VMPFC)): the orbital prefrontal areas (a) code the hedonic quality of choice options, (b) are critical for maintaining associative information about expected outcomes in representational memory so that it can be compared and integrated with information about internal states and current goals, (c) serve as a store of implicitly acquired linkages between factual knowledge and bio-regulatory states, including those that constitute feelings and emotions, (d) serve as a detector of potential content that is derived from the critical aspects of the input, that is, the gist information, (e) are crucially involved in the integration of emotional signals in the decision-making process, and (f) may specialize in integrating the external and internal environment. In the last part of this contribution, we try to bring together these varying but related approaches and propose a preliminary working hypothesis with regard to the role of orbital prefrontal areas in decision making.

Keywords: orbitofrontal cortex; decision making; relative reward value; biasing affective signal

Introduction

Some of the most significant breakthroughs in cognitive neuroscience have arisen from the study of brain-damaged patients. For example, Broca's (1861) description of the language-production deficit in patient "Tan" or Scoville and Milner's (1957) description of a declarative memory acquisition in patient "H.M." paved the way for identifying the neural substrates of language and memory. Similarly, Esslinger and Damasio's (1985) description of profound real-life decision-making impairments and acquired disturbances in personality in patient "EVR" led to the assumption that the ventromedial prefrontal cortex (VMPFC) including parts of the orbitofrontal cortex (OFC) can be considered the neural substrate of social behavior in general, and decision-making behavior in particular. "EVR" has been characterized to show gross impairments in his ability to make advantageous decisions in his personal and social life (Esslinger and Damasio, 1985). For example, despite his

*Corresponding author.
Tel.: +49-221-4726-255; Fax: +49-221-4726-298;
E-mail: volz@nf.mpg.de

DOI: 10.1016/S0079-6123(09)01306-5

stable and successful marriage, "EVR" divorced his wife shortly after the resection of the bilateral orbitofrontal meningioma and entered into a short-lived marriage. Although "EVR" has been described to have a keen business sense, within months of the operation, he engaged in a number of highly speculative ventures eventually leading to bankruptcy. Before the resection "EVR" was characterized as an accomplished professional, yet after the resection he lived in a sheltered environment. Remarkably, on tests of language, memory, reasoning, and general knowledge, "EVR" showed an intact intelligence. This pattern of deficits in social and personal decision making with concomitant intact intelligence has repeatedly been observed in subsequent patients with damage to the VMPFC and OFC (Tranel, 2002). These findings suggest that not only the fund of knowledge concerning appropriate social conduct seems to be impaired but also the ability to distinguish the consequences of one's own actions as being advantageous or disadvantageous.

This assumption has been supported by observations in patients suffering from drug addiction. These patients show an exaggerated motivation to procure and administer the drug regardless of its adverse consequences. Based on imaging data and data from animal models, it has been proposed that repeated exposure to drugs of abuse significantly change the function and structure of the OFC (Schoenbaum et al., 2006). More specifically, a disruption of the function of the striato-thalamo-orbitofrontal circuit has been postulated to be central for the addictive state (Volkow and Fowler, 2000).

Accordingly, clinical observations of patients with malfunction of the OFC suggest this area (or pathways traversing this region) to be necessarily involved in decision-making processes, especially in the perception and evaluation of choice options. Since valuable insights as to the function of a specific brain area can be inferred from its anatomical location and connectivity to other brain areas, these data will shortly be outlined. Afterwards, we will review neuroscientific results concerning the function of the OFC in decision making.

Connectivity data of the OFC

The OFC, located above the orbits of the eyes, is a part of the prefrontal cortex (see Fig. 1). Results from axonal tracer studies revealed at least two functionally distinct but inter-related networks, the orbital and the medial prefrontal network, though the latter also includes areas on the orbital surface (i.e., areas along the gyrus rectus and a caudolateral orbital region; Price, 2006). The orbital network is special in that it receives input from the cortical areas associated with most of the sensory systems, including olfaction, taste/visceral afferents, somatic sensation, and vision (Price, 2006). In addition to the sensory inputs, the orbital network has specific connections with the thalamus and the striatum and furthermore is connected with a number of limbic structures, some of which are assigned a key role in emotional processes (including the amygdala, hippocampus, entorhinal cortex, and parahippocampal gyrus; Price, 2006). Based on these features, Barbas and Zikopoulos (2006) have suggested that the orbitofrontal areas specialize in integrating the external and internal environment therewith enabling to determine the emotional value of events and their conscious

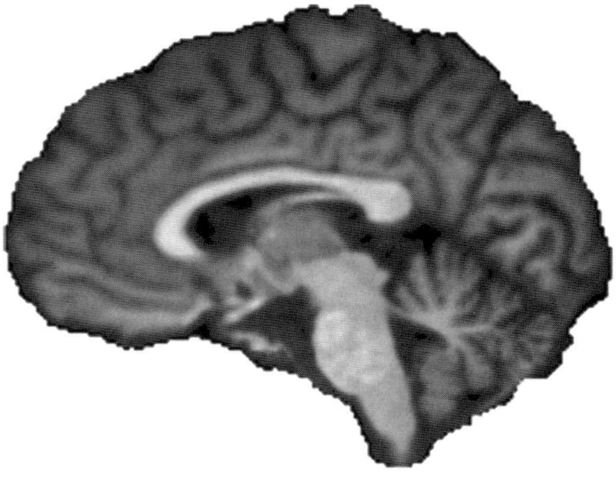

Fig. 1. Basic prefrontal regions shown on a sagittal view of an individual brain normalized and aligned to the Talairach stereotactic space. Area shown in light blue illustrates orbitofrontal cortex (OFC), purple ventromedial prefrontal cortex (VMPFC), and green medial prefrontal cortex. (See Color Plate 6.1 in color plate section.)

appreciation. In contrast, studying the intrinsic cortico-cortical connections of the medial prefrontal network revealed no obvious sensory inputs to this network. Instead, the medial prefrontal network is distinguished by outputs to the visceral control structures in the hypothalamus and brainstem, hence being involved in the cortical modulation of visceral function (Barbas and Zikopoulos, 2006; Price, 2006). The strong bidirectional connections between the orbital and medial prefrontal network may promote a "continuously updated dialogue between these functionally distinct cortices in behavior" (Barbas and Zikopoulos, 2006, p. 81).

Neuroscientific results concerning the putative function of the OFC

Numerous studies in humans as well as in nonhuman primates and rats suggest that the OFC codes the reward value of stimuli across the range of sensory modalities, including the somatosensory, auditory, and visual domains (Blood et al., 1999; O'Doherty et al., 2003b; Rolls et al., 2003b). In most studies, reward was considered as a stimulus that can act as a positive reinforcer. However, in any given experiment, "a reward may or may not be used to reinforce anything" (Montague and Berns, 2002, p. 267). In human imaging studies, reward has mostly been operationalized as appetitive stimuli (food, water, drugs, odors) or money.

For example, when subjects were presented with pleasant (sweet) and unpleasant (salt) taste stimuli, significant OFC activation was observed (O'Doherty et al., 2001b). The activation foci for positive and negative affective valence revealed to be distinct. Likewise, when subjects were presented with pleasant and unpleasant odor stimuli, activation within the OFC was observed (Gottfried et al., 2002a; Anderson et al., 2003; Rolls et al., 2003a). The OFC activation correlated with valence but was independent of intensity (Anderson et al., 2003). Furthermore, when asking participants which stimuli they find rewarding, and letting them evaluate the stimuli according to their preference, the evoked activity within the OFC showed a clear correspondence with the subjective ratings. Accordingly, the OFC seems not only to discriminate between rewards and nonrewards, but also to correlate with the subjective experience of sensory pleasure (Kringelbach et al., 2003; O'Doherty et al., 2003b). Activation within the OFC was shown to respond also to abstract reinforcers, such as money, which are not bound to a sensory modality (O'Doherty et al., 2001a; Rolls et al., 2008). The finding that reward processing is a major function of the OFC has been paralleled in nonhuman primates. Results from neurophysiological studies in monkeys using different versions of the (delay) go/no-go task showed that OFC neurons distinguished between reward and nonreward, between reward and punishment, between different rewards, and between different volumes of the same reward (for an overview see Schultz and Tremblay, 2006).[1]

Based on the finding that different sensory pathways project to definable subregions within the OFC, a functional segregation through sensory modalities seems to be obvious. Yet, a good deal of imaging data revealed a unitary representation of reward value across modalities within the anterior medial OFC (O'Doherty et al., 2001b, 2003a, b; Rolls et al., 2003a). A functional segregation though revealed as a function of reward modality, that is, the medial OFC seems to be crucially involved in rewards and the lateral OFC in punishments. For example, several studies showed that pleasant odors activate the medial OFC, whereas aversive odors activate the lateral OFC (Gottfried et al., 2002b; Anderson et al., 2003; Rolls et al., 2003a). This medial-lateral distinction has also been shown for abstract reinforcers: O'Doherty et al. (2001a) used a probabilistic reversal learning task in which participants could

[1]As outlined above, brain areas do not function in isolation but are embedded in a wider network. Hence, besides the OFC there are also other reward centers in the brain including the ventral and dorsal striatum, which show additional movement-related activity not found in the OFC; the amygdala, for which only limited information on reward processes is available; and the dopamine neurons, which differ from the OFC in processing so-called reward prediction errors (Schultz and Tremblay, 2006). To describe the division of labor of these areas in detail would go beyond the scope of this article.

win or lose money.[2] That is, participants received a probabilistically determined monetary reward when choosing the correct stimulus and a (probabilistically determined) monetary loss when choosing the incorrect stimulus. The authors found a medial versus lateral distinction within the OFC, such that monetary rewards activated the medial OFC and monetary punishments the lateral OFC. Notably, a correspondence was observed between OFC activation and the magnitude of money won or lost on single trials. In particular, medial OFC activation correlated positively with the amount of money won on single trials, whereas lateral OFC correlated positively with the amount of money lost on single trials. Interestingly, a medial versus lateral distinction as a function of reward value was also observed even when the reward component was not in the focus of the study, that is, not essential for the task to be accomplished. For example, O'Doherty et al. (2003b) presented their participants with face stimuli high and low in attractiveness. Participants' task was to judge the gender of the respective stimulus. Apparently, this task does not require the assessment of attractiveness to solve the gender task. Nevertheless, more attractive-face stimuli activated the medial OFC, whereas less-attractive face stimuli activated the lateral OFC while participants performed the unrelated gender task. Another functional magnetic resonance imaging (fMRI) study on facial attractiveness and its relation to sexual orientation revealed a three-way interaction effect between stimulus gender, beauty, and sexual preference within the medial anterior OFC (Ishai, 2007). That is, in heterosexual women and homosexual men, attractive male faces elicited stronger activation within the OFC than attractive female faces, whereas in heterosexual men and homosexual women the inverse activation pattern in the medial anterior OFC was observed. All in all, the suggestion of the medial-lateral trend in the OFC for reward representation is supported by a comprehensive meta-analysis by Kringelbach and Rolls (2004).

It has to be noted however, that there are several studies not supporting the medial-lateral dissociation within the OFC as a function of reward value (Elliott et al., 2003; O'Doherty et al., 2003a). Common to studies challenging a simple valence hypothesis is that they used complex gambling or decision-making paradigms, which may explain the disparate findings. Yet, it remains to be studied which conditions are causing the difference.

Together, neuroscientific results suggest the orbitofrontal areas to be crucially involved in the representation of rewards of choice options. According to the terminology of research in judgment and decision making, the OFC may encode the subjectively expected utility of a choice option. It is general knowledge in research on judgment and decision making that decisions are context-sensitive in that the overall subjectively expected utility of an option is highly dependent on the consequences of the alternatives (Hastie and Dawes, 2001; Jungermann et al., 2005, see also Chapter 2: On the relativity of athletic performance: a comparison perspective on performance judgments in sports). That is, a particular choice alternative appears different to a decision maker when considered in different contexts. Specifically, a choice alternative may be evaluated more or less desirable when presented in different choice sets. A number of imaging studies found that the OFC accounts for such context effects, in that activation within the OFC encodes the relative rather than the absolute value of rewards. For example, Small et al. (2001) recorded the hemodynamic activity while participants consumed chocolate to satiety. The authors observed activation within the medial OFC early in the experiment when the chocolate was highly rewarding. Yet, later in the experiment, when the chocolate was eaten to satiety, and hence the reward value went from being pleasant to aversive, activation was observed within the lateral OFC. Subsequent studies revealed congruent findings showing OFC responses to pleasant tastes and smells to depend

[2] In common reversal learning paradigms behavior is altered on reversing reinforcement contingencies. Accordingly, such paradigms may conceptually be subdivided in acquiring stimulus-reinforcement associations and subsequently alter behavior by switching to new associations as stimulus-reinforcement contingencies reverse.

on the subjective state, that is, hungry versus satiated (Gottfried et al., 2002b; Kringelbach et al., 2003). The proposition that the OFC encodes the relative rather than the absolute reward value has recently been confirmed for abstract monetary rewards (Elliott et al., 2008). In their study, Elliott and colleagues demonstrated that the same perceptual stimulus elicited differential activation within the medial OFC depending on the learned reward history and the current context: activation within medial OFC was observed for a medium-rewarding stimulus when paired with a low-rewarding stimulus but not when the same medium-rewarding stimulus was paired with a high-rewarding stimulus. The finding of differential OFC activity to the same perceptual input in dependence of the relative reward value has been shown in monkeys before (Tremblay and Schultz, 1999): activation within orbitofrontal neurons was higher for a piece of apple when it was presented alternatively to cereal but not when the piece of apple was presented alternatively to a raisin. All in all, these findings suggest that OFC activation reflects the integration of simple associative information about the reward value of an option with nonassociative variables such as current context, probability of the outcome, internal state, and subsequent plans (Roesch and Schoenbaum, 2006).

Although not common, there are situations in which the decision maker does not only learn about the outcome of the chosen option but also about the outcome of the unchosen option. For example, when choosing between two gambles (in the casino), one immediately learns whether or not one won the jackpot. In case the consequences of the unchosen option are much better than the ones of the chosen option, the decision maker feels regret. Conversely, when the consequences of the chosen option are much better than the ones of the unchosen option the decision maker is delighted. These utility post-decision emotions, which are named counterfactual emotions, have received special attention from decision research since they considerably influence the evaluation of a choice option (Bell, 1982; Loomes and Sudgen, 1982, 1987a). To avoid regret in decision making, people often tend to ignore what would have happened once committed to a course of action. Despite progress being made in studying and incorporating such emotions into decision-making theories, findings suggest that "regret theory" is only valid under very specific conditions (Loomes and Sudgen, 1987b; Loomes, 1988). Anyhow, since the OFC has been suggested to be critical for holding associative information about reward value in representational memory and for comparing it to other available options in the choice set, it is rather straightforward to investigate the involvement of the OFC on the experience of regret. Camille and colleagues (2004) conducted an elegant study on the involvement of the OFC in the experience of regret. Healthy participants, patients with OFC lesions, and patients with lesions in other parts of the prefrontal cortex were presented with a choice between two risky gambles with a monetary reward. In a partial feedback condition participants only learned about the outcome of the chosen gamble, whereas in a complete feedback condition participants learned about both outcomes, the one of the chosen and the unchosen gamble. After each gamble, participants had to rate their effect from sad to happy. Results revealed that specifically patients with lesions to the OFC do not experience regret whatsoever since their emotions were not at all modulated by the knowledge of the outcome of the rejected alternative. Likewise, skin conductance responses in these patients confirmed a lack of emotional reaction to the outcome of the unchosen gamble. In contrast, healthy controls and control patients showed a clear regret effect in that a reward made them happier and a loss made them less sad when they learned that they avoided a large punishment. Congruent, feedback about the outcome of the unchosen gamble enhanced skin conductance responses as compared to emotional reactivity in the partial feedback condition. The findings by Camille and colleagues convincingly illustrate context effects on OFC activation and support the suggestion that the OFC plays a crucial role in the generation and use of subjective outcome expectancies, taking into account internal states (including emotional variables), current context, and future goals.

The finding that OFC-damaged patients show a lack of emotional reactivity in decision-making tasks is in accordance with previous results (Bechara et al., 1994). In this context, Damasio and colleagues (1991) put forward the so-called "somatic marker hypothesis," attempting to understand the striking emotional and everyday decision-making deficits observed in patients with lesions to the OFC and VMPFC. According to the somatic marker hypothesis, the observed decision-making deficits are attributable to an inability to use emotion-based biasing signals generated from the body ("somatic markers") when evaluating different response options (Damasio, 1994; Bechara et al., 2002; Damasio, 2004). These somatic markers are assumed to be integrated in higher brain regions, particularly in the VMPFC. Based on connectional features, this functional assumption seems highly plausible: orbital prefrontal network areas receive visceral afferents, among others, and are strongly connected with medial prefrontal network areas, which in turn have rich direct and indirect and highly efficient networks with central autonomic pathways extending as far as spinal autonomic centers.

Evidence for the somatic marker hypothesis is mainly based on performance in only one single experimental paradigm, namely the "Iowa gambling task." In this task, participants sample from four decks of cards whereby each card wins or looses money. Unbeknown to participants, two decks are disadvantageous decks revealing high rewards and occasional high penalties, whereas the other two decks are advantageous decks revealing low rewards and no penalties. Participants' task is to win as much money as possible. Therefore, participants have to formulate a plan of action that is advantageous in the long run, which means to forgo a high reward in favor of a high payoff at the end of the experimental session. Healthy participants initially choose from high-reward decks, but later bias their choices to the low-reward decks. At first, this change in behavior cannot consciously be justified, rather participants report basing their choices on a "gut" (instinctive) feeling until penalties become apparent. This feeling of rightness has been suggested to be reflected in anticipatory skin conductance response activity in response to a possible outcome of the action (e.g., Bechara et al., 2000). OFC/VMPFC-damaged patients alike prefer the high-reward decks at the beginning. Yet, they fail to modify their responding, which is suggested to reflect the inability to generate outcome expectancies required to compare the relative value of the preferred and the unpreferred outcome. Importantly, these patients fail to generate anticipatory skin conductance responses before picking a card (Bechara et al., 2002). This pattern of verbal reports and concomitant autonomic responses, led to the assumption that healthy participants pass through a "hunch-phase" in their learning of the advantageous strategy. During this phase, they show signs of having implicitly learned the rule at an intuitive level, while still being unable to verbalize it (de Vries et al., 2008). Other examples of such kinds of feelings or intuitions, often experienced to guide our decision behavior in everyday life, are feelings of knowing, feelings of rightness, feelings of coherence, or feelings of preference (Price and Norman, 2008). What these feelings have in common is "their metacognitive informational content and the lack of conscious access to the antecedents of the feelings" (Price and Norman, 2008, p. 31). In this context, Price and Norman (2008) distinguish between the consciousness of feelings — we are very much aware of our instinctive feelings — and then nonconsciousness of their antecedents — we do not have detailed conscious access to the searched-for information itself, the relative accessibility of relevant information, or the familiarity of memory cues and the like. On top of this, these kinds of instinctive feelings specifically draw on the OFC. Schnyer et al. (2005) investigated healthy participants working on an episodic feeling-of-knowing paradigm. In this task, participants were asked to make predictions about the retrievability of final words that completed studied sentences. They then had to select the correct completion from among the alternatives. Activation within the VMPFC/OFC was critically involved in accurate feeling-of-knowing-judgments. Moreover, the activation was modulated by retrieval rating. Congruent, when Schnyer and colleagues (2004) tested patients with lesions to the frontal cortex (including the VMPFC/OFC) with the same episodic feeling-of-knowing task, they observed a clear impairment in the accuracy of the prospective feeling-of-knowing judgments in these patients. Bar and colleagues (2006), as well as Volz and von Cramon (2006), both investigated the neural correlates of feelings of coherence by using a visual object recognition task. In these paradigms, object recognition was made difficult by presenting participants with blurred images of objects. Both studies revealed differential activation unique to successful object recognition in the anterior medial OFC. By using both functional magnetic resonance imaging (fMRI) and magnetoencephalography, Bar and colleagues (2006) discovered that the OFC activation associated with successful object recognition develops 50 ms *before* activation in fusiform areas is observed, which in turn are known to be crucially involved in object recognition processes. Accordingly, information from the OFC is projected to recognition-related visual areas. Bar and colleagues interpret these dynamics as top-down facilitation of object recognition. In this context, it has to be noted that the OFC is not a region traditionally viewed as a visual area. Volz et al. (2008) repeated their study on feelings of coherence in the auditory domain and again observed significant activation within anterior medial OFC for subjective (auditory) object recognition, that is, the subjective perception of coherent sounds. Together, all these "gut" feelings or intuitions, that are characterized by a dissociation between "knowing" the correct answer and a lack of conscious access to the antecedents of such feelings, crucially draw on the OFC.

Given the neuroscientific results so far, there are several different but related interpretations as to the putative function of the OFC/VMPFC: the orbital prefrontal areas:

(a) code the hedonic quality of choice options (e.g., O'Doherty, 2007),
(b) are critical for maintaining associative information about expected outcomes in representational memory so that it can be compared and integrated with information about internal states and current goals (e.g., Roesch and Schoenbaum, 2006; Schoenbaum et al., 2006),
(c) serve as a store of implicitly acquired linkages between factual knowledge and bio-regulatory states, including those that constitute feelings and emotions (e.g., Damasio et al., 1991; Tranel, 2002; Damasio, 2004),
(d) serve as a detector of potential content that is derived from the critical aspects of the input, that is, the gist information (e.g., Bar et al., 2006; Volz and von Cramon, 2006),
(e) are crucially involved in the integration of emotional signals in the decision-making process (e.g., Camille et al., 2004; Schnyer et al., 2004), and
(f) may specialize in integrating the external and internal environment (e.g., Barbas and Zikopoulos, 2006).

Proposition of how the OFC contributes to decision making

Although ambitious, we try to bring together these varying but related approaches: the major empirical finding is that the OFC codes the hedonic quality of a choice option be it an object or a behavioral strategy (compare interpretation (a)). The stimuli that were used in those studies supporting this view were obviously positive or negative, for example, pleasant odor stimuli or unpleasant odor stimuli presented without a relevant context and mostly without the need for a decision.[3] Generally, the perception of an option depends on internal (e.g., goals, internal state)

[3]We adhere to the definition of decision according to the literature on judgment and decision making: a decision is a response to a situation that is composed of three parts: (1) there is more than one possible course of action under consideration, (2) the decision maker can form expectations concerning future events and outcomes following from each course of action; expectations are often described in terms of probabilities or degrees of confidence, and (3) consequences, associated with the possible outcomes, that can be assessed on an evaluative continuum reflecting personal values and current goals (Hastie and Dawes, 2001; Jungermann et al., 2005).

and external factors (e.g., context in which the options are presented, probabilities of the consequences of each option); together both factors constitute the context of a decision problem. As such, the hedonic quality of a choice option can never be absolute but is highly context-sensitive (compare interpretations (b) and (f)). To appreciate an option accurately, the individual must rely on past experience with the option under consideration or with a related option. In decision situations such as in the Iowa gambling task, past experience such as the number of winnings may exceed cognitive capacities, in other decision situations the searched-for information may not even be accessible to consciousness, as for example cue familiarity or retrieval strength. Hence, we assume that past experience is mainly conveyed by affect, which is certainly accessible to consciousness (compare interpretations (c) and (e)). In turn, these affective reactions toward an option depend on the current goals and hence these are assumed to reflect the entire stream of prior goal-relevant experiences, which may be considered gist information (compare interpretation (d)). The resulting affective orientation establishes the crucial link between past experience and future decisions. How exactly affective reactions may influence decision-making behavior is highly elaborated in "preference theory," an affect-based approach to recurrent decision making put forward by Betsch (2005).

Taken together, the OFC seems to code for the relative reward value (or subjective expected utility) of choice options in simple decision-making situations and to provide biasing affective signals in sophisticated decision-making situations. In the former type of decision situations, the OFC may compute an ongoing valuation of rewards, punishments, and their predictors across a broad domain of stimuli (Montague and Berns, 2002). In such situations, the relations between rewards, punishments, and their predictors are either given or easily inferable by thorough feedback analysis. For example, in reversal learning tasks, contingencies are straightforward and the appropriate strategy has only to be changed once reinforcement contingencies and therewith the reward value of the options change. In contrast, in sophisticated decision situations when the appropriate response cannot be selected using simple associations, the OFC is suggested to provide affective reactions toward the (behavioral) option under consideration reflecting the entire stream of prior experiences. We suggest that affective signals are either in demand when cognitive capacities are limited or when the gist information (the antecedents of the affective signals) is not consciously accessible. Based on the sentence "preferences need no inferences" (Zajonc, 1980), we assume OFC triggered affective responses to be available as soon as a decision problem is encountered. Thereby this kind of default-mode operation requires only a minimum of mental resources and cannot be terminated by volitional dictum (cf. Betsch, 2005).

To illustrate the latter proposition, we give an example from the visual domain. Previous research showed that normal visual object identification ("what") depends on occipito-temporal regions, whereas object location ("where") requires parietal cortical areas (Ungerleider and Mishkin, 1982). The dual visual system view was then combined with physiological data. Information about form and color delivered by so-called P-cells was considered to project ultimately to the what-hierarchy of visual areas ending in the temporal lobe, whereas information about motion delivered by so-called M-cells was considered to project ultimately to the where-hierarchy of visual areas ending in the parietal lobe. The pathway providing information from so-called K-cells was ignored, since it did not fit well with the dual visual system view and is only recently an object of interest (Hendry and Reid, 2000; Xu et al., 2001). K-cells have been found to share many achromatic spatial and temporal properties in common with both M- and P-cells but are limited in the message they send, and hence it is difficult to grade them according to system. One possibility is, however, that K-cells provide *the* critical information important for both motion and object identification in parallel to the processing within the P- and M-cells. Hence, information from K-cells may be relied on only when information from P- and M-cells is not available. Specifically, K-cells are the only cell-group that projects directly to the extrastriate cortex (V2, V4), which may be the reason why this information is not accessible to consciousness. Because of this, K-cells have recently been suggested to sustain some visual behavior in the absence of visual cortex (V1), and thus are assigned a crucial role in blindsight (Hendry and Reid, 2000; Vakalopoulos, 2005). Another example how gist information may be constructed comes from research on implicit attitude formation (Betsch et al., 2001), which suggests evaluative judgments to reflect the total value of prior encounters (Betsch et al., 2003).

Concluding, especially in social situations where there are no simple rules (about reward contingencies) but rather an immense number of contextual cues determining which information is relevant at any particular time, affective signals seem to be vital to judge advantageously. This may explain why "EVR" showed gross impairments especially in his ability to make advantageous decisions in his personal and social life.

Abbreviations

fMRI	functional magnetic resonance imaging
OFC	orbitofrontal cortex
VMPFC	ventromedial prefrontal cortex

References

Anderson, A. K., Christoff, K., Stappen, I., Panitz, D., Ghahremani, D. G., Glover, G., Gabrieli, J. D., & Sobel, N. (2003). Dissociated neural representations of intensity and valence in human olfaction. *Nature Neuroscience*, *6*, 196–202.

Bar, M., Kassam, K. S., Ghuman, A. S., Boshyan, J., Schmid, A. M., Dale, A. M., Hämäläinen, M. S., Marinkovic, K., Schacter, D. L., Rosen, B. R., & Halgren, E. (2006). Top-down facilitation of visual recognition. *Proceedings of the National Academy of Sciences of the United States of America*, *103*, 449–454.

Barbas, H., & Zikopoulos, B. (2006). Sequential and parallel circuits for emotional processing in primate orbitofrontal cortex. In D. H. Zald & S. L. Rauch (Eds.), *The orbitofrontal cortex* (pp. 57–91). New York: Oxford University Press.

Bechara, A., Damasio, A. R., Damasio, H., & Anderson, S. W. (1994). Insensitivity to future consequences following damage to human prefrontal cortex. *Cognition*, *50*, 7–15.

Bechara, A., Tranel, D., & Damasio, A. R. (2002). The somatic marker hypothesis and decision-making. In F. Boller & J. Grafman (Eds.), *Handbook of neuropsychology* ((2nd ed.), pp. 117–143). Amsterdam: Elsevier.

Bechara, A., Tranel, D., & Damasio, H. (2000). Characterization of the decision-making deficit of patients with ventromedial prefrontal cortex lesions. *Brain*, *123* (Pt. 11), 2189–2202.

Bell, D. E. (1982). Regret in decision making under uncertainty. *Operations Research*, *30*, 961–981.

Betsch, T. (2005). Preference theory: an affect-based approach to recurrent decision making. In T. Betsch & S. Haberstroh (Eds.), *The routines of decision making* (pp. 39–65). Mahwah, NJ: Lawrence Erlbaum Associates.

Betsch, T., Hoffmann, K., Hoffrage, U., & Plessner, H. (2003). Intuition beyond recognition: when less familiar events are liked more. *Experimental Psychology*, *50*, 49–54.

Betsch, T., Plessner, H., Schwieren, C., & Gütig, R. (2001). I like it but I don't know why: a value-account approach to implicit attitude formation. *Personality and Social Psychology Bulletin*, *27*, 242–253.

Blood, A. J., Zatorre, R. J., Bermudez, P., & Evans, A. C. (1999). Emotional responses to pleasant and unpleasant music correlate with activity in paralimbic brain regions. *Nature Neuroscience*, *2*, 382–387.

Broca, P. (1861). Remarques sur le siege de la faculte du langage articule, suivies d'une observation d'aphemie (perte de la parole) [Comments regarding the seat of the faculty of spoken language, followed by an observation of aphemia (loss of speech)]. *Bulletin Societe Anatomique de Paris*, *6*, 330–357.

Camille, N., Coricelli, G., Sallet, J., Pradat-Diehl, P., Duhamel, J. R., & Sirigu, A. (2004). The involvement of the orbitofrontal cortex in the experience of regret. *Science*, *304*, 1167–1170.

Damasio, A. R. (1994). *Descartes error: emotion, reason and the human brain*. New York: Avon.

Damasio, A. R. (2004). William James and the modern neurobiology of emotion. In D. Evans & P. Cruse (Eds.), *Emotion, evolution and rationality* (pp. 3–14). Oxford: Oxford University Press.

Damasio, A. R., Tranel, D., & Damasio, H. (1991). Somatic markers and the guidance of behavior: theory and preliminary testing. In H. S. Levin, H. M. Eisenberg, & A. L. Benton (Eds.), *Frontal lobe function and dysfunction* (pp. 217–229). New York: Oxford University Press.

de Vries, M., Holland, R. W., & Witteman, C. L. M. (2008). In the winning mood: affect in the Iowa gambling task. *Judgment and Decision Making*, *3*, 42–50.

Elliott, R., Agnew, Z., & Deakin, J. F. (2008). Medial orbitofrontal cortex codes relative rather than absolute value of financial rewards in humans. *European Journal of Neuroscience*, *27*, 2213–2218.

Elliott, R., Newman, J. L., Longe, O. A., & Deakin, J. F. (2003) Differential response patterns in the striatum and orbitofrontal cortex to financial reward in humans: a parametric functional magnetic resonance imaging study. *Journal of Neuroscience*, *23*, 303–307.

Esslinger, P. J., & Damasio, A. R. (1985). Severe disturbance of higher cognition after bilateral frontal lobe ablation: patient EVR. *Neurology*, *35*, 1731–1741.

Gottfried, J. A., Deichmann, R., Winston, J. S., & Dolan, R. J. (2002a). Functional heterogeneity in human olfactory cortex: an event-related functional magnetic resonance imaging study. *Journal of Neuroscience*, *22*, 10819–10828.

Gottfried, J. A., O'Doherty, J., & Dolan, R. J. (2002b). Appetitive and aversive olfactory learning in humans studied using event-related functional magnetic resonance imaging. *Journal of Neuroscience*, *22*, 10829–10837.

Hastie, R., & Dawes, R. M. (2001). *Rational choice in an uncertain world. The psychology of judgment and decision making*. Thousand Oaks, CA: Sage Publications.

Hendry, S. H. C., & Reid, R. C. (2000). The koniocellular pathway in primate vision. *Annual Review of Neuroscience*, *23*, 127–153.

Ishai, A. (2007). Sex, beauty and the orbitofrontal cortex. *International Journal of Psychophysiology*, *63*, 181–185.

Jungermann, H., Pfister, H.-R., & Fischer, K. (2005). *Die Psychologie der Entscheidung: Eine Einführung [The psychology of decision: an introduction]*. Heidelberg: Spektrum Akademischer Verlag.

Kringelbach, M. L., O'Doherty, J., Rolls, E. T., & Andrews, C. (2003). Activation of the human orbitofrontal cortex to a liquid food stimulus is correlated with its subjective pleasantness. *Cerebral Cortex*, *13*, 1064–1071.

Kringelbach, M. L., & Rolls, E. T. (2004). The functional neuroanatomy of the human orbitofrontal cortex: evidence from neuroimaging and neuropsychology. *Progress in Neurobiology*, *72*, 341–372.

Loomes, G. (1988). Further evidence of the impact of regret and disappointment in choice under uncertainty. *Econometrica*, *55*, 47–62.

Loomes, G., & Sudgen, R. (1982). Regret theory: an alternative theory of rational choice under uncertainty. *Economic Journal*, *92*, 805–824.

Loomes, G., & Sudgen, R. (1987a). Some implications of a more general form of regret theory. *Journal of Economic Theory*, *41*, 270–287.

Loomes, G., & Sudgen, R. (1987b). Testing for regret and disappointment in choice under uncertainty. *Economic Journal*, *97*, 118–129.

Montague, P. R., & Berns, G. S. (2002). Neural economics and the biological substrates of valuation. *Neuron*, *36*, 265–284.

O'Doherty, J., Critchley, H., Deichmann, R., & Dolan, R. J. (2003a). Dissociating valence of outcome from behavioral control in human orbital and ventral prefrontal cortices. *Journal of Neuroscience*, *23*, 7931–7939.

O'Doherty, J., Kringelbach, M. L., Rolls, E. T., Hornak, J., & Andrews, C. (2001a). Abstract reward and punishment representations in the human orbitofrontal cortex. *Nature Neuroscience*, *4*, 95–102.

O'Doherty, J., Rolls, E. T., Francis, R., Bowtell, R., & McGlone, F. (2001b). Representation of pleasant and aversive taste in the human brain. *Journal of Neurophysiology*, *85*, 1315–1321.

O'Doherty, J., Winston, J., Critchley, H., Perrett, D., Burt, D. M., & Dolan, R. J. (2003b). Beauty in a smile: the role of medial orbitofrontal cortex in facial attractiveness. *Neuropsychologia*, *41*, 147–155.

O'Doherty, J. P. (2007). Lights, camembert, action! The role of human orbitofrontal cortex in encoding stimuli, rewards, and choices. *Annals of the New York Academy of Sciences*, *1121*, 254–272.

Price, J. L. (2006). Connections of the orbital cortex. In D. H. Zald & S. L. Rauch (Eds.), *The orbitofrontal cortex* (pp. 39–55). New York: Oxford University Press.

Price, M. C., & Norman, E. (2008). Intuitive decisions on the fringe of consciousness: are they conscious and does it matter? *Judgment and Decision Making*, *3*, 28–41.

Roesch, M., & Schoenbaum, G. (2006). From associations to expectancies: orbitofrontal cortex as gateway between limbic system and representational memory. In D. H. Zald & S. L. Rauch (Eds.), *The orbitofrontal cortex* (pp. 199–235). New York: Oxford University Press.

Rolls, E. T., Kringelbach, M. L., & de Araujo, I. E. (2003a). Different representations of pleasant and unpleasant odors in the human brain. *European Journal of Neuroscience*, *18*, 695–703.

Rolls, E. T., McCabe, C., & Redoute, J. (2008). Expected value, reward outcome, and temporal difference error representations in a probabilistic decision task. *Cerebral Cortex*, *18*, 652–663.

Rolls, E. T., O'Doherty, J., Kringelbach, M. L., Francis, S., Bowtell, R., & McGlone, F. (2003b). Representation of pleasant and painful touch in the human orbitofrontal and cingulate cortices. *Cerebral Cortex*, *13*, 308–317.

Schnyer, D. M., Nicholls, L., & Verfaellie, M. (2005). The role of VMPC in metamemorial judgments of content retrievability. *Journal of Cognitive Neuroscience*, *17*, 832–846.

Schnyer, D. M., Verfaellie, M., Alexander, M. P., Lafleche, G., Nicholls, L., & Kaszniak, A. W. (2004). A role for right medial prefrontal cortex in accurate feeling of knowing judgments: evidence from patients with lesions to frontal cortex. *Neuropsychologia*, *42*, 957–966.

Schoenbaum, G., Roesch, M. R., & Stalnaker, T. A. (2006). Orbitofrontal cortex, decision-making, and drug addiction. *Trends in Neuroscience*, *29*, 116–124.

Schultz, W., & Tremblay, L. (2006). Involvement of primate orbitofrontal neurons in reward, uncertainty, and learning. In D. H. Zald & S. L. Rauch (Eds.), *The orbitofrontal cortex* (pp. 173–198). New York: Oxford University Press.

Scoville, W. B., & Milner, B. (1957). Loss of recent memory after bilateral hippocampal lesions. *Journal of Neuropsychiatry and Clinical Neurosciences*, *12*, 103–113.

Small, D. M., Zatorre, R. J., Dagher, A., Evans, A. C., & Jones-Gotman, M. (2001). Changes in brain activity related to eating chocolate: from pleasure to aversion. *Brain*, *124*, 1720–1733.

Tranel, D. (2002). Emotion, decision making, and the ventromedial prefontal cortex. In D. T. Stuss & R. T. Knight

(Eds.), *Principles of frontal lobe function* (pp. 338–353). New York: Oxford University Press.

Tremblay, L., & Schultz, W. (1999). Relative reward preference in primate orbitofrontal cortex. *Nature*, *398*, 704–708.

Ungerleider, L. G., & Mishkin, M. (1982). Two cortical visual systems. In D. J. Ingle, M. A. Goodale, & R. J. W. Mansfield (Eds.), *Analysis of visual behavior* (pp. 549–586). Cambridge: MIT Press.

Vakalopoulos, C. (2005). A theory of blindsight — the anatomy of the unconscious: a proposal for the koniocellular projections and intralaminar thalamus. *Medical Hypotheses*, *65*, 1183–1190.

Volkow, N. D., & Fowler, J. S. (2000). Addiction, a disease of compulsion and drive: involvement of the orbitofrontal cortex. *Cerebral Cortex*, *10*, 318–325.

Volz, K. G., Rubsamen, R., & von Cramon, D. Y. (2008). Cortical regions activated by the subjective sense of perceptual coherence of environmental sounds: a proposal for a neuroscience of intuition. *Cognitive, Affective & Behavioral Neuroscience*, *8*, 318–328.

Volz, K. G., & von Cramon, D. Y. (2006). What neuroscience can tell about intuitive processes in the context of perceptual discovery. *Journal of Cognitive Neuroscience*, *18*, 2077–2087.

Xu, X., Ichida, J. M., Allison, J. D., Boyd, J. D., Bonds, A. B., & Casagrande, V. A. (2001). A comparison of koniocellular, magnocellular and parvocellular receptive field properties in the lateral geniculate nucleus of the owl monkey. *Journal of Physiology*, *531*, 203–218.

Zajonc, R. B. (1980). Feeling and thinking: preference need no inferences. *American Psychologist*, *35*, 151–175.

M. Raab et al. (Eds.)
Progress in Brain Research, Vol. 174
ISSN 0079-6123

CHAPTER 7

Perceiving the intentions of others: how do skilled performers make anticipation judgments?

A. Mark Williams*

Research Institute for Sport and Exercise Sciences, Liverpool John Moores University, Liverpool, United Kingdom

Abstract: In this chapter, the processes underpinning anticipation skill in dynamic, sport tasks are reviewed. A number of perceptual-cognitive skills that contribute to our ability to perceive the intentions of opponents are highlighted, such as the ability to: (a) pick up advance information from an opponent's postural orientation, (b) identify stimulus familiarity in patterns of play, (c) explore the visual scene in an efficient and effective manner, and (d) ascertain the most likely options facing an opponent at any given moment in time. A framework is presented to illustrate how these different perceptual-cognitive skills interact in a continuous and dynamic manner during performance. Finally, an attempt is made to illustrate how the interaction between these perceptual-cognitive skills, as well as the nature of the link between perception and action, may be influenced by player, task, and situational constraints.

Keywords: perception; cognition; expertise; sport

Introduction

In most daily activities, there is continual need to perceive the intentions of others. This need is particularly evident in competitive sport, where the ability to "read" opponents' intentions, often while simultaneously disguising one's own intentions, is crucial to performance, particularly at the elite level (Reilly et al., 2000). Moreover, in fast ball sports like tennis or squash, as well as in team sports such as soccer and handball, the need to perceive the intentions of others occurs under severe time pressure. For example, when returning serve in tennis or attempting to save a penalty kick in soccer the duration of ball flight is often shorter than the combined duration of player response time and movement time. In the latter situations, the need to perceive the intentions of an opponent is intrinsically interwoven with the requirement to make predictive judgments early, often solely based on information available prior to racket–ball or foot–ball contact. The ability to anticipate is therefore paramount to successful performance in such domains, providing additional time for performers to choose an appropriate course of action and initiate a response.

The aim in this chapter is to highlight how skilled performers make judgments about the intentions of opponents in dynamic, time-constrained environments. The main focus is on how players anticipate an opponent's potential mode of action and the approximate future location of the ball rather than on how players coordinate and control their actions during interception and action. An attempt is made to

*Corresponding author.
Tel.: +44-151-231-4489; Fax: +44-151-231-4353;
E-mail: m.williams@ljmu.ac.uk

DOI: 10.1016/S0079-6123(09)01307-7

synthesize contemporary research that has attempted to highlight the key perceptual-cognitive skills underlying anticipation skill in sport. These skills include: (a) the ability to pick up advance information from an opponent's postural orientation, (b) a capacity to identify familiar patterns of play or sequences within dynamic team sports, (c) greater efficiency and effectiveness in moving the eyes around the display in an effort to extract information from relevant display areas, and (d) the capability to accurately predict the likely choice options open to an opponent at any given moment. A framework is presented to illustrate how these perceptual-cognitive skills interact in a continuous, dynamic, and parallel fashion during performance. Finally, an attempt is made to briefly illustrate how the interaction between these perceptual-cognitive skills, as well as the nature of the link between perception and action, may be influenced by different player, task, and situational constraints (see also Chapter 8: The bidirectional links between decision making, perception, and action).

Perceptual-cognitive skills underpinning anticipation judgments

A brief overview is provided of some of the perceptual-cognitive skills underlying anticipation judgments.

Advance cue utilization

Numerous published reports exist to suggest that skilled performers are able to pick up information from an opponent's postural orientation in the moments leading up to a key event such as foot–ball or racket–ball contact. This ability to pick up advance (early) visual cues is typically evaluated using a film-based temporal occlusion approach, where the action is filmed from the same viewing perspective as in the competitive situation and then played back to the athlete with varying amounts of advance and ball flight information. The skilled performer's superiority over his/her less skilled counterpart is most apparent at earlier occlusion conditions (i.e., pre- rather than post-ball flight cues). Although researchers initially employed this method in tennis (e.g., see Jones and Miles, 1978), they subsequently extended this finding to other sports including squash (Abernethy and Russell, 1987), soccer (Williams and Burwitz, 1993), and field-hockey (Salmela and Fiorito, 1979). Moreover, a myriad of different techniques has been employed to verify the findings derived using the film-based occlusion technique such as high-speed film analysis and the use of liquid crystal occlusion glasses (e.g., see Müller and Abernethy, 2006; for a detailed review, see Hodges et al., 2007; Williams and Ward, 2007).

Most recently, researchers have endeavored to ascertain the specific postural cues that performers pick up when attempting to anticipate an opponent's intentions. Several techniques have been used to address this issue including film-based spatial occlusion, eye movement registration, and verbal protocol analysis (e.g., see Williams and Davids, 1998; Müller et al., 2006; Vaeyens et al., 2007a, b; McRobert et al., in press). Another approach that has been used with increasing popularity has been to present stimuli as point-light or stick-figure images. Such displays are typically constructed by attaching reflective markers to the major joints of the body, recording the body in motion, then displaying these markers alone against a black background. The intention is to remove background and contextual information and to present movement in its simplest terms (Johansson, 1973; Cutting and Proffitt, 1982). A typical stick-figure representation for a tennis forehand shot is presented in Fig. 1.

Several researchers have proposed that the effective pick up of relative motion (captured by sequences of point-light displays) is an essential component of anticipation skill in fast-ball sports (e.g., see Abernethy et al., 2001; Ward et al., 2002). Performers appear to determine opponents' intentions based on their perception of the relative motion between specific bodily features, rather than via the extraction of information from more superficial features or an isolated area or cue.

A more recent extension to the point-light display method was provided by Huys and

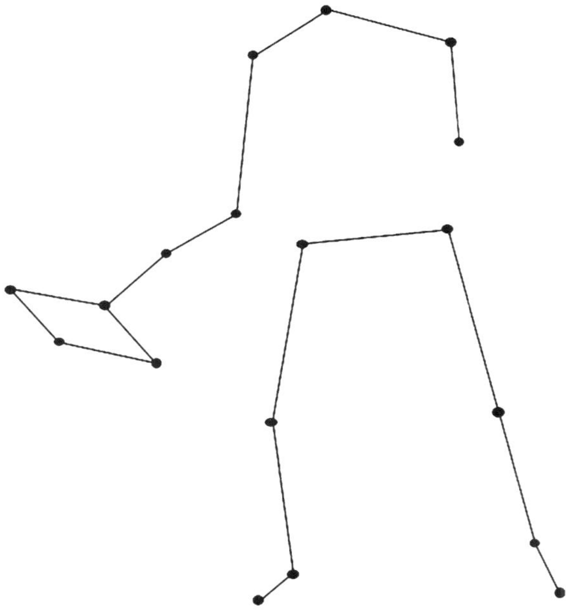

Fig. 1. An example of a stick-figure image of a tennis forehand stroke.

colleagues (e.g., Huys et al., 2008, in press; Williams et al., in press). These authors created stick-figure displays after initially running principle component analyses on kinematic data gathered from several tennis players performing forehand drive shots to different areas of the court. The intention was to describe each shot based on a smaller number of components or modes. These modes were then manipulated in order to interchange, neutralize, or occlude the dynamic information presented at different body regions (+ racket) when players performed crosscourt and inside-out forehand drive shots in tennis. The findings imply that skilled players employ a more "global" than "local" perceptual strategy when compared with less skilled counterparts, being able to extract relevant information from a number of different body areas simultaneously when attempting to anticipate their opponents' intentions. This latter conclusion would support the notion that skilled performers demonstrate greater perceptual flexibility or redundancy than less skilled athletes (Williams and Ward, 2007).

Identifying stimulus familiarity

The ability to identify familiarity in sequences of play is usually inferred from performance on recall or recognition paradigms imported from cognitive psychology (see Chase and Simon, 1973; Goldin, 1978). In the recall paradigm, participants are required to recall players' positions after viewing structured (i.e., a typical offensive move) and unstructured (e.g., players warming up before a match) filmed sequences of play lasting a few seconds. Performance is evaluated by comparing each player's actual position with that recalled by participants. A skill effect is observed on structured sequences only (e.g., see Williams and Davids, 1995; Ward and Williams, 2003).

In the recognition paradigm, players are presented with film clips involving structured and unstructured sequences of play lasting a few seconds. In a subsequent recognition phase, players are presented with a similar sample of sequences some of which were shown in the earlier viewing phase and others of which are novel. Players are required to indicate which sequences are new and which were presented earlier. Performance is determined by the accuracy with which players are able to recognize previously viewed clips. Skilled players are more accurate than less-skilled players in recognizing structured sequences only (e.g., see Williams and Davids, 1995; Ward and Williams, 2003; Smeeton et al., 2004).

In recent years, the focus has been on identifying "what" information players use to make these stimulus familiarity judgments and "when" does this information emerge in dynamic, action sequences. For example, Williams et al. (2006) examined the relative importance of superficial display features and relational information between players when recognizing sequences of play. Skilled and less-skilled soccer players completed a recognition test where sequences of play were presented under both film and point-light display conditions. In the point-light display condition, player positions and movements were highlighted as colored dots against a black background with the playing area represented as white lines, whereas more superficial features such as the color of players' uniforms, postural cues, or the condition of the playing surface and other environmental effects were removed.

Although skilled players' recognition accuracy suffered under point-light compared with film conditions, the decrement was less marked than for less-skilled players and, most importantly, the skill main effect was maintained across viewing conditions. The findings are presented in Fig. 2. Skilled performers are able to detect similarity across patterns of play based upon structural relations (e.g., positions and movements of players) and the higher-order predicates they convey (e.g., tactical significance of these relations between players), whereas less-skilled players are reliant on more superficial structural features (Gentner and Markman, 1998).

In a follow-up experiment, Williams et al. (2006) used a spatial occlusion technique in an attempt to identify whether some players were more important than others were when identifying sequences. A decrement in performance was observed when the two central attackers and their corresponding defensive markers were occluded, compared with performance under a control condition involving the removal of two peripheral offensive players and their corresponding defensive markers. This effect was most pronounced in the skilled compared with less-skilled players. The same question was examined by North et al. (in press, in preparation). who, in two separate experiments, collected eye movements and verbal reports as players attempted to make familiarity-based judgments. Skilled players recorded a higher number of fixations on central attacking players and made more fixation transitions from the ball to an attacking feature and vice versa when compared with less-skilled players. Moreover, skilled players made significantly more verbalizations involving the movements of offensive players "off the ball" than less-skilled players, with the central attacking players being particularly important for the skilled players when sequences were presented in point-light format.

In another experiment, Williams and North (in press) examined whether these familiarity judgments were made based on information from the players' positions or their movements. Skill and less-skilled soccer players were initially presented with 3 s sequences representing either dynamic patterns of play or a static image from the final frame of such a sequence. In the recognition phase, both new and old static images and

Fig. 2. Mean response accuracy scores and standard deviations for the skilled and less-skilled participants on the film endpoint-light display conditions (data from Williams et al., 2006).

dynamic film sequences were presented. Participants made familiarity judgments as to whether or not they had viewed the static and dynamic stimuli. The skilled players were more accurate when recognizing structured sequences presented as dynamic rather than static images, whereas no differences were observed across the two display formats for less-skilled players. It appears that skilled players rely on motion information rather than the positional orientation of specific players when recognizing sequences of play.

North and Williams (2008) attempted to identify whether crucial time windows exist for the extraction of relative motion information from key players when viewing dynamic sequences. They manipulated viewing duration during the recognition phase using three groups of skilled players. In the viewing phase the players viewed sequences of play high and low in structure each 5 s in duration, whereas, in the recognition phase, new and old sequences of play were presented with viewing durations that showed either the final 1, 3, or 5 s. The skilled players were significantly more accurate in the 3-s condition when compared with the 1- and 5-s condition when viewing structured sequences, whereas no significant differences in response accuracy were reported across the three viewing conditions on sequences low in structure. Findings suggest that in relatively chaotic sports such as soccer, structure emerges only in the final moments preceding an important attacking event such as a penetrative pass.

Although several researchers have proposed that the ability to recognize patterns of play is central to anticipation skill in team sports (e.g., Abernethy et al., 1994; Williams and Davids, 1995; Abernethy et al., 2005), this issue has not been adequately addressed in the literature. North et al. (in press, in preparation) found no significant correlations between anticipation and recognition performance for skilled or less-skilled players, respectively, although when the two groups were collapsed a moderate correlation between anticipation accuracy and recognition sensitivity was obtained. In two separate experiments, involving an analysis of eye movement and verbal report data, evidence was presented to support the argument that recognition processes underpin anticipation judgments. For example, the skilled participants showed no differences in fixation transitions between key features across the two tasks, indicating a broad, relation-based perceptual strategy. However, when instructed to anticipate rather than recognize film clips, participants verbalized more stimuli, actions, and cognitions; fixated on more locations; showed an increase in the number of fixations; and spent less time fixating the ball and more time viewing the offensive team. Although both anticipation and recognition tasks stimulate complex retrieval structures, the processes involved in activating these structures do differ to some degree. While recognition may be involved in anticipation, the latter skill appears more complex, invoking different and more refined retrieval structures.

Visual search behaviors

The ability of performers to pick up advance visual cues or to identify patterns of play is partly determined by the manner in which they move their eyes around the display in an attempt to extract the most pertinent information. The visual search behaviors employed in dynamic sport settings are typically recorded using a head-mounted, corneal reflection technique (see Williams et al., 1999). The system works by detecting the location of the pupil and corneal reflection and then superimposing a cursor or crosshair onto a video-taped image of the scene. These data can then be analyzed to ascertain areas of interest as well as the number and frequency of eye fixations employed during scene perception. Overall, this body of work suggests that skilled performers fixate on different areas of the display, for varying periods of time and in a different sequence or order than their less-skilled counterparts. The differences between skill groups are assumed to be indicative of more refined selective attention processes and enhanced task-specific knowledge structures (Henderson, 2003).

In much of the early work researchers presented stimuli using static slides (e.g., see Bard and Fleury, 1976) and/or focused on relatively closed skill situations such as the penalty kick in soccer (e.g., see Tyldesley et al., 1982). Technological advancements have facilitated a progressive shift towards the analysis of gaze behaviors using dynamic film simulations (e.g., see Williams and Davids, 1998; Helsen and Starkes, 1999; Vaeyens et al., 2007a, b) and field-based test protocols (e.g., see Martell and Vickers, 2004; Vickers and Williams, 2007). Although differences in fixation location between skilled and less-skilled performers have been reported in the majority of papers, occasionally no differences between skill groups are observed, implying that variations in performance are due either to more effective use of the para-fovea and visual periphery and/or to the quality of information extracted (e.g., see Williams and Davids, 1998).

In a review paper, Williams and colleagues (e.g., Williams et al., 2004) illustrated how the visual behaviors employed by athletes varied significantly based on task constraints such as the number of players involved in the sequence (e.g., 3 vs. 3, 5 vs. 5) and the participant's specific functional role or goal (defender vs. attacker) (e.g., see also Vaeyens et al., 2007a, b). Moreover, evidence was presented to illustrate how factors related to the individual such as anxiety and physiological fatigue influence the specific gaze behaviors employed during task performance (see Williams and Elliott, 1999; Vickers and Williams, 2007). The challenge for the performer is to ascertain how the different components of the visual system (e.g., fovea, para-fovea, and periphery) can be used most effectively to deal with the specific requirements of the task.

Situational probabilities

The perceptual-cognitive skills highlighted above reflect the skilled performer's superior ability to process contextual information present in the display as the action unfolds. However, in addition to this ability, skilled performers developed more accurate and refined expectations of likely event probabilities than their less-skilled counterparts did. Historically, this issue has been examined experimentally by manipulating stimulus probabilities in laboratory-based choice reaction time paradigms (for a review see Williams et al., 1999). However, in more recent years there have been a few attempts to develop more representative paradigms to examine the role of situational probabilities in sport settings (e.g., see Paull and Glencross, 1997; Ward and Williams, 2003; Crognier and Féry, 2005).

Ward and Williams (2003) asked groups of elite and subelite soccer players to assign probability values to the "best passing options" available to a player in possession of the ball. Film sequences were paused immediately prior to the player passing the ball and participants were required to highlight likely passing options. The elite players were better than the subelite group at identifying players who were in the best position to receive the ball and were more accurate in assigning an appropriate probability to players in threatening and nonthreatening positions, as determined by a panel of expert coaches. In contrast, subelite players were less efficient in their selection of critical and noncritical players and were not as adept at assigning a hierarchy of probabilities to likely events.

McRobert and colleagues (in press) used a novel variation of this paradigm to illustrate how cricket batters develop accurate expectations or situational models that enable them to predict the type of delivery that they are likely to face from a bowler. Participants played a stroke in response to life-size filmed representations of cricket bowlers. In one condition, the batters viewed 36 random deliveries from 10 different bowlers, whereas in a second condition batters were presented with an entire over (i.e., six consecutive deliveries) from six bowlers who had delivered only one ball each in the first condition. The batter's performance on the final delivery from each six-ball over in the second condition was compared with their performance on the same delivery when presented as an independent trial in the first condition. Participants improved their anticipation accuracy when additional contextual information was presented compared to the random viewing condition. Moreover, progressive changes in visual search behaviors and verbal reports gleaned from participants were observed across deliveries. The batters spent more time fixating central (i.e., head–shoulder region) rather than peripheral (i.e., ball–hand) areas and articulated more higher-order predictive and planning statements rather than less-refined cognition and evaluation statements as the degree of contextual information presented increased.

These findings illustrate how batters are able to buildup a more refined contextual model of a bowler over a series of deliveries. The batters pick up idiosyncrasies in bowling technique over successive deliveries and modify their visual search behavior to focus gaze on more subtle and informative postural cues. Moreover, the batters refine their expectations of the type of delivery that a bowler is more or less likely to perform, guiding and refining their search for relevant contextual information as the bowler approaches the wicket.

How do these perceptual-cognitive skills interact when making anticipation judgments? Why there is more to anticipation than meets the eye?

Although there is significant experimental evidence to illustrate the importance of each of these perceptual-cognitive skills when making anticipatory judgments, there have been no reported attempts to examine how these skills interact with each other in a dynamic and evolving manner to facilitate appropriate anticipation in the competitive setting (cf. Williams and Ward, 2007). How does the relative importance of each of these perceptual-cognitive skills vary across different tasks and situations? In certain situations it is conceivable that players may rely exclusively on the ability to process information from an opponent's postural orientation, whereas in others players may make an anticipatory judgment based solely on perceived event probabilities. What factors influence the role that the different perceptual-cognitive skills play in the anticipation process? The potential interaction between these perceptual-cognitive skills and constraints related to the task, situation, and performer is illustrated in Fig. 3.

The figure, which is based on the constraints-based model proposed by Newell (1986), is intended to convey the notion that the relative importance of these perceptual-cognitive skills may vary based on a range of constraints related to the task, situation, and performer. In relation to task constraints, it is likely that the ability to identify stimulus familiarity in patterns of play is crucial in team ball games such as soccer, hockey, rugby, and basketball. In contrast, in racket sports, such as tennis, squash, and badminton, situational probabilities and the ability to extract information from an opponent's postural orientation may be more important than the ability to identify patterns of play.

Within a specific task, situational constraints also influence the importance and relevance of each perceptual-cognitive skill. For example, the position of the player(s) on the field/court may determine the influence of each perceptual-cognitive skill (i.e., half-way line vs. edge of penalty area in soccer; back-court vs. at the net in tennis) as well as the distance between the player and the opponent(s)/ball. The above factors often influence the time available to formulate a response as well as the costs and benefits associated with accurate and inaccurate judgments. The emphasis on one perceptual-cognitive skill over another may also be affected by the score and time remaining in the match. If a tennis player is match point down on an opponent's first serve, he/she may choose to make an anticipatory judgment almost solely on the probabilities associated with the opponent playing a certain type of serve. In contrast, on the second serve mid-way through the set, the same player may alter his/her strategy in an effort to rely more so on advance information arising from the opponent's postural orientation. These strategies have different cost–benefit ratios and the strategy selected may be influenced by several factors, some of which are not directly associated with the specific context (e.g., the strategy selected may be very different when faced with a lower or higher ranked opponent since the impact of winning/losing alters).

Finally, factors associated with the player such as physiological and emotional state influence how information is extracted from the display. For example, when anxious, players may narrow their focus of attention, potentially reducing the

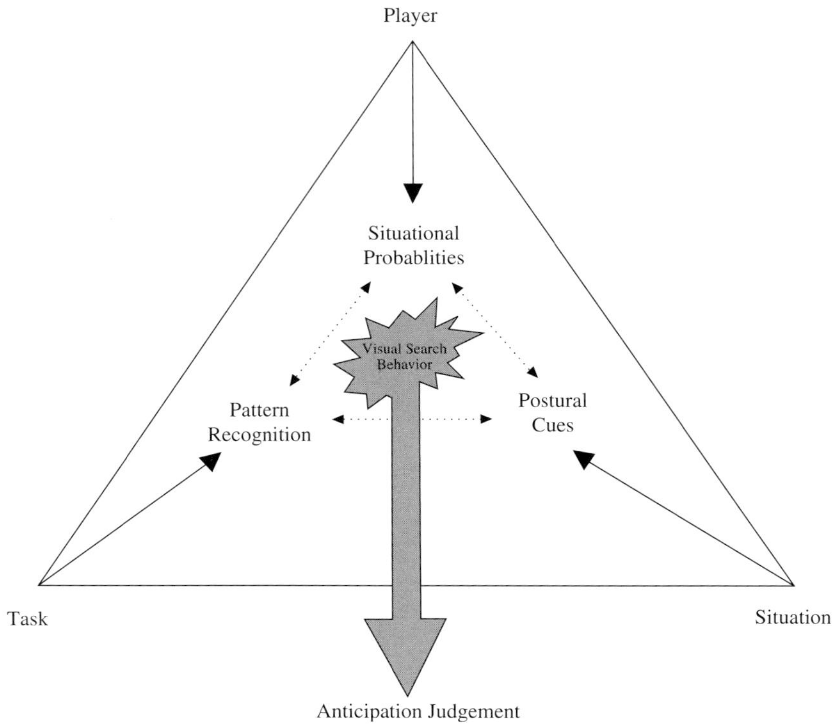

Fig. 3. The interactive relationship between various perceptual-cognitive skills and constraints related to the task, situation, and player when making anticipation judgments (adapted from Williams and Ward, 2007).

capacity to use peripheral vision and influencing the relative importance of different sources of information (i.e., a shift from focusing on the runs and positions of players "off the ball" to the extraction of information from the player in possession of the ball). Similarly, when fatigued, players may have to rely on situational probabilities to anticipate an opponent's intentions since attention may be focused internally on how best to cope rather than externally on the positions and movements of players. A key issue is that the importance of these perceptual-cognitive skills is likely to be highly dynamic, varying considerably from one situation to the next depending on a range of factors or constraints (see also Chapter 8: The bidirectional links between decision making, perception, and action). Unfortunately, thus far, few, if any, researchers have examined how the importance of these perceptual-cognitive skills varies across these different constraints.

The reality of the matter is that researchers have largely treated any interaction between perceptual-cognitive skills as a potential confound, designing paradigms that have enabled the importance of each skill to be examined in isolation from the other. It is therefore conceivable that the importance of some of these skills has actually been overemphasized in comparison with their value in the actual competitive setting. For example, when asking participants to respond to film clips involving forehand drive shots in tennis, the only information available emanates from the postural orientation of the opponent, whereas in the actual competitive setting there is continual access to other sources of contextual information such as shot probabilities and on-court location.

Finally, a broader question relates to whether we have even overemphasized the importance of anticipation skill in high-level sport by the very nature of the paradigms that have been employed. Our paradigms typically constrain performers to make an anticipation judgment in response to a film or on-court scenario, whereas there is evidence to suggest that top-level performers anticipate far less than is commonly assumed (see James et al., 2006). In competition, the costs of anticipating incorrectly may be far too great, constraining players to anticipate only infrequently and in situations where the chances of success are high or the costs associated with failure palatable. Research is needed to identify more specifically when anticipation may be important in sport. For example, in tennis, do players anticipate most often when returning serve, when volleying at the net, or when engaging in an extended backcourt rally? Is anticipation more important when facing an opponent of lower, comparable, or higher skill level? Does the importance of anticipation vary across early or late stages of a game (e.g., when at deuce or 0-40 down) or match (e.g., when 3-3 or in a tie-break)? Providing answers to such questions are essential if we are to create realistic field- or laboratory-based simulations to measure and evaluate the mechanisms underpinning anticipation skill (Williams and Ericsson, 2005).

Perception, cognition, and action: integration and differentiation

Although the focus in this chapter has been almost exclusively on the question of how performers make anticipatory judgments, it is acknowledged that players' intentions are often tightly coupled with their ability to implement the chosen course of action physically. It would therefore be difficult to disagree with the general conceptualization that action influences perception and visa versa or that perception, cognition, and action are tightly interlinked and interwoven (see Williams and Ericsson, 2007). In many situations, these processes are inevitably part of the same performance rubric. For example, returning serve in volleyball, catching a fly ball in baseball, or defending the basket in a one-on-one situation in basketball are all situations where there is likely to be reciprocal mutuality between perception and action (Gibson, 1979).

In other contexts, however, the degree of coupling may not be so tight and, in such situations, there may be obvious benefits to ensure differentiation between these processes. For example, a rugby player who anticipates that the ball will be thrown to his/her immediate opponent in the line-out may decide not to challenge for the ball but to form a maul by linking with his/her team-mates in an effort to pressurize the opposition forwards. Similarly, in soccer, a central defender may anticipate that the attacking midfield player will play the ball out to a team-mate positioned on the wing, but even though there is a chance of intercepting this pass, the decision may be to remain in a central area of the pitch so as to avoid being pulled out of position. In tennis doubles, a player may move as if expecting an opponent to execute a crosscourt drive in order to deceive an opponent into playing a down-the-line stroke to his/her team-mate who is well placed to perform a winning volley at the net. In these scenarios, the natural links between perception and action are mediated by over-riding strategic and tactical considerations, which, in turn, influence our intentions.

An important observation is that while the majority of scientists have focused considerable effort in attempting to identifying the mechanisms underpinning skilled "perception" as well as "perception and action" in sport, relatively few have attempted to examine how cognition frequently acts as a bridge between perception and action. There are few published reports focusing on how strategic and tactical issues constrain the nature of the links between perception and action. Verbal protocol analysis provides one potentially very useful approach to bridge this gap (e.g., see McPherson and Kernodle, 2003; McRobert et al., in press).

Conclusions

The aim in this chapter was to provide a brief overview of some of the important perceptual-cognitive skills that govern our ability to make successful anticipatory judgments in sport. The main question addressed was how do skilled performers anticipate the intentions of opponents? The importance of a number of perceptual-cognitive skills were highlighted, such as the ability to: (a) pick up advance information from an opponent's postural orientation, (b) identify stimulus familiarity in patterns of play, (c) explore the visual scene in an efficient and effective manner, and (d) ascertain the most likely options facing an opponent at any given moment in time.

Three additional issues were highlighted. First, these perceptual-cognitive skills are not mutually exclusive and most likely function in an integrated and overlapping manner during performance. The relative importance of each skill and how this varies across different situations and tasks has yet to be adequately examined. Anticipation skill is likely to involve far more than the sum of these different perceptual-cognitive skills.

Second, we still know relatively little about the "when," "where," and "what" of anticipation. When during a match is anticipation important? Does it vary in importance across different areas of the field/court? What specific sources of information do players use to make anticipation judgments? Clearly, the above factors likely interact, adding a further layer of complexity that needs to be unraveled.

Third, perception, cognition, and action interact in a continuous, dynamic, and parallel manner during performance. However, the links between these components vary markedly based on the demands of competition. Perception, cognition, and action are often very tightly interlinked providing an abundance of information to guide performance in dynamic tasks. At other times, these elements may be less tightly coupled or even strongly differentiated, particularly in sports that are heavily governed by tactical and strategic factors. In these latter situations, the challenge is to identify how cognition acts as an effective bridge between perception and action. Our efforts to advance knowledge relating to these three issues may require us to be creative in designing paradigms that mimic the demands of actual competition, while at the same time, providing sufficient experimental control so as to tease out the complex interactions between perception, cognition, and action in sport.

References

Abernethy, B., Baker, J., & Cote, J. (2005). Transfer of pattern recall skills may contribute to the development of sport expertise. *Applied Cognitive Psychology*, *19*, 705–718.

Abernethy, B., Gill, D. P., Parks, S. L., & Packer, S. T. (2001). Expertise and the perception of kinematic and situational probability information. *Perception*, *30*, 233–252.

Abernethy, B., Neal, R. J., & Koning, P. (1994). Visual-perceptual and cognitive differences between expert, intermediate, and novice snooker players. *Applied Cognitive Psychology*, *8*, 185–211.

Abernethy, B., & Russell, D. G. (1987). Expert-novice differences in an applied selective attention task. *Journal of Sport Psychology*, *9*, 326–345.

Bard, C., & Fleury, M. (1976). Analysis of visual search activity during sport problem situations. *Journal of Human Movement Studies*, *3*, 214–222.

Chase, W. G., & Simon, A. S. (1973). Perception in chess. *Cognitive Psychology*, *4*, 55–81.

Crognier, L., & Féry, Y. (2005). Effect of tactical initiative on predicting passing shots in tennis. *Applied Cognitive Psychology*, *19*, 1–13.

Cutting, J. E., & Proffitt, D. R. (1982). The minimum principle and the perception of absolute, common, and relative motions. *Cognitive Psychology*, *14*, 211–246.

Gentner, D., & Markman, A. B. (1998). Structure mapping in analogy and similarity. In P. Thagard (Ed.), *Mind readings* (pp. 127–156). Cambridge, MA: MIT Press.

Gibson, J. J. (1979). *The ecological approach to visual perception*. Boston: Houghton Mifflin.

Goldin, S. E. (1978). Effects of orienting tasks on recognition of chess positions. *American Journal of Psychology*, *91*, 659–671.

Helsen, W. F., & Starkes, J. L. (1999). A multidimensional approach to skilled perception and performance in sport. *Applied Cognitive Psychology*, *13*, 1–27.

Henderson, J. M. (2003). Human gaze control during real-world scene perception. *Trends in Cognitive Sciences*, *7*(11), 498–504.

Hodges, N. J., Huys, R., & Starkes, J. L. (2007). Methodological review and evaluation of research in expert performance in sport. In G. Tenenbaum & R. C. Eklund (Eds.), *Handbook of sport psychology* (3rd ed, pp. 161–183). New York: Wiley.

Huys, R., Cañal-Bruland, R., Hagemann, N., Beek, P. J., Smeeton, N. J., & Williams, A. M. (in press). 'Global' information pick up underpins anticipation skill of tennis shot direction. *Journal of Motor Behavior*.

Huys, R., Smeeton, N. J., Hodges, N. J., Beek, P. J., & Williams, A. M. (2008). On the dynamical information underlying anticipation skill. *Perception & Psychophysics*, *70*, 1207–1216.

James, N., Caudrelier, T., & Murray, S. (2006). The use of anticipation by elite squash players. *Journal of Sports Sciences*, *23*, 1249–1250.

Johansson, G. (1973). Visual perception of biological motion and a model for its analysis. *Perception & Psychophysics, 14*, 201–211.

Jones, C. M., & Miles, T. R. (1978). Use of advance cues in predicting the flight of a lawn tennis ball. *Journal of Human Movement Studies, 4*, 231–235.

Martell, S. G., & Vickers, J. N. (2004). Gaze characteristics of elite and near-elite athletes in ice hockey defensive tactics. *Human Movement Science, 22*, 689–712.

McPherson, S. L., & Kernodle, M. W. (2003). Tactics, the neglected attribute of expertise: problem representations and performance skill in tennis. In J. L. Starkes & K. A. Ericsson (Eds.), *Expert performance in sports: advances in research on sport expertise* (pp. 137–168). Champaign, IL: Human Kinetics.

McRobert, A., Williams, A. M., Ward, P., & Eccles, D. (in press). Perceptual-cognitive mechanisms underpinning expertise: the effects of task constraints. *Ergonomics*.

Müller, S., & Abernethy, B. (2006). Batting with occluded vision: an in situ examination of the information pick-up and interceptive skills of high- and low-skilled cricket batsmen. *Journal of Science and Medicine in Sport, 9*, 446–458.

Müller, S., Abernethy, B., & Farrow, D. (2006). How do world-class cricket batsmen anticipate a bowler's intention? *Quarterly Journal of Experimental Psychology, 59*(12), 2162–2186.

Newell, K. M. (1986). Constraints on the development of coordination. In M. G. Wade & H. T. A. Whiting (Eds.), *Motor development in children: aspects of coordination and control* (pp. 341–361). Amsterdam: Martin Nijhoff.

North, J. S., & Williams, A. M. (2008). Identifying the critical time period for information extraction when recognizing sequences of play. *Research Quarterly for Exercise and Sport, 79*, 268–273.

North, J. S., Williams, A. M., Ward, P. and Ericsson, K. A. (in preparation). Identifying the critical information sources to skilled anticipation and recognition using retrospective verbal reports.

North, J. S., Williams, A. M., Ward, P., Hodges, N. J. and Ericsson, K. A. (in press). Perceiving patterns in dynamic action sequences: The relationship between anticipation and pattern recognition skill. *Applied Cognitive Psychology*.

Paull, G., & Glencross, D. (1997). Expert perception and decision making in baseball. *International Journal of Sport Psychology, 28*, 35–56.

Reilly, T., Williams, A. M., Nevill, A., & Franks, A. (2000). A multidisciplinary approach to talent identification in soccer. *Journal of Sports Sciences, 18*, 695–702.

Salmela, J. H., & Fiorito, P. (1979). Visual cues in ice hockey goaltending. *Canadian Journal of Applied Sport Sciences, 4*, 56–59.

Smeeton, N., Ward, P., & Williams, A. M. (2004). Transfer of perceptual skill in sport. *Journal of Sports Sciences, 19*(2), 3–9.

Tyldesley, D. A., Bootsma, R. J., & Bomhoff, G. T. (1982). Skill level and eye movement patterns in a sport orientated reaction time task. In H. Rieder, H. Mechling, & K. Reischle (Eds.), *Proceedings of an international symposium on motor behaviour: contribution to learning in sport* (pp. 290–296). Cologne: Hofmann.

Vaeyens, R., Lenoir, M., Williams, A. M., Mazyn, L., & Philippaerts, R. M. (2007a). The effects of task constraints on visual search behavior and decision-making skill in youth soccer players. *Journal of Sport & Exercise Psychology, 29*(2), 156–175.

Vaeyens, R., Lenoir, M., Williams, A. M., Mazyn, L., & Philippaerts, R. M. (2007b). Visual search behavior and decision-making skill in soccer. *Journal of Motor Behavior, 39*(5), 395–408.

Vickers, J., & Williams, A. M. (2007). Why some choke and others don't! *Journal of Motor Behavior, 39*(5), 381–394.

Ward, P., & Williams, A. M. (2003). Perceptual and cognitive skill development in soccer: the multidimensional nature of expert performance. *Journal of Sport & Exercise Psychology, 25*(1), 93–111.

Ward, P., Williams, A. M., & Bennett, S. J. (2002). Visual search and biological motion perception in tennis. *Research Quarterly, 73*(1), 107–112.

Williams, A. M., & Burwitz, K. (1993). Advance cue utilization in soccer. In T. P. Reilly, J. Clarys, & A. Stibbe (Eds.), *Science and football II* (pp. 239–244). London: E and F N Spon.

Williams, A. M., & Davids, K. (1995). Declarative knowledge in sport: a byproduct of experience or a characteristic of expertise? *Journal of Sport & Exercise Psychology, 7*(3), 259–275.

Williams, A. M., & Davids, K. (1998). Visual search strategy, selective attention, and expertise in soccer. *Research Quarterly, 69*(2), 111–128.

Williams, A. M., Davids, K., & Williams, J. G. (1999). *Visual perception and action in sport*. London: E and F N Spon.

Williams, A. M., & Elliott, D. (1999). Anxiety and visual search strategy in karate. *Journal of Sport & Exercise Psychology, 21*, 362–375.

Williams, A. M., & Ericsson, K. A. (2005). Some considerations when applying the expert performance approach in sport. *Human Movement Science, 24*, 283–307.

Williams, A. M., & Ericsson, K. A. (2007). Perception, cognition, action and skilled performance. *Journal of Motor Behavior, 39*(5), 338–340.

Williams, A. M., Hodges, N. J., North, J. S., & Barton, G. (2006). Perceiving patterns of play in dynamic sport tasks: identifying the essential information underlying skilled performance. *Perception, 35*, 317–332.

Williams, A. M., Huys, R., Canal, R., & Hagemann, N. (in press). The dynamical information underpinning anticipation skill. Human Movement Science.

Williams, A. M., Janelle, C. M., & Davids, K. (2004). Constraints on the search for visual information in sport. *International Journal of Sport and Exercise Psychology, 2*, 301–318.

Williams A. M., & North, J. S. (in press). Identifying the minimal essential information underlying pattern recognition. In H. Ripoll, M. Raab and D. Arajuo (Eds.), *Perspectives on cognition and action*. New York: Nova Science Publishing Inc.

Williams, A. M., & Ward, P. (2007). Perceptual-cognitive expertise in sport: exploring new horizons. In G. Tenenbaum & R. C. Eklund (Eds.), *Handbook of sport psychology* (3rd ed., pp. 203–223). New York: Wiley.

M. Raab et al. (Eds.)
Progress in Brain Research, Vol. 174
ISSN 0079-6123

CHAPTER 8

The bidirectional links between decision making, perception, and action

Rita F. de Oliveira[1,*], Lysann Damisch[2], Ernst-Joachim Hossner[3], Raôul R. D. Oudejans[4], Markus Raab[1], Kirsten G. Volz[5] and A. Mark Williams[6]

[1]*Institute of Psychology, German Sport University Cologne, Cologne, Germany*
[2]*Department of Psychology, University of Cologne, Cologne, Germany*
[3]*Department of Sport Science, Liverpool Hope University, Liverpool, United Kingdom*
[4]*Research Institute MOVE, Faculty of Human Movement Sciences, VU University Amsterdam, Amsterdam, The Netherlands*
[5]*Max Planck Institute for Neurological Research, Cologne, Germany*
[6]*Research Institute for Sport and Exercise Sciences, Liverpool John Moores University, Liverpool, United Kingdom*

Abstract: In different scientific disciplines, decision making is studied by examining how options are perceived or generated in isolated situations. In this chapter we explore the benefits of an interdisciplinary approach to understanding option perception within human behavior. As a consequence of the contributions from different disciplines, we present a framework that describes the bidirectional links between decision making, perception, and action. We will argue that, given their interdependency, the study of decision making profits greatly from a more integrated view of the situations it studies. The framework also illustrates how constraints may influence these bidirectional links. Finally, we use this framework to convey new ideas for experimental paradigms, data interpretation, and applications.

Keywords: decision making; perception; action; skill; constraints; interdisciplinarity

Introduction

There is a multitude of options available to each individual that span different levels of abstraction and different contexts of behavior. Many such options are cognitively derived and thought over in more or less lengthy decision-making processes, whereas other options are given or are immediately available from the context (e.g., whether or not to buy a house, and whether to have red or white wine). If the task is to choose between a given set of options, decision making is restricted to the options presented (see Chapter 6: How the orbitofrontal cortex contributes to decision making. A view from neuroscience). However, when the task involves the perception of options, decision making depends heavily on the information available (see Chapter 5: How do people perceive and generate options?). Thus, while decision making

*Corresponding author.
Tel.: +49-221-4982-5721; Fax: +49-221-4982-8320;
E-mail: r.oliveira@dshs-koeln.de

DOI: 10.1016/S0079-6123(09)01308-9

does occur in closed settings, it is more often influenced by new perceptual information, by the execution of actions, and by various constraints.

In this chapter, we explore the benefits of an interdisciplinary approach to understanding option perception within overt human behavior. In the different scientific disciplines decision making is often studied in an insular manner. We present an alternative framework that accommodates findings from different disciplines in order to explore their commonalities and amplify their contributions. The framework describes the bidirectional links between decision making, perception, and action. We will argue that these links are so prominent that attempting to study decision making without taking them into account might prove restricted. The framework also illustrates how constraints may influence these bidirectional links. Finally, we use the framework to convey new ideas for experimental paradigms, data interpretation, and applications. Before presenting our framework, we first briefly describe current paradigms.

Current paradigms and different perspectives

Different disciplinary approaches typically focus on one or two domains of human behavior. For instance, those interested in decision making, focus on how options are generated based on previous knowledge, neural scientists can explore how choice behaviors are neurally implemented, ecological psychologists might focus on what information can guide decision making, motor control theorists focus on the implementation of solely one option, cognitive psychologists may be concerned with anticipatory processes, and social psychologists can inform how others influence one's decision making.

These approaches focus on one or two components of human behavior in isolation and consequently, within each discipline there are models from which to derive specific hypotheses that may be tested in a controlled and systematic fashion. However, while such models can be tested and refined through well-controlled research, the disciplinary isolation may preclude a better understanding of decision making as a whole. In this sense, the current dominant approach used in most fields is monodisciplinary and reductionist. To illustrate, while the study of decision making is valuable per se, there are clear benefits in considering how a person picks up given perceptual information from the environment (see Chapter 5: How do people perceive and generate options? and Chapter 7: Perceiving the intentions of others: how do skilled performers make anticipation judgments?), or how the current emotional state might influence the decision-making process (see Chapter 3: A cognitive movement scientist's view on the link between thought and action: insights from the "Badische Zimmer" metaphor and Chapter 4: Perceiving and moving in sports and other high-pressure contexts).

In addition, for reasons of experimental control and ease of measurement, the tasks selected for study are often discrete and static. For instance in neurosciences, the majority of researchers use experimental paradigms of action observation or action imagery to handle the realization of action using brain-imaging techniques. More recently, virtual reality paradigms have been used that illicit neural activation related to effective motor output, for instance, in stroke patients (e.g., August et al., 2006), but the realization of actions under imaging paradigms remains restricted. Even in the study of inherently complex and dynamic situations, like sport, researchers typically select one single discrete task for scrutiny. The soccer penalty kick, the basketball free throw, or the return of serve in tennis arguably constitute the most exciting moments of a match, but they do not capture the complexity of the continuous decision making that happens during game play. Although behavioral studies can in principle handle the complexity and dynamics of sport, there is a paucity of research into longer duration and more dynamic situations such as the counter-attack in soccer or the dribble penetration in basketball.

For similar reasons, the processes under study are often seen as being consecutive in nature. In other words, perception, decision making, and action are seen as following one another in an orderly pattern. For instance, a basketball player dribbling down the court sees an opponent closing in and a team-mate appearing to the right. Being close to the basket the player perceives the options to shoot at the basket and pass the ball to the team-mate. That particular team-mate has missed the last two shots so our player decides to shoot instead of pass. This is a standard example that lends itself to the study of how perception influences decision making, and how the decision influences action in a consecutive manner. However, this example does not correspond to a standard situation within game play. In a more realistic example, the opponent would move closer, compromising the player's decision to shoot and prompting the player's action to protect the ball by slightly turning to the left. This action would disclose another available team-mate (to the left) to whom the player would be happy to pass the ball. In this more realistic example, new perceptions compromise the link between decision and action, perception prompts action, and action discloses new perceptions. Although relevant to the full understanding of decision making, these bidirectional influences of perception and action are often neglected.

By fostering an interdisciplinary discussion, the limitations and capabilities of each discipline become more evident. As stated above, the capabilities of each discipline are the robustness of its models and their testability, whereas they are reductionist and frequently treat behavior as discrete, static, and consecutive. To explore the interdisciplinary contributions to the understanding of decision making, we employ a framework that expresses the bidirectional links between decision making, perception, and action. We also consider the influence of various constraints in shaping these bidirectional links so as to present a more complete description of human behavior.

Exploring the bidirectional links between decision making, perception, and action

We explored a framework of bidirectional links between decision making, perception, and action, each of these links being dynamic, and strongly constrained by characteristics of the person, task, and environment (Newell, 1986). Figure 1 represents the four main characteristics of the framework: bidirectional, nonconsecutive, dynamic, and under the influence of constrains. First, decision making, perception, and action are linked with bidirectional arrows that represent the reciprocal nature of their interactions (see Chapter 3: A cognitive movement scientist's view on the link between thought and action: insights from the "Badische Zimmer" metaphor, for a discussion about the format of these links). Second, they are presented in a circle to reflect the nonconsecutive nature of their interaction. In other words, decision can follow from perception and result in action but any other direction of interaction occurs as well. Third, its dynamic nature is highlighted because the whole framework develops over time along two different scales. On the short-term scale, the framework is dynamic because perception does not end when a decision is made or an action performed; these processes are continuous and influence one another. On the long-term scale, the links between decision making, perception, and action are strengthened over time through repetition and learning, such that, for instance, a given decision prompts the most efficient execution. Finally, the figure also shows constraints, in the center of the circular interactions, to illustrate that behavior is heavily impacted by the characteristics of the task, environment, and person. In sum, the framework attempts to increase awareness of the bidirectional, dynamic, and continuous interactions between decision making, perception, and action, and illustrates the influential role of different constraints upon these interactions. Next, we explain the four characteristics of the framework in more detail with reference to appropriate examples, many of which can be revisited in other chapters of this volume.

Bidirectional links between decision making, perception, and action

We start this description with the bidirectional link between decision making and perception. One good illustration of these links is the relativity of performance judgments in gymnastics

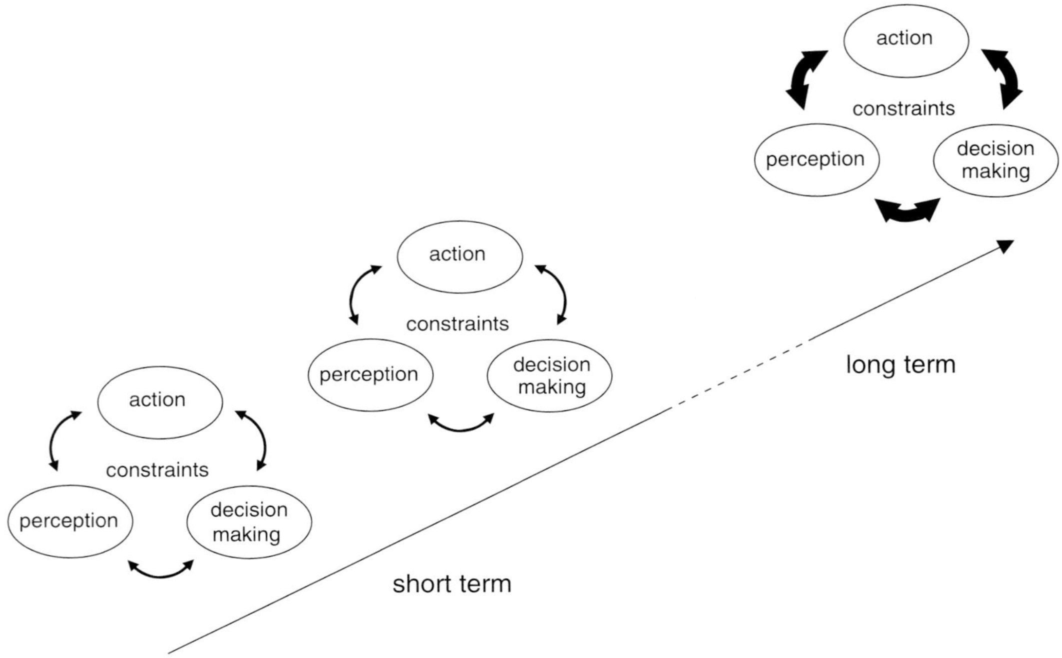

Fig. 1. Framework of the bidirectional links between decision making, perception, and action. Circular bidirectional arrows represent the interaction between elements. Decision making, perception, and action are arranged in a circle to represent their nonconsecutive interaction. In addition, constraints heavily impact on these interactions and are thus represented within the cycles. The dynamic nature of the framework is represented by the straight arrow that runs throughout cycles (short-term), and by the tricking of the circular arrows (long-term).

(see Chapter 2: On the relativity of athletic performance: a comparison perspective on performance judgments in sports). While the perception of similarities between two athletes imparts upon decisions of their relative quality (perception–decision-making link), making a decision on the relative quality of two athletes is likely to codetermine the perception of their future performance (decision-making–perception link). In another example, the decision to pick apples from a tree will likely direct perception to information about ripeness (e.g., how red apples look; decision-making–perception link), while perceiving red apples on a tree may also trigger the decision to pick apples (perception–decision-making link).

Decision making is also linked to action in a bidirectional fashion. For instance in handball, the preference to select a pass to a team-mate on the left or on the right is dynamically influenced by action. The decision to pass the ball to a team-mate on the left may directly lead to the execution of that pass (decision-making–action link), however, if the player is faster or more accurate at passing to his/her right-hand side this action preference may directly guide the decision (action–decision-making link). Thus, decisions may lead to actions, but actions may also change the preference states of the player (see Chapter 5: How do people perceive and generate options?).

For completeness, we address the bidirectional link between perception and action that has been laid out in the context of direct perception (Gibson, 1979/1986; see also Chapter 4: Perceiving and moving in sports and other high-pressure contexts; Chapter 5: How do people perceive and generate options?). Contrary to information-processing theories, direct perception claims that meaningful information is available in the environment (e.g., light reflected from objects) and is specific to the objects, events, and properties in the environment. Hence, simply picking up this information results in perceiving what one can do with and in the environment (Gibson, 1979/1986). A crucial characteristic of this pick-up of information is that it is often an active process in which action is as important to perception as perception is to action. For instance, the action of looking at a target brings about information that is specific to target location, which in turn allows the person to act on the target (de Oliveira et al., 2009; see Chapter 5: How do people perceive and generate options?). Thus, in interacting with the environment, action implementation brings about relevant perceptions (action–perception link) while perceptual information can directly guide actions (perception–action link).

Nonconsecutive nature of decision making, perception, and action

As it is probably evident from the description of the framework so far, decision making, perception, and action do not occur in a neat consecutive manner. Any given decision can lead to new perceptions or new actions, or the current action can lead to new perceptions or new decisions. The idea is not that perception, decision making, and action do not follow from each other because they are independent processes, but that these processes are not consecutive in a deterministic sense. In other words, a decision-making process (e.g., whether to study or go out) may lead, and often does lead, to an action (e.g., to study), however, the link between decision and action may also be interrupted by new perceptions (e.g., the relevant book is missing) or by unexpected constraints (e.g., sudden back pain). Similarly, a decision-making process may be abruptly terminated by the perception of new action possibilities. For instance in handball, players generate their best allocation options after the first few fixations of video footage, but in a more dynamic system that preference would change over time due to the perception of other options for action (see Chapter 5: How do people perceive and generate options?).

Dynamic nature of decision making, perception, and action

The interplay between decision making, perception, and action occurs at different temporally continuous scales. For instance, a novice tennis player may notice that, on a clay surface, the opponent is able to slide and therefore cover large distances in returning the ball. Thus, the novice player may explore the sliding properties of the clay in his/her decisions for ball placement and for the actions of returning the ball. A more experienced player will have learned and integrated all properties of the clay surface. Such calibration to the properties of the surface will allow the experienced player to perceive, from early information, where he/she must intercept the ball or where he/she must place the ball on the opposite field. In other words, players learn to use the bouncing properties of the clay, so as to calibrate the links between perception and action in their playing decisions. Within the framework, there are reciprocal links between decision, perception, and action that evolve in short-term adaptations as exemplified with the short-term adaptation of the novice player described above. Importantly, over time and repetition, some of the specific links are used repeatedly and thus strengthened (as represented by the thicker arrows in Fig. 1) permitting the player to learn. Thus, the long-term dynamics of the framework refers to the strengthening of connections that occurs through learning as exemplified by the experienced tennis player. Continued practice may enhance performance through changes in cortical organization, suggesting that new synaptic connections might have formed (see Chapter 6: How the orbitofrontal cortex contributes to decision making. A view from neuroscience).

Constraints on the bidirectional links between perception, action, and decision making

In studying the influence that constraints have on decision making, perception, and action, Proffitt (2006) asserted that the person's goal, physiologic state, and emotions all influence the way he/she sees and interacts with the world. He found that, for instance, people report a hill slant to be steeper when they are tired, and the height of a balcony to be higher when they stand on it and are afraid of heights. The author used these examples to illustrate that perceptions are embodied in the sense that they relate the person and the task goals to action possibilities and their costs. Notably, Newell (1986) expounded on the role of constraints on the optimal pattern of control and coordination. In his constraints-led perspective, he identifies constraints related to the organism (e.g., the person), the task, and the environment. In the framework of the dynamic bidirectional links between decision making, perception, and action, constraints play a relevant role because they can influence each of the processes as well as the links between them. Hence, we turn now to the constraints, as set out by Newell.

Organismic constraints refer to characteristics of the person. These characteristics may be relatively stable, such as height, weight, or the properties of the musculoskeletal system, or transient such as fatigue, satiety, or anxiety. For instance, increased anxiety has been shown to influence perception and hence decision making (see Chapter 7: Perceiving the intentions of others: how do skilled performers make anticipation judgments?). Compared to low-anxiety conditions, novice climbers under high anxiety perceived smaller reaching height, selected a larger number of holds, performed slower, and needed more time to detect and use task-relevant information (see Chapter 4: Perceiving and moving in sports and other high-pressure contexts). The authors argue that by practicing under high levels of anxiety the bidirectional links can be recalibrated to minimize the impairing effect of this transient organismic constraint. Another example of transient and dynamic organismic constraints stems from research on satiety and chocolate (see Chapter 6: How the orbitofrontal cortex contributes to decision making. A view from neuroscience). Small and collaborators (2001) measured the hemodynamic activity while participants ate chocolate to satiety. At the beginning of the experiment, the chocolate was highly rewarding and accompanied by activation within the medial orbitofrontal cortex, an area that has been suggested to reflect the reward value of stimuli, but when chocolate was eaten to satiety the reward value of the chocolate went from being pleasant to aversive and activation within the medial orbitofrontal cortex was no longer observed. These results show that non-associative information about the internal state of the organism and associative information about reward value are integrated in a dynamic manner.

Task constraints refer to the goal and rules of the task, and the eventual use of equipment. In sport, the rules of the game provide very specific sets of constraints on the actions that are possible and allowed. In soccer, one of the rules is that field players cannot touch the ball with their hands, thus when intercepting a ball, decision making is bounded by this task constraint. Similarly, the player will act as to avoid ball–hand contact and will also perceive information to this effect. Task constraints may also be exploited for perception and action. For instance in basketball shooting, experienced players appear to use the official height of the basket (a task constraint) to locate the basket accurately (de Oliveira et al., 2009). When the player stands in the field, the basket is at an angle of elevation in respect to the player. This angular elevation can specify the location of the basket if, and only if, the basket is set at a given constant height.

Environmental constraints refer to those factors that are external to the person and not specific to the task, such as ambient light and temperature, the presence of spectators, or the playing surface. For instance in soccer, the presence of crowd noise can significantly interfere with referees' decisions. Compared with referees viewing games in silence, those watching them with background crowd noise were more uncertain in their decisions. In addition, watching games with noisy spectators related to referees awarding fewer fouls against the home team. Thus, the presence of spectators and the noise they make, thought to be at the core of the home advantage phenomenon, constrains the links between referees' decisions, perceptions, and actions. The surface where a game is played constitutes another environmental constraint. For instance in the Grand

Slam tennis tournament, grass is used in Wimbledon, clay in the French Open, and hard courts in the US and Australian Open. These three courts differ in slipperiness, friction, and absorption, such that the same ball bouncing off different surfaces will have differing characteristics of velocity, angle, and spin. Players must adapt, or calibrate, their decisions and style of play to the surfaces they play on. If the player adapts, the same visual situation will prompt different decisions as well as different perceptions and actions depending on the courts.

Consequences for research paradigms, data interpretation, and application

The framework presented in the preceding sections describes the dynamic bidirectional links between decision making, perception, and action, and illustrates how constraints may influence these links. While this framework can accommodate the research findings of different disciplines, it also has the potential to amplify their contributions. This amplification can be effected by modifying experimental paradigms, broadening the scope of data interpretation, and probing new opportunities for applications.

Applications of the framework to experimental paradigms

Knowledge about human decision making has evolved through the accumulation of research findings in accord with established theories. It follows that researchers address a given set of questions using similar methodologies. As mentioned earlier, disciplinary approaches have the potential to generate precise and uncontaminated results but at the same time they can be reductionist and that may hamper a more complete understanding of human decision making.

One way of examining decision making is to ask participants to watch a video of a game situation and ask them, at some point, what the options are and which option they would choose to perform. This line of research reduces the problem of option perception and decision making to the realm of visual perception. The paradigm informs about the cognitive capabilities of participants in deciding upon game situations and deriving or anticipating future scenarios. However, it puts aside the fact that the perception of action possibilities is not solely in the realm of perception but also in the realm of action and under the stringent influence of constraints. The participant can decide correctly that the best option in a given video situation is to execute a long pass to a team-mate on the opposite side of the field, whereas in reality the participant is incapable of performing such a pass with accuracy. The framework that we present here highlights the bidirectional link between decision making, perception, and action, and also the role of constraints such as skill level. As such, it brings awareness to potential caveats linked with well-established methodologies while at the same time giving insight into ways of improvement. There are two ways in which researchers could rely on a more encompassing approach to the study of decision making. Clearly, the first is a more interdisciplinarity approach in deriving research questions and devising research methods. For example, in deriving research questions for studying decision making one can ask how options are perceived, or how exploratory actions bring about relevant perception. As to devising effective methodologies, some examples include the combined use of neuroimaging techniques and behavioral methods to investigate decision making (see Chapter 6: How the orbitofrontal cortex contributes to decision making. A view from neuroscience), or the combination of visual occlusion and kinematic analysis to investigate visual perception (see Chapter 4: Perceiving and moving in sports and other high-pressure contexts; Chapter 7: Perceiving the intentions of others: how do skilled performers make anticipation judgments?).

The second way in which researchers could rely on a more encompassing approach to the study of decision making is the inclusion of question-driven research. Whereas theory-driven research, which is common within disciplinary approaches, restricts questions, methods, and interpretations, more question-driven research has the potential to treat decision making in its full complexity. For instance, in tackling the question of how children cross busy streets, one might want to consider gaze behavior taking into account the children's low point of observation (visual perception); the velocity and consistency of walking or running (action); the timing and sequencing of gaze and gait initiation (interaction between perception and action); and the familiarity of the children with the properties of the road's surface (constraints). All these factors can be considered when examining how children cross roads. However, they are not considered as easily when studying decision making in road crossing.

Applications of the framework to data interpretation

In the same way, data interpretation is often tainted by the theoretical approach or discipline from which results are interpreted. This is because the manner in which an element works in isolation may differ markedly from the manner in which the same element works within an embedded multilevel framework. For instance, the participant asked to report on the best option after viewing the footage of a game, may be reporting on three different aspects. The first is the best option in absolute terms, which includes a long and accurate pass to a team-mate on the opposite side of the field. The second is the best option for the player and team he/she has been watching on the video. The third is the best option he/she would be able to perform. In the latter situation, the participant would still have to make inferences about the playing surface, the wind conditions, or the fitness level of the team-mates and opponents when deciding for the long pass across the field. Obviously, researchers involved in this type of experiments are well aware of such caveats when interpreting their data yet only seldom are these caveats made explicit.

Applications of the framework to interventions

In light of the importance of the bidirectional influences between decision making, perception, and action, the final issue is how to develop and refine these links through interventions. Again, the prototypical approach has been to isolate and train elements of the framework in isolation. For example, cognitive psychologists have developed interventions that focus solely on developing the knowledge base that may underpin skilled anticipation. Other interventions aim at instilling novices with the gaze pattern of experts. Yet another approach is to train technical skills in environments where the demands on decision making and perception are reduced.

Clearly, each of these approaches has its merits but limit applicability to the context in which the participant has to perform, or to the current characteristics of the participant. For instance, novice and older drivers have a relatively narrow field of vision such that they are oblivious to much of what happens in their visual periphery. With a narrow field of vision, these drivers may not react timely to bikes approaching at a crossing, cars preparing to overtake, or animals running across the road. In other words, a limited peripheral vision is not only inefficient but also a danger to the driver, instructor, and others. However, training the novice driver to attend to stimuli in the periphery without coupling this specific training to the decisions and actions habitual to driving would likely cause more accidents than prevent them. The narrow attention focus of the novice driver is coupled to the decisions and actions that he/she has to complete and should thus be trained in a coupled manner.

Conclusion

The issues of decision making, perception, and action have been analyzed from different disciplinary perspectives over the years and much knowledge has been gathered. Notwithstanding the advances from individual disciplines in understanding human decision making, it is fruitful to integrate the knowledge and methodologies in a framework that can amplify their contributions. We have described a framework of bidirectional links between decision making, perception, and action, including the role of different constraints. This dynamic framework can be understood as a interdisciplinarity view on decision making.

In principle, there is information in the environment that can guide decision making, as well as perception and actions for a particular person in a particular state. Some states, or organismic constraints, increase the probability for the perception of some options that would otherwise not be evident. For instance, thirst prompts the decision to drink water and increases the probability of the person perceiving a fountain, a restaurant, or a snack shop as viable options. Yet, relevant information can also be derived from cognitive processes including decision making itself. When deciding what motorcycle to buy, one can think of which one is cheaper, cooler, more convenient, and safer, and through this process perceive the entirely new option that fits all criteria which is to buy a bicycle. Thus, the information predominantly used in decision making may rest in the several elements of the framework we described earlier and/or on their interaction. Moreover, options can be perceived through the continuous and dynamic interaction between the elements in the framework.

References

August, K., Lewis, J. A., Chandar, G., Merians, A., Biswal, B., & Adamovich, S. (2006). fMRI analysis of neural mechanisms underlying rehabilitation in virtual reality: activating secondary motor areas. *Conference Proceedings of the IEEE Engineering in Medicine and Biology Society, 1*, 3692–3695.

de Oliveira, R. F., Oudejans, R. R. D., & Beek, P. J. (2009). Experts appear to use angle of elevation information in basketball shooting. *Journal of Experimental Psychology — Human Perception and Performance, 35*.

Gibson, J. J. (1979/1986). *The ecological approach to visual perception*. Hillsdale, NJ: Lawrence Erlbaum Associates.

Newell, K. M. (1986). Constraints on the development of coordination. In M. G. Wade & H. T. A. Whiting (Eds.), *Motor development in children: aspects of coordination and control* (pp. 341–361). Amsterdam: Martin Nijhoff.

Proffitt, D. R. (2006). Embodied perception and the economy of action. *Perspectives on Psychological Science, 1*, 110–122.

Small, D. M., Zatorre, R. J., Dagher, A., Evans, A. C., & Jones-Gotman, M. (2001). Changes in brain activity related to eating chocolate. *Brain, 124*, 1720–1733.

SECTION II

How are these Options Evaluated and an Intended Course of Action Selected?

M. Raab et al. (Eds.)
Progress in Brain Research, Vol. 174
ISSN 0079-6123

CHAPTER 9

(Ir)rationality in action: do soccer players and goalkeepers fail to learn how to best perform during a penalty kick?

Michael Bar-Eli[1,*], Ofer H. Azar[1] and Yotam Lurie[2]

[1]Department of Business Administration, Guildford Glazer School of Business and Management, Ben-Gurion University of the Negev, Beer Sheva, Israel
[2]Department of Management, Guildford Glazer School of Business and Management, Ben-Gurion University of the Negev, Beer Sheva, Israel

Abstract: This chapter discusses penalty kicks in soccer, interpreted within the framework of behavioral economics. We present two behaviors of professional soccer players during penalty kicks that seem nonoptimal, and possibly indicate biases in decision making. We ask whether, despite the huge incentives involved in professional soccer and the possibility of learning through feedback from the outcomes of previous penalty kicks, goalkeepers and penalty kickers are not optimizing their actions. We suggest that they do indeed learn to optimize, but have a different utility function; goalkeepers are not solely interested in minimizing the chances of the goal, and kickers are not solely interested in maximizing these chances.

We believe that, in general, in cases where decision makers have the ability to learn through feedback about the outcome of their actions but exhibit behavior that seems nonoptimal, it is possible that they do optimize, but that their utility function is different from the one assumed. We propose that such decision makers' behavior be reconceived as "socially rational," in the sense that their social environment seems to be incorporated into their utility functions. Finally, the concept of "socio-emotional rationality" is suggested to account for possible implications in future studies of motion–cognition interactions.

Keywords: social rationality; behavioral economics; soccer; penalty; decision making

Introduction

In their message to the participants of the interdisciplinary workshop on Motion and Cognition, Raab et al. (2008) stated that the broadest goal of the workshop is "to gain a better understanding of individuals' decision-making processes when complex movements are part of the behavior," and "to provide insights into decision making that can also be extended to other situations… unlike the "standard" economic utility models." Furthermore, they stated that the rationale of the workshop is, among others, to find out whether "models in the judgment and decision-making area in cognitive science" are "applicable to real actions in real environments," and ask whether "as a consequence of knowing the underlying mechanisms of choice, can training schedules

*Corresponding author.
Tel.: +972-8-6472208; Fax: +972-8-6477691;
E-mail: mbareli@som.bgu.ac.il

DOI: 10.1016/S0079-6123(09)01309-0

or other instructional aspects be altered as one of the long-standing goals to improve performance?"

In their message, Raab et al. (2008) requested that our team discuss the implications of our (applied) decision-making research conducted in sport (which is closely related to the field of economics) on the "feedback of decisional outcomes for subsequent situations," that is, how feedback from states of the environment result from the produced action. In essence, it was stated that among all issues discussed in the workshop, "this question will be perhaps the most concerned with interactions between motor and cognitive systems, because… responses and adaptations to this feedback by both the motor system and the cognitive system result in learning."

For the past decade, our team has been investigating different aspects of penalty kicks in soccer. In this chapter, we will briefly present two of these studies and demonstrate how the behavior of both goalkeepers and kickers during penalty kicks in soccer can be used to address some of the above-mentioned issues, when theorized and analyzed within the framework of behavioral economics. We will also discuss some possible broader implications for the question of human rationality.

Behavioral economics

One of the popular topics in economics in the 2000s is behavioral economics. Not long ago, most economists did not consider adding insights from psychology and/or sociology to economic analysis a worthwhile endeavor, and consequently, behavioral economics was a peripheral area that received very little attention. Recently, however, the field has gained more recognition, and today psychological and sociological factors, such as decision-making biases, are believed to influence economic behavior in many important ways (for review see Rabin, 1998; Kahneman, 2003). For example, awarding the John Bates Clark Medal (considered to be the second highest prize in economics after the Nobel Prize) to economist Matthew Rabin in 2001, and the Nobel Prize in economics to the psychologist Daniel Kahneman (and Vernon Smith) in 2002, demonstrates the increased recognition of behavioral economics as an important research area.

Traditional economic theory (e.g., Samuelson and Nordhaus, 2004) assumes that people have well-defined preferences, and these can be represented by a utility function. People then maximize their utility, subject to budget constraints. Moreover, it was usually assumed that economic agents are selfish and care only about their own well-being or the well-being of their own household. In models that added uncertainty to the environment, maximizing utility was replaced by maximizing expected utility, using the probabilities of the different possible future states.

While this model of economic decision making was helpful in many areas (including sports; e.g., see Leeds and von Allmen, 2005; Fort, 2006; Eschenfelder and Li, 2007), and still continues to be a standard model in economics, it was also shown in numerous experiments and empirical studies that often decision makers deviate from these assumptions in important ways, and that this might affect not only the individual, but even firm strategy and market outcomes as well (e.g., see Azar, 2004a). Consequently, in recent years many economic models have become more flexible about the assumptions they make regarding economic behavior, for example, allowing the economic agent to be concerned about social norms (Azar, 2004b, 2005) or to be inequality averse (Fehr and Schmidt, 1999).

There are several major reasons why people might behave differently from the assumed behavior of traditional economic theory. One reason is that they are prone to biases in their judgment and decision-making processes, and therefore they may not optimize correctly. For example, people might have systematic biases in their perception of probabilities (Gilovich et al., 2002), or they may consider relative price differences even when only absolute price differences matter (denoted "relative thinking," see Azar, 2007). Another reason is that the task may be too complex for people to accomplish, in particular when time constraints and stress might prevent the performer from accurately processing all the information and reaching optimal decisions (Payne and Bettman, 2004). Finally, the preferences of people do not always match those assumed in traditional economic theory. Evidence for this has been found, not only in economics, but also in finance (Glaser et al., 2004) and accounting (Kotchetova and Salterio, 2004).

In this chapter, we propose that when decision makers exhibit behaviors that seem nonoptimal, despite having enough opportunities to practice the task, receive feedback about the outcome, and thus learn to perform it better, it should be carefully examined whether the preferences of the decision makers are different from the ones assumed. More specifically, we suggest that while it seems at first that the only thing that should matter to soccer players during penalty kicks is whether a goal is scored or not, they in fact also care about how the outcome came about and how it is perceived, and this could lead to what seems at first to be nonoptimal behavior. In what follows, we demonstrate how our research on penalty kicks in soccer is related to this observation, and how behaviors that seem nonoptimal at first can be not only rationalized in the derogatory Freudian sense but actually reconceived as optimal and thus rational. Then, we explain why these behaviors consequently exist, despite the ample learning and feedback that professional goalkeepers and players have on previous penalty kicks.

Penalty kicks in soccer

The topic of judgment and decision making under uncertainty has recently attracted the attention of sport/exercise psychologists (Bar-Eli and Raab, 2006). Penalty kicks in soccer create a simple but interesting real-life example of a decision-making task under uncertainty. According to Palacios-Huerta (2003), the time it takes the ball to get from the penalty mark to the goal is about 0.3 s; this would imply that because of the short distance between the ball and the goal, and because of the high speed at which the ball is kicked, the goalkeeper usually cannot afford to wait until he/she sees clearly in what direction the ball is kicked. Therefore, the goalkeeper has to make a decision whether to jump to one side or to stay in the center, and this decision is made at about the same time that the kicker chooses where to direct the kick.

The goalkeeper has to choose whether to jump to the right or the left, or to stay in the center, in order to minimize the risk of a goal being scored, under conditions of uncertainty regarding the direction and height of the ball. The goalkeeper may use his/her knowledge of the directional distribution of penalty kicks in general, the past behavior of the kicker, and cues that might be obtained from the kicker's behavior and approach of the ball, to help him/her decide to which side to jump, if at all. Similarly, the kicker has to decide how to take the shot, facing uncertainty regarding what the goalkeeper will do.

Needless to say, the importance of the goalkeeper's and kicker's performances in professional soccer during penalty kicks is crucial. Not only does the outcome of the game often depend on the goalkeeper's and kicker's performances, but also the issue of the huge amounts of money involved can be raised: teams can make large amounts of money by winning and climbing to the next stage of a tournament, players receive bonuses for successful games, and the kicker's and goalkeeper's reputation and, thus, future earnings also depend on their performance, to give just a few examples. Since players in top clubs earn hundreds of thousands and often millions of dollars annually (see Kupfer, 2006 for review), it is obvious that the kicker's and goalkeeper's performances (e.g., in penalty kicks) have highly significant monetary implications.

Consequently, the penalty kick represents a decision problem with major incentives, making it very intriguing to determine whether expert players — who are encouraged through hefty financial rewards to do their best during the few seconds of a penalty kick — deviate from rational decision making. Numerous studies have examined cognitive processes such as anticipation, cue utilization, and response time of goalkeepers during penalty kicks (e.g., Morris and Burwitz, 1989; McMorris et al., 1993; Williams and Burwitz, 1993; McMorris et al., 1995; McMorris and Colenso, 1996; Savelsbergh et al., 2002, 2005). However, the question of whether goalkeepers and kickers act optimally or reveal biased decision making is still under-researched. Accordingly, Bar-Eli et al. (2007) investigated this problem, and found that goalkeepers jump to their right or left in 94% of the penalty kicks, although given the distribution of kicks, it is optimal for the goalkeeper to stay in the center. Bar-Eli and Azar (2009), although not directly addressing the issue of biased decision making, found that the most difficult penalty kicks to stop are the ones that reach the upper third part of the goal, and nevertheless only 13% of penalty kicks reach this area. We discuss these two findings in more detail below.

Goalkeeper behavior

In Bar-Eli et al. (2007), the behavior of goalkeepers during penalty kicks was examined. The authors collected video recordings of 311 penalty kicks from actual games in top male soccer leagues and championships worldwide, and asked three judges to determine, using a diagram of the goal's area, to which part of the goal the ball was kicked, to which direction the goalkeeper jumped (if at all), and whether he/she stopped the ball. After excluding kicks on which there was significant disagreement among the judges and kicks that missed the goal frame, 286 remained and were used for the analysis. The main statistics in the data are summarized in Table 1. (It should be noted that when we mention right or left, it is from the goalkeeper's perspective; therefore, a kick to the left actually means that the kicker shot the ball to his/her right, and vice versa.)

As Table 1 shows, about 29% of the kicks reached the central third of the goal, but the goalkeeper chose to stay in the center only in about 6% of the cases. This behavior is even more puzzling and striking when we take into account that the chances of the goalkeeper to stop a kick when his/her choice matches the direction of the kick is much higher in the center than at the sides (60% vs. 25–30%; Table 1). Consequently, as the table suggests, the chances of stopping a kick are higher when the goalkeeper stays in the center than when he/she jumps to one of the sides. Nevertheless, goalkeepers almost always (in 94% of the kicks) jump to the right or left instead of staying in the center.

Bar-Eli et al. (2007) suggest that the reason for this nonoptimal behavior is the so-called "action bias." Because the norm is that goalkeepers should "do something" (i.e., jump) during penalty kicks, norm theory (Kahneman and Miller, 1986) implies that a goal scored yields feelings for the goalkeeper that are worse following inaction (staying in the center) than following action (jumping), leading to a bias for action. The more frequently investigated "omission bias" (a bias in favor of inaction; see Ritov and Baron, 1990, 1992, 1995) was reversed in that study, because the norm among goalkeepers was to act rather than to choose inaction. A survey conducted with 32 top professional goalkeepers supported the claim that jumping is the norm and that a goal scored caused worse feelings for the goalkeeper if it was the result of his decision to stay in the center.

Traditional economic theory would imply that the goalkeeper's behavior is optimal when it minimizes the chances of the other team to score

Table 1. Penalty kick outcomes

	Left	Center	Right
Goalkeeper's choice[a]	49.3%	6.3%	44.4%
Goalkeeper's chances to stop the kick overall[b]	14.2%	33.3%	12.6%
Goalkeeper's chances to stop if the goalkeeper and kicker choose the same direction[c]	29.6%	60.0%	25.4%
Kicker's choice[d]	32.2%	28.7%	39.2%

[a]Presents the percentage of cases in which the goalkeeper chose to jump left, right, or stay in the center.
[b]Presents the fraction of kicks that were stopped following each of the goalkeeper's possible actions, regardless of the kick's direction.
[c]Presents the fraction of kicks that were stopped when both the goalkeeper and the kicker chose this direction (i.e., the goalkeeper chose the correct direction — the one that matched the kick direction).
[d]Presents the distribution of the kicks' direction.

a goal. This suggests that the goalkeeper should stay in the center and not jump as long as the kickers do not change their kicking strategy. However, from the goalkeeper's perspective, the documented behavior of almost always jumping might be optimal if his/her utility function includes not only the score outcome (goal or not), but also other ingredients (e.g., if he/she felt worse following a scored goal when he/she did not jump than when he/she jumped); indeed, the survey of the top goalkeepers mentioned above supports this idea. It is also possible that observers of the goalkeeper, including soccer managers who can affect the goalkeeper's career and earnings in the future, are biased in their evaluation of the goalkeeper's actions. That is, they may view a goalkeeper who stays in the center and gets a goal as acting less professionally than a goalkeeper who jumps to one of the sides and gets a goal. If that is the case, then this also justifies, from the goalkeeper's perspective, the decision to almost always jump, even though this does not maximize the team's expected score (because it does not minimize the chances of a goal).

One could even go a step further and make the conjecture that even from the team's perspective, minimizing the chances to get a goal is not the entire component of the utility function. Teams are also economic organizations, and, as such, they want to attract fans to their games. While fans are more attracted by successful teams, they may also be concerned with how the team is playing, and may view "heroic jumps" to save a penalty kick as more attractive than a goalkeeper who just stays in the center. Therefore, it is possible that the goalkeeper's behavior, which at first seems irrational, is rational not only from his/her own perspective, but also because it matches the team's preferences.

This discussion may suggest why, despite the huge monetary incentives that are involved in professional soccer, and despite the fact that the highly experienced goalkeepers included in our study have had ample opportunities for learning and receiving feedback about the prospects of their possible actions, the goalkeepers still only rarely choose the optimal action, that is, the one that minimizes the chances of a goal — to stay in the center. According to this reasoning, it is not that they do not learn how to optimize their behavior despite having learning and feedback opportunities; their behavior is consistent with preferences that are different from merely minimizing the chances of a goal, and the goalkeepers' behavior might indeed optimize their utility, once these preferences are taken into account.

Kicker behavior

Bar-Eli and Azar (2009) used the set of penalty kicks reported in Bar-Eli et al. (2007) but augmented it by also considering the height of the kick and not only its horizontal direction. As opposed to Bar-Eli et al. (2007), who focused on the goalkeeper's strategy and whether he/she exhibited decision-making biases, Bar-Eli and Azar (2009) examined the optimal strategy for the kicker in penalty kicks. They found that approximately 13% of the kicks reached the upper third of the goal, 30% the central part, and 57% the lower part. None of the kicks kicked to the upper part was stopped, compared to 13% stopping chances of the kicks to the central part and 20% to the lower part. Bar-Eli and Azar concluded that the optimal shooting strategy is to aim the ball to the upper part (in particular to the upper two corners), and that with proper training, the missing rates should be low enough to justify applying such a strategy.

The major factor that may deter kickers from kicking to the upper part of the goal is probably that this would increase the chances of missing the entire goal frame. However, the scoring chances of a flat kick, which is shot to the lower third of the goal, are at most 80% (according to the data, the goalkeeper has about a 20% chance to stop a ball that is shot to the lower-third inside the goal frame, but there is also an additional probability that the kick will be shot to the goalpost, or miss the goal frame altogether). Therefore, if the chances of a kick that is kicked to the upper third to miss the goal frame are lower than 20%, this strategy should be considered better than kicking to the lower third. It seems reasonable that, with proper training, a less than 20% missing rate can be achieved; yet, we see kickers shooting much more often (over four times more frequently) to the lower part than to the upper part of the goal (obviously, there is also the alternative of kicking to the central third, but to simplify the discussion we are focusing on the upper and lower parts).

Once again, at first glance these results are puzzling. Kickers have ample opportunity to practice penalty kicks, to practice various alternative shots, and to receive feedback (e.g., by seeing whether they and/or their colleagues scored a goal or not) and learn how to improve their performance of these kicks. Why, despite this, and even though they have huge incentives to perform penalty kicks optimally in real games, do they rarely shoot to the upper part of the goal, where the goalkeeper's chances to stop the ball are negligible?

This question can probably be answered by the preferences of kickers, which are not necessarily only to maximize the chances to score a goal (even though from the team's perspective this should be the most important issue). For example, it is possible that from the kicker's perspective, missing the goal frame is considered worse than kicking a ball that the goalkeeper succeeds in stopping. The reason could be that in the first case it is clear that the missed goal is solely the kicker's fault, while in the second case the outcome (of no goal) may also be attributed (e.g., by different observers) to the goalkeeper's ability to stop the kick, and not only to the kicker's inability to kick well. Consequently, kickers might avoid shooting to the upper part. If kickers' behavior is indeed motivated not only by maximizing the chances of scoring a goal but also by other preferences, it should not be surprising that learning does not solve this seemingly nonoptimal behavior. It is not that the kickers do not learn how to optimize their performance in kicking penalties; they do optimize, but not a classic "economic" utility function (i.e., maximizing the chances of scoring). Instead, their utility function also reflects their significant disutility from missing the goal frame, which is higher than their disutility from a kick that the goalkeeper stops.

Social rationality

We present two behaviors of professional soccer players during penalty kicks that seem at first to be nonoptimal, and possibly indicate biases in decision making. One is that goalkeepers almost always jump to one of the sides, even though staying in the center is the strategy that minimizes the chances of a goal, given the distribution of the direction of kicks. The second is that kickers shoot to the upper third of the goal in only 13% of the cases, even though with proper training the chances of scoring such a kick are extremely high (i.e., this is the action that maximizes the chances of scoring a goal).

The question we ask is whether it is *reasonable*, despite the huge incentives involved in professional soccer, and despite the possibility of receiving feedback from the outcomes of previous penalty kicks (including ones in training) and consequently of learning, that soccer players will still not optimize their actions. We suggest that the answer might be that the players are reasonable, but they are reasonable in the sense that they optimize a different utility function than what was expected — the goalkeepers are not only interested in minimizing the chances of a goal being scored, and the kickers are not only interested in maximizing these chances of scoring.

In the case of goalkeepers, their utility also depends on whether they have jumped or not — they feel worse when a goal is scored if they did not jump than if they did jump. The reason for this might be that it is considered a more professional and correct action to jump (perhaps because this is what goalkeepers are expected to do in penalty kicks), or that goalkeepers jump because they want to appear as though they are trying hard to stop the ball; in sports, trying your best is an important value. Similarly, in the kickers' case, their behavior can be utility maximizing if they view missing the goal frame as a worse outcome than shooting a kick that the goalkeeper stops.

The interaction between the kicker and the goalkeeper can be described as follows: both the kicker and goalkeeper know quite well that the hardest areas for the goalkeeper to stop a penalty are the two upper corners of the goal

(Bar-Eli and Azar, 2009); that is, the optimal strategy for the kicker would be to shoot the ball to one of the two upper corners accurately. In this case, no matter what the goalkeeper does — the outcome of the kick would depend (almost) only on the kicker's performance. However, because of the kicker's utility function, which includes the perception of missing the goal frame as being a worse outcome than shooting a kick that the goalkeeper will eventually stop, the kicker prefers not to shoot to the upper corners. As a result, an interaction between the kicker and the goalkeeper takes place, because in most of the goal's areas (except for the two upper corners), the outcome of the kick will depend on the performance of both the kicker and the goalkeeper; this interaction can be best described in game-theoretical terms (e.g., Chiappori et al., 2002; Palacis-Huerta, 2003).

Because of this interaction or game between the kicker and the goalkeeper, an analysis of goalkeepers' behavior and possible decision-making biases — particularly the action bias (see Bar-Eli et al., 2007), and our above argument concerning the goalkeeper's possible different utility function (i.e., one that includes also their will to appear as trying hard to stop the ball) are both relevant. However, in reference to the major principles of motor control and learning (Schmidt and Lee, 2005), we believe that with sufficient and proper training of kicking penalties to the top corners of the goal (i.e., according to the most updated training principles; e.g., see Blumenstein et al., 2007 for review), the kickers' percentage of success in performing such kicks could be substantially increased, resulting in a goal being scored regardless of the goalkeepers' performance (because it is nearly impossible to stop a penalty kick shot to the upper two corners). Therefore, only the kickers' "irrationality" enables the kicker–goalkeeper game, as well as the consequent "irrational" behavior of goalkeepers, to take place.

The more general lesson we propose is that in cases where we observe decision makers who receive feedback about the outcome of their actions and thus have the ability to learn, and yet exhibit behavior that seems nonoptimal, one should carefully consider their assumed utility function. It is possible that the decision makers do optimize, but they optimize a utility function that is different from the one assumed, and therefore it might seem, incorrectly, that they behave irrationally and do not maximize utility.

A concept that is closely related to this lesson is that of "social rationality." More specifically, in his seminal book, Gigerenzer (2000, p. 202) strongly challenges the idea that "formal axioms and rules of choice can define rational behavior without referring to factors external to choice behavior," that is, that such axioms and rules can be imposed a priori as context-independent, general purpose yardsticks of rationality (see Chapter 10: Getting around: making fast and frugal navigation decisions, for further treatment of rationality). In line with other studies (e.g., Simon, 1986; Elster, 1990; Sen, 1993), he argued that "these principles are incomplete as behavioral norms in the sense that their normative validity depends on the social context of the behavior, such as social objectives, values, and motivations." Without taking into account the social context, they are limited, confusing, and insufficient for defining rational behavior.

To demonstrate this idea, Gigerenzer (2000) used, among others, the phenomenon of "probability matching" (for reviews see Gallistel, 1990; Vulkan, 2000), according to which even after many rounds in an experiment (and therefore an ample opportunity for learning), as the decisions become more rational (i.e., choosing the option that more often has the higher probability of success), people still occasionally choose the non-optimal action. This is irrational if people only try to maximize their earnings in the experiment, which is the usual assumption in experimental economics. But if people's utility also depends on other factors, such as how boring the experiment is, and if it is more interesting for people to participate in an experiment where occasionally one chooses something different and does not always perform the same action, then it might be rational and utility maximizing for the participant to occasionally choose the option with the smaller probability of success, even though this might decrease the participant's expected earnings. As Gigerenzer (2000, p. 206) puts it, "the maximizing principle does not capture the distinction between the individual in social isolation and in social interaction… Whether an organism performs in isolation or in the context of other organisms can determine, among other things, whether maximization is entailed as an optimal choice rule" (e.g., for evolutionary reasons; see Gallistel, 1990).

Viewed from this perspective, the goalkeepers and penalty kickers in our studies may be defined as "socially rational" in the sense that their social environment seems to be incorporated into their utility functions. In other words, social factors such as spectators, team-mates, and the coach (for review see Jowett and Lavallee, 2007), whose perception by the athletes was found to have a substantial effect on their performance (Bar-Eli, 1997), may play a major role in understanding the utility functions of penalty kickers (e.g., "I don't want my miss — if it occurs — to be attributed by my spectators/team-mates/coach only to my inability, and therefore, I'll avoid shooting to the upper part of the goal") and goalkeepers (e.g., "I want to present myself to my observers as more professional — i.e., as someone who actively tries hard to do something — and therefore, I'll always jump"). Indeed, self-presentational considerations were found to be crucial in explaining people's behavior in physical activity contexts (Martin-Ginis et al., 2007), with individuals' mental representations of significant others substantially affecting the cognitive processes which underlie judgment and decision making in sport (Eccles and Tenenbaum, 2007).

Beyond social rationality: psychophilosophical considerations

Performance is defined as "goal-directed behavior" (e.g., by action theory; see Frese and Zapf, 1994; Schack and Hackfort, 2007). In order to maximize performance, athletes usually attempt to optimize something — be it, for example, their movement (Schack and Bar-Eli, 2007) or their arousal state (Gould and Carson, 2007). Thus, the pursuit of excellence in sport requires athletes to achieve performance "maximization through optimization" — a principle that is central among the major aspects of human rationality (Bar-Eli et al., 1999, 2006). Moreover, sport psychology has been provided with rational models, such as the Bayesian approach (for a review see Tenenbaum and Bar-Eli, 1993), and/or optimization aims derived from the domain of operations research in management science (Mehrez et al., 2006; Sinuany-Stern et al., 2006), which can be used to aid in optimizing the decision makers' thought processes required for performance maximization.

Such models reflect rationality in its instrumental sense, which has to do with the effectiveness of one's application of means towards the accomplishment of a certain goal (Weber, 1919/1946). Instrumental rationality and/or reasoning are reflected, for example, in the current literature on expert sport performance (for review see Starkes and Ericsson, 2003), and on the professionalization of managerial processes in organized elite sport (Slack and Parent, 2006). One might argue that in principle "social rationality" reflects an instrumental conception of rationality, with the utility function of both the goalkeeper and the penalty kicker being more complex than just to score or to stop the ball (i.e., maximizing chances of scoring as opposed to minimizing chances of a goal being scored). From the standpoint of the goalkeeper, it is preferred to have the goal scored while trying virtuously to stop the ball by jumping, than to have a goal scored while merely standing straight and effortlessly in the middle of the goal frame, because it is believed that an athlete must at least try to perform with power and grace. Similarly, from the standpoint of the kicker, the negative value of kicking the penalty totally out of bounds (i.e., a full miss of the goal frame) is a much worse outcome than kicking a penalty kick which is stopped by the goalkeeper, because it is believed that an athlete should prefer a solid effort (even if unsuccessful) to a complete failure which might even be perceived as shameful.

However, by modifying the utility function in order to be in line with the instrumental conception of reason, one misses the deeper ramifications of the abovementioned example for understanding the connection between human reason, emotion, decision making, and action. In many facets of human life, it is not only the outcome that matters; but also how (i.e., by which route) the outcome was reached: for example, how active or passive was the agent, what were the agent's intentions, and what did the agent actively do (or refrain from doing) in order to reach the specific outcome. What one *did* and what one *did not* do in order to reach the outcome can be more significant than the outcome itself. The point is that human beings are not just utility-maximizing consequentialists; rather, emotions, social relations, and environmental considerations play an important role in people's decisions and actions. Thus, to drive the point home, within a loaded emotional situation, where pride and money are at stake and tension is high, the goalkeeper is an athlete and is expected to perform as an athlete. The goalkeeper is also a member of a team that has been making a tremendous effort for their fans and other stakeholders. As such, beyond maximizing a certain result per se, the goalkeeper is expected behave as an athlete and represent his/her entire team. As such the goalkeeper has to put up a fight and make a serious effort at stopping the penalty kick.

The "trolley problem" is an ethical thought experiment which might be used to clarify this matter; it was first introduced by Foot (1978, reprinted in 2002), and later richly elaborated upon by Thomson (1985) and Unger (1996). The trolley problem has also been a significant feature in Greene's (Greene, 2004; Greene et al., 2001) neuroscientific approach to questions of rational decision making. The relevance of the trolley problem to the current issue is not in the kind of situation it describes, but that by means of this moral dilemma we can demonstrate that our judgment and decision-making process takes into account more than just the optimal outcome.

According to the original version of the trolley problem suggested in 1978 (see Foot, 2002), you are driving a runaway trolley that is headed for five people who will be killed if it proceeds on its present course. The only way to save them is to hit a switch that will turn the trolley onto an alternate track where it will kill one person instead of five. Should you turn the trolley in order to save five at the expense of one? Most people say yes.

An alternative version of the trolley Problem (Thompson, 1985) is called the "footbridge dilemma." As before, a trolley threatens to kill five people. You are standing next to a very fat stranger on a bridge. Your only way to stop the trolley is to push him over the bridge onto the track, killing him but saving five others. Should you proceed? Most people say no.

Though the outcome of the two cases can be similar, attempts to find relevant moral distinctions between these two cases, in order elucidate the rationale behind the agents' decision-making process, have traditionally focused on the agent's intention: in the first case one does not intend harm toward anyone and harming is just a side effect. Greene (2004), in taking a neuroscientific approach to the trolley problem, examined the brain's response to such decisions using functional magnetic resonance imaging (fMRI). His findings were that in cases where the dilemma was of a more personal nature, such as the footbridge dilemma, where one's decision involved a more active role of authorship, then the decision process involves a more social-emotional decision process. In contrast, if the dilemma was of a more impersonal nature, it involved a more cognitive decision process (Greene, 2004). More specifically, when participants considered personal moral dilemmas, the medial prefrontal cortex, posterior cingulate/precuneus, and superior temporal sulcus/temperoparietal junction exhibited increased activity, but when participants considered impersonal moral dilemmas, "cognitive" brain areas associated with abstract reasoning and problem solving exhibited increased activity. Thus, this study distinguished between judgments based on cognitive processes and judgments based on emotional processes, which enables us to suggest an extension of the very meaning of human rationality.

Conclusion: towards socio-emotional rationality

In the introduction to this chapter, we briefly summarized the broadest goal and the rationale of the workshop, as presented by Raab et al. (2008). We believe that our abovementioned two studies on penalty kicks in soccer successfully demonstrated how behaviors of both goalkeepers and kickers in this situation can be used to address some of the major issues raised by Raab et al. (2008), when analyzed and theorized within the framework of behavioral economics. Gigerenzer's (2000) concept of "social rationality" seems to be adequate in capturing the behavior of goalkeepers and kickers during penalty kicks in soccer, mainly because it enables the incorporation of these actors' social environment into their respective utility functions. Although in principle this can be presented as an extended version of the "instrumental rationality" concept, by manipulating and extending the utility function, this way of capturing the goalkeeper's and kicker's behavior misses the deeper ramifications of the abovementioned example for understanding the connection between human reason, emotion, decision making, and action.

We suggest that in order to conceive better the relationship between motion and cognition in terms of the connections among psychology, economics, and sport, a concept of human rationality should be used, which will go beyond social rationality in the instrumental sense. In line with our discussion above, we suggest that the concept of "social rationality" be extended to include not only the social and environmental context but also emotional elements pertaining to the agent's character and role. Emotions have been of increasing interest to sport/exercise psychologists (e.g., Vallerand and Blanchard, 2000; Hanin, 2007), for example within the action theory (e.g., Kuhl, 1994; Schack and Hackfort, 2007) and the psychophysiological approach (e.g., Blumenstein et al., 2002; Hatfield and Kerick, 2007). We propose a concept of "socio-emotional rationality," that will extend the psychophilosophical study of human rationality to include social and emotional components as well, and may provide a significant breakthrough and to a better understanding of complex motion–cognition relationships.

References

Azar, O. H. (2004a). *Psychological motivations and biases in economic behavior and their effects on markets and firm strategy.* Doctoral dissertation. Evanston, IL: Northwestern University.

Azar, O. H. (2004b). What sustains social norms and how they evolve? The case of Tipping. *Journal of Economic Behavior & Organization, 54*, 49–64.

Azar, O. H. (2005). The social norm of tipping: does it improve social welfare? *Journal of Economics, 85*, 141–173.

Azar, O. H. (2007). Relative thinking theory. *Journal of Socio-Economics, 36*, 1–14.

Bar-Eli, M. (1997). Psychological performance crisis in competition, 1984–1996: a review. *European Yearbook of Sport Psychology, 1*, 73–112.

Bar-Eli, M., & Azar, O. H. (2009). Penalty kicks in soccer: an empirical analysis of shooting strategies and goalkeepers' preferences. *Soccer and Society, 10*, 183–191.

Bar-Eli, M., Azar, O. H., Ritov, I., Keidar-Levin, Y., & Schein, G. (2007). Action bias among elite soccer goalkeepers: the case of penalty kicks. *Journal of Economic Psychology, 28*, 606–621.

Bar-Eli, M., Lowengart, O., Master-Barak, M., Oreg, S., Goldenberg, J., Epstein, S., & Fosbury, R. D. (2006). Developing peak performers in sport: optimization versus creativity. In D. Hackfort & G. Tenenbaum (Eds.), *Perspectives on sport and exercise psychology series: Vol. 1. Essential processes for attaining peak performance* (pp. 158–177). Oxford: Meyer and Meyer.

Bar-Eli, M., Lurie, Y., & Breivik, G. (1999). Rationality in sport: a psychophilosophical approach. In R. Lidor & M. Bar-Eli (Eds.), *Sport psychology: linking theory and practice* (pp. 35–58). Morgantown, WV: Fitness Information Technology.

Bar-Eli, M., & Raab, M. (Eds.). (2006). Judgment and decision making in sport and exercise: a special issue. *Psychology of Sport and Exercise, 7*(6), 519–688.

Blumenstein, B., Bar-Eli, M., & Tenenbaum, G. (Eds.). (2002). *Brain and body in sport and exercise: biofeedback applications in performance enhancement.* New York: Wiley.

Blumenstein, B., Lidor, R., & Tenenbaum, G. (Eds.). (2007). *Perspectives on sport and exercise psychology series: Vol. 2. Psychology of sport training.* Oxford: Meyer and Meyer.

Chiappori, P. A., Levitt, S., & Groseclose, T. (2002). Testing mixed-strategy equilibria when players are heterogeneous: the case of penalty kicks in soccer. *American Economic Review, 92*, 1138–1151.

Eccles, D. W., & Tenenbaum, G. (2007). A social-cognitive perspective on team functioning in sport. In G. Tenenbaum & R. C. Eklund (Eds.), *Handbook of sport psychology* (3rd ed., pp. 264–283). New York: Wiley.

Elster, J. (1990). When rationality fails. In K. S. Cook & M. Levi (Eds.), *The limits of rationality* (pp. 19–51). Chicago, IL: University of Chicago Press.

Eschenfelder, M. J., & Li, M. (2007). *Economics of sport* (2nd ed.). Morgantown, WV: Fitness Information Technology.

Fehr, E., & Schmidt, K. M. (1999). A theory of fairness, competition and cooperation. *Quarterly Journal of Economics, 114*, 817–868.

Foot, P. (2002). The problem of abortion and the doctrine of the double effect. In P. Foot (Ed.), *Virtues and vices and other essays in moral philosophy* (pp. 19–33). Oxford: Blackwell.

Fort, R. D. (2006). *Sports economics* (2nd ed.). Upper Saddle River, NJ: Prentice Hall.

Frese, M., & Zapf, D. (1994). Action as the core of work psychology: a German approach. In H. C. Triandis, M. D. Dunnette, & L. M. Haugh (Eds.), *Handbook of industrial and organizational psychology* (pp. 271–340). Palo Alto, CA: Consulting Psychologist.

Gallistel, C. R. (1990). *The organization of learning*. Cambridge, MA: MIT Press.

Gigerenzer, G. (2000). *Adaptive thinking: rationality in the real world*. Oxford: Oxford University Press.

Gilovich, T., Griffin, D., & Kahneman, D. (2002). *Heuristics and biases*. Cambridge: Cambridge University Press.

Glaser, M., Nöth, M., & Weber, M. (2004). Behavioral finance. In D. J. Koehler & N. Harvey (Eds.), *Blackwell handbook of judgment and decision making* (pp. 527–546). Malden, MA: Blackwell.

Gould, D., & Carson, S. (2007). Psychological preparation in sport. In B. Blumenstein, R. Lidor, & G. Tenenbaum (Eds.), *Perspectives on sport and exercise psychology: Vol. 2. psychology of sport training* (pp. 115–136). Oxford: Meyer and Meyer.

Greene, J. D. (2004). The neural bases of cognitive conflict and control in moral judgment. *Neuron, 44*, 389–400.

Greene, J. D., Sommerville, R. B., Nystrom, L. E., Darley, J. M., & Cohen, J. D. (2001). An MRI investigation of emotional engagement in moral judgment. *Science, 293*, 2105–2108.

Hanin, Y. L. (2007). Emotions in sport: current issues and perspectives. In G. Tenenbaum & R. C. Eklund (Eds.), *Handbook of sport psychology* (3rd ed., pp. 31–58). Hoboken, NJ: Wiley.

Hatfield, B. D., & Kerick, S. E. (2007). The psychology of superior sport performance: a cognitive and affective neuroscience perspective. In G. Tenenbaum & R. C. Eklund (Eds.), *Handbook of sport psychology* (3rd ed., pp. 84–109). Hoboken, NJ: Wiley.

Jowett, S., & Lavallee, D. (Eds.). (2007). *Social psychology in sport*. Champaign, IL: Human Kinetics.

Kahneman, D. (2003). Maps of bounded rationality: psychology for behavioral economics. *American Economic Review, 93*, 1449–1475.

Kahneman, D., & Miller, D. T. (1986). Norm theory: comparing reality to its alternatives. *Psychological Review, 93*, 136–153.

Kotchetova, N., & Salterio, S. (2004). Judgment and decision-making accounting research: a quest to improve the production, certification and use of accounting information. In D. J. Koehler & N. Harvey (Eds.), *Blackwell handbook of judgment and decision making* (pp. 547–566). Malden, MA: Blackwell.

Kuhl, J. (1994). A theory of action and state orientations. In J. Kuhl & J. Beckmann (Eds.), *Volition and personality* (pp. 9–46). South Lower Saxony, Germany: Hogrefe and Huber.

Kupfer, F. A. T. (2006). *Erfolgreiches Fussballclub Management* [Successful management of football clubs]. Göttingen: Die Werkstatt.

Leeds, M., & von Allmen, P. (2005). *The economics of sport* (2nd ed.). Boston, MA: Addison-Wesley.

Martin-Ginis, K. A., Lindwall, M., & Prapavessis, H. (2007). Who cares what other people think? Self-presentation in exercise and sport. In G. Tenenbaum & R. C. Eklund (Eds.), *Handbook of sport psychology* (3rd ed., pp. 136–157). New York: Wiley.

McMorris, T., & Colenso, S. (1996). Anticipation of professional goalkeepers when facing right- and left-footed penalty kicks. *Perceptual and Motor Skills, 82*, 931–934.

McMorris, T., Copeman, D., Corcoran, D., Saunders, G., & Potter, S. (1993). Anticipation of soccer goalkeepers facing penalty kicks. In T. P. Reilly, J. Clarys, & A. Stibbe (Eds.), *Science and football II* (pp. 250–253). London: E and F N Spon.

McMorris, T., Hauxwell, B., & Holder, T. (1995). Anticipation of soccer goalkeepers when facing penalty kicks to the right and left of the goal using different kicking techniques. In W. K. Simpson, A. D. LeUnes, & J. S. Picou (Eds.), *Applied research in coaching and athletics, annual* (pp. 32–43). Boston, MA: American Press.

Mehrez, A., Friedman, L., Sinuany-Stern, Z., & Bar-Eli, M. (2006). Optimal threshold in multi-stage competitions. *International Journal of Sport Management and Marketing, 1*, 215–238.

Morris, A., & Burwitz, L. (1989). Anticipation and movement strategies in elite soccer goalkeepers at penalty kicks. *Journal of Sports Sciences, 7*, 79–80.

Palacios-Huerta, I. (2003). Professionals play minimax. *Review of Economic Studies, 70*, 395–415.

Payne, J. W., & Bettman, J. R. (2004). Walking with the scarecrow: the information-processing approach to decision research. In D. J. Koehler & N. Harvey (Eds.), *Blackwell handbook of judgment and decision making* (pp. 110–132). Malden, MA: Blackwell.

Rabin, M. (1998). Psychology and economics. *Journal of Economic Literature, 36*, 11–46.

Ritov, I., & Baron, J. (1990). Reluctance to vaccinate: omission bias and ambiguity. *Journal of Behavioral Decision Making, 3*, 263–277.

Ritov, I., & Baron, J. (1992). Status-quo and omission biases. *Journal of Risk and Uncertainty, 5*, 49–61.

Ritov, I., & Baron, J. (1995). Outcome knowledge, regret, and omission bias. *Organizational Behavior and Human Decision Processes, 64*, 119–127.

Samuelson, P. A., & Nordhaus, W. D. (2004). *Economics* (18th ed.). New York: McGraw-Hill/Irwin.

Savelsbergh, G. J. P., van der Kamp, J., Williams, A. M., & Ward, P. (2005). Anticipation and visual search behavior in expert soccer goal keepers. *Ergonomics, 48*, 1686–1697.

Savelsbergh, G. J. P., Williams, A. M., van der Kamp, J., & Ward, P. (2002). Visual search, anticipation and expertise in soccer goalkeepers. *Journal of Sports Sciences, 20*, 279–287.

Schack, T., & Bar-Eli, M. (2007). Psychological factors of technical preparation. In B. Blumenstein, R. Lidor, & G. Tenenbaum (Eds.), *Perspectives on sport and exercise psychology: Vol. 2. Psychology of sport training* (pp. 62–103). Oxford: Meyer and Meyer.

Schack, T., & Hackfort, D. (2007). Action-theory approach to applied sport psychology. In G. Tenenbaum & R. C. Eklund (Eds.), *Handbook of sport psychology* (3 rd ed., pp. 332–351). New York: Wiley.

Schmidt, R. A., & Lee, T. D. (2005). *Motor control and learning: a behavioral emphasis* (4 th ed.). Champaign, IL: Human Kinetics.

Sen, A. (1993). Internal consistency of choice. *Econometrica*, *61*, 495–521.

Simon, H. A. (1986). Rationality in psychology and economics. In R. Hogarth & M. Reder (Eds.), *Rational choice: the contrast between economics and psychology* (pp. 25–40). Chicago, IL: University of Chicago Press.

Sinuany-Stern, Z., Israeli, Y., & Bar-Eli, M. (2006). Application of the analytic hierarchy process for the evaluation of basketball teams. *International Journal of Sport Management and Marketing*, *1*, 193–207.

Slack, T., & Parent, M. M. (2006). *Understanding sport organizations: the application of organization theory* (3rd ed.). Champaign, IL: Human Kinetics.

Starkes, J. L., & Ericsson, K. A. (Eds.). (2003). *Expert performance in sports: advances in research in sport expertise*. Champaign, IL: Human Kinetics.

Tenenbaum, G., & Bar-Eli, M. (1993). Decision making in sport: a cognitive perspective. In R. N. Singer, M. Murphey, & L. K. Tennant (Eds.), *Handbook of research on sport psychology* (pp. 171–192). New York: Macmillan.

Thomson, J. J. (1985). The trolley problem. *Yale Law Journal*, *94*, 1395–1415.

Unger, P. (1996). *Living high and letting die*. Oxford: Oxford University Press.

Vallerand, R. J., & Blanchard, C. M. (2000). The study of emotion in sport and exercise. In Y. L. Hanin (Ed.), *Emotions in sport* (pp. 3–37). Champaign, IL: Human Kinetics.

Vulkan, N. (2000). An economist's perspective on probability matching. *Journal of Economic Surveys*, *14*, 101–118.

Weber, M. (1919/1946). Politics as a vocation. In H. H. Gerth & C. W. Mills (Eds.), *From Max Weber: essays in sociology* (pp. 77–156). New York: Oxford University Press.

Williams, A. M., & Burwitz, L. (1993). Advance cue utilization in soccer. In T. Reilly, J. Clarys, & A. Stibbe (Eds.), *Science and football, II* (pp. 239–243). London: E and F N Spon.

M. Raab et al. (Eds.)
Progress in Brain Research, Vol. 174
ISSN 0079-6123

CHAPTER 10

Getting around: making fast and frugal navigation decisions

Juliet A. Conlin*

Center for Adaptive Behavior and Cognition, Max Planck Institute for Human Development, Berlin, Germany

Abstract: The spatial environments in which humans and animals navigate tend to be complex and dynamic, and are often characterized by information shortage. Getting around is a fundamental, adaptive problem for many individuals, but one they are capable of solving with remarkable speed and accuracy. In this chapter, it is argued that what on occasion appears to be a complex behavior based on internal representations is often the outcome of simple mechanisms — termed fast and frugal heuristics — and their interaction with the environment. Thus, the spatial navigation decisions made by individuals can be characterized as "good enough" rather than optimal, and are often crucially dependent on the environments in which they operate. This chapter reviews approaches from navigation research that differ in their emphasis on environment versus internal spatial representations in explaining movement behavior and offers an explanation of human movement behavior in terms of relatively simple mechanisms that can exploit properties of the task environment. The main contention put forward here is that a large part of explanatory power would be lost by focusing merely on the underlying cognitive processes without reference to the structure of the environment(s) in which those processes operate.

Keywords: spatial navigation; heuristics; task environment; movement behavior

Introduction

The 2008 Center for Interdisciplinary Research (ZiF) interdisciplinary workshop entitled "*Mind and motion: the bidirectional link between thought and action*" (29–31 May 2008; Bielefeld, Germany) took as its overarching topic the investigation of decision making as intimately rooted in real action; that is, human decision making as principally embodied in the physical environment(s) the individual is situated in. This chapter addresses the particular question of what drives the selection of a course of action in movement behavior among a set of choices, given the physical, psychological and physiological constraints faced by the navigator.

The spatial environments in which humans and animals navigate tend to be complex and dynamic, and are often characterized by information shortage. Getting around is a fundamental, adaptive problem for many living organisms, examples of which include avian migration across continents and foraging for food, or even evacuating a building in an emergency, negotiating an unfamiliar public transport system, or shopping in a supermarket under time pressure. In very few cases do individuals have perfect information,

*Corresponding author.
Tel.: +49-30-8240-6694; Fax: +49-30-8240-6394;
E-mail: conlin@mpib-berlin.mpg.de

DOI: 10.1016/S0079-6123(09)01310-7

such as knowledge about the location of a food source, or the exact details of a particular route; that is, they are often not in a position to make "optimal" navigation decisions. However, despite many apparent constraints, individuals are capable of negotiating challenging environments with remarkable speed and accuracy – making navigation decisions that are "good enough" – suggesting that they are likely to have developed efficient methods for dealing with spatial information and translocation.

Consider, for example, the Saharan desert ant *Cataglyphis fortis*. This ant is truly an impressive navigator: it travels along a tortuous foraging route, often across barren landscapes devoid of landmarks, and yet returns to the nest along a straight line (e.g., Wehner, 2003). How is such a feat accomplished? One explanation is that while traveling, the ant computes and stores the trajectory of the outward journey, and then calculates a return vector on the basis of this representation using something akin to a trigonometric or other nonlinear function. However, this act would appear amazingly complex in view of the size of the ant's brain (approximately 0.1 mg in a 10 mg ant). Instead, it has been suggested that rather than using complex cognition, the ant has simple mechanisms available that act on features of its environment that allow it to keep track of its location in relation to home (e.g., Ziegler and Wehner, 1997).

In this chapter, I will outline an approach to decision-making behavior that similarly goes beyond purely "internalistic" explications; one that takes into account not just the cognitive processes of the individual, but also the task environment in which behavior takes place. According to this approach, what on occasion appears to be a complex behavior based on internal representations is often the outcome of simple mechanisms – termed fast and frugal heuristics – and their interaction with the environment (Gigerenzer et al., 1999).

Overview

This chapter is divided into four sections. The first section introduces the major theoretical foundations for the decision-making approach described here. Section two reviews approaches from navigation research that differ in their emphasis on environment versus internal spatial representations in explaining movement behavior. The third section outlines some applications of the fast and frugal approach to debates specific to navigation behavior, concluding with an exploratory consideration of potential areas of further investigation and application. The main argument put forward is that – as with any other real-world task that demands rapid, accurate decision making – human movement behavior and navigation cannot be adequately described, explained, or predicted by focusing mainly on complex cognitive calculations that act on internal representations. Instead, it is argued that much of human decision making takes place in the form of relatively simple mechanisms that form an "adaptive toolbox," which can exploit properties of the task environment (e.g., Gigerenzer et al., 1999). Thus, a large piece of explanatory power would be lost by focusing merely on the underlying cognitive processes without reference to the structure of the environment(s) in which those processes operate.

Theoretical foundations of the adaptive toolbox

Bounded rationality

The adaptive toolbox approach to decision making draws largely on Herbert A. Simon's (e.g., 1956) notion of bounded rationality. Bounded rationality stands as a challenge to the concept of unbounded rationality, according to which humans make decisions (or at least, how they *should* make decisions) by gathering all available information, weighting the pieces of information according to their relevance, and then integrating the weighted information in order to arrive at an optimal decision (e.g., Edwards and Fasolo, 2001). A view of bounded rationality according to Simon (e.g., 1956), in contrast, assumes that *actual* decision making takes place within the bounds of time, information, and computational ability that the task environment and the cognitive system affords. Indeed, many real-life decisions, including route choices, are so complex as to be computationally intractable, even if the decision maker had limitless time and cognitive power (e.g., Gigerenzer, 2004).

When foraging on patchily distributed resources, for example, animals have to select a route that passes through several different sites that contain the preferred resources, before returning to their starting point. This task, the success of which has obvious implications in terms of food intake, exposure to predators, and mating success, can be likened to a notorious optimization problem known as the "traveling salesperson problem" (TSP), which consists of finding the shortest closed path through a set of points. The TSP belongs to a class of problems that are computationally intractable, that is, problems for which no algorithm is known to exist that solves the problem in polynomial time; in other words, the shortest tour cannot necessarily be found during the lifetime of the individual (Graham et al., 2002).

The computational intractability is illustrated by the fact that the number of possible solutions is given by $(n\text{-}1)!/2$. Exhaustive search of a 15 node problem would take millions of years given a computer program that could find solutions at a rate of 1000 per second. However, despite the apparent intractability of this task, animals — both human and nonhuman — can find tours that are close to the best known solution quickly and with little apparent effort. For example, both chimps and vervet monkeys display search patterns for randomly placed hidden foods that were close to the shortest route (Menzel, 1973; Cramer and Gallistel, 1997, respectively).

MacGregor and colleagues (e.g., MacGregor and Ormerod, 1996; MacGregor et al., 2000, 2004) have investigated human performance on the TSP, and found that naïve adults reliably outperform simple computer-based algorithms. How do they do this? The authors argue that the solutions produced are the result of low-level neurocomputational processes that have evolved to cope with the requirements of vision (based on Gestalt principles like "good form" or "*Prägnanz*"), which in turn provide the underpinnings for a "convex-hull" heuristic.

Simon (1990) used the metaphor of a pair of scissors to illustrate his view of bounded rationality, according to which one blade represents the characteristics of the task environment, and the other blade represents the cognitive capabilities of the decision maker. He argued that to study computational capabilities of the decision maker without reference to the characteristics of the task environment would be like trying to use a pair of scissors to cut using only one blade.

Ecological rationality

The notion of ecological rationality brings environmental structure back into bounded rationality. Ecologically rational behavior refers to that which emerges when the structure of environmental information — be it physical, biological, social, or cultural — fits with the cognitive mechanisms of the organism (e.g., Todd and Gigerenzer, 2007). The main contentions of an ecological rationality perspective are that (1) the rationality of human decision-making behavior should be evaluated according to how successfully it performs in the real world, (2) the underlying inference and decision-making processes cannot be studied effectively without reference to the environment(s) in which they operate, and (3) adaptive, ecologically rational behavior emerges at the point where the mechanisms of the mind correspond to and exploit the structure of the environment.

According to the adaptive toolbox research program, the mind is viewed as containing a collection of specialized mechanisms, or heuristics, that evolution has built into the human mind, designed for the specific domains of inference and reasoning about the world (see also Cosmides and Tooby, 1996; Payne et al., 1993). This collection of mechanisms has been termed the "adaptive toolbox," a metaphor that captures the notion that different domains of thought require different specialized tools instead of one universal tool, and that these tools are psychological adaptations to the environment(s) in which they are applied. The research program takes ecological rationality as its normative framework to the extent that by designing computational models of cognitive mechanisms, it can specify what environments are appropriate for which mechanisms and vice versa. It provides a systematic investigation of human decision making in terms of simple heuristics, that is decision-making strategies that are fast, frugal, and accurate when the cognitive mechanisms can exploit particular informational structures in the environment (e.g., Conlin and Gigerenzer, 2009).

Indeed, it would appear plausible that if relevant environments reliably include certain informational resources, evolution may not duplicate them internally (cf. Brooks, 1999; Clark, 2001). The introductory example of the *Cataglyphis fortis* echoes a metaphor used by Simon in his seminal book entitled "*The sciences of the artificial*" (1969). In his "ant on the beach" metaphor, Simon describes an ant traveling across an irregular beach with a constantly changing surface due to the random action of waves and wind. The ant's trajectory across the beach is a highly complicated sequence of twists and turns, as the ant must negotiate all kinds of obstacles that stand in its way; a trajectory, as Simon argues, that is far too complex to be the result of the ant's cognition. Instead, the apparent complexity of the trajectory is the result of the *interaction of the ant with its environment*. Thus, the navigation behavior displayed by the ant is neither a property of the ant itself, nor a property of the environment alone. In the case of the TSP described above, it is the fit between the structure of information in the environment and the organism's evolved perceptual processes that can give rise to good solutions for computationally intractable problems.

The adaptive toolbox research program investigates the mechanisms that exploit the structure of such environments to provide quick, accurate, and computationally inexpensive solutions to guide everyday decision problems. Although is has long been assumed and borne out empirically that humans use heuristics in a variety of everyday tasks (e.g., see the comprehensive heuristics-and-biases program by Kahneman, Tversky, and colleagues; e.g., Tversky and Kahneman, 1974, 1983; Kahneman et al., 1982), the fast and frugal heuristics program takes issue with claims that heuristics are generally "second best" solutions, or that they are used because of cognitive limitations.

There is a fast growing body of experimental evidence that shows that these heuristics are successful, not merely because they provide "quick and dirty" solutions to tasks faced by limited capacity cognitive systems, but that their accuracy is often based on their simplicity (e.g., Gigerenzer and Goldstein, 1996; Gigerenzer et al., 1999; Goldstein and Gigerenzer, 2002; Martignon and Hoffrage, 2002; Johnson and Raab, 2003; Bröder and Schiffer, 2006; Hogarth and Karelaia, 2006; Katsikopoulos and Martignon, 2006). These findings support the idea that individuals do not always need complex internal representations to act successfully. This is presumably the case when the cognitive system can rely on the fact that relevant task environments contain certain informational resources.

This alternative view of rationality can also be applied to the social domain. In Chapter 9 of this volume (entitled: (Ir)rationality in action: do soccer players and goalkeepers fail to learn how to best perform during a penalty kick?), Bar-Eli, Azar and Lurie provide an insightful explanation of goalkeepers' and soccer players' behavior during penalty kicks in terms of social rationality: what may appear to be nonoptimal or irrational behavior can be reframed as rational behavior once the social environments of the actors, and the resulting social constraints such as the expectations of the onlooker's, are taken into account.

The role of internal representations and environmental structure in navigation

Cognitive maps

Psychological research into how people make navigation decisions or move from one place to another has very often focused on the role of internal spatial representations in guiding such behavior (e.g., Huttenlocher, Hedges, & Duncan, 1991; Wang and Spelke, 2002). Human spatial memory has been the focus of intense research interest in psychology for many decades, not least since Tolman (1948) published his seminal paper, in which he coined the phrase "cognitive map" to refer to an internal representation of the environment that indicates the routes, paths, and environmental relationships that an animal uses in making decisions about where to move. These maps were believed to underlie navigation performance in human and nonhuman animals, and, according to Tolman (1948), their essential feature was the ability to make such novel shortcuts between two points, because this ability suggests the existence of a robust representation of a spatial environment that allows for multiple routes to the same goal (e.g., if one has a representation of the route A to B, and another of the route A to C, then one should be able to generate the novel route B to C).

In more recent experimental paradigms, much of the research into human spatial memory has provided support for the idea that spatial representations are organized in a very particular way. The predominant view is that spatial memory contains nested levels of detail, such that representations of environmental space are organized hierarchically, in which representations of smaller-scale environments are stored at progressively lower levels (e.g., Hirtle and Jonides, 1985; McNamara, 1986; Wiener and Mallot, 2003). For example, one might hold in memory individual representations of a building, apartments within that building, rooms within the apartments, and so on.

According to one class of accounts, a consequence of this hierarchical structuring is that humans show reliable and systematic distortions when making inferences about spatial relationships. For example, in their influential paper, Stevens and Coupe (1978) demonstrated that people made systematic errors in judging geographical relations between two locations that were housed in different political "units." The researchers asked students in San Diego to indicate the compass direction between San Diego (in California, United States of America) and Reno (in Nevada, United States of America). The participants in their experiment judged Reno to be east of San Diego, even though it is actually further west, an erroneous judgment that resulted — according to Stevens and Coupe (1978) — from the participants' knowledge that Nevada is east of California. Other studies have shown that people underestimate within-region distances in spatial judgments (e.g., McNamara, 1986), or exaggerate distances between objects separated by barriers, both physical (e.g., Thorndyke, 1981; Newcombe and Liben, 1982), and psychological (Carbon and Leder, 2005), in small-scale, local environments (McNamara, 1986) as well as large-scale, global environments (Friedman and Montello, 2006).

But how meaningful is it to interpret such effects in terms of "systematic errors"? Indeed, while the structure, content, and form of spatial memory have been studied extensively, there is little emphasis in the psychological literature on the role of the environmental structure in shaping spatial judgments and navigation.

Behaviorally relevant properties of space

Spatial environments that humans and animals navigate in can be characterized in various ways, for example, in terms of the distribution of spatial information, or the stability and reliability of various cues (e.g., landmarks) available for navigation. Biologists and mathematical ecologists, for example, have pointed out that spatial distributions in the environment tend to be clumped or clustered, rather than uniform or random (e.g., Pielou, 1977; Hubbell, 1979), because living organisms grow, reproduce, and disperse in spatially autocorrelated ways (Hills, 2006).

Both natural and modern environments often show predictable kinds of distribution of spatial stimuli and (informational) resources in the environment. For example, the majority of stream organisms tend to have an aggregated or clumped distribution (Hildrew and Townsend, 1976), and urination by ungulates creates clusters of particular species of grasses and shrubs, creating preferential grazing sites for native herbivores (Day and Detling, 1990). A particular type of spatial foraging heuristic known as area-restricted search (ARS) capitalizes on exactly such distributions. When foraging animals encounter patchily distributed food items, many alter their search path in a way that limits the search area (e.g., Benedix, 1993), increasing foraging efficiency.

Similarly in modern, urban environments, one can rely on certain spatial distributions of resources: for example, gas stations tend to be located on corners, and automatic telling machines (ATM) tend to be clustered in shopping districts, rather than randomly or uniformly distributed across a city.

Thus, it appears that the cognitive system need not explicitly represent the spatial environment in each and every detail; rather, it might instead exploit behaviorally relevant properties of space, many of which are likely to be stable (such as the large-scale geometry of landscapes, or particular resource distributions), resulting in spatial representations based on regularities in uncertain environments.

Representation-free behavior

But are internal representations necessary at all to explain and predict movement behavior? Taking a somewhat radical view, cognitive scientist Rodney Brooks claimed, "*...there need be no explicit representation of either the world or the intentions of the system to generate intelligent behaviors...*" (Brooks, 1999, p. 90). That is, intelligence, as expressed by complex behavior, "emerges" from the interaction of a few simple behaviors, rather than complex representations. For example, a robot whose simple behaviors include collision avoidance and motion toward a moving object will appear to stalk the object, pausing whenever it gets too close. One famous example is Brooks' robot Herbert, whose task was to search desks and tables for empty soda cans in a crowded laboratory. The robot's seemingly goal-directed spatial behavior emerged from the interaction of about 15 simple behavioral routines, rather than powerful sensing capacities and detailed advanced planning. That is, Herbert required no detailed inner representation or model of the environment to perform the task successfully.

Further support for this view comes from the movement of the bluefish tuna (Clark, 1998). According to its physiology alone, this fish is physically incapable (too weak, in fact, by a factor of seven) to swim as fast as it does. Fluid dynamicists Triantafyllou and Triantafyllou (1995) suggest that the paradoxical speeds the tuna attains can be explained by the way the fish actively creates and exploits its natural environment. That is, the fish uses naturally occurring swirls and eddies to increase its speed, and uses its tail to create additional vortices and pressure gradients, which can then be exploited for acceleration and so on. One might also consider soaring birds, who take advantage of updrafts generated when a steady wind strikes a hill or cliff, and can thereby maintain or increase their altitude without flapping their wings and wasting precious energy resources.

Patterns of movement behavior

Under certain conditions, human pedestrian flows can form collective patterns of motion, much like fluids or streams (Helbing et al., 2005). Following this view, certain aspects of human movement behavior – analogous to the tuna's swimming movement – can be explained solely in terms of physical environmental constraints, with no reference to internal representations or individual cognitive processes. Research on pedestrian crowd dynamics, for example, has shown that zigzag-shaped geometries can reduce the pushing-related buildup of pressure in panicking crowds (Helbing et al., 2005). Similarly, strategically placed "obstacles" can stabilize flow patterns and make them more fluid by acting in a way similar to wave breakers, absorbing pressure in a crowd buildup (Escobar and de la Rosa, 2003).

Such studies typically investigate how much of human movement behavior can be predicted and explained "for free" – that is, without invoking the need for particular cognitive representations or individual goals – on the basis of space-related constraints. One method of formally describing and analyzing environments in order to predict movement behavior comes from a line of research known as space syntax analysis (e.g., Hillier and Hanson, 1984). This type of analysis typically represents and quantifies characteristics of built environments by redescribing continuous space as a connected set of discrete units. From this it is possible to describe the "intelligibility" of a space using maps and graphs, that is, the ease of navigability of particular environments in terms of the correlation between global and local properties of those environments (e.g., Penn et al., 1998). Bafna (2003), for example, describes how a regular grid map like Manhattan shows very *low* intelligibility, whereas a labyrinthine street network found in a traditional market town, such as Mytilini in Greece, shows high intelligibility (Peponis et al., 1989). This appears surprising given the apparent simplicity of Manhattan's grid structure when viewed as a map, but Bafna (2003) goes on to report that participants actually navigate a market town more efficiently, and that, conversely, orientation in a grid network is extremely difficult in the absence of additional information such as numbered streets.

It appears, then, that a formal description of the spatial properties of environments is useful in predicting human movement behavior. In general though, the space syntax theory and analysis literature is silent about how or whether the cognitive system represents space. One possibility, then, is that the notion of enduring internal cognitive representations, like those presumed to be held in a cognitive map, is not necessary at all. There is undoubtedly much value in the idea that the form of an organism's embodiment constrains the way it interacts with the environment, and empirical evidence from a navigation task supports the crucial role for body-based information from full physical movement during navigational search (Ruddle and Lessels, 2006). However, spatial thinking about large-scale environments is necessarily an "off-line" cognitive activity, and one that requires the ability to generate and manipulate mental representations of spaces we cannot currently perceive.

Thus, while most real-world behavior takes place in complex environments and can exploit the possibility of interaction with features of the external world, it is implausible to consider spatial thinking as being completely free of cognitive representations. The following section describes in more detail the fast-and-frugal heuristics research program that investigates ecologically rational behavior as arising from the fit between cognition and environment.

Bringing cognition and environment together

As outlined earlier, fast and frugal decisions are generally based on mechanisms that arise from the interaction between a decision maker and his/her environment(s). This casts doubt on the idea that accurate navigation depends on internal, enduring representations (such as a cognitive map) of those environments. Rather, the adaptive toolbox approach would favor a view according to which navigation decisions are made on the basis of mechanisms arising from transient, dynamic representations that are formed in relation to the current environmental conditions, much like Simon's "ant on the beach." This perspective is echoed in comparative research into human and animal navigation: Wang and Spelke (2002, p. 376) report a review of research that provides evidence that human nonhuman animals navigate using representations that are "momentary rather than enduring, egocentric rather than geocentric, and limited in the environmental information they capture." According to the authors, humans navigate by establishing a set of target locations in the immediate environment and continuously computing the positions of these targets relative to themselves as they move, using various internal and external perceptual cues. It would seem, therefore, that navigation decisions need not internally represent all available environmental information; in fact, once a spatial cue (e.g., a landmark) has fulfilled its orientation purpose, it would seem maladaptive to continue to store this information internally.

To date, empirical investigations into to how people make navigation decisions given limited time, computational power, and information using a the fast-and-frugal heuristics approach remain sparse. However, given the nature of navigation and large-scale environment movement decisions that must be made by human and nonhuman animals on a regular basis, the fast-and-frugal heuristics approach appears to be a promising route in modeling actual behavior in the real world. For example, recent research suggests that for animals that forage or travel in groups, making movement decisions often depends on social interactions among group members (Couzin et al., 2005). It has been suggested that structures in and from social environments are among the most important in decision making (Todd and Gigerenzer, 2007). A fast-and-frugal heuristics approach provides a useful starting point for the investigation of group movement and navigation, to address the question, for example, about the cues the individuals rely on in social interactions to make decisions. This approach may also be useful in describing as yet unspecified heuristics, such as whether group travel decisions are made using a "follow the leader" or "follow the majority" heuristic.

Eccles et al. (2002) have investigated the use of heuristics in orienteering. This research suggests the use of different heuristics by individuals, depending on whether they were novice or expert orienteers. Further application of a fast-and-frugal heuristics approach to orienteering might attempt to determine such heuristics' ecological rationality; that is, the fit between the particular environmental structure and the structure of the heuristic.

Conclusion

The main contentions of the fast-and-frugal heuristics approach — namely that decision making is often (although not always) the result of simple mechanisms that exploit the structures of the environment — demand a strong focus not merely on the cognitive system, but also on the environment in which the cognitive system acts. This focus is addressed by the concept of ecological rationality. Importantly, the normative question asked by an ecological rationality approach is to specify the environments in which heuristics are successful. A comprehensive understanding of navigation decisions and movement behavior will require insights from a variety of disciplines, such as psychology, geography, artificial intelligence, and behavioral ecology, not least in order to import meaningful ways to describe psychologically relevant aspects of spatial structure.

References

Bafna, S. (2003). Space syntax: a brief introduction to its logic and analytical techniques. *Environment and Behavior, 35*, 17–29.

Benedix, J. H. (1993). Area-restricted search by the plains pocket gopher (*Geomys bursarius*) in tallgrass prairie habitat. *Behavioral Ecology, 4*, 318–324.

Bröder, A., & Schiffer, S. (2006). Stimulus format and working memory in fast and frugal strategy selection. *Journal of Behavioral Decision Making, 19*, 361–380.

Brooks, R. (1999). *Cambrian intelligence: the early history of the new AI*. Cambridge, MA: MIT Press.

Carbon, C.-C., & Leder, H. (2005). The wall inside the brain: overestimation of distances crossing the former Iron Curtain. *Psychonomic Bulletin & Review, 12*, 746–750.

Clark, A. (1998). Where brain, body, and world collide. *Journal of the American Academy of Arts and Sciences, 127*, 257–280.

Clark, A. (2001). Reasons, robots and the extended mind. *Mind & Language, 16*, 121–145.

Conlin, J. A. & Gigerenzer, G. (2009). The ins and outs of ecological rationality. Manuscript in preparation.

Cosmides, L., & Tooby, J. (1996). Are humans good intuitive statisticians after all? Rethinking some conclusions from the literature on judgment under uncertainty. *Cognition, 58*, 1–73.

Couzin, I. D., Krause, J., Franks, N. R., & Levin, S. A. (2005). Effective leadership and decision-making in animal groups on the move. *Nature, 433*, 513–516.

Cramer, A. E., & Gallistel, C. R. (1997). Vervet monkeys as travelling salesmen. *Nature, 387*, 464.

Day, T. A., & Detling, J. K. (1990). Grassland patch dynamics and herbivore grazing preference following urine deposition. *Ecology, 71*, 180–188.

Eccles, D. W., Walsh, S. E., & Ingledew, D. K. (2002). The use of heuristics during route planning by expert and novice orienteers. *Journal of Sports Science, 20*, 327–337.

Edwards, W., & Fasolo, B. (2001). Decision technology. *Annual Review of Psychology, 52*, 581–606.

Escobar, R. A., & de la Rosa, A. (2003). Architectural design for the survival optimization of panicking fleeing victims. In *Proceedings of the 7th European Conference on Artificial Life (ECAL)* (14–17 September), Dortmund, Germany.

Friedman, A., & Montello, D. R. (2006). Global-scale location and distance estimates: Common representations and strategies in absolute and relative judgments. *Journal of Experimental Psychology — Learning Memory and Cognition, 32*, 333–346.

Gigerenzer, G. (2004). Fast and frugal heuristics: the tools of bounded rationality. In D. J. Koehler & N. Harvey (Eds.), *Blackwell handbook of judgment and decision making* (pp. 62–88). Oxford: Blackwell.

Gigerenzer, G., & Goldstein, D. G. (1996). Reasoning the fast and frugal way: models of bounded rationality. *Psychological Review, 103*, 650–669.

Gigerenzer, G., Todd, P. M., & *the ABC Research Group*. (1999). *Simple heuristics that make us smart*. Oxford: Oxford University Press.

Goldstein, D. G., & Gigerenzer, G. (2002). Models of ecological rationality: the recognition heuristic. *Psychological Review*, *109*, 75–90.

Graham, S. M., Joshi, A., & Pizlo, Z. (2002). The traveling salesman problem: a hierarchical model. *Memory & Cognition*, *28*, 1191–1204.

Helbing, D., Buzna, L., Johansson, A., & Werner, T. (2005). Self-organized pedestrian crowd dynamics: experiments, simulations, and design solutions. *Transportation Science*, *39*, 1–24.

Hildrew, A. G., & Townsend, C. R. (1976). The distribution of two predators and their prey in an iron rich stream. *Journal of Animal Ecology*, *45*, 41–57.

Hillier, B., & Hanson, J. (1984). *The social logic of space*. Cambridge: Cambridge University Press.

Hills, T. T. (2006). Animal foraging and the evolution of goal-directed cognition. *Cognitive Science*, *30*, 3–41.

Hirtle, S. C., & Jonides, J. (1985). Evidence of hierarchies in cognitive maps. *Memory & Cognition*, *13*, 208–217.

Hogarth, R. M., & Karelaia, N. (2006). "Take-the-best" and other simple strategies: why and when they work "well" with binary cues. *Theory and Decision*, *61*, 205–249.

Hubbell, S. P. (1979). Tree dispersion, abundance, and diversity in a tropical dry forest. *Science*, *203*, 1299–1309.

Huttenlocher, J., Hedges, L. V., & Duncan, S. (1991). Categories and particulars: prototype effects in estimating spatial location. *Psychological Review*, *98*, 352–376.

Johnson, J. G., & Raab, M. (2003). Take the first: option-generation and resulting choices. *Organizational Behavior and Human Decision Processes*, *91*, 215–229.

Kahneman, D., Slovic, P., & Tversky, A. (1982). *Judgment under uncertainty: heuristics and biases*. Cambridge: Cambridge University Press.

Katsikopoulos, K., & Martignon, L. (2006). Naïve heuristics for paired comparisons: some results on their relative accuracy. *Journal of Mathematical Psychology*, *50*, 488–494.

MacGregor, J. N., Chronicle, E. P., & Ormerod, T. C. (2004). Convex hull crossing avoidance? Solution heuristics in the travelling salesperson problem. *Memory & Cognition*, *32*, 260–270.

MacGregor, J. N., & Ormerod, T. C. (1996). Human performance on the travelling salesman problem. *Perception & Psychophysics*, *58*, 527–539.

MacGregor, J. N., Ormerod, T. C., & Chronicle, E. P. (2000). A model of human performance on the travelling salesperson problem. *Memory & Cognition*, *28*, 1183–1190.

Martignon, L., & Hoffrage, U. (2002). Fast, frugal and fit: simple heuristics for paired comparison. *Theory and Decision*, *52*, 29–71.

McNamara, T. P. (1986). Mental representations of spatial relations. *Cognitive Psychology*, *18*, 87–121.

Menzel, E. (1973). Chimpanzee spatial memory organization. *Science*, *182*, 943–945.

Newcombe, N., & Liben, L. S. (1982). Barrier effects in the cognitive maps of children and adults. *Journal of Experimental Child Psychology*, *34*, 46–58.

Payne, J. W., Bettman, J. R., & Johnson, E. J. (1993). *The adaptive decision maker*. New York: Cambridge University Press.

Penn, A., Hillier, B., Banister, D., & Xu, J. (1998). Configurational modeling of urban movement networks. *Environment and Planning B: Planning and Design*, *25*, 59–84.

Peponis, J., Hajinikoloaou, E., Livieratos, C., & Fatouros, D. A. (1989). The spatial core of urban culture. *Ekistics*, *334*, 43–55.

Pielou, E. C. (1977). *Mathematical ecology*. New York: Wiley.

Ruddle, R. A., & Lessels, S. (2006). For efficient navigational search, humans require full physical movement, but not a rich visual scene. *Psychological Science*, *17*, 460–465.

Simon, H. A. (1956). Rational choice and the structure of environments. *Psychological Science*, *63*, 129–138.

Simon, H. A. (1969). *The sciences of the artificial*. Cambridge, MA: MIT Press.

Simon, H. A. (1990). Invariants of human behavior. *Annual Review of Psychology*, *41*, 1–19.

Stevens, A., & Coupe, P. (1978). Distortions in judged spatial relations. *Cognitive Psychology*, *10*, 422–437.

Thorndyke, P. (1981). Distance estimation from cognitive maps. *Cognitive Psychology*, *13*, 526–550.

Todd, P. M., & Gigerenzer, G. (2007). Environments that make us smart: ecological rationality. *Current Directions in Psychological Science*, *16*, 167–171.

Tolman, E. C. (1948). Cognitive maps in rats and men. *Psychological Review*, *55*, 189–208.

Triantafyllou, M., & Triantafyllou, G. (1995). An efficient swimming machine. *Scientific American*, *272*, 64–71.

Tversky, A., & Kahneman, D. (1974). Judgment under uncertainty: heuristics and biases. *Science*, *185*, 1124–1131.

Tversky, A., & Kahneman, D. (1983). Extensional versus intuitive reasoning: the conjunction fallacy in probability judgment. *Psychological Review*, *90*, 293–315.

Wang, R. F., & Spelke, E. S. (2002). Human spatial representations: insights from animals. *Trends in Cognitive Sciences*, *6*, 376–382.

Wehner, R. (2003). Desert ant navigation: how miniature brains solve complex tasks. *Journal of Comparative Physiology A*, *189*, 579–588.

Wiener, J., & Mallot, H. (2003). Fine-to-coarse route planning and navigation in regionalized environments. *Spatial Cognition and Computation*, *3*, 331–358.

Ziegler, P. E., & Wehner, R. (1997). Time-courses of memory decay in vector-based and landmark-based systems of navigation in desert ants *Cataglyphis fortis*. *Journal of Comparative Physiology A*, *181*, 13–20.

M. Raab et al. (Eds.)
Progress in Brain Research, Vol. 174
ISSN 0079-6123

CHAPTER 11

Crossmodal interaction in speeded responses: time window of integration model

Adele Diederich[1,*] and Hans Colonius[2]

[1]School of Humanities and Social Sciences, Jacobs University, Bremen, Germany
[2]Department of Psychology, Oldenburg University, Oldenburg, Germany

Abstract: Saccadic reaction time (SRT) to a visual stimulus tends to be faster when an auditory and/or somatosensory stimulus is presented in close temporal or spatial proximity, even when participants are instructed to ignore the accessory input (focused attention task). The time course of SRT as a function of stimulus onset asynchrony (SOA) is consistent with the time-window-of-integration (TWIN) model assuming a peripheral stage of parallel processing in separate sensory channels followed by a secondary stage of multisensory integration. TWIN has been shown to account for effects of the spatial configuration of the stimuli, for the effect of increasing the number of nontargets presented together with the target, for a possible warning effect of the nontarget, for effects of increasing the intensity of the nontarget, and for the effect of background noise on multisensory integration. Moreover, it has been able to accommodate some effects of aging on multisensory integration.

There is empirical support for TWIN's tenet of the separability between spatial and temporal factors on multisensory integration. Besides presenting many features of TWIN within the context of crossmodal interaction modeling efforts, some possible directions on how the TWIN framework could serve to elucidate the link between perception and action are shown.

Keywords: multisensory integration; warning effect; time-window-of-integration; response time; saccadic eye movement

Introduction

Adaptive behavior depends on the ability of the perceptual system to deliver information about ongoing events in the environment rapidly. This information typically arrives via different sensory channels and has to be integrated to produce a coherent internal representation of the outside world. Our daily experience provides many examples for successful multisensory integration. We identify objects by visual information such as shape, color, or size but also by auditory or tactile information, for example a singing bird, a fire engine, velvet. A simple act like drinking sweet hot green tea from a Styrofoam cup involves integrating information from various modalities such as vision, taste, olfaction, and touch. Sensory information arriving from various modalities needs to be processed in parallel for adequate action.

Action execution also depends on crossmodal integration. For instance, kicking a ball requires

*Corresponding author.
Tel.: +49-421-200-3431; Fax: +49-421-200-3303;
E-mail: a.diederich@jacobs-university.de

DOI: 10.1016/S0079-6123(09)01311-9

the integration of visual, proprioceptive, and tactile information, and performance may critically depend on how well this information is integrated across the different reference frames. In keeping with the ball game example, motion of the body in physical space, or even a simple movement of the head, will in turn modify the incoming crossmodal stream of information requiring yet another crossmodal integration process. Obviously, action execution and crossmodal integration are closely intertwined components of a bidirectional dynamic process involving in an organism adapting to the environment.

Studying multisensory integration is an emerging field in neuroscience strongly influencing behavioral scientists traditionally investigating crossmodal interaction. Neurophysiologists are predominantly interested in where and how different sensory modalities are integrated in the brain. For example, they found multisensory neurons in various animals, for example, in the superior colliculus (SC) of the cat. The neurons in the deep layers of the superior colliculus (DLSC) are important in the control of eye movements and, in general, for directing exogenous attention. In fact, a large majority of multisensory neurons in cat DLSC show an enhanced response to particular combinations of visual, auditory, and tactile stimuli relative to the best modality-specific responses. For instance, Meredith and Stein (1986) found that a visual stimulus evoked responses on only six of 16 trials in a bimodal neuron and these responses were composed of few impulses. An auditory stimulus also evoked weak and unreliable responses. However, their combination produced a vigorous response on every trial. The mean number of impulses was increased by up to 1207% compared to the most effective unimodal stimulus. Information about stimulus location is represented topographically within the structure by an orderly arrangement of neurons according to the location of their respective receptive fields (RFs). The spatial register among the different sensory maps is formed by the multisensory neurons whose different RFs are in register with one another (for a review see Stein and Meredith, 1993). In addition, the SC contains a motor map composed of output neurons coding appropriate eye movements (Sparks, 1986), that is, the locus of activity in the motor map encodes a movement command that reorients the eyes (and the head) a given distance in a particular direction. Thus, if the goal is to translate a sensory target into an appropriate motor command, the alignment of the visual, auditory, and tactile maps to each other and to the motor map is critical.

In the behavioral sciences and, in particular, in the psychological literature, numerous effects of crossmodal stimulation on perception and action have been described over more than a century (cf. Welch and Warren, 1986). Prominent examples are the McGurk effect (McGurk and MacDonald, 1976), the ventriloquism effect (Howard and Templeton, 1966; for the effect in humans and monkeys, e.g., Wood and Recanzone, 2004), synesthesia, and many more (see Calvert et al., 2004, for an overview). While many phenomena, like synesthetic experiences, can only be described at a qualitative level, other crossmodal effects are amenable to standard psychophysical measurement techniques.

One such measure is response time (RT). It is a ubiquitous measure used in experimental psychology for more than 150 years to investigate hypotheses about the mental and motor processes leading to the generation of a response. Thus, it comes as no surprise that some of the first psychological studies of intersensory interaction employed RT as a measure to assess the effect of combining stimuli from different modalities and of varying their intensities (Todd, 1912). RT analysis becomes most powerful in uncovering underlying processes when it is based on explicit quantitative hypotheses (Luce, 1986). One simple reason for the preponderance of response time as a measure of crossmodal effects is that, whatever happens within an organism processing stimulus information from several sensory modalities, it must unfold over time. One of the main experimental findings in behavioral studies is that the occurrence of crossmodal effects critically depends on the temporal arrangement of the stimulus sets. For example, the speed-up of the RT to a visual stimulus resulting from presenting, say, an accessory auditory stimulus typically becomes greatest when the visual stimulus precedes the auditory by an interval that equals the difference in RT between response to the visual stimulus alone and the auditory stimulus alone (Hershenson, 1962). For instance, assuming a mean RT to a visual stimulus of 160 ms and to an auditory of 120 ms, the largest amount of facilitation is expected when the visual stimulus precedes the auditory by 160 ms − 120 ms = 40 ms. Interestingly, temporal integration rules similar to this one have been discerned for the responses at the level of individual multisensory neurons as well. Specifically, bimodal cells in cat superior colliculus (SC) show maximal levels of response enhancement when the peak discharge periods evoked by each modality overlap in time (Meredith et al., 1987). This overlap is often correlated not with simultaneous stimulus presentation but with differences in stimulus presentation that match their latency differences. Since multisensory SC neurons project to premotor areas of the brainstem that control orientation of the eyes, pinnae, and head, the temporal relationships of stimulus complexes that control the activity of these neurons should ultimately contribute to orientation behavior as well. Of course, individual neurons differ in their temporal response properties, and behavior is the result of a large number of neurons at different sites of the brain. Nonetheless, it is a distinct possibility that the temporal integration rules observed in behavioral experiments can ultimately be deduced from certain principles holding at the level of the individual neuron or of an assembly of neurons.

In the following, we describe mathematical models of RT that have been developed to explain and to predict crossmodal stimulus effects resulting from a manipulation of their spatial-temporal configuration and of physical stimulus dimensions like intensity. One feature of the models is that they are based on a few, relatively simple principles — if not mathematically, then at least conceptually — so that they certainly do not reflect the full complexity of the underlying multisensory processes. Note that this should be considered a virtue rather than a defect at this stage: these models can be tested rigorously and, hopefully, converge on those principles that constitute a valid framework for a more detailed account to be developed as more evidence is collected.

Redundant target versus focused attention paradigm

Two different experimental paradigms have been utilized to measure RT to a crossmodal stimulus set. In the *redundant target* (also known as *divided-attention*) paradigm, stimuli from different modalities (e.g., a tone and a light) are presented simultaneously or with certain stimulus onset asynchrony (SOA), and the participant is instructed to respond to the stimulus detected first. Typically, the time to respond in the crossmodal condition (e.g., a tone plus a light) is faster than in either of the unimodal conditions (e.g., a tone or a light). In the *focused attention* paradigm, crossmodal stimulus sets are presented in the same manner but now participants are instructed to respond only to the onset of a stimulus from a specifically defined target modality (e.g., such as the visual), and to ignore the remaining nontarget stimulus (e.g., the tactile or the auditory). In the latter setting, when a stimulus of a nontarget modality (e.g., a tone), appears before the visual target at some spatial disparity, there is no *overt* response to the tone if the participant is following the task instructions. Nevertheless, the nontarget stimulus has been shown to modulate the response to the target stimulus: depending on the exact spatiotemporal configuration of target and nontarget, the effect can be a speed-up or an inhibition of RT (for an overview see, for example, Diederich and Colonius, 2004a, b).

Note that the distinction between the redundant target and the focused attention paradigm is not only an interesting experimental variation as such, but it may also provide an important theoretical perspective. In fact, since the stimuli can be chosen to be physically identical in both paradigms, any differences observed in the corresponding responses would have to be due to the instructions only, thereby giving the researcher an opportunity to separate top-down from bottom-up processes in the multisensory integration mechanism.

Separate activation versus coactivation models

Usually, observable RT is divided into two additive components, that is,

$$RT = S + R \quad (1)$$

where the random variable S denotes stimulus processing time proper, and R is base time random variable denoting response preparation, motor delay, etc. Here, RT will mostly be considered at the level of expectation only, instead of the entire probability distribution. This simplifies the discussion significantly; in particular, the expected value of R, $E[R]$, can be treated as a constant which, when added, often does not modify the main results of interest and which will, therefore, often not be mentioned explicitly. Numerous models have been suggested to account for the crossmodal effects in RTs. Two different approaches are commonly distinguished, *separate activation* versus *coactivation* models. Separate activation models, also known as *race models*, assume (1) that presenting a crossmodal stimulus produces parallel, separate activation in different sensory channels that build to the level at which they can produce a response, and (2) that the response is triggered by the signal that reaches that level first. Assuming statistical variability in the channel processing times, separate activation models predict faster average RT to crossmodal stimuli than to unimodal stimuli because the average of the winner's processing time is smaller than the average processing time in each single channel ("statistical facilitation", cf. Raab, 1962).

For example, let T, V, and A denote the random processing time for a tactile, visual, and auditory stimulus, respectively. Then,

$$E[\min(T, V, A)], \leq \min(E[T], E[V], E[A]) \quad (2)$$

where $E[\ ]$ indicates the expected value of the random variables (*Jensen's inequality*, cf. Billingsley, 1979). Thus, statistical facilitation predicts faster mean responses to crossmodal stimuli as a statistical phenomenon without assuming an additional neural mechanism.

A more general test of separate activation models proposed by Miller (1982) has become the standard tool to assess whether statistical facilitation suffices to explain the response speed-up of crossmodal stimulus presentations. For three modalities, the test is based on the inequality

$$P(\min\{T, V, A\} \leq t) \leq P(T \leq t) + P(V \leq t) + P(A \leq t) \quad (3)$$

with the same notation as above. This inequality puts an upper bound on the facilitation produced by trimodal stimuli. However, since it is the sum of three probabilities approaching a value of three for increasing values of t, this bound may not always be of use in testing separate activation. Interestingly, the upper bound can be sharpened in various ways (cf. Diederich, 1992). In any event, Miller's test has often been shown to be violated in crossmodal experiments with two modalities, and the amount of violation of the inequality has been interpreted as a measure of the strength of neural coactivation (e.g., see Colonius and Diederich, 2006).

As an alternative approach, *coactivation models* assume that activation raised in different sensory channels by presenting crossmodal stimuli is combined to satisfy a single criterion for response initiation. Coactivation models predict faster average RT to multiple stimuli compared to single stimuli because the combined activation reaches that criterion faster. Assume that presentation of a stimulus triggers a sequence of "events" occurring randomly over time. In a neurophysiological context, these events are typically interpreted as spikings of a neuron, for example, but the model can be formulated at a more abstract level (cf. Tuckwell, 1989). The only relevant property of the events is their time of occurrence, and all information about the stimulus is contained in the time course of the events. For example, the rate of the event sequence, that is, the mean number of events per unit time interval, is typically thought to be related to signal intensity. Let $N(t)$ denote the number of events that have occurred by time t after stimulus presentation. *Counter models* assume that $N(t)$ has some internal representation registering the number of events over the course of time.

interaction $\Pr[I]$ and the amount and sign of interaction (Δ). According to the assumptions, the first factor depends on the temporal configuration of the stimuli (SOA), whereas the second factor depends on their spatial configuration. Note that this *separation of temporal and spatial factors* is in accordance with the definition of the window of integration: the incidence of multisensory integration hinges upon the stimuli to occur in temporal closeness, whereas the amount and sign of interaction (Δ) is modulated by spatial proximity, reaching from enhancement for neighboring stimuli to possible inhibition for distant stimuli.

Predictions

The TWIN model makes a number of empirical predictions. First, the amount of crossmodal interaction should depend on the SOA between the stimuli. Indeed, the effect tends to be most prominent when there is some characteristic temporal asynchrony between the stimuli (Frens et al., 1995; Colonius and Arndt, 2001). Within the model, this simply means that a stimulus with faster peripheral processing has to be delayed (in terms of SOA) in such a way that the arrival times of both stimuli have a higher probability of falling into the window of integration.

Second, the probability of interaction, $\Pr[I]$, should depend on unimodal features that affect the speed of processing in the first stage, such as stimulus intensity or eccentricity. For example, if a stimulus from one modality is very strong compared to the intensity of the other stimulus, the peripheral processing speed of the former stimulus will increase, and the chances that both peripheral processes terminate within the time window will be smaller (assuming simultaneous stimulus presentations). The resulting low value of $\Pr[I]$ is in line with the empirical observation that a very strong target signal will effectively rule out any further reduction of RT by adding a stimulus from another modality (e.g., Corneil et al., 2002).

Finally, the amount of multisensory integration (Δ) and its direction (facilitation or inhibition) occurring in the second stage depend on crossmodal features of the stimulus set, in particular spatial disparity and laterality (laterality here refers to whether or not all stimuli appear in the same hemisphere). Crossmodal features cannot have an influence on first-stage processing time since the modalities are yet being processed in separate pathways.

Integration rule assumptions

More specific predictions require an explication of the rules governing the window-of-integration mechanism in specific task requirements.

(5) *Focused attention task:* when the task is to orient toward the target stimulus ignoring stimuli from other modalities, the first stage terminates when the target peripheral process ends. Multisensory integration occurs only if the nontarget stimulus wins the race in the first stage opening a time window such that the termination of the target peripheral process is enclosed in the window.

In other words, in the focused attention situation, the window of integration is "opened" only by activity triggered by the nontarget stimulus, but first-stage duration is determined by the target stimulus processing time regardless of the identity of the winner of the race.

(6) *Redundant target task:* when the task is to orient toward the first stimulus detected regardless of modality, the first-stage duration is defined by the winner's peripheral processing time, and the window of integration is opened by whichever stimulus wins the race.

From these assumptions, further predictions concerning the effects of varying stimulus intensity follow. Take, for example, a focused attention task with a visual target and an auditory nontarget stimulus. Increasing the intensity of the visual stimulus will speed up visual peripheral processing (up to some minimum level) thereby increasing the chances for the visual target to win the race. Thus, the probability that the window of integration opens decreases, predicting less multisensory integration. Increasing the intensity of the nontarget auditory stimulus, in contrast, leads to the opposite prediction: the auditory stimulus will have a better chance to win the race and to open the window of integration, hence predicting more multisensory integration to occur on average.

Deriving the probability of interaction in TWIN

The race in the first stage of the model is made explicit by assigning independent non-negative random variables V and A to the peripheral processing times for the visual target and, for example, an auditory nontarget stimulus, respectively. With τ as SOA value and ω_I as integration window width parameter, the time window of integration assumption in the case of a focused attention experiment is equivalent to the (stochastic) event I, for example,

$$I = \{A + \tau < V < A + \tau + \omega\}$$

Thus, the probability of integration to occur, $P(I)$, is a function of both τ and ω, and it can be determined numerically once the distribution functions of A and V have been specified.

Exponential distribution are postulated, for simplicity, for the peripheral processing time V for a visual target and A for an auditory nontarget with parameters λ_V and λ_A, respectively. That is,

$$f_V(t) = \lambda_V e^{-\lambda_V t}$$

$$f_A(t) = \lambda_A e^{-\lambda_A t}$$

for $t \geq 0$, and $f_V(t) = f_A(t) \equiv 0$ for $t < 0$. The corresponding distribution functions are referred to by $F_V(t)$ and $F_A(t)$.

In order to compute

$$\begin{aligned} P(I) &= Pr(A + \tau < V < A + \tau + \omega) \\ &= \int_0^\infty f_A(a)\{F_V(a + \tau + \omega) - F_V(a + \tau)\}da \end{aligned}$$

it is necessary to distinguish three cases for the magnitude of $\tau + \omega$ resulting in the following expressions:

(i) $\tau < \tau + \omega < 0$

$$P(I) = \frac{\lambda_V}{\lambda_V + \lambda_A} e^{\lambda_A \tau}\left(-1 + e^{\lambda_A \omega}\right)$$

(ii) $\tau < 0 < \tau + \omega$

$$P(I) = \frac{1}{\lambda_V + \lambda_A}\left\{\lambda_A\left(1 - e^{-\lambda_V(\omega+\tau)}\right) + \lambda_V\left(1 - e^{\lambda_A \tau}\right)\right\}$$

(iii) $0 < \tau < \tau + \omega$

$$P(I) = \frac{\lambda_A}{\lambda_V + \lambda_A}\left\{e^{-\lambda_V \tau} - e^{-\lambda_V(\omega+\tau)}\right\}$$

Figures 2 and 3 depict the probability of integration as a function of the parameters λ and ω, respectively.

In the case of a redundant target experiment with both a visual and an auditory target, for example, the probability of integration is defined by both target processes finishing within the time window, that is,

$$I = \{\max(V, A + \tau) < \min(V, A + \tau) + \omega\},$$

and expressions similar to the above can be derived under the exponential assumption. More elaborate assumptions can be introduced with more than two target modalities.

Extending the TWIN model

The basic TWIN model can be extended in various ways. We consider two directions. First, in the focused attention case, empirical observations suggest that the nontarget may play an additional role as a warning cue modulating the RT under certain conditions. Second, the mechanism operating at the second processing stage determining the amount multisensory integration may be specified.

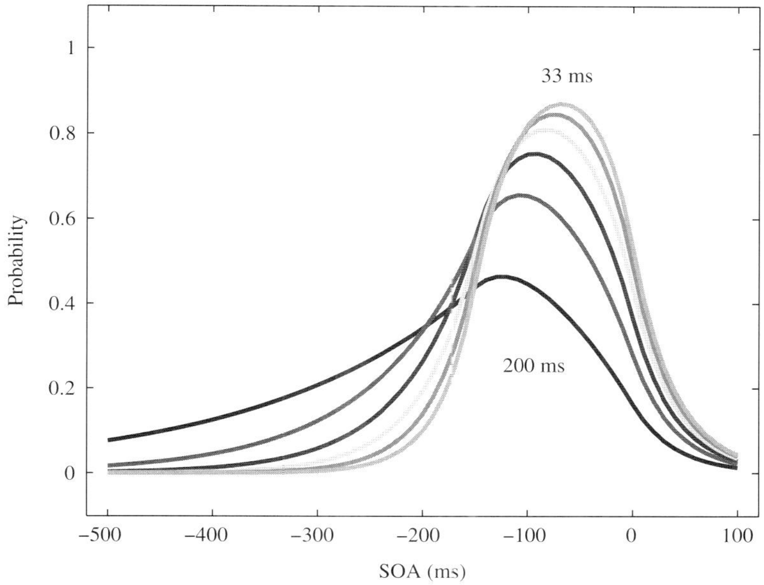

Fig. 2. Probability of integration as a function of the intensity parameter $\lambda_{nontarget}$ of the nontarget, ranging from $1/\lambda = 200$ ms to $1/\lambda = 33$ ms.

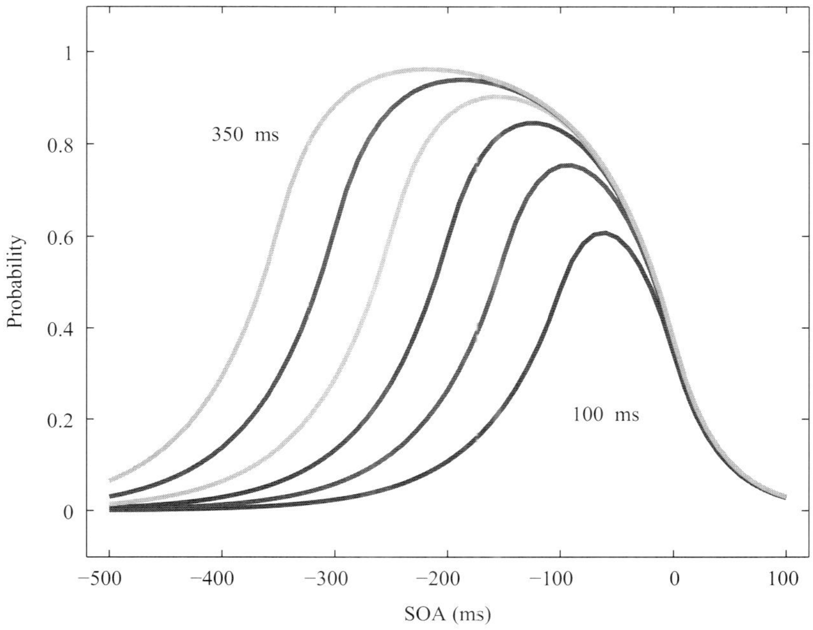

Fig. 3. Probability of integration as a function window width ω ranging from 100 ms to 350 ms in steps of 50 ms.

Warning mechanism: TWIN-W

Although estimates for the time window of integration vary somewhat across subjects and task specifics, the 200 ms width emerges in several studies (e.g., Eimer, 2001). In contrast, when the nontarget occurs at an earlier point in time (SOA of 200 ms or more before the target), a substantial decrease of RT compared to the unimodal condition has still been observed in our study (Diederich and Colonius, 2007a). This decrease, however, no longer depended on whether target and nontarget appeared at ipsi- or contralateral positions, thereby supporting the hypothesis that the nontarget plays the role of a *spatially unspecific* alerting or warning cue for the upcoming target whenever the SOA is large enough. Note that the hypothesis of increased crossmodal processing triggered by an alerting or warning cue had already been advanced in Nickerson (1973) who referred to it as "preparation enhancement." In the eye movement literature, the effects of a warning signal have been studied primarily in the context of explaining the gap effect, that is, the latency to initiate a saccade to an eccentric target is reduced by extinguishing the fixation stimulus prior to the target onset (Reuter-Lorenz et al., 1991; Klein and Kingstone, 1993). An early study on the effect of auditory or visual warning signals on saccade latency, but without considering multisensory integration effects, was conducted by Ross and Ross (1981). Here, the dual role of the nontarget — inducing multisensory integration that is governed by the above mentioned spatiotemporal rules on the one hand and acting as a spatially unspecific crossmodal warning cue on the other hand — will be taken into account by an extension of TWIN that yields an estimate of the relative contribution of either mechanism for any specific SOA value.

(7) *Assumption of warning mechanism:* if the nontarget wins the processing race in the first stage by a wide enough margin, then subsequent processing will be facilitated or inhibited without dependence on the spatial configuration of the stimuli.

The occurrence of warning depends on intramodal characteristics of the target and the nontarget such as modality or intensity. For instance, an intense auditory nontarget may have a higher chance to win the race with a headstart (i.e., advantage) compared to a weak tactile nontarget.

The warning mechanism of the nontarget is triggered whenever the nontarget wins the race by a certain margin or headstart γ_A and, thus, its occurrence corresponds to the event:

$$W = \{A + \tau + \gamma_A < V\}.$$

The probability of warning to occur, $P(W)$, is a function of both τ and γ_A, and its value can be determined numerically as soon as the distribution functions of A and V have been specified. If the headstart γ_A is large enough for the integration window to close again, this implies:

$$\gamma_A > \omega \geq 0 \text{ and, therefore, } P(I \cap W) = 0$$

Assuming exponential distributions again, Fig. 4 shows the probability of integration for a window width of $\omega = 150$ ms and the probability of warning with various values of the headstart parameter γ.

The next step is to compute expected total RT for the unimodal and crossmodal conditions. From the two-stage assumption, total RT in the crossmodal condition can again be written as a sum of two random variables:

$$RT_{crossmodal} = S_1 + S_2 \qquad (12)$$

where S_1 and S_2 refer, as before, to the first- and second-stage processing time, respectively. For the expected saccadic RT in the crossmodal condition then follows:

$$\begin{aligned}
\mathrm{E}[RT_{crossmodal}] &= \mathrm{E}[S_1] + \mathrm{E}[S_2] \\
&= \mathrm{E}[S_1] + P(I) \cdot \mathrm{E}[S_2|I] \\
&\quad + P(W) \cdot \mathrm{E}[S_2|W] + \{1 - P(I) \\
&\quad - P(W)\} \cdot \mathrm{E}[S_2|I^c \cap W^c] \\
&= \mathrm{E}[S_1] + \mathrm{E}[S_2|I^c \cap W^c] \\
&\quad - P(I) \cdot \{\mathrm{E}[S_2|I^c \cap W^c] - \mathrm{E}[S_2|I]\} \\
&\quad - P(W) \cdot \{\mathrm{E}[S_2|I^c \cap W^c] \\
&\quad - \mathrm{E}[S_2|W]\}
\end{aligned}$$

where $\mathrm{E}[S_2|I]$, $\mathrm{E}[S_2|W]$, and $\mathrm{E}[S_2|I^c \cap W^c]$ denote the expected second-stage processing time

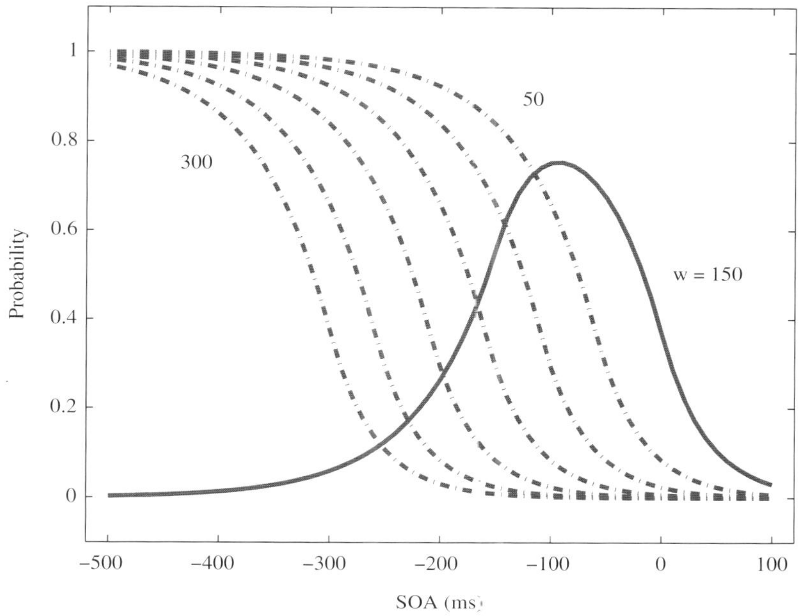

Fig. 4. Probability of integration (solid line) with window width $\omega = 150$ ms and probability of warning (dashed lines) with headstart γ ranging from 50 ms to 300 ms in steps of 50 ms.

conditioned on interaction occurring (I), warning occurring (W), or neither of them occurring ($I^c \cap W^c$), respectively (I^c, W^c represent the complement of events I, W). Setting:

$$\Delta \equiv \mathrm{E}[S_2|I^c \cap W^c] - \mathrm{E}[S_2|I]$$

$$\kappa \equiv \mathrm{E}[S_2|I^c \cap W^c] - \mathrm{E}[S_2|W]$$

this becomes:

$$\mathrm{E}[RT_{crossmodal}] = \mathrm{E}[S_1] + \mathrm{E}[S_2|I^c \cap W^c] - P(I) \cdot \Delta - \mathrm{P}(W) \cdot \kappa \quad (13)$$

In the unimodal condition, no integration or warning is possible. Thus,

$$\mathrm{E}[RT_{unimodal}] = \mathrm{E}[S_1] + \mathrm{E}[S_2|I^c \cap W^c]$$

and we arrive at a simple expression for the combined effect of multisensory integration and warning, crossmodal interaction (CI),

$$\begin{aligned} \mathrm{CI} &\equiv \mathrm{E}[RT_{unimodal}] - \mathrm{E}[RT_{crossmodal}] \\ &= P(I) \cdot \Delta + P(W) \cdot \kappa \end{aligned} \quad (14)$$

Note that Δ and κ can separately take on positive or negative values (or zero) depending on whether multisensory integration and warning have a facilitative or inhibitory effect.

The basic assumptions of TWIN-W imply that for a given spatial configuration and nontarget modality there are no sign reversals or changes in magnitude of Δ or κ across all SOA values. In contrast, both the probability of integration $P(I)$ and the probability of warning $P(W)$ do change with SOA. In particular, when the nontarget is presented very late relative to the target (large positive SOA), its chances of winning the race against the target and thus opening the window of integration become very small. When it is presented rather early (large negative SOA), it is likely to win the race and to open the window, but the window may be closed by the time the target arrives. Again, the probability of integration, $P(I)$, is small. Therefore, the largest integration effects are expected for some mid-range SOA values. In contrast, the probability of warning $P(W)$ decreases monotonically with SOA: the later the nontarget is presented, the smaller are its chances to win the race against the target with some headstart γ. Assuming exponential distributions once again, Figure 5 shows predicted mean saccadic reaction time (SRT) as a function of SOA under various conditions of warning and/or integration occurring.

It is interesting to note that this difference in how $P(I)$ and $P(W)$ should depend on SOA is, in principle, empirically testable without any distributional assumptions by manipulating the conditions of the experiment. Specifically, if target and nontarget are presented in two distinct spatial conditions, ipsilateral and contralateral, for example, one would expect Δ to take on two different values, Δ_i and Δ_c, whereas $P(W) \cdot \kappa$, the expected nonspatial warning effect, should remain the same under both conditions. Subtracting the corresponding crossmodal interaction terms then gives, after canceling the warning effect terms (Eq. 14),

$$\mathrm{CI}_i - \mathrm{CI}_c = P(I) \cdot (\Delta_i - \Delta_c) \tag{15}$$

an expression that should yield the same qualitative behavior, as a function of SOA, as $P(I)$.

For further details and empirical tests of TWIN-W, we refer to Diederich and Colonius (2008a, b) and to Diederich et al. (2008).

Explicating the second stage of TWIN

Spatial stimulus configurations affect the second processing stage (see above). Spatially aligned stimuli may facilitate a response, whereas spatially disparate stimuli have no effect or may even inhibit a response. So far, no specific distributional assumptions about the second stage have been made. Colonius and Arndt (2001) assumed a normal distribution leading to an ex-Gaussian overall distribution. An alternative approach is to incorporate a sequential sampling process as proposed in Eq. (5).

Future directions for TWIN

The bidirectional links between action and perception could be made more explicit in the TWIN model framework. In fact, the separation of spatial and temporal factors determining crossmodal effects and the distinction between intersensory and warning effects as afforded by TWIN, could help to discern the perceptual and attentional consequences of a specific action. For example, moving the body in space, or even a simple turn of the head, may transform the spatial configuration of a crossmodal stimulus complex from a contralateral to an ipsilateral condition (relative to the observer) leading to specific changes in the amount of crossmodal effects, possibly transforming enhancement into inhibition or vice versa. In a similar manner, moving closer toward the source of an acoustical nontarget may increase its potential to act as an unspecific warning signal while, simultaneously, lowering its chances to play a role as a spatially specific part of a crossmodal stimulus configuration.

Another direction would be to extend the TWIN framework to account for choice response times and choice frequencies for two alternative choice problems as well (see Chapter 12: Embodied cognition of movement decisions: a computational modeling approach). To this end, the second stage of the TWIN model may be framed as a decision process. The idea is as follows. For a binary choice, for example between options A and B, a sequential sampling model is assumed. The options in this context could be actions such as reaching to the left or right, or moving the eyes to a target or not. Upon presentation of the choice options and completion of the first stage, the decision maker sequentially samples information from the stimulus display over time. The small increments of evidence sampled at any moment in time are such that they either favor option A or option B. Assume that a positive increment indicates an increment favoring option A and a negative increment is an increment favoring option B. The evidence is accumulated from one moment in time to the next by summing the current state with the new increment. This process continues until the magnitude of the cumulative

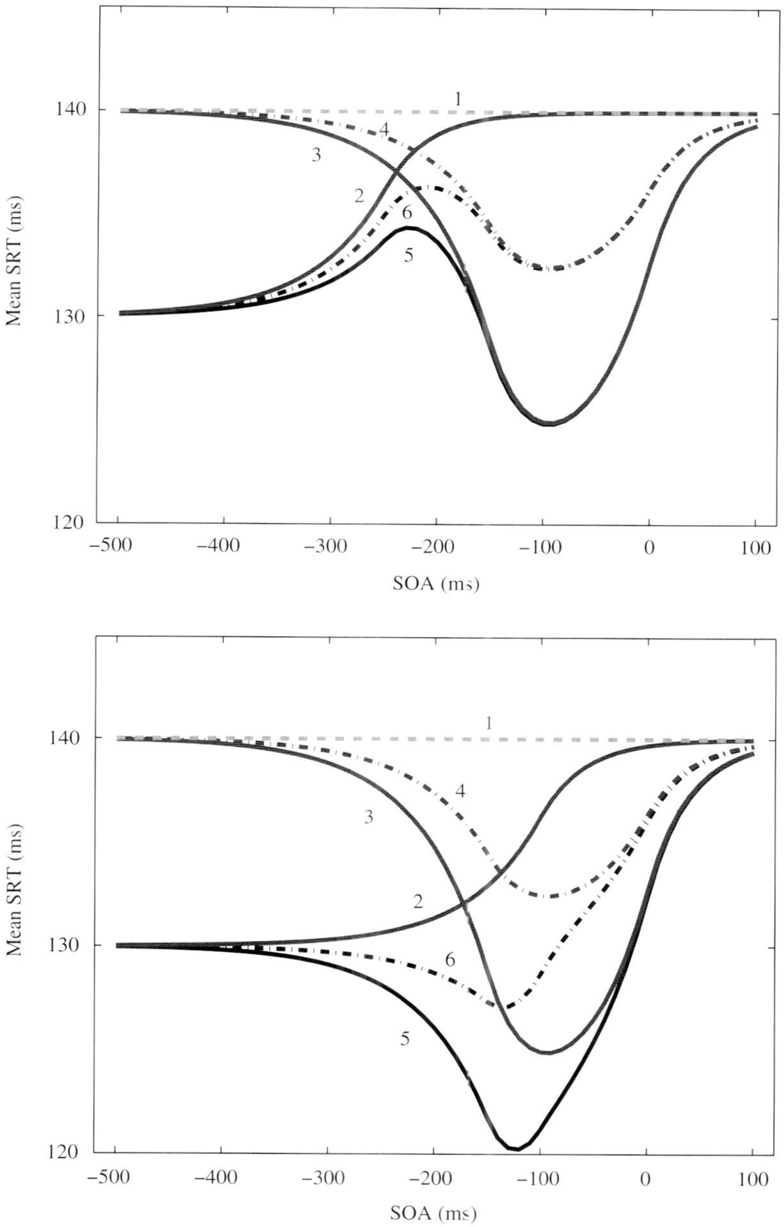

Fig. 5. Predicted mean saccadic reaction time (SRT) as a function of stimulus onset asynchrony (SOA). In the upper panel, $\gamma > \omega$; in the lower panel, $\gamma < \omega$. The horizontal (1) line indicates the predicted mean SRT to the target only. The remaining lines indicate the predicted mean SRT when target and nontarget are presented, as a function of SOA. Line 2 refers to the mean SRT when only warning takes place. Lines 3 and 4 refer to the predicted mean SRT when bimodal stimuli were presented ipsi- (3, solid) and contralaterally (4, dashed) and only integration occurs. The black lines refer to the predicted mean SRT when bimodal stimuli were presented ipsi- (5, solid) and contralaterally (6, dashed) and both integration and warning occur.

evidence exceeds a threshold criterion. The process stops and option A is chosen as soon as the accumulated evidence reaches a criterion value for choosing A, or it stops and chooses option B as soon as the accumulated evidence reaches a criterion value for choosing B. The probability of choosing A over B is determined by the accumulation process reaching the threshold for A before reaching the threshold for B. The criterion is assumed to be set by the decision maker prior to the decision task (e.g., see Diederich and Busemeyer, 2006; Diederich, 2008 for various models within this framework). Here, two criteria (two absorbing boundaries) of the process are assumed, one for initiating an A response and one for initiating a B response, while in the *multichannel diffusion* model stated above only one criterion (one absorbing boundary and one reflecting boundary) was assumed.

Within the context of the focused attention or redundant target paradigm, where subjects do not have to make a choice between alternative responses, an obvious interpretation for the options would be a choice between the decision to respond and the decision not to respond, given stimuli are presented at threshold levels or within a noisy background.

References

Bell, A. H., Meredith, A., Van Opstal, A. J., & Munoz, D. P. (2005). Crossmodal integration in the primate superior colliculus underlying the preparation and initiation of saccadic eye movements. *Journal of Neurophysiology*, *93*, 3659–3673.

Billingsley, P. (1979). *Probability and measure*. New York: Wiley.

Calvert, G., Spence, C., & Stein, B. E. (Eds.). (2004). *Handbook of multisensory processes*. Cambridge, MA: MIT Press.

Colonius, H., & Arndt, P. (2001). A two-stage model for visual–auditory interaction in saccadic latencies. *Perception & Psychophysics*, *63*, 126–147.

Colonius, H., & Diederich, A. (2004). Multisensory interaction in saccadic reaction time: a time-window-of-integration model. *Journal of Cognitive Neuroscience*, *16*, 1000–1009.

Colonius, H., & Diederich, A. (2006). Race model inequality: interpreting a geometric measure of the amount of violation. *Psychological Review*, *113*, 148–154.

Corneil, B. D., Van Wanrooij, M., Munoz, D. P., & Van Opstal, A. J. (2002). Auditory–visual interactions subserving goal-directed saccades in a complex scene. *Journal of Neurophysiology*, *88*, 438–454.

Diederich, A. (1992). Probability inequalities for testing separate activation models of divided attention. *Perception & Psychophysics*, *52*, 714–716.

Diederich, A. (1995). Intersensory facilitation of reaction time: evaluation of counter and diffusion coactivation models. *Journal of Mathematical Psychology*, *39*, 197–215.

Diederich, A. (2008). A further test on sequential sampling models accounting for payoff effects on response bias in perceptual decision tasks. *Perception & Psychophysics*, *70*, 229–256.

Diederich, A., & Busemeyer, J. R. (2006). Modeling the effects of payoff on response bias in a perceptual discrimination task: threshold-bound, drift-rate-change, or two-stage-processing hypothesis. *Perception & Psychophysics*, *68*, 194–207.

Diederich, A., & Colonius, H. (1991). A further test of the superposition model for the redundant-signals effect in bimodal detection. *Perception & Psychophysics*, *50*, 83–86.

Diederich, A., & Colonius, H. (2004a). Modeling the time course of multisensory interaction in manual and saccadic responses. In G. Calvert, C. Spence, & B. E. Stein (Eds.), *Handbook of multisensory processes*. Cambridge, MA: MIT Press.

Diederich, A., & Colonius, H. (2004b). Bimodal and trimodal multisensory enhancement: effects of stimulus onset and intensity on reaction time. *Perception & Psychophysics*, *66*, 1388–1404.

Diederich, A., & Colonius, H. (2007a). Why two "distractors" are better than one: modeling the effect of nontarget auditory and tactile stimuli on visual saccadic reaction time. *Experimental Brain Research*, *179*, 43–54.

Diederich, A., & Colonius, H. (2007b). Modeling spatial effects in visual–tactile saccadic reaction time. *Perception & Psychophysics*, *69*, 56–67.

Diederich, A., & Colonius, H. (2008a). Crossmodal interaction in saccadic reaction time: separating multisensory from warning effects in the time window of integration model. *Experimental Brain Research*, *186*, 1–22.

Diederich, A., & Colonius, H. (2008b). When a high-intensity "distractor" is better then a low-intensity one: Modeling the effect of an auditory or tactile nontarget stimulus on visual saccadic reaction time. *Brain Research*, *1242*, 219–230.

Diederich, A., Colonius, H., Bockhorst, D., & Tabeling, S. (2003). Visual–tactile spatial interaction in saccade generation. *Experimental Brain Research*, *148*, 328–337.

Diederich, A., Colonius, H., & Schomburg, A. (2008). Assessing age-related multisensory enhancement with the time-window-of-integration model. *Neuropsychologia*, *46*, 2556–2562.

Eimer, M. (2001). Crossmodal links in spatial attention between vision, audition, and touch: evidence from event-related brain potentials. *Neuropsychologia*, *39*, 1292–1303.

Frens, M. A., Van Opstal, A. J., & Van der Willigen, R. F. (1995). Spatial and temporal factors determine auditory–visual interactions in human saccadic eye movements. *Perception & Psychophysics*, *57*, 802–816.

Hershenson, M. (1962). Reaction time as a measure of intersensory facilitation. *Journal of Experimental Psychology, 63*, 289–293.

Howard, I. P., & Templeton, W. B. (1996). *Human spatial orientation*. London: Wiley.

Klein, R., & Kingstone, A. (1993). Why do visual offsets reduce saccadic latencies? *Behavioral and Brain Sciences, 16*, 583–584.

Luce, R. D. (1986). *Response times: their role in inferring elementary mental organization*. New York: Oxford University Press.

McGurk, H., & MacDonald, J. (1976). Hearing lips and reading voices. *Nature, 264*, 746–748.

Meredith, M. A. (2002). On the neural basis for multisensory convergence: a brief overview. *Cognitive Brain Research, 14*, 31–40.

Meredith, M. A., Nemitz, J. W., & Stein, B. E. (1987). Determinants of multisensory integration in superior colliculus neurons. I. Temporal factors. *Journal of Neuroscience, 10*, 3215–3229.

Meredith, M. A., & Stein, B. E. (1986). Visual, auditory, and somatosensory convergence on cells in superior colliculus results in multisensory integration. *Journal of Neurophysiology, 56*, 640–662.

Miller, J. O. (1982). Divided attention: evidence for coactivation with redundant signals. *Cognitive Psychology, 14*, 247–279.

Nickerson, R. S. (1973). Intersensory facilitation of reaction time: energy summation or preparation enhancement. *Psychological Review, 80*, 489–509.

Raab, D. H. (1962). Statistical facilitation of simple reaction times. *Transactions of the New York Academy of Sciences, 24*, 574–590.

Reuter-Lorenz, P. A., Hughes, H. C., & Fendrich, R. (1991). The reduction of saccadic latency by prior offset of the fixation point: an analysis of the gap effect. *Perception & Psychophysics, 49*, 167–175.

Ricciardi, L. M. (1977). *Diffusion processes and related topics in biology*. Berlin: Springer-Verlag.

Ross, S. M., & Ross, L. E. (1981). Saccade latency and warning signals: effects of auditory and visual stimulus onset and offset. *Perception & Psychophysics, 29*, 429–437.

Schwarz, W. (1989). A new model to explain the redundant-signal effect. *Perception & Psychophysics, 46*, 498–500.

Schwarz, W. (1994). Diffusion, superposition, and the redundant-targets effect. *Journal of Mathematical Psychology, 38*, 504–520.

Sparks, D. L. (1986). Translation of sensory signals into commands for control of saccadic eye movements: role of primate superior colliculus. *Physiological Reviews, 66*, 116–177.

Spence, C., & Squire, S. (2003). Multisensory integration: maintaining the perception of synchrony. *Current Biology, 13*, R519–R521.

Stein, B. E., & Meredith, M. A. (1993). *The merging of the senses*. Cambridge MA: MIT Press.

Todd, J. W. (1912). Reaction to multiple stimuli. In R. S. Woodworth (Ed.), *Archives of psychology, No. 25. Columbia contributions to philosophy and psychology* (Vol. XXI, No. 8). New York: The Science Press.

Tuckwell, H. C. (1989). *Elementary applications of probability theory*. London: Chapman & Hall.

Van Wassenhove, V., Grant, K. W., & Poeppel, D. (2007). Temporal window of integration in auditory–visual speech perception. *Neuropsychologia, 45*, 598–607.

Welch, R. B., & Warren, D. H. (1986). Intersensory interactions. In K. R. Boff, L. Kaufmann, & J. P. Thomas (Eds.), *Handbook of perception and human performance, vol. 1: sensory processes and perception* (pp. 25-1–25-36). New York: Wiley.

Wood, T. W., & Recanzone, G. H. (2004). Cross-modal interactions evidenced by the ventriloquism effect in humans and monkeys. In G. Calvert, C. Spence, & B. E. Stein (Eds.), *Handbook of multisensory processes*. Cambridge, MA: MIT Press.

M. Raab et al. (Eds.)
Progress in Brain Research, Vol. 174
ISSN 0079-6123

CHAPTER 12

Embodied cognition of movement decisions: a computational modeling approach

Joseph G. Johnson*

Department of Psychology, Miami University, Oxford, OH, USA

Abstract: This chapter presents a cognitive computational view of decision making as the search for, and accumulation of, evidence for options under consideration. It is based on existing models that have been successful in traditional decision tasks involving preferential choice. The model assumes shifting attention over time that determines momentary inputs to an evolving preference state. In this chapter, the cognitive model is extended to illustrate how links from the motor system may be incorporated. These links can basically be categorized into one of three influences: modifying the subjective evaluation of choice options, restricting attention, and altering the options that are to be found in the choice set. The implications for the formal model are introduced and preliminary evidence is drawn from the extant literature.

Keywords: attention; decision making; motor system

Introduction

Each contributor to this volume recognizes the importance of the link between the cognitive and motor systems. In practice, however, we scientists as a whole often take a reductionist approach and focus on our own specializations, assuming we can easily integrate our research into the larger schema if and when it is necessary. For example, as a cognitive psychologist, I find myself studying how the brain may process information to produce a course of action. However, rarely am I interested in how that course of action becomes physically implemented. This becomes problematic when one realizes that the other components of the system — in this case, the system being the human agent — reciprocally influence one another, and thus a complete understanding is only possible when they are considered jointly. Not to underestimate the daunting realities of such a comprehensive approach, this chapter instead aims for a more modest goal. In particular, I will outline the relevant cognitive processes that are involved with the processing of information. Then, I will offer suggestions for how the motor system can be represented as a coupled influence on these processing assumptions. Throughout, I will tend to focus on movement decisions involving the gross motor system (as opposed to saccadic decisions or key presses) to make more apparent the strong connections between motion and cognition.

Cognitive components of ball sports

I will begin with a short, focused primer on the relevant cognitive processes that I assume to

*Corresponding author.
Tel.: +1-513-529-4161; Fax: +1-513-529-2420;
E-mail: johnsojg@muohio.edu

DOI: 10.1016/S0079-6123(09)01312-0

underlie overt behavior in movement decisions. This will provide a sort of road map, not only for the remainder of the current discussion but also for the implementation of the integrative approach that I am advocating. After introducing these topics, we will be able to see how they can be formally modeled as the mental precursors for movement decisions.

Attention is the first component of cognition that will be essential for understanding athlete behavior. Attention serves as the "gatekeeper" of the mind, serving as a filter that determines what information is actively processed at any given moment (e.g., retained in "working memory;" Broadbent, 1958; Baddeley and Hitch, 1974; see Knudsen, 2007, for a review in a neuroscientific context). Our multiple senses are perpetually bombarded with input, requiring a mechanism for focusing mental efforts on some subset of immediately relevant information for subsequent processing. It is important in the context of the current discussion to realize that information comes not only from senses interacting with the world, such as vision and audition, but also proprioception such as kinesthetic and vestibular senses. Attention is what allows the athlete to hear the voice of a coach over the roar of a crowd, or to focus on the movements of team-mates setting up a play or defenders rotating positions, or to consciously modify his/her hand or arm position to perfect the topspin on a return in tennis.

Closely related to attention is the perception of the information that is currently attended. Information does not just passively enter our minds, but it is shaped in large part by our expectations, experiences, and other inherent biases. In other words, the information conveyed by our senses may be objectively defined by physical properties such as hue, pitch, or direction of motion, but our subjective interpretation of this information is what becomes the basis of thought. Decades (indeed centuries) of work in psychophysics has examined this relationship, which suggests decreasing marginal subjective response with increasing objective stimulus magnitude, summarized by the Weber-Fechner Law (see also Stevens, 1957). In other words, a constant increase in stimulus magnitude will be more subjectively impactful if it occurs at low intensities — a candle appears brighter in a cave than outside on a sunny day, and the first punch in a boxing match is likely more painful than the twenty-first.

What purpose does this influx of information serve? That is, what are the cognitive goals associated with movement decisions? Answering this question is simply a matter of working backwards in a sense, determining what cognitive operations are required to produce the behaviors that constitute a "successful" movement. To ground some of these concepts, it will be instructive to use a running example, such as an athletic performance. The continuous stream of an athletic contest is actually composed of a series of discrete actions, the aggregate of the choices of the athletes engaged in the sport. What is a half of soccer, really; how is it best described? By a halftime score of 1-0? No, this conveys very little information about what has taken place. In fact, it is a period of 45 minutes during which unfolds a constant series of running, passing, shooting, diving, sliding, celebrating, etc. by 22 (or more) individuals. To understand this half of play, we need to understand the contribution of each action, and to understand a single action from this series, for example, the lob pass from a midfielder to a forward, we can decompose the action into its cognitive antecedents. Specifically, driven by *attention* to different information aspects, any action can be examined as the *generation* of possible options, the *deliberation* among these options, and the ultimate *choice* of a single option.

Consider the situation facing the midfielder, who currently has the ball and dribbles across the midfield line. At this point, he/she must advance the play, and the cognitive processes that do so evolve in a sequence of events. First, he/she must survey the field and ascertain any relevant information, such as defender positions and the dynamic movements of his/her team-mates. Additional information is attended as well, ranging from relevant information from long-term memory — such as the preferences of his/her center forward and striker in receiving passes and shooting — to immediate context information such as the number of penalties on the opposing defenders and the time remaining in the half.

This attended information is then used to generate possible options — such as a lob pass to a forward, a crossfield pass to a wing player, or continuing to dribble up the sideline. Note that these options may not necessarily be explicitly generated and verbalizable at any given moment, and also that they depend largely on the (perception of the) attended information. Nevertheless, from among this set of potential options a decision is made, presumably requiring some level of cognitive processing. Perhaps a simple, repeatedly rehearsed "if–then" rule, based on pattern matching, is almost automatically enacted; or maybe a systematic analysis of the possible options reveals a clear "best choice" and results in a more explicit overt choice.

Any single choice, or action, is not performed and then lost in the chronicles of a play-by-play summary. That is, an athletic contest is indeed a series, a configural *Gestalt* that is more than the sum of its parts, something more than a collection of independent choices. Instead, these choices are decidedly *dependent*, with one affecting the next. Furthermore, each individual choice is evaluated — and by more than just tens of thousands of screaming critics. Each individual must assess the functional outcome of his/her actions, and thereby learn about his/her successes or failures. Cognitively, performance feedback becomes the impetus for modifying future behavior, through modifying future option generation, deliberation, and choice strategies. A poor choice in one instance is less likely to be generated as a viable option in future instances, less likely to be favored during deliberation even if it is considered, and less likely to be chosen even if it is momentarily favored.

Motoric influences on cognition

In an abstract sense, and in sterile laboratory conditions, these concepts of attention, perception, option generation, deliberation, choice, outcome assessment, and learning have been studied for decades by cognitive psychologists. However, there is a huge discrepancy between the study of learning shape and color patterns by undergraduates and the learning of successful shots on goal by highly motivated athletes in sports. Not only is the athletic domain different (i.e., realistic), and the athlete more emotionally involved, but the *physical* immersion of the athlete in the athletic contest suggests the importance of the physical position and movement. Recently, a successful research paradigm in naturalistic decision making has emerged that addresses some of the deficiencies of laboratory research (Zsambok and Klein, 1997). This work does involve decision agents in their real environments, but has not necessarily highlighted the role of physical embedment.

This is a critical point because although the discussion thus far has described the cognitive components that lead to observable action, the link is really bidirectional. In particular, there are a number of findings that suggest we as theorists must acknowledge the simple fact that a decision is ultimately one of movement. Work on cognitive tuning has shown that indeed the cognitive processes described above can be greatly influenced by the position of the body's muscles and limbs (e.g., Friedman and Förster, 2002). Furthermore, obvious influences stem from factors such as physical orientation: if one is facing the left side of the field, then information from this direction is more salient and thus more influential in subsequent deliberation, and options are more likely to be generated within this restricted range.

Perhaps most importantly, especially in situations such as athletic contests, what one would cognitively wish to perform is not necessarily attainable physically. Due to constraints on one individual's abilities, perhaps the "best" solution or decision in a given situation is beyond the skill level of the individual (or sometimes, any individual). Therefore, even though one may know what the best choice is, it may not correspond to an option that is available to the specific decision maker. Maybe an opponent in tennis has immense trouble handling backhand returns, but if I am incapable of producing a decent backhand return then this option is not viable, even if I know that it would be the "best" against this opponent. In sum, I conceptualize the influence of the motor system *during* decision-making deliberation as being manifest in one or more of three primary ways: (a) priming or modifying the subjective evaluation or perception of courses of action, as in cognitive tuning; (b) restricting one's momentary focus of attention, based on physical orientation; and (c) altering the options that are to be found in the choice set, or at least those that are seriously considered to be enacted.

Finally, it is important to acknowledge the performance of the motor system *after* cognitive processes have produced a "winner" or intended course of action. Cognitive models rarely consider the direct translation of thought into action. That is, although a cognitive model may predict which option is favored as a result of cognitive operations — such as the careful weighing of pros and cons, or simply the "gut" reaction (i.e., instinct) that leads one to prefer a specific option — the physical implementation of this choice is seen as a foregone conclusion. It is typically assumed that cognitive decisions directly and infallibly produce the corresponding action. However, a ball is not passed, kicked, hit, or thrown simply by willing it to happen, but rather as the result of physical action. Thus, the motor system can be seen as taking a (cognitive) input and producing the physical output. This process is also prone to unique sources of error — the playmaker may overshoot the pass to one team-mate, resulting in possession by another (unchosen) team-mate. Granted, this still assumes a "privileged" status of the cognitive system and relegates the motor system to a serially secondary process that is undoubtedly too simplistic. Other approaches assume a more direct role of the motor system (and even downplay the cognitive role altogether in presuming perception–action coupling, see Chapter 4: Perceiving and moving in sports and other high-pressure contexts). Future extensions to the framework introduced here will need to better specify the bidirectional nature of these links and the more central role played by the motor system.

The remainder of the chapter will introduce a formal approach to incorporating these motoric influences on decision behavior, with the caveat that any attempts made here are exploratory. In particular, I will outline a general framework for modeling decision making that has been very successful in traditional (laboratory) decision tasks. Then, I will detail two distinct extensions to this framework to accommodate the two key notions introduced here: (a) the explicit influences of the motor system on the cognitive processing of information; and (b) the subsequent influences upon the observed decision (overt action) attributable to the motor system. It is a challenging task to incorporate these important components, but one that will lead to a more comprehensive view of athlete behavior and other movement decisions.

Formal modeling of human movement decisions

Aristotle is often credited with the first popular model of planetary/stellar motion, which placed the earth at the center of the solar system and suggested spherical planetary/stellar orbits. Because this model was unable to account for several observable phenomena, it required extensive modification. This led to the development of Ptolemy's rather complicated geocentric model (with input from Hipparchus), requiring 13 books to present fully. This mathematical model required several specific geometric devices to explain observed motions. It was Copernicus, circa 1543, who advanced the notion of a sun-centered (heliocentric) model. This model provided a much simpler and parsimonious explanation for the observed data by focusing on a wholly different approach. It was the Copernican model that was expanded on by Galileo, Kepler, and Newton to become what we know today to be the correct description of planetary motion. Similarly, I advocate a Copernican revolution of sorts — more properly a computational revolution — in the study of human decision making.

In the field of decision making, the evolution of contemporary models can similarly be traced by examining the failure of popular models in accounting for aspects of behavioral data. Each failure (e.g., "bias") spurred subsequent modification of the basic model (expected utility theory) to accommodate the "anomalous" empirical results. However, the general approach of the basic model has been retained, resulting in a present-day patchwork of mechanisms built in to the basic model to explain mounting evidence against expected utility computations. Metaphorically, decision researchers are still clinging to the geocentric (algebraic) model rather than adopting a more parsimonious heliocentric (computational) approach.

The "basic model," expected utility theory, is based on an algebraic calculation of evidence in favor of competing courses of action. Specifically, theories in this tradition specify a utility function that transforms objective values (e.g., monetary outcomes of gambles) into subjective values, called utilities; a weighting function that transforms objective event probabilities (e.g., chance of each gamble outcome) into subjective assessments, or decision weights; and rules for utilizing the transformed information. Typically, these rules involve combination (multiplication) of weight and utility for a given outcome or consequence, as well as integration (addition) of weighted utilities in computing a holistic value for each possible alternative or action. The option with the highest holistic value is then chosen. The most popular current incarnations of the basic model are termed rank-dependent utility (RDU) models, such as prospect theory (Kahneman and Tversky, 1979; Tversky and Kahneman, 1992).

In contrast, computational models formally describe the transformation of information into action, not just the relations among inputs and outputs, and thus produce precise, quantitative, testable predictions about mental processes. Cognitive modeling, in particular, has enjoyed a recent surge of popularity. The "cognitive revolution" during the last half of the last century has permeated much of psychology, promoting cognitive mechanisms to describe behavior. In particular, there has been an increase in attention to the information processing that underlies human behaviors, in contrast to the behaviorist viewpoint of the first half of the century. That is, rather than simply viewing behavior as conditioned responses, or matching of situations to actions, the cognitive processing that drives these responses is taken into consideration. The increased interest in cognitive modeling is due in large part to the success these models have enjoyed across domains outside of mainstream cognitive psychology (i.e., beyond memory, language, categorization, etc.). This advance is not yet apparent to the same degree in examining decision making and other behaviors with motor consequences.

In decision making in particular, computational models are only beginning to become the "state of the art" in a field long dominated by utility theories and assumptions of human rationality and adherence to the laws of probability. Next, I will describe a modeling framework that is arguably the most successful in accounting for empirical results in the decision-making literature. These sequential sampling models have been applied to binary choices (Busemeyer and Townsend, 1993); multiattribute decisions (Diederich, 1997), multialternative settings (Roe et al., 2001); influences of motivational and drive states on decision making (Busemeyer et al., 2002); decisions under time pressure (Diederich, 2003); other response modes such as prices (Johnson and Busemeyer, 2005); and many more (see Busemeyer and Johnson, 2004, 2008, for reviews). Furthermore, this same class of models has been successful across many content domains in cognitive psychology, including perceptual discrimination (Link and Heath, 1975), recognition memory (Ratcliff, 1978), probabilistic inference (Wallsten and Barton, 1982), and others.

Sequential sampling model representation

Sequential sampling models assume that deliberation during a decision occurs at some subconscious level, rather than as an exhaustive and calculated assessment of the benefits and drawbacks of each option. That is, in contrast to the most popular conceptualizations of choice (utility theories), it is unlikely that athletes compute expected values during an athletic contest. As an alternative to this view of "economic" decision making, sequential sampling models suggest that information is sampled over time, which results in increases or decreases in the relative preference for each option.

First, the sequential sampling model allows for a non-neutral *initial preference*, meaning there may be preference for a particular option before any task-relevant information is considered. The midfielder may exhibit some favoritism for a particular team-mate, regardless of the specific situation. From this point, information is sampled (attended) over the course of deliberation. At one moment, the midfielder may be focused on the need to score a goal and consider the scoring potential of different actions, at the next moment he/she may be focused on playing conservatively to retain possession of the ball.

Psychologically, the sequential sampling model assumes that the attended information brings to mind affective reactions to each option, largely based on previous experiences (if available) and/or implicit predictions of potential outcomes. If the midfielder considers defender distances, and one team-mate is closely guarded, this may produce a negative reaction towards passing to this team-mate based on recalled instances of turnovers or the predicted possibility of a turnover. If he/she considers the fact that his/her team is down with little time remaining, then passing to team-mates in scoring position will be evaluated positively. Affective *valences* such as these are produced for each option, at each moment in time, and are integrated over time to derive a *preference state* for each option. The evolution of preference states proceeds as additional information is considered over the course of deliberation. At some point an option must be selected — after all, the midfielder must decide what to do at some point, or stand near the midfield line paralyzed with inaction! Sequential sampling models introduce a *threshold*, or level at which an option is considered "good enough," to determine choice. As preferences for each relevant option accumulate, the midfielder eventually must decide that the preference for one single option is strong enough to deserve action. This model has accounted for a variety of findings that have challenged other decision models (Busemeyer and Johnson, 2008) and has been specifically applied to sports tasks (Johnson, 2006).

The intuitive model description above can be precisely modeled as a dynamic system to afford quantitative predictions. Formally, I will here follow the presentation of Roe et al. (2001) that allows for any number of options, described by any number of attributes (see also Diederich and Busemeyer, 2003, for an excellent practical tutorial on how to apply these models to data). Assume a decision maker, such as our midfielder, is considering some m number of actions (e.g., lob pass to center forward), each described by n attributes (e.g., safety/conservativeness, scoring potential, adherence to game plan, etc.). These may be represented as an $m \times n$ matrix, $\mathbf{M}$, where the "value" of option i on the jth attribute is found at m_{ij}. For example, if A = "lob pass to center forward," and B = "dribble to the right," then perhaps A has a higher scoring potential ($m_{\text{A,scoring}} > m_{\text{B,scoring}}$) whereas the latter is less risky ($m_{\text{A,safety}} < m_{\text{B,safety}}$). For mathematical tractability when dealing with attributes that may vary in range, we typically assume that each column of $\mathbf{M}$ is divided by the maximum value in that column. This makes the contribution of attributes uniform that may otherwise vary greatly. For example, attributes for a new car decision may include price, which is measured in tens of thousands, as well as fuel economy in liters/kilometer, which is measured by values less than one!

I propose a significant extension to this representation that is especially relevant to dynamic situations such as movement decisions in general, and athletics in particular. Whereas Roe et al. (2001) introduce the $\mathbf{M}$ matrix as static over the course of the decision task, I propose relaxing this assumption of time-homogeneity and allow for $\mathbf{M}(t)$. Specifically, the dimensionality of $\mathbf{M}(t)$ may change over time as new options are considered and added to the choice set. In contrast to laboratory tasks where the choice options are a closed set explicitly presented to the participant, in real situations potential actions must often be generated "on the fly" over time. For example, rather than having a preconceived set of options in mind, a playmaker dynamically generates these options as he/she scans the field during a play and advances the ball up the field. Option generation has not received considerable attention in decision making and thus has not entered into formal models (but see Gettys et al., 1987; Klein et al., 1995; Johnson and Raab, 2003; and Thomas et al., 2008 for notable exceptions). Here, I simply assume that the $1 \times n$ vector of attribute values for an option is concatenated to the choice set matrix $\mathbf{M}(t)$ at the time t when it is generated. It is beyond the scope of this chapter to detail the option generation process proper, detailing which options are generated and when (but see Johnson and Raab, 2003; Raab and Johnson, 2007a, b for our work on this topic).

The sequential sampling models described here assume that these values are evaluated relatively, rather than absolutely. That is, an action with a very high scoring potential will appear very favorable compared to an action with a low scoring potential, but only slightly better than an action with a scoring potential that is similar. This relative comparison, or contrast operator, is performed mathematically with an $m \times m$ matrix $\mathbf{C}$ that typically takes the form of ones along the main diagonal, and $-(1/m-1)$ as all off-diagonal elements. In other words, when we take the matrix product $\mathbf{C} \cdot \mathbf{M}(t)$ it converts the value of action i on attribute j from its absolute value to a value that is scaled by the average of all other actions $k \neq i$ on attribute j. Hereafter, we assume this contrast operator has been applied and will simply refer to the product $\mathbf{C} \cdot \mathbf{M}(t)$ as $\mathbf{M}(t)$.

Sequential sampling models do not assume that all the information (i.e., attributes) for each potential action are simultaneously weighed and considered. Rather, they describe the shifts in attention across different pieces of information or attributes over time. Typically, they assume that at any given moment, attention focuses on a single attribute in an all-or-none fashion. This is modeled by an $n \times 1$ attention weight vector $\mathbf{W}(t)$, which models current attention to attribute k as $w_k(t) = 1$, $w_j(t) = 0$, for all $j \neq k$. This may be a simplifying assumption, based on the ability of working memory to process multiple pieces of information, and the debates found in an entire literature on divided attention. In any case, we retain this assumption for the moment, but acknowledge the possibility that multiple nonzero elements could exist in $\mathbf{W}(t)$, representing the proportion of attention to each attribute at each moment, with $\Sigma\, \mathbf{W}(t) = 1$.

The mechanism for these momentary shifts in attention varies across sequential sampling models. Busemeyer and Townsend (1993) and Roe et al. (2001) make the simplifying assumption that the focus of attention — that is, the location of the "1" element in $\mathbf{W}(t)$ — changes stochastically over time based on the relative importance or "weight" of each attribute. For example, if scoring potential is the most important attribute, and furthermore is equally as important as all other attributes combined, then this would be formally modeled as $\Pr[w_{scoring}(t) = 1] = 0.50$, for all t. Diederich (1997) has developed sequential sampling models that specify a particular (rather than stochastic) order by which attributes are considered. Especially intriguing is the possibility of measuring overt visual attention as a proxy for covert attention to be input to Diederich's (1997) models; the use of eye-tracking methods offer promising potential in this pursuit (Raab and Johnson, 2007a, b; Johnson and Raab, 2008).

Johnson and Busemeyer (2008) have developed a computational model of the attention-switching processes assumed to operate for people in more tightly controlled (although more abstract) experimental settings, involving choices in the laboratory among sets of gambles. However, the same basic principles can be applied to the practical domain of movement decisions in athletics. Essentially, the model suggests that dynamic patterns of attention can be wholly specified by considering (1) what attribute is first considered, and (2) the conditional probability of attending to each attribute, given the current focus of attention. Formally, this suggests attention switching is a Markov process defined by transitions in attention over time. Application to any task simply requires specifying the probability that each piece of information is initially considered, and the conditional transition probabilities.

In the soccer example, the first attributes considered can be based on factors such as: immediate context — for example, if it is late in the second half and one's team is trailing, then scoring potential is more likely to be considered first, or a rapidly approaching defender may trigger initial thought of safe passing options; perceptual salience — attributes that are more prominent are likely to be considered first; or previous experience — past situations, especially those with successful outcomes or those frequently occurring (e.g., during training), may prompt initial consideration of specific attributes. Then, the conditional probability of considering the next attribute could depend on factors such as the degree of similarity between attributes, or specific attentional patterns acquired during training (e.g., the order of a quarterback's "reads" in American football).

At this point, we have specified the attributes that describe each option, $\mathbf{M}(t)$, as well as the mechanism of shifting attention across these attributes, $\mathbf{W}(t)$. Simple matrix multiplication of $\mathbf{M}(t) \cdot \mathbf{W}(t) = \mathbf{V}(t)$ produces an $m \times 1$ vector of the relative attribute values that are considered at moment t, collectively referred to as the momentary *valence*. This describes the subjective assessment of each option, relative to other options, at any given moment in time based on the currently attended attribute. As attention shifts over time among attributes, the momentary valence changes as well. At one moment attention may be focused on scoring a game-winning goal, in which case those options with a high scoring potential will be evaluated more favorably, and the momentary valence at that point will reflect this. At the next moment, perhaps attention shifts to the need to retain possession of the ball to prevent a game-winning goal by the other team, in which case those options with higher "safety" or less riskiness will be evaluated more favorably in $\mathbf{V}(t)$. As the momentary valence changes over time, sequential sampling models assume that these are collected and accumulated into a momentary preference state, $\mathbf{P}(t)$. In particular, I assume the *preference state* at time t is a simple linear combination of the previous preference state and the current valence input: $\mathbf{P}(t) = \mathbf{SP}(t{-}1){+}\mathbf{V}(t)$, where $\mathbf{S}$ is an $m \times m$ matrix that allows for growth/decay of the previous preference state, as well as dependencies across options (see Roe et al., 2001, for a discussion of $\mathbf{S}$, including psychological interpretations).

I have now described how one's preference over a set of options in a movement decision evolves over time, driven by shifting attention to different attributes of the options. To specify the model fully, I need only determine the beginning and end of this process. In particular, the initial state of the model, or the *initial bias* of the decision maker prior to any information acquisition, is represented as an $m \times 1$ vector, $\mathbf{P}(0) = \mathbf{z}$. For example, if there is no initial preference for any options, then all $z_i = 0$. Alternatively, if the midfielder has a tendency to "dribble first, pass later," then that could be modeled by a higher vale for z_{dribble} than any other option. Perhaps the midfielder has a favored forward player to whom he/she has a strong rapport and a marked predisposition for passing; in this case, the option of passing to that player might have an elevated z_i relative to other options.

Finally, a method must be used to end deliberation. That is, I have described how the preference state changes over time, but at some point a decision must be made and action must be taken, or the midfielder will find himself/herself constantly thinking and never acting! Intuitively, there is typically no need to process attribute information exhaustively during a decision. Especially for dynamic situations such as the midfielder's, the information could readily change and thus there could arguably be a functionally infinite amount of potential information. To prevent paralyzing indecision, sequential sampling models specify a *threshold* preference level, or a level of preference which is "good enough" to justify selecting an option. Formally, a free parameter θ denotes the necessary preference whereby $P_i(t) > \theta$ produces a choice of option i at time t. Although this value is typically held constant (e.g., Busemeyer and Townsend, 1993), one could imagine situations where it may decrease over the course of deliberation, or be defined as a relative rather than absolute value.

Incorporating motor system influences on cognition

The previous section introduced a formal representation of movement decisions via a computational (sequential sampling) model. This model has been applied to many "purely cognitive" decisions where the only required response was a key press or a mouse click. How could — or *should* — the model be modified to reflect the realities of an agent that is situated physically in a decision situation? Recall that I advocated for three primary routes by which the motor system could directly impact the cognitive decision-making apparatus: (1) changes in the subjective perception of value; (2) changes in attentional focus; and (3) changing the actions in the choice set. I now discuss how to incorporate each of these factors in turn.

First, the motor system may be responsible for changes in the perception of the attributes of the choice options. For example, if the motor system is fatigued, then perhaps this changes the perception of attributes associated with some options. A long lob pass would be perceived as a riskier maneuver if the midfielder knew that his/her body might not physically be able to produce such a pass. Poor calibration during a given contest may lower the midfielder's confidence in his/her shooting ability, and thus lower the scoring potential associated with any direct shots on goal. A more provocative method for formally incorporating the influence of the motor system is to assume that the motor system itself contains attributes. That is, although the current **M** is assumed to be perceptual, this is not a requirement or a restriction. Various attributes that could define an option relevant to the motor system, such as physical effort required or likelihood of proper physical implementation, could be collected as distinct entries (columns) in **M**. In this case, motoric influences such as fatigue would be represented independently from other considerations, meaning that the subjective assessment of physical effort required to enact an option would be modified, but the unconditional scoring potential of the option would not. The differences between these formal representations would become apparent based on how attentional shifts proceed. For the former case, where the motor system directly changes the option's "perceptual" attributes such as scoring potential, then any attention to this attribute would involve a motoric tempering of the attribute value and thus the momentary valence. In the latter case, attention to perceptual attributes would leave the valence unaffected by the motor system, and only explicit attention to motoric attributes could produce an influence.

Second, the motor system could directly impact shifting attention, the driving force of the sequential sampling model. For example, perhaps fatigue does not only diminish values (either perceptual or motoric), but it may also increase the likelihood of attending to these values. Assume for a moment that we represent motoric dimensions independently in **M**, such as the physical effort to enact option i as $m_{i,\text{effort}}$. Early in an athletic contest, the midfielder may pay very little attention to the effort required to produce a certain movement, such as a long lob pass; however, after running for 80 minutes this may be a much more salient dimension on the midfielder's mind. In this case, $\Pr[w_{\text{effort}}(t) = 1]$ would be much larger at the end of the contest than at the beginning. Changes in attentional focus based on physical constraints could also make some options more likely to be considered than others. For example if the midfielder is facing to the left then one might expect greater assessment of options that are on the left — although, of course, knowledge of unseen players' positions and habits would not preclude other possibilities. In any case, this could be performed in the model by selectively "zeroing out" or greatly diminishing values on a given row of $\mathbf{M}(t)$ at a given moment that do not match the momentary physical orientation. Johnson and Raab (2008) formally model these sorts of spatial dependencies in visual attention in the context of a sampling model to predict choices in handball.

Third, and also in line with this notion of modifying rows of $\mathbf{M}(t)$, is the addition or deletion of rows within $\mathbf{M}(t)$ due to physical impossibility. This would formally restrict cognitive appraisal of options to those options that are able to be instantiated physically, obviating the potential paradox of preferring or selecting an option that cannot be carried out. Even if an option i is generated at time t when facing in one direction, if the midfielder is in a different position and orientation at time t' which makes this option physically unfeasible, then we would assume the corresponding row of values $m_{i,\cdot}(t') = 0$. If at a later point t'' this action could again be completed, then $m_{i,\cdot}(t'')$ would return to their original values. Although mathematically the addition of a row due to (cognitive) option generation would produce the same result as this physical "reacquisition" of a potential action, only the physical constraints are assumed to result in the deletion or "zeroing out" of values in $\mathbf{M}(t)$.

There are several auxiliary assumptions that could be relaxed in the sequential sampling model to accommodate the unique nature of movement decisions in real environments. For example, perhaps attention does not shift among attributes, but across *options*. In other words, $\mathbf{W}(t)$ would become an $m \times 1$ column vector that would indicate the current option under consideration. This would make more concrete some of the other assumptions of motoric influence as well, such as increased attention to physically congruent options. Physical fatigue or other factors may adjust the decision threshold bound θ as well, such as by requiring less support or accumulated preference for an option before action is initiated. The possibilities outlined in this section, as well as others, are intriguing avenues for future work in using sequential sampling models for movement decisions. Ideally, one could perform model comparisons to determine which candidate implementations are most successful at reproducing choices and response times of real movement decisions (see Raab and Johnson, 2004, for an analogous quantitative application of the sequential sampling model to test alternative hypotheses for decisions in basketball).

Motor system realization of cognitive intentions

The previous section detailed how to incorporate motoric influences on deliberation formally. However, it did not provide any insight into how the (cognitively) selected action was implemented. That is, although the attainment of a threshold level of cognitive preference for an option may dictate which action is preferred, and when, it does not describe how this action is physically implemented, or how long this action production takes. One can imagine additional influences during this stage as well that may produce an action distinctly different than the one intended. Especially in behavioral science, where we only have access to observed actions, we typically assume that those observations reveal the intentions of the agent. However, this need not always be the case. The motor system can exhibit its own characteristic sources of error that produce significant deviations from expected or planned behavior. The tennis star never intends to hit a ball 5 cm beyond the edge line, and the action "shoot ball 1 m over cross bar" was probably not the first action to reach a decision threshold during a soccer player's penalty kick deliberation. Only by appreciating this fact of the motor system (at least), and ultimately modeling it explicitly (at best), can we hope to truly capture in an explanatory framework decisions involving complex, coordinated movements. This is the biggest challenge facing a formal model, which for now will regrettably have to be relegated to a simple ε appended to the cognitive model.

Bridging the mind–body gap

The examples from previous work surveyed above illustrate a steady production of studies and modeling endeavors that are helping us to understand the cognitive processes underlying movement decisions better. These processes are summarized and illustrated in Fig. 1. Options are generated dynamically, adding options to the choice set $\mathbf{M}(t)$ as time elapses. Each option (larger circle) is conceptually decomposed into a collection of its relevant attributes (smaller circles, $m_{i,j}$). At any moment in time, attention is focused on some aspect or attribute of each choice option, according to the attentional dynamics in $\mathbf{W}(t)$ described earlier. This would result typically in a common feature across options receiving attention (as illustrated by the dark lines in Fig. 1), but could also be represented by all features of one option receiving attention, as proposed in the model extension suggesting attention shifts across options (rows in $\mathbf{M}(t)$). The current focus of

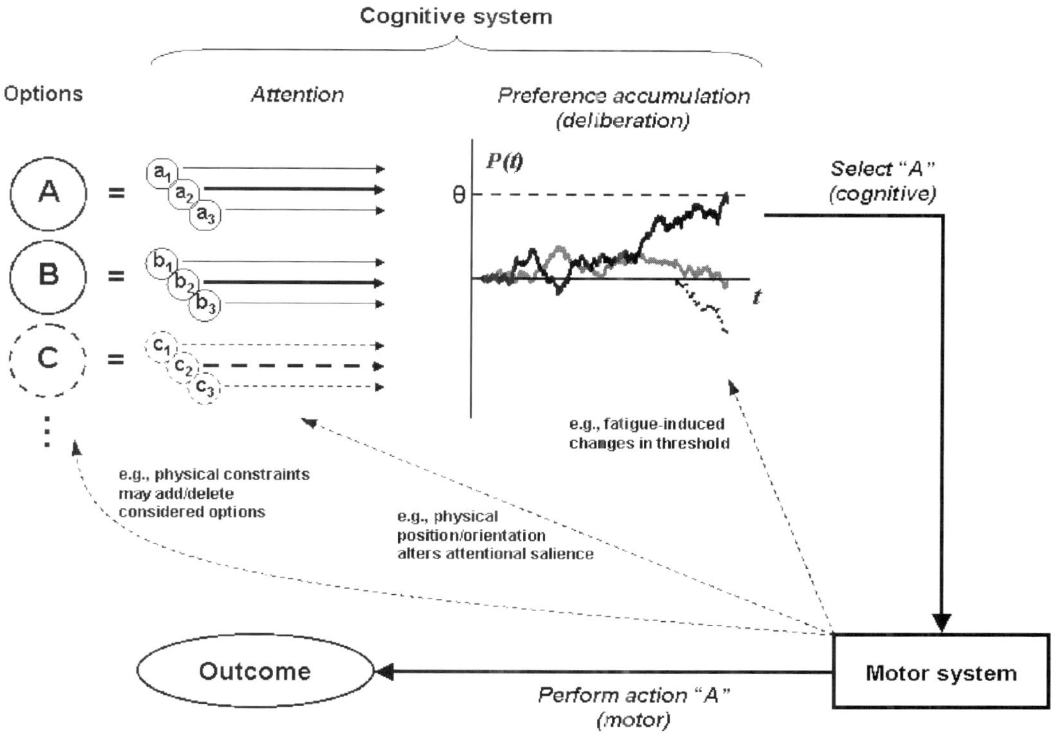

Fig. 1. Integrated model of cognitive and motor systems regulating athlete behavior. Option generation appends possible choice alternatives to the choice set, such as the newly generated option "C." Attention to specific attributes of these options (currently shown as attention to the second attribute of each option, represented as dark lines) drives preference accumulation over time for each option. The first option to reach a threshold level of preference (here, option "A") is selected as the output of the cognitive system. This selection drives the motor system to enact the movements associated with choice of the option. The resulting outcome is used for learning, which affects subsequent cognitive and motor performance. Note that influences (dashed lines) between the cognitive and motor system, and within the cognitive system, also appear. See text and related references for details of each component.

attention determines how the preference for each option increases (or decreases) during that instant, shown in the plots of option preferences, $P_i(t)$, against time, t. For example, in Fig. 1, the illustrated preference accumulation process suggests that during roughly the second half of deliberation, the attention allocation was towards information that favored option "A," shown as the dark trajectory. This is inferred because this preference trajectory increased more than preference for option "B," the gray trajectory, or option "C," the dotted trajectory. The option "C" was generated during the final moments of deliberation, and as a result this preference trajectory does not have a history earlier in the deliberation process. Option "A" is the first to reach the preference threshold (the dashed, horizontal line θ), resulting in choice of "A" and a corresponding response time prediction.

An important consideration in Fig. 1 is that the (cognitive) choice of "A" does not necessarily result in initiation of action that supports this choice. That is, although the midfielder may decide that passing to the striker is the best alternative, enacting the physical motions to bring about or realize this choice is not a deterministic extension of the cognitive appraisal, as mentioned earlier in this chapter. Issues such as motor memory, muscle fatigue, and calibration error, among a host of others factors, can result in an outcome that does not correspond to the one preferred by the cognitive system; these influences are the purview of movement scientists. Only the input is specified in the current iteration of the model, with the motor mapping of the cognitive input to the physical outcome to be more precisely detailed.

One facet that is important but withheld from Fig. 1 (for clarity) is the role of feedback (which would emanate from the "Outcome" node to the motor and cognitive systems). Feedback leads to cognitive adjustment, including the likelihood of generating specific options in subsequent situations, or the guidance of attention, or the operation of the mechanisms of deliberation (e.g., initial preference for different options, shown in the plot in Fig. 1 as being equal and zero). However, feedback can also influence subsequent motor system performance, such as "fine-tuning" of one's shot in basketball. While these types of feedback extensions are unidirectional, perhaps the most important form of bidirectional links in the context of the current discussion is the ongoing crosstalk between the motor system and the cognitive system. This feature is manifest most explicitly in the guidance of attention. Movement or body position in a specific direction will obviously bias attention in the given direction, for example. However, there are more subtle influences on cognition as well, as evidenced by work on motor ontology and cognitive tuning (e.g., Friedman and Förster, 2002; Gallese and Metzinger, 2003).

The mechanisms of motor ontology (i.e., how the motor system constrains high-level mental phenomena, such as through goal representation) are not yet fully understood, and a formal model fitting the relevant data is still missing. For example, under the influence of a negative signal a decision might be reached later, and fewer options might be generated, compared to the influence of a positive feedback signal (see Raab and Green, 2005 for empirical evidence). In the model framework for deliberation introduced earlier, these influences are most predictive in their effect on initial preferences. In a context where a fast decision is made, it is not surprising that initial preferences (i.e., beginning preference closer to threshold) has the highest impact because less external and internal information is integrated. That is, the preference formation is highly influenced by the initial bias value and to a smaller amount by the signal values. In this framework, in short, proprioceptive feedback signals the cognitive system and thereby alters the preference of options.

Conclusions

I have provided an introduction and brief survey of the use and usefulness of cognitive modeling of behavior in general, and decision making in particular, to movement behaviors. The cognitive component of athletes' decision behavior is an important consideration that is only recently becoming appreciated. However, the role of cognition in ball sports obviously cannot be removed from the situation in which it is embedded, nor from the physical system which produces the relevant behaviors (motions). Thus, it is important to consider models of athlete behavior from a perspective that recognizes the interactive links between the body, mind, and environment. Recent work on "motor cognition" has begun to make strides in this direction as well.

I have presented initial attempts at formally modeling the cognitive processes that give rise to movement behavior, as well as the relationship between cognitive and physical systems. Theoretically, I advocate the use of sequential sampling models that have been successful in explaining cognitive aspects of decision making. Here, I showed how to extend these models to incorporate motoric influences. Another key advantage of these models is their ability to be cast in terms of potential neural substrates and thereby link to the neuroscientific evidence on decision making (see Schall, 2004, and Chapter 23: Juggling with the brain — thought and action in the human motor system, for excellent reviews; see Chapter 17: Perceptual decision making: a bidirectional link between mind and motion, for neuroscientific treatment of perceptual decision making; see Busemeyer et al., 2006, for a discussion specific to sequential sampling models).

Empirically, I have selectively investigated the relevant cognitive process including how movement options are generated and selected (Johnson and Raab, 2003); how models of mental deliberation incorporating individual differences and cognitive traits can predict movement decisions (Raab and Johnson, 2004); general applications and advantages of cognitive modeling to movement decisions in sports (Johnson, 2006); and methods for illuminating these cognitive processes through measurements of visual attention via eye tracking (Raab and Johnson, 2007a, b; Johnson and Raab, 2008). This chapter has begun to integrate these efforts in hopes of ultimately developing a model that considers jointly proprioceptive information, perceptual inputs, and cognitive processing in models of movement behavior. Through continued integration and refinement, I am confident that cognitive models, when connected bidirectionally to our knowledge of the human body, will provide a comprehensive understanding of movement behaviors.

References

Baddeley, A., & Hitch, G. (1974). Working memory. In G. A. Bower (Ed.), *Recent advances in learning and motivation* Vol. 8, (pp. 47–90). New York: Academic Press.

Broadbent, D. E. (1958). *Perception and communication*. New York: Oxford University Press.

Busemeyer, J. R., Jessup, R. K., Johnson, J. G., & Townsend, J. T. (2006). Building bridges between neural models and complex decision making behavior. *Neural Networks*, *19*, 1047–1058.

Busemeyer, J. R., & Johnson, J. G. (2004). Computational models of decision making. In D. Koehler & N. Harvey (Eds.), *Blackwell handbook of judgment and decision making* (pp. 133–154). Oxford: Blackwell.

Busemeyer, J. R., & Johnson, J. G. (2008). Micro-process models of decision making. In R. Sun (Ed.), *Cambridge handbook of computational psychology* (pp. 302–321). Cambridge: Cambridge University Press.

Busemeyer, J. R., & Townsend, J. T. (1993). Decision field theory: a dynamic-cognitive approach to decision making in an uncertain environment. *Psychological Review*, *100*, 432–459.

Busemeyer, J. R., Townsend, J. T., & Stout, J. C. (2002). Motivational underpinnings of utility in decision making: decision field theory analysis of deprivation and satiation. In S. Moore (Ed.), *Emotional cognition* (pp. 197–218). Amsterdam: John Benjamins.

Diederich, A. (1997). Dynamic stochastic models for decision making under time constraints. *Journal of Mathematical Psychology*, *41*, 260–274.

Diederich, A. (2003). MDFT account of decision making under time pressure. *Psychonomic Bulletin & Review*, *10*, 157–166.

Diederich, A., & Busemeyer, J. R. (2003). Simple matrix methods for analyzing diffusion models of choice probability, choice response time, and simple response time. *Journal of Mathematical Psychology*, *47*, 304–322.

Friedman, R., & Förster, J. (2002). The influence of approach and avoidance motor actions on creative cognition. *Journal of Experimental Social Psychology*, *38*, 41–55.

Gallese, V., & Metzinger, T. (2003). Motor ontology: the representational reality of goals, actions and selves. *Philosophical Psychology*, *16*, 365–388.

Gettys, C. F., Pliske, R. M., Manning, C., & Casey, J. T. (1987). An evaluation of human act generation performance. *Organizational Behavior and Human Decision Processes*, *39*, 23–51.

Johnson, J. G. (2006). Cognitive modeling of decision making in sports. *Psychology of Sport and Exercise*, *7*, 631–652.

Johnson, J. G., & Busemeyer, J. R. (2005). A dynamic, stochastic, computational model of preference reversal phenomena. *Psychological Review*, *112*, 841–861.

Johnson, J. G., and Busemeyer, J. R. (2008). A computational model of the process generating decision weights.

Johnson, J. G., & Raab, M. (2003). Take the first: option-generation and resulting choices. *Organizational Behavior and Human Decision Processes*, *91*(2), 215–229.

Johnson, J. G., and Raab, M. (2008). In the eye of the beholder: Modeling intention from attention.

Kahneman, D., & Tversky, A. (1979). Prospect theory: an analysis of decision under risk. *Econometrica*, *47*, 263–291.

Klein, G., Wolf, S., Militello, L., & Zsambok, C. (1995). Characteristics of skilled option generation in chess. *Organizational Behavior and Human Decision Processes*, *62*(1), 63–69.

Knudsen, E. I. (2007). Fundamental components of attention. *Annual Review of Neuroscience*, *30*, 57–78.

Link, S. W., & Heath, R. A. (1975). A sequential theory of psychological discrimination. *Psychometrika*, *40*, 77–105.

Raab, M., & Green, N. (2005). Motion as input: a functional explanation of movement effects on cognitive processes. *Perceptual and Motor Skills*, *100*, 333–348.

Raab, M., & Johnson, J. G. (2004). Individual differences of action–orientation for risk-taking in sports. *Research Quarterly for Exercise and Sport*, *75*(3), 326–336.

Raab, M., & Johnson, J. G. (2007a). Expertise-based differences in search and option generation strategies. *Journal of Experimental Psychology — Applied*, *13*, 158–170.

Raab, M., & Johnson, J. G. (2007b). Implicit learning as a means to intuitive decision making in sports. In H. Plessner, C. Betsch, & T. Betsch (Eds.), *A new look on intuition in judgment and decision making* (pp. 119–133). Mahwah, NJ: Lawrence Erlbaum Associates.

Ratcliff, R. (1978). A theory of memory retrieval. *Psychological Review*, *85*, 59–108.

Roe, R. M., Busemeyer, J. R., & Townsend, J. T. (2001). Multialternative decision field theory: a dynamic connectionist model of decision making. *Psychological Review*, *108*, 370–392.

Schall, J. D. (2004). On building a bridge between brain and behavior. *Annual Review of Psychology*, *55*, 23–50.

Stevens, S. S. (1957). On the psychophysical law. *Psychological Review*, *64*, 153–181.

Thomas, R. P., Dougherty, M. R., Sprenger, A., & Harbison, J. I. (2008). Diagnostic hypothesis generation and human judgment. *Psychological Review*, *115*, 155–185.

Tversky, A., & Kahneman, D. (1992). Advances in prospect theory: cumulative representation of uncertainty. *Journal of Risk and Uncertainty*, *5*, 297–323.

Wallsten, T. S., & Barton, C. (1982). Processing probabilistic multidimensional information for decisions. *Journal of Experimental Psychology — Learning Memory and Cognition*, *8*, 361–384.

Zsambok, C. A., & Klein, G. (Eds.). (1997). *Naturalistic decision making*. Mahwah, NJ: Elsevier.

M. Raab et al. (Eds.)
Progress in Brain Research, Vol. 174
ISSN 0079-6123

CHAPTER 13

A multiple-cue learning approach as the basis for understanding and improving soccer referees' decision making

Henning Plessner[1,*], Geoffrey Schweizer[2], Ralf Brand[2] and David O'Hare[3]

[1]Institute of Psychology 1, University of Leipzig, Leipzig, Germany
[2]Department of Sport and Exercise Psychology, University of Potsdam, Potsdam, Germany
[3]Department of Psychology, University of Otago, Dunedin, New Zealand

Abstract: A significant proportion of all referee decisions during a soccer match are about fouls and misconduct. We argue that most of these decisions can be considered as a perceptual-categorization task in which the referee has to categorize a set of features into two discrete classes (foul/no-foul). Due to the dynamic nature of tackling situations in football, these features share a probabilistic rather that a deterministic relationship with the decision criteria. Accordingly, these processes can be studied on the basis of a multiple-cue learning framework as proposed by Brunswick (1955), which focuses among others on how people learn from repeated exposure to probabilistic information. Such learning processes have been studied on a wide range of tasks, but until now not (to our knowledge) in the area of judging sport performance. We suggest that decision accuracy of referees can be improved by creating a learning environment that fits the requirements of this theoretical perspective.

Keywords: multiple-cue learning; social judgment theory; soccer referees

Introduction

The present volume aims at the understanding of embodied decision making, that is, the processes that underlie peoples' decided actions while interacting with a complex and dynamic environment (Wilson, 2002). In contrast to cognitive sciences' traditional view of the mind as an abstract information processor, this perspective highlights the significance of the mind's connections to the outside world. Thus, perceptual and motor systems are considered to be highly relevant for the understanding of central cognitive processes, for example, decision making. By dividing embodied decision making into three parts, this chapter is dedicated to the central decision-making process of selecting a particular course of action from among a set of choices. There are different routes that can be taken when addressing such a complex problem. The most obvious one, at least from an academic perspective, is to follow a good theory, such as the "decision field theory" (Busemeyer and Townsend, 1993) and to discuss its implications for the understanding of embodied decision making in various contexts (Chapter 12: Embodied cognition

*Corresponding author.
Tel.: +49-341-973-5964; Fax: +49-341-973-5969;
E-mail: plessner@uni-leipzig.de

DOI: 10.1016/S0079-6123(09)01313-2

of movement decisions: a computational modeling approach). The present chapter, however, follows a slightly different road by starting with an applied problem of embodied decision making, that is, the soccer referees' task of deciding between foul and no-foul in ambiguous tackling situations. A suitable theoretical framework (social judgment theory, e.g., Hammond et al., 1975), which conceives expert behavior as product of the dynamic relationship between an active performer and his/her functionally defined environments (e.g., Araújo, 2007), will then be considered.

While decision making by athletes appears to be a prototype of embodied cognition (Araújo et al., 2006; Araújo and Kirlik, 2008), for example, by sharing many of its key features such as time pressure, action orientation, dynamical transactions, and motor responses, this is less obvious for the decisions of soccer referees. Of course, they are involved in a dynamic physical environment almost to the same extent as the players. However, in contrast to players' actions, soccer referees' decisions are determined by judgment rather than by decision processes (Plessner and Haar, 2006). That means, once a situation has been judged by a referee, his/her decision follows almost automatically from the (internalized) rules of the game. The assumed consequences (option evaluations) of this decision should be of minor influence (apart from certain aspects of game management, e.g., Brand et al., 2009). On the other hand, these are still *real* decisions that people execute in a dynamic environment while interacting also on a physical level. Although they may not be typical for decisions that abundantly demand an embodied cognition perspective, in our opinion they clearly belong to this class of actions. That is why we think that uncovering the processes that underlie soccer referees' foul/no-foul decisions may be of interest for the broader class of embodied decision-making situations as well. After all, our analysis will reveal the central role of the decision maker's learned representation of an environment's probabilistic relationships between variables as a course for his/her action in *natural* situations (cf. Klein, 1999). The applied problem with which we start, however, is the notion that soccer referees' decisions seem to be inevitably error prone (e.g., Brand et al., 2008).

Foul/no-foul decisions by soccer referees

It is mid-February, the media called out a Super-Sunday in the German soccer league, because there were four games with a direct duel of top teams of the first and second division. By and large, one would expect these games to become decided by performance differences between the competing teams. After all, soccer is a game where performance is supposed to be measured mainly by a scoring system (Stefani, 1998). However, if one believes the newspaper coverage after the games, the teams' performances were only partly responsible for the final outcome of the games. Almost all of the interviewed players and coaches of the unsuccessful teams tend to blame the respective referees rather than their opponents' strength. If they take a look in the scientific literature they would find quite some support for their claim of imperfect referee decisions. For example, soccer referees have been found to be inconsistent in general (Dawson et al., 2007) and to be systematically influenced by the home crowd (Nevill et al., 2002; Sutter and Kocher, 2004), the players' reputation (Jones et al., 2002), the teams' origin (Messner and Schmid, 2007), the passage of the game (Unkelbach and Memmert, 2008), and their own prior decisions (Plessner and Betsch, 2001).

Only one published study (Helsen and Bultynck, 2004) has categorized the observable decision-making activity of soccer referees. In this study the observable decisions of officials at the Euro 2000 championship were categorized into "ball in and out of play" decisions (average of 75 per game) and "foul detection and management" (e.g., awarding a direct free kick; showing a yellow card), which averaged 49 decisions per game. Of course this does not count the decisions not to interfere with play. Given that every event involving player contact has to be judged foul or no-foul, there may well be at least four or five times as many nonobservable decisions not to interfere as there are observable decisions, meaning that referees are making as many as 200–250 foul/no-foul decisions per game. For an overview of the demands of soccer refereeing see Mascarenhas et al. (2006).

Foul decisions are covered by Law 12 of the Laws of the Game (Federation Internationale de Football Association, 2007). This provides a detailed list of offences (e.g., kicking an opponent) and behaviors that should be penalized with sanctions of different degrees of severity (e.g., a direct free kick). While some of the listed offences, such as spitting at an opponent, refer to clearly observable criteria, others refer to more complex judgments that must be inferred from the observed behavior of one or more players, for example, to play in a *dangerous* manner. The only offence for which intentionality must be considered is that of handball. Although judgments about fouls and misconduct are frequently treated in public as judgments about matters of fact, it is actually very difficult for most critical situations (e.g., a tackle) to determine the *correct* decision. For example, recent studies have found agreement rates between match referees and expert referee panels on such decisions to be only between 70% (Fuller et al., 2004) and 79% (Andersen et al., 2004), so there is clearly a level of ambiguity that exists in these situations. In addition, the viewing perspective of referees is frequently imperfect (Oudejans et al., 2000) and, therefore, some relevant judgment criteria may be not directly observable in a given decision situation. For example, assistant referees are generally not exactly in line with the second last defender and the referee may be some distance from play or have their line of sight occluded by other players. Decisions have to be made nevertheless with a minimum of delay.

In terms of human cognition, the application of Law 12 can be conceived as a perceptual categorization task in which the decision maker has to categorize a diverse set of features into two discrete classes (in this case foul or no-foul). Because this happens mostly under time pressure, referees' responses are more likely to result from automatic (intuitive) than from deliberative processes. What can referees do to improve their decision-making skills concerning Law 12? Typically, referees are recommended to elaborate their knowledge on the game's rules continuously: they watch and analyze videotaped games at home, meet several times per season and discuss situations, and they receive instruction as to why one situation has to be considered foul play and another situation not (for an overview see MacMahon and Plessner, 2008). However, such training methods are suboptimal in at least one important aspect: they may lay a necessary foundation of relevant (declarative) knowledge on *how* situations should be judged, but none of them addresses the decision-making *process* and its *proceduralization*. Hence, the basic idea of our work is that the process of decision making should be trained, and that it is not enough to just know the reasons why a decision is correct or incorrect. Our goal is to design an effective training method to be used by expert referees in order to systematically improve their decision-making skills. However, in order to do so we must first understand the cognitive processes that underlie referees' decision making in the dynamic environment of a soccer game.

In general, how can people make decisions under time pressure when not all relevant criteria are accessible? Among others, they may use criteria that are observable, even if their validity is imperfect (Kahneman et al., 1982). In the referee context, this is evident, for example, in an experimental study by Nevill et al. (2002). They investigated whether crowd noise has an influence on soccer referees' decisions concerning potential foul situations and, thus, the contribution to the phenomenon of a home advantage. They assumed that referees have learned to use crowd noise as a decision cue because in general it would serve as a useful indicator for the severity of the foul. However, because the reaction of a crowd is usually biased against the away team, the use of this knowledge would be inappropriate. In the experiment, referees had to assess various challenges videotaped from a match in the English Premier League. Half of them observed the video with the original crowd noise audible, whereas the other half viewed the video in silence. The presence or absence of crowd noise had an effect on decisions made by the referees. Most importantly, referees who viewed challenges in the noise condition called significantly fewer fouls on the home team than those observing the video in silence, thus constituting the effect of a home advantage. The authors concluded that this effect might be partly due to judgment processes in which the salient, yet potentially biased, judgment of the crowd is used as an additional decision *cue*. The usage of multiple cues in making decisions has been studied extensively within the framework of social judgment theory (SJT), which will be described next.

A multiple-cue probabilistic learning framework

So called multiple-cue approaches to judgment and decision making generally highlight people's ability to make decisions by assessing multiple features of the decision-making situation. In contrast to single-cue approaches (e.g., Gigerenzer and Goldstein, 1996) they do not generally assume that people have to rely on single properties of complex situations (while neglecting all others, including expert knowledge) in order to perform successfully. In addition, there is some recent evidence that they seem to be more successful than single-cue approaches in describing people's default decision processes under time pressure (Glöckner, 2008). Thus, they rather match the task that (expert) referees encounter when making foul/no-foul decisions. Surprisingly, the value of multiple-cue approaches for the understanding of expert performance in sport has been recognized just recently (Araújo and Kirlik, 2008). Nowadays several slightly different approaches to multiple-cue judgment exist, but they all stem from Brunswik's idea of probabilistic functionalism (Brunswik, 1955; Goldstein, 2004).

According to Brunswik, people's ultimate goal is called *achievement* (Doherty and Kurz, 1996; Goldstein, 2004). Achievement is the basis for survival and procreation and depends on people's ability to perceive their respective environments as accurately as possible. The problem that arises with achievement is that people usually do not have direct access to the "true state of the world" (called *distal variables* or *criteria*), they have to infer it from visible features of the environment (called *proximal variables* or *cues*). Importantly, these cues are equivocal and probabilistic in nature, meaning that their relations to both distal variables and their perceptions are not deterministic but expressed by correlations. These concepts are prominently illustrated in the lens model (Doherty and Kurz, 1996; Goldstein, 2004).

But Brunswik's ideas are not restricted to the domain of perception (Hammond, 1955). SJT applies the concepts illustrated in the lens model to research on human judgments, especially in social situations (Hammond et al., 1975). The Brunswikian lens and SJT came to notable prominence particularly in the domain of medical judgment (Wigton, 1996). The main idea of SJT, closely related to Brunswik's ideas, is the following: people have to judge certain *distal variables* (e.g., illnesses). As they have no access to this variable itself, they have to rely on accessible *proximal variables* or *cues* instead (e.g., symptoms of the disease). These cues are correlated with the distal variable. As people learn the identity of the relevant cues and the relationships of the cues to the distal variables, the quality of their *judgment* improves (e.g., more correct diagnoses). This improvement is expressed by an ascending correlation r_a (*achievement*) between distal variables and judgments. One of the advantages of lens model research is that this correlation can be divided into several components, among these cue-criterion correlations (*ecological validities*) and cue-judgment correlations (*cue utilization coefficients*), thereby providing more comprehensive insight into human judgment than by investigating achievement only (Cooksey, 1996; Goldstein, 2004).

But how are people able to learn cue-criterion correlations? This is exactly the question research on multiple-cue probability learning (MCPL; Holzworth, 2001) deals with. Numerous studies investigate issues as the importance of appropriate feedback for learning (Klayman, 1988), the dependency of learning on certain task characteristics or the significance of insight into the learning process (Lagnado et al., 2006). Although disagreeing on details most researchers agree that the optimal way of learning probabilistic relationships between cues and outcomes is heavily dependant on task characteristics such as type of cues (binary vs. continuous), representation format (verbal vs. pictorial), the role of the learner (observing vs. intervening), the number of cues presented or the type of relations between cues and criteria (linear vs. u-shaped) (Brehmer et al., 1974; Bröder and Schiffer, 2003; Enkvist et al., 2006).

A question that is also heavily discussed is if human beings (1) *are able* to engage in multiple-cue decision making and (2) *in fact do* engage in multiple-cue decision making. The arguments against humans basing their decisions on multiple cues primarily rely on limited cognitive capacities, stating that people's capacities are not sufficient for taking multiple cues into account, at least not within a limited amount of time (Gigerenzer and Goldstein, 1996). But the aforementioned questions are, at least to some degree, empirical ones, and as such the position that people are incapable of taking multiple cues into account cannot be supported. Substantial evidence exists showing that human beings are able to engage in multiple-cue decision making (Bröder and Schiffer, 2003; Enkvist et al., 2006; Lagnado et al., 2006; Newell et al., 2007; Glöckner, 2008). To what extent people use multiple cues instead of single cues is dependent on characteristics of the decision-making situation again (e.g., representation format, Bröder and Schiffer, 2003). Most importantly, the argument that usage of multiple cues affords too much time to be functional for decisions under time pressure does not hold if one takes into account the possibility of automatic processes in decision making (Glöckner, 2008). Automatic or intuitive decision making allows for the simultaneous (parallel) processing of numerous features within very short amounts of time.

According to the previous sections, soccer referees' decision making can be described as (1) primarily automatic (or intuitive) rather than deliberative and (2) based on multiple rather than on single cues. A training program created to improve the quality of referees' decisions therefore needs to center around these two fundamentals: the acquisition of correlations between *multiple cues and criteria* via *automatic learning processes*.

But how can one successfully learn cue-criterion correlations in an automatic manner? First, Hogarth (2001, 2008) emphasizes the importance of appropriate feedback for automatic learning processes. He defines a good intuition as something that has been acquired in a kind learning environment. Kind learning environments provide observable, relevant, and accurate feedback to persons acting within these environments.

Second, the learning environment must provide learners with access to cues that are relevant as well as irrelevant for the decision to be made. While the relevant (diagnostic) cues should covary with the judgment criteria the irrelevant cues are supposed not to. Transfer from the learning environment to real-life situations is only possible if real-life cues are properly represented in the learning situations. Otherwise, learners acquire knowledge on cue-criterion relations that are not valid for real-life decisions.

In the following section, we briefly describe the decision-making training program SET (*S*chiedsrichter-*E*ntscheidungs-*T*raining, i.e., referees' decision-making training), which aims to improve elite soccer referees' decisions in possible foul situations (for more details see Brand et al., 2009).

Training soccer referees' decision making

SET was constructed according to the assumptions described in the previous section. SET is an internet device that basically consists of a database and an online training module. The database contains a large number of video sequences. These sequences show possible foul situations, usually contact situations between two or more players. Also stored in the database are additional facts for each sequence, such as foul category (e.g., kicking, pushing, or holding), source of the video (e.g., recorded for television or for research purposes), and most importantly the correct decision to be made (as defined by the German Soccer Association's referee board, based on careful video analyses).

Referees participating in the training program get access to the online platform. They regularly participate in training sessions via the internet. During each session they are shown several videos. Each video is stopped after the critical incident to be judged, usually a contact between two players. The participant indicates his/her decision (foul or no-foul) by clicking the mouse. If he/she decides to call foul, he/she is subsequently asked to indicate the appropriate sanction (e.g., direct free kick, yellow card, red card). Afterwards, participants receive feedback as to the correctness of their decision.

Feedback is automatically generated via an online comparison between the participant's decision and the information stored in the database. Importantly, when creating a training session, researchers (or training operators) can vary the session's parameters systematically via a user-friendly interface. For example, a researcher could select the videos to be shown, the foul categories to be trained, the time pressure to be imposed on the participants, and, most importantly, the feedback mode.

Because appropriate feedback is so crucial for the acquisition of good intuitions (Hogarth, 2008), the feedback function is of prime importance within the SET program. Researchers can therefore change various feedback properties according to theoretical concerns, such as delay of feedback (immediate vs. delayed), feedback content (e.g., outcome feedback vs. feedback containing additional information), or feedback presentation (e.g., single feedback for the latest decision only vs. feedback for several decisions at the same time). Although these factors have been addressed in various studies (e.g., Mathan and Koedinger, 2005), so far it is unclear which of the previous results can be applied to the complex process of judging sport performance. Thus, SET is not only a training program, but a research tool at the same time, as it allows the investigation of various still unanswered questions related to improving decision-making quality of referees, such as if delayed or immediate feedback is optimal for learning success or if learning is enhanced or hindered by the provision of additional information. In addition, systematic variation of observable (proximal) cues (e.g., crowd noise, Nevill et al., 2002) within the video material provides the opportunity of assessing individuals' cue utilization coefficients. Among others, this would allow the researcher to explore the extent to which irrelevant cues are used.

As mentioned above, learners must be provided with representative cues for the real life situation in order to achieve transfer from training to real life. Therefore, videos in the SET database are carefully selected from recordings of real-life soccer matches from different tournaments (German Bundesliga, UEFA Champions League, New Zealand Football Championship, 2006 FIFA World Cup). Video items have to meet high quality standards to be included in the database. In the project's starting phase, out of a sample of more than 2000 videos only a little over 200 were included. Thereby, we have tried to ensure the training situations' validity for referees' decision making on the football pitch. One possible limitation is that the video clips are all taken from external pitch-side cameras. With modern technology it is technically possible to obtain good quality video from the referee's perspective by having the referee wear a light-weight head-mounted video camera. We have trialed this technology in game situations with some success. Petit and Ripoll (2008) found that soccer players made faster, more accurate decisions to pass or not pass the ball in video simulations showing the players' perspective rather than the traditional external camera perspective. This has obvious implications for the training of referee decision making.

So far, we have conducted four empirical studies in order to validate assumptions on the optimal design of training sessions (research in progress). Preliminary results indicate that immediate feedback (right or wrong, as classified by the Deutscher Fußball-Bund (DFB) referees' board) after each decision most effectively increases decision accuracy. Further research is planned that includes the assessment of referees' cue-judgment correlations in order to understand which cues are currently used to which extent. This would be a first step to not only improve referees' decision-making accuracy in general but also to create a training environment that specifically immunizes them against the undesired influence of irrelevant cues, such as crowd noise (Nevill et al., 2002) or the color of a player's shirt (Frank and Gilovich, 1988).

Conclusions

We started with the assumption that soccer referees' task of deciding between foul and no-foul in ambiguous tackling situations can be considered as a subclass of embodied cognition that shares most, although not all, features of this concept (Wilson, 2002). The applied problem in this area is that many biases have been identified in the corresponding literature and there is a natural interest by referees to improve their decision making. In order to build up a useful training tool for referees we chose SJT and the idea of MCPL as a suitable theoretical framework that helps to understand the processes that underlie soccer referees' decision making and the development of corresponding skills. With this approach we emphasize the central role of perceptual categorization processes in *natural* decision making (cf. Klein, 1999). Moreover, referring to classic SJT and the lens model accentuates the significance of an expert decision maker's representation of environmental variables and their probabilistic relationships as a course for his/her action (e.g., Araújo and Kirlik, 2008). In our opinion, this framework incorporates the essential elements upon which a broader framework of embodied decision making can be built.

References

Andersen, T. E., Engebretsen, L., & Bahr, R. (2004). Rule violations as a cause of injuries in male Norwegian professional football: are the referees doing their job? *American Journal of Sports Medicine*, *32*(Suppl. 1), 62S–68S.

Araújo, D. (2007). Promotion ecologies where performers exhibit expert interactions. *International Journal of Sport Psychology*, *38*(1), 73–77.

Araújo, D., Davids, K., & Hristovski, R. (2006). The ecological dynamics of decision making in sport. *Psychology of Sport and Exercise*, *7*(6), 653–676.

Araújo, D., & Kirlik, A. (2008). Towards an ecological approach to visual anticipation for expert performance in sport. *International Journal of Sport Psychology*, *39*(2), 157–165.

Brand, R., Plessner, H., & Unkelbach, C. (2008). Basic psychological processes underlying referees' decision-making. In P. Anderson, P. Ayton, & C. Schmidt (Eds.), *Myths and facts about football: the economics and psychology of the world's greatest sport* (pp. 173–190). Oxford: Cambridge Scholars Press.

Brand, R., Schweizer, G., & Plessner, H. (2009). Conceptual considerations about the development of a decision-making training method for expert soccer referees. In: D. Araujo, H. Ripoll, & M. Raab, (Eds.), *Perspectives on cognition and action in sport* (pp. 181–190). Hauppauge, NY: Nova Science.

Brehmer, B., Kuylenstierna, J., & Liljergren, J.-E. (1974). Effects of function form and cue validity on the subjects' hypotheses in probabilistic inference tasks. *Organizational Behavior and Human Performance*, *11*(3), 338–354.

Bröder, A., & Schiffer, S. (2003). Take the best versus simultaneous feature matching: Probabilistic inferences from memory and effects of representation format. *Journal of Experimental Psychology — General*, *132*(2), 277–293.

Brunswik, E. (1955). Representative design and probabilistic theory in a functional psychology. *Psychological Review*, *62*(3), 193–217.

Busemeyer, J. R., & Townsend, J. T. (1993). Decision field theory: a dynamic-cognitive approach to decision making in an uncertain environment. *Psychological Review*, *100*(3), 432–459.

Cooksey, R. W. (1996). The methodology of social judgment theory. *Thinking and Reasoning*, *2*(2&3), 141–173.

Dawson, P., Dobson, S., Goddard, J., & Wilson, J. (2007). Are football referees really biased and inconsistent? Evidence on the incidence of disciplinary sanction in the English Premier League. *Journal of the Royal Statistical Society Series A: Statistics in Society*, *170*(1), 231–250.

Doherty, M. E., & Kurz, E. M. (1996). Social judgment theory. *Thinking and Reasoning*, *2*(2&3), 109–140.

Enkvist, T., Newell, B. R., Juslin, P., & Olsson, H. (2006). On the role of causal intervention in multiple-cue judgment: positive and negative effects on learning. *Journal of Experimental Psychology — Learning Memory and Cognition*, *32*(1), 163–179.

Federation Internationale de Football Association. (2007). *Laws of the game 2007/2008*. Zürich, Switzerland: Federation Internationale de Football Association.

Frank, M. G., & Gilovich, T. (1988). The dark side of self- and social perception: Black uniforms and aggression in professional sports. *Journal of Personality and Social Psychology*, *54*(1), 74–85.

Fuller, C. W., Junge, A., & Dvorak, J. (2004). An assessment of football referees' decisions in incidents leading to player injuries. *American Journal of Sports Medicine*, *32*(Suppl. 1), 17S–22S.

Gigerenzer, G., & Goldstein, D. G. (1996). Reasoning the fast and frugal way: models of bounded rationality. *Psychological Review*, *103*(4), 650–669.

Glöckner, A. (2008). Does intuition beat fast and frugal heuristics? A systematic empirical analysis. In H. Plessner, C. Betsch, & T. Betsch (Eds.), *Intuition in judgment and decision making* (pp. 91–105). Mahwah, NJ: Lawrence Erlbaum Associates.

Goldstein, W. M. (2004). Social judgment theory: applying and extending Brunswik's probabilistic functionalism. In D. J. Koehler & N. Harvey (Eds.), *Blackwell handbook of judgment and decision making* (pp. 37–61). Malden, MA: Blackwell Publishing.

Hammond, K. R. (1955). Probabilistic functionalism and the clinical method. *Psychological Review*, *62*(1), 255–262.

Hammond, K. R., Stewart, T. R., Brehmer, B., & Steinmann, D. (1975). Social judgment theory. In M. F. Kaplan & S. Schwartz (Eds.), *Human judgment and decision processes* (pp. 271–312). New York: Academic Press.

Helsen, W., & Bultynck, J. B. (2004). Physical and perceptual-cognitive demands of top-class refereeing in association football. *Journal of Sports Science*, *22*(2), 179–189.

Hogarth, R. (2001). *Educating intuition*. Chicago: The University of Chicago Press.

Hogarth, R. (2008). On the learning of intuition. In H. Plessner, C. Betsch, & T. Betsch (Eds.), *Intuition in judgment and decision making* (pp. 91–105). Mahwah, NJ: Lawrence Erlbaum Associates.

Holzworth, R. J. (2001). Multiple cue probability learning. In K. R. Hammond & T. R. Stewart (Eds.), *The essential Brunswik: beginnings, explications, applications* (pp. 348–359). Oxford: Oxford University Press.

Jones, M. V., Paull, G. C., & Erskine, J. (2002). The impact of a team's reputation on the decisions of association football referees. *Journal of Sports Science*, *20*(12), 991–1000.

Kahneman, D., Slovic, P., & Tversky, A. (1982). *Judgment under uncertainty: heuristics and biases*. Cambridge: Cambridge University Press.

Klayman, J. (1988). On the how and why (not) of learning from outcomes. In B. Brehmer & C. R. B. Joyce (Eds.), *Human judgment: the SJT view* (pp. 115–162). Amsterdam: North-Holland Elsevier.

Klein, G. (1999). *Sources of power: how people make decisions*. Cambridge: MIT Press.

Lagnado, D. A., Newell, B. R., Kahan, S., & Shanks, D. R. (2006). Insight and strategy in multiple-cue learning. *Journal of Experimental Psychology — General*, *135*(2), 162–183.

MacMahon, C., & Plessner, H. (2008). The sports official in research and practice. In D. Farrow, J. Baker, & C. MacMahon (Eds.), *Developing sport expertise: researchers and coaches put theory into practice* (pp. 172–188). London: Routledge.

Mascarenhas, D. R. D., O'Hare, D., & Plessner, H. (2006). The psychological and performance demands of association football refereeing. *International Journal of Sport Psychology*, *37*, 99–120.

Mathan, S. A., & Koedinger, K. R. (2005). Fostering the intelligent novice: learning from errors with metacognitive tutoring. *Educational Psychologist*, *40*(4), 257–265.

Messner, C., & Schmid, B. (2007). Über die Schwierigkeit, unparteiische Entscheidungen zu fällen: Schiedsrichter bevorzugen Fußballteams ihrer Kultur [About the difficulty to make impartial decisions: referees favor culturally familiar football teams]. *Zeitschrift für Sozialpsychologie*, *38*(2), 105–110.

Nevill, A. M., Balmer, N. J., & Williams, A. M. (2002). The influence of crowd noise and experience upon refereeing decisions in football. *Psychology of Sport and Exercise*, *3*(4), 261–272.

Newell, B. R., Lagnado, D. A., & Shanks, D. R. (2007). Challenging the role of implicit processes in probabilistic category learning. *Psychonomic Bulletin & Review*, *14*(3), 505–511.

Oudejans, R. R. D., Verheijen, R., Bakker, F. C., Gerrits, J. C., Steinbrückner, M., & Beek, P. J. (2000). Errors in judging 'offside' in football. *Nature*, *404*(6773), 33.

Petit, J.-P., & Ripoll, H. (2008). Scene perceptions and decision making in sport simulation: a masked priming investigation. *International Journal of Sport Psychology*, *39*(1), 1–19.

Plessner, H., & Betsch, T. (2001). Sequential effects in important referee decisions: The case of penalties in soccer. *Journal of Sport & Exercise Psychology*, *23*(3), 200–205.

Plessner, H., & Haar, T. (2006). Sports performance judgments from a social cognition perspective. *Psychology of Sport and Exercise*, *7*(6), 555–575.

Stefani, R. (1998). Predicting outcomes. In J. Bennett (Ed.), *Statistics in sport* (pp. 249–275). London: Arnold.

Sutter, M., & Kocher, M. G. (2004). Favoritism of agents – The case of referees' home bias. *Journal of Economic Psychology*, *25*(4), 461–469.

Unkelbach, C., & Memmert, D. (2008). Game-management, context-effects, and calibration: the case of yellow cards in soccer. *Journal of Sport & Exercise Psychology*, *30*(1), 95–109.

Wigton, R. S. (1996). Social judgement theory and medical judgment. *Thinking and Reasoning*, *2*(2&3), 175–190.

Wilson, M. (2002). Six views of embodied cognition. *Psychonomic Bulletin & Review*, *9*(4), 625–636.

M. Raab et al. (Eds.)
Progress in Brain Research, Vol. 174
ISSN 0079-6123

CHAPTER 14

A conceptual framework for studying emotions–cognitions–performance linkage under conditions that vary in perceived pressure

Gershon Tenenbaum[1,*], Bradley D. Hatfield[2], Robert C. Eklund[1], William M. Land[1], Luis Calmeiro[3], Selen Razon[1] and Thomas Schack[4]

[1]Department of Educational Psychology and Learning Systems, Florida State University, Tallahassee, FL, USA
[2]Department of Kinesiology, University of Maryland, College Park, MD, USA
[3]School of Social and Health Sciences, University of Abertay, Dundee, United Kingdom
[4]Neurocognition and Action Research Group and Center of Excellence "Cognitive Interaction Technology" (CITEC) and Research Institute for Cognition and Robotics (CoR-Lab), Bielefeld University, Bielefeld, Germany

Abstract: A unified conceptual framework, which integrates the structural components of human performance, such as emotional processes (i.e., feelings, mood), cognitive processes and structures (e.g., knowledge architecture, long-term working memory), motor processes (coordination, endurance), and the neurophysiologic basis of these structural components (i.e., activation of cortical areas) is introduced. Recent developments in the cognitive, neurological, expertise, and emotion sciences provide a sound evidence for this conceptualization. The unified conceptual framework enables a better understanding of human performance, and allows generating applications, which share scientific validity.

Keywords: mental representation; emotions; brain activity; cognitive processes; action

Introduction

Many performers exhibit high-level performance in practice, but sometimes struggle under stressful/competitive conditions (see Beilock and Gray, 2007 for extensive review). Though motor skills and mental representations of these skills are inherited and learned, the performer's use of them alters under emotional/mental and temporal pressure. How does the emotional–cognitive–motor linkage change under pressure? What are the underlying mechanisms that permit or prevent an efficient course of action? Questions such as these cannot be answered with confidence at this stage. Though sound theories and extensive research have been devoted to explore this linkage, almost all efforts have not taken an integrative approach to answer these questions. In this chapter, a conceptual framework is provided offering an integrative approach to study mental and motor operations under emotion-invoking pressure conditions.

There is a substantial interest in exploring a framework of human performance. Yet, foundations for the experimental analysis of the structure

*Corresponding author.
Tel.: 850-644-8791; Fax: 850-644-8776;
E-mail: tenenbau@coe.fsu.edu

DOI: 10.1016/S0079-6123(09)01314-4

and functioning of human capabilities to perform have only been established in recent years as new methods in cognitive neuroscience and movement science have developed. Our aim in this chapter is to compile results from various fields, such as intelligence, problem solving, emotions and emotion regulation, action and the motor system, and expertise, and to make comprehensive claims regarding the underlying mechanisms of human performance. Once these mechanisms are uncovered, ecologic applications can be made to enhance performance. So far, the structural components of human performance, such as emotional processes (i.e., feelings, mood), cognitive processes and structures (e.g., knowledge architecture, long-term working memory), motor processes (coordination, endurance), and the neurophysiologic basis of these structural components (i.e., activation of cortical areas) have been studied independently. We integrate current understandings into a unified theoretical framework to enable a better understanding of human performance.

Basic assumptions and concepts

Every conscious action made by humans is a consequence of response selection, whether intentional or unintentional. By definition, response selection indicates adaptive behavior based upon the capacity to solve problems. This "behavioral effectiveness" is directed by cognitive processes and mental operation. The effectiveness of these processes consists of the richness and variety of perceptions processed at a given time; that is, the system capacity to encode (store and represent) and access (retrieve) information relevant to the task being performed (Tenenbaum, 2003). From an information-processing perspective, motor behaviors consist of encoding relevant environmental cues through the utilization of attentional strategies, processing the information through an ongoing interaction between working memory and long-term memory, making an action-related decision, and executing the action while leaving room for refinements and modifications. Under pressure, changes in each of these components are seen. These changes are sequential in nature (i.e., begin with the perceptual components, continue with the cognitive components, and end with the motor system). Taking a complimentary perspective, the cognitive construction of human actions consists of four levels, each with a designated function (Schack, 2004a, b; see Table 1). The function of the *mental control* level (level IV) is sketched for voluntary movement regulation, and the coding or the anticipated outcome of movement. The *mental representations* level (level III) predominantly forms a cognitive benchmark for the mental control level (i.e., level IV). It is organized conceptually, and is responsible for transferring the anticipated action outcome into a model of the movement structure it requires. Because an action is "no chain of details, but a structure subdivided into details" (Bernstein, 1988, p. 27, translated), movement organization must have a working model of this structure at its disposal. The corresponding abilities for using movement representations have been acquired stepwise during technical preparation. These movement representations hold the knowledge that relates directly to performance. However, the model also clearly reveals that these representations are functionally embedded in further levels and components of action organization. Therefore, the functioning of the lower levels (levels I and II) is sketched as well. The level of *sensorimotor control* is linked directly to the

Table 1. Levels of action organization (Schack, 2004b)

Code	Level	Main function	Subfunction	Means
IV	Mental control	Regulation	Volitional initiation Control strategies	Symbols; strategies
III	Mental representation	Representation	Effect-oriented adjustment	Basic action concepts
II	Sensorimotor representation	Representation	Spatiotemporal adjustment	Perceptual effect representations
I	Sensorimotor control	Regulation	Automation	Functional systems; Basic reflexes

environment. In contrast to the level of mental control (level IV), which is *induced intentionally*, the level of sensorimotor control (level I) is *induced perceptually*.

For the stability of practical performance, all four levels are of significance. Yet, if excellent performance is shown under neutral conditions but collapses under pressure, one should assume that *though cognitive and motor systems are established, the performer's access to these systems under pressure is impaired*. It is furthermore plausible that there is another system besides the predominantly cognitive motor system, which, among other things, is accountable for the control of emotional processes (i.e., self-regulation and mental control). For working purposes, reliance on three independent, but inter-related systems is made:

- Cognitive structure components and processes (cognitive appraisal)
- Emotional system
- Self-regulation structure (e.g., various strategies are applied here: emotional control, motivation control, attention control, etc.).

When performance is satisfactory under neutral (nonpressure) conditions, all levels (see Table 1) function harmoniously and optimally.

Current state of the art on the emotions–cognitions–performance linkage

Affect, emotions, and links to cognition

According to Rosenberg (1998), affective traits are stable predispositions, which determine emotional responses, and reflect individual differences in emotional reactivity. Similar to affective traits, moods determine the threshold of occurrence of a given emotion. The classification of emotions as "positive" (e.g., pride, pleasure) or "negative" (e.g., anxiety, anger) is frequently based on hedonic tone (i.e., pleasant vs. unpleasant). Therefore, positive emotions are mistakenly thought to be good or desirable, while negative emotions are thought to be bad or undesirable (Plutchik, 2003). However, such categorization does not have a functional basis. Emotions motivate, organize, and guide perception, thought processes, and mobilize action towards a behavioral purpose. Positive or pleasant emotions are believed to have a role of enlarging the behavioral and thought repertoire to face a given situation (Fredrickson, 1998). A wider repertoire allows the individuals to create and develop new solutions, and enhance personal resources. Such an enhancement is associated with several processes: (1) expansion of attention focus, (2) expansion of cognitive processes, (3) expansion of action possibilities, by inducing unusual and diverse responses that add to subjects' personal resources, (4) expansion of intellectual resources, and (5) expansion of social resources (Fredrickson, 1998). The effects of positive emotions are also relevant for performers who operate under stressful conditions as they may increase the spectrum of relevant information attended to, encourage the utilization of novel and unpredictable strategies and actions, prompt the development of mastery, and enhance team interactions. These factors increase the likelihood of successful performance. By verifying the broader functions of emotions, it is possible to generate their positive effects while reducing their negative potential (Damásio, 1994).

Negative emotions, such as anxiety, should be understood not only as having an undermining effect on performance, but also as a process of adaptation to specific events or a motivating factor for a particular action. A performer's anxiety in a particular moment of an action can reinforce his/her sensitivity to dangerous situations, prompting adoption of defensive strategies, and engaging in more realistic decision making. Carver and Scheier's (1988) control process model of anxiety and performance holds that anxiety can have both facilitative and debilitative effects on performance depending on subjects' expectancy of being able to cope with anxiety and complete the action. Support for this contention in sport comes from the work of Jones and colleagues, where highly skilled swimmers (Jones et al., 1994) and cricketers (Jones and Swain, 1995) interpreted both cognitive and somatic anxiety symptoms as more facilitative to performance. Swimmers who had positive expectancies of goal attainment interpreted anxiety as more facilitative than swimmers who had negative expectations of goal attainment (Jones and Hanton, 1996). Thus, cognitive anxiety can improve motivation and facilitate appropriate attention focus (Jones et al., 1993).

When movements were performed following exposure to emotional stimuli, active defensive circuitry resulted in faster, but more variable, voluntary movements. In the study of Coombes et al. (2006), exposure to unpleasant images led to an increase in mean force production of a sustained voluntary movement more than exposure to pleasant stimuli, as well as acceleration of central processing times. Furthermore, the length of exposure to affective stimuli mediated speed and accuracy of motor performance; compared to pleasant stimuli, unpleasant stimuli led to either increased error or increased speed. According to biphasic theory of emotion (Lang et al., 1997), the wide array of emotion can be classified by two dimensions; *valence* (i.e., appetitive or defensive), and *arousal intensity*. When performing a task, the two affective dimensions are activated by the functional brain, priming physiologic adaptations and mental representations, which correspond with the environmental context. Thus, the nature of the task, the context, and the emotion eliciting stimuli all mediate the production of movement (Coombes et al., 2006).

The relationship between emotion and cognition from an action-oriented perspective is depicted in Fig. 1. The appraisal of events, action effects, or stimuli in the environment is the first cognitive process in action organization in the model. The result of appraisal is not only stored in memory, but is of central meaning for evoking emotions. The stimuli and appraisal-dependent emotions are stored in memory as specific elements of cognitive event profiles. Furthermore, emotions are functionally linked with the beginning and maintaining of motivation. One stimulus may produce not only one, but several types of motivation.

At this stage, the level of mental control comes into play. Processing at this level starts with the decision about a relevant action. The result of this decision-making process is the intention to reach specific action effects. Based on this intention an action plan is created and the mental control processing runs to a module, which is responsible for action-execution. This module is linked to the level of sensorimotor control (see Table 1), and includes all motor components necessary for production of goal-directed action effects. The kind and quality of action effects are important information for the action system. But if the action effects are not congruent to the intended outcomes, or not valid to cope with the actual situation, the appraisal system will read an insufficient action, and will evoke negative emotions. In case of problems in action realization (e.g., the real situation in competition is much more difficult than the expected one), mental control processing must run through another path. Now, the performer must use action strategies such as control of attention, control of emotion, or motivational control. Such strategies are supported by inner speech, and are used to stabilize action realization. Thus, if the performer lacks such strategies, he/she has no tools to control the action in a befitting manner. If the performer lacks mental control, he/she will not realize his/her intentions and will lack intended action effects. This kind of information is negatively valued by the appraisal system, and influences the development of emotions dramatically. Thus, one important link between emotion and information storage is caused by the representation of emotionally induced action effects in long-term memory. From this point of view, emotions are a part of information storage in general.

Emotions can be viewed as memory units (Bower, 1981). They are components linked to the memory system that facilitate access to mental representations associated with targets of judgment (Forgas, 1991). Due to prior associations, innate and learned environmental situations activate particular emotion nodes stored in memory. This activation spreads throughout neuronal circuits to mental representations of events associated with that emotion influencing encoding and retrieval of material, as well as the valence of judgments of people, events, objects, and behaviors (Bower, 1991). Emotions are activated by experiencing the emotions again, or by activation of any

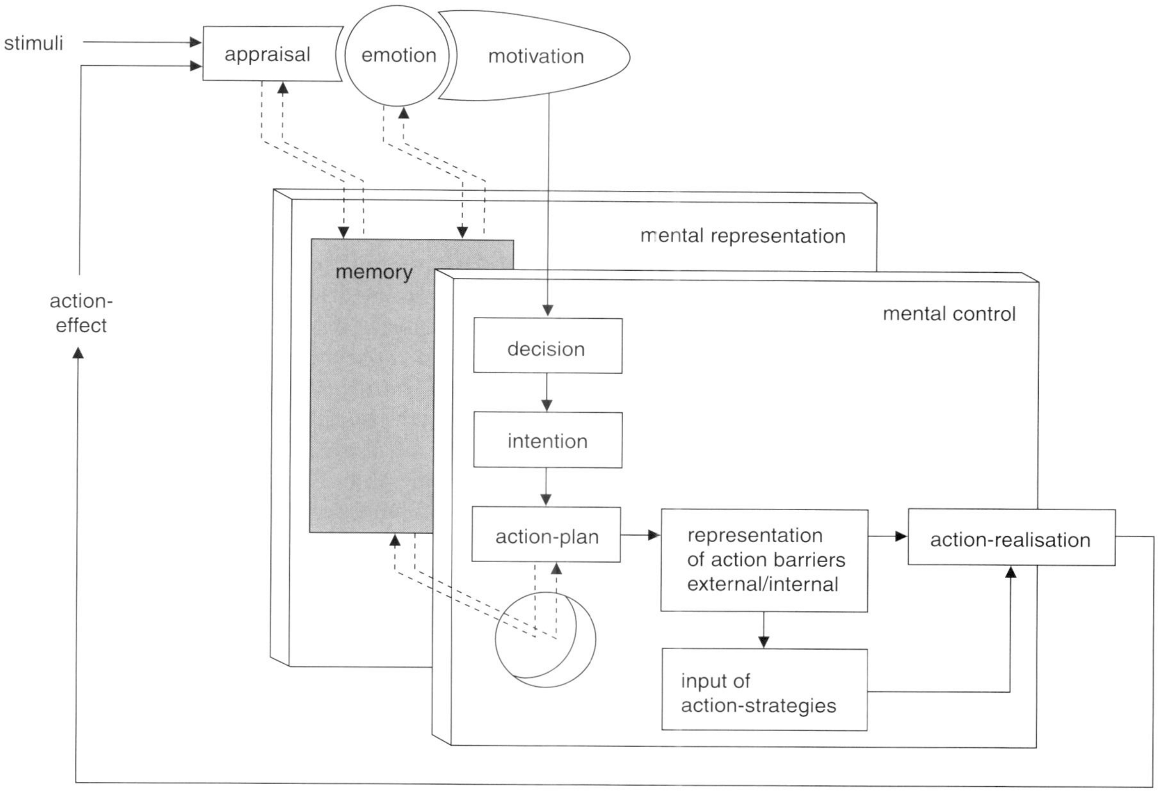

Fig. 1. Emotion–cognition–linkage model.

of their links (Barry et al., 2004). The stronger the activation of particular emotional nodes is, the greater the mood-congruent effect. When emotions are strongly activated emotion-congruent constructs (e.g., concepts, words, themes, and rules of inference) become primed and available for use bringing into readiness certain perceptual categories, themes, or ways of interpreting the world congruent with current emotional states. In affect-priming terms, an emotion node spreads activation throughout the memories to which it is connected, increasing the chance that those memories will be retrieved (Bower, 1981).

Emotions are stored in memory as functional elements of event files. Changes of emotional states trigger the related event files (see Fig. 1), and activates related structures of mental representation. This is important for capturing human performance because perception and action are based on the same representation structures (see coding theory; Prinz, 2005). Such an emotion-based activation of representation affects the perception of the actual situation, and the focus of attention. Therefore, there is the risk that a vicious circle (i.e., a no-win situation) between threatening stimuli, and negative emotional states, which activates event files, representation structures, and focuses attention on threatening stimuli, with performance decrement as a probable result.

The model proposed in Fig. 1 enhances capturing the complex interaction between emotion, judgment, and decision making. After perceiving a situation or a stimulus-configuration, appraisal processes play an important role for the formation of emotions. The appraisal of a particular situation is based on different memory traces. In case of positive experience and a high congruence between intended and actual action, positive-oriented memory sources become the reference point for appraisal. The result of such an appraisal process could be in the form of a positive mood. The positive mood affects the generation of motivation in this situation, and influences the process of decision making (see Fig. 1).

In turn, emotions influence the judgments about situations both directly and indirectly. The direct influence is based on the development of different types of motivation, which result in appraisal-based emotions. The question in this stage of action processing is, how much and what type of motivation is generated in a situation? The indirect influence has to do with the informational quality of emotions, and is a reference point for decision making and judgments. Thus, one may decide more holistically, or nonanalytically, in case of positive mood. From this point of view, the direct influence of emotions has to do with the content of decision making, especially the number and quality of motivation, while the indirect influence of emotions is defined by the type of decision making (holistic vs. analytic mode of decision making). This emotion-based influence on mental control modes is valid for action organization, because the formation of emotions is grounded on memory-based appraisal processes (see Fig. 1). To conclude, emotions are important information sources for individuals' decision making and mental control. For instance, concerning a special situation, positive mood is an indicator for successful coping, and an effective action organization in the past. This information reduces the probability of errors in action, or the probability for threat and danger.

As can be noted in the model (see Table 1), processes of mental regulations are required for a high capacity of information processing and mental control. They take place at the lower sensorimotor levels, and have to do with the use of basic routines and sensorimotor representation structures. At the level of mental control it is important to generate valid plans and to control attention and emotions in case of unexpected external events in the action organization. Basic regulation must generate and implement such tools like routines at the lower level, and implement control strategies at the mental control level. If such steps are taken, the action system contains sufficient capacity for the regulation of the actual action.

Affect can also be used as input for cognitive processing and behavior. Positive affective states inform subjects that the environment is safe, and no specific action is required (Schwarz and Bless, 1991). According to Schwarz and Bless, people experiencing positive affects also benefit from greater cognitive flexibility, being able to establish more word associations, and find more relationships or differences between concepts than controls. Subjects in negative moods are constrained by the identified problematic situation, thus access to and application of diverse knowledge structures are less flexible, and performance decline is more probable. Thus, emotions can be seen as preceding or resulting from an action as well as "accompanying" it, and creating a *coping* style, which must be considered when performing an action, particularly under pressure conditions.

The model shows a functional relationship between the level of mental control and emotions (Fig. 1). From this point of view, the development of emotions is functionally related to the actual generated action effects. Of importance for the formation of emotions is the difference between intended and actual effects, and the appraisal of a difference. For this reason, the model is in accordance with specific emotion theories. Mandler (1979, 1985) assumes that the abort of a planned action can be seen as a central reason for anxiety development. Based on Mandler's approach, task-relevant stimuli are perceived through the increased interruption of the action. In turn, these stimuli demand their own attention resources, and therefore disturb action performance. *Thus, anxiety could not only be the reason of a performance interruption, but also stringently seen as its consequence.* The models (see Table 1 and Fig. 1) consists of the assumption that the interruption of activated performance plans and the increasing inability to subordinate the action performance to an action program are attributed both emotionally and negatively. Particularly, through the collapse of solid connections within the system of mental control, the inner speech loses the function of an instrument of action control (see Schack, 1997). The result is a functional damage of mental control; a mental control deficit. One can deduce three reasons for the development of such a mental control deficit including the involved emotional reactions from the model.

The first reason is the development of an *inappropriate action* plan. On the basis of the former, specifically negative, experience within an action space, and therefore motivated by specific information–evaluation mechanisms in the appraisal system, a biased scheme of the internal and external action conditions is underlying the process. Subsequently, reference stimuli are selected, which match this appropriate scheme, and influence the information processing disadvantageously. Already before selecting an action, one must rely on a cognitive scheme, which was established through former experience, which suggests the recourse to stereotypical action sequences (e.g., avoidance, dysfunctional internal dialogue, etc.). The second reason of the mental control deficit issue is directly related to the first reason, and refers to the *deficient availability of action control strategies*. A shortcoming of aim-oriented control of the action performance (e.g., failure of coping or motivation control), can lead to termination of the initiated action and the collapse of the mental control.

The third reason of such an action control deficit is related to a *collapse of the control anticipation*. With decreases of the anticipatory system to perform and control an action, and cope with incidental difficulties when performing an action, the effect of negative emotions increases within the process of action argumentation and action performance (Bandura, 1991; Carver and Scheier, 1991). Finally, affective reactions can inhibit an organized course of the interaction of cognitive processes during the action construction, and prevent the building of solid connections of mental control (Schack, 1997).

The notion of appraisal and coping

People make decisions based on an evaluation or appraisal process through which the transaction is judged to be essential in maintaining, enhancing, or hindering well-being (Lazarus, 1999). When encountering a stressful condition, people appraise the resources they can use to cope with the situational demands. *Primary appraisal* consists of judgments concerning the relevance of what is happening in the transaction to one's values, goal commitments, beliefs about self and the world, and situational intentions. *Secondary appraisal, which may precede primary appraisal in some cases*, refers to a cognitive–evaluative process that is focused on *what can be done about a stressful transaction*. This type of appraisal refers to subjects' evaluation of such factors as *agency*, *future expectancies*, and *coping options*. Positive emotions arise in the presence of favorable conditions, but negative emotions follow under unfavorable ones.

Generally, as it is assumed in the model, the triggering of emotional states is bound to appraisal processes, whereupon the necessary information evaluation results from the evaluation criteria. Phylogenetically predispositioned (i.e., subcortical) evaluation mechanisms are related to an immediate appraisal system, whereas experience-dependent (ontogenetically experienced) evaluation mechanisms are assigned to a (cortically attached) memory-dependent appraisal system. The former mechanisms are connected to parts of the limbic system, and therefore can generate emotional reactions virtually concurrently when getting information, which is liable to a primary (subcortical) evaluation. If information relies on an individual memory inventory, which can only be evaluated through experience-based representations, then the triggering of emotional reactions takes place after the cognitive processing. Here, both the experience the performers have gathered in certain situations, and the coping strategies they have at their disposal are of vital importance to the quality of action generation and control.

Coping

Coping refers to "constantly changing cognitive and behavioral efforts to manage specific external and/or internal demands that are appraised as taxing or exceeding the resources of the person"

(Lazarus and Folkman, 1984, p. 141). Thus, coping is an ongoing process that takes into account the "fit between what one does, the requirements of the conditions being faced, and one's individual needs" (Lazarus, 1999, p. 80). Lazarus considers that good coping consists not only of choosing the most adequate strategy, but also being flexible about abandoning an effective strategy for another, even more effective one. Coping helps subjects to deal with *aspects of a problem* (i.e., *problem focused*), and with the *emotions* associated with it (i.e., *emotion focused*). Endler and Parker (1990) added *avoidance-coping* as another dimension.

Appraisals and *coping* are considered to be associated with emotions (Lazarus, 1999). In order for a coping response to be executed, one has to evaluate what is at stake (i.e., primary appraisal), and whether one can cope with the situation (i.e., secondary appraisal). In contrast, efficient coping processes also influence and change appraisals and the nature of the transaction. As a threatening person–environment relationship is reappraised, a new relational meaning of the stressful encounter is constructed (Folkman and Lazarus, 1990), and different emotions arise.

The conceptual framework presented here extends Lazarus's (1999) concept, as it is more suitable to the motor domain. It assumes that two basic forms of performance regulation are necessary to secure smooth performance: a *process regulation* and a *basic regulation*. While the latter regulates emotional–motivational basic processes, the former accounts for the precise execution of the action goals. Process regulation builds on basic regulation; basic regulation creates the necessary conditions for process regulation. As known from many situations in sport, some athletes falter in competition despite being able to perform well under training conditions. It is argued here that these athletes show good process regulation under low pressure, but they fail to do likewise under high pressure. In such cases, it is possible to diagnose difficulties in basic regulation. Basic regulation stabilizes the action system under different pressure conditions. Thus, the goal of basic regulation is to stabilize the emotional and motivational processes in every situation of action organization. Therefore, basis regulation uses different tools at every level of action organization.

At the level of sensorimotor control, routines are helpful to operate an optimal fit among anticipation of perceptual action effects, perceptual input, and behavior units for stabilizing emotions. Routines provide an athlete with a consistent set of behaviors that can be specifically activated in moments of intense pressure and in those situations when it is difficult to become motivated. Routines give performers a sense of security and the confidence that they can rise up to meet the challenge. At the level of mental control, well-functioning performers are using a toolbox of strategies to control emotion and motivation. These strategies are based on inner speech (i.e., self-instructions), and are considered important psychological tools for coping with adversity (Schack, 1997).

In contrast to basic regulation, the process regulation is a procedure responsible for controlling the actual action organization. Process regulation activates and focuses modules like, for instance, attention and motor decisions making for reaching goal-directed action effects. Basic regulation and process regulation are not separated in reality. Furthermore, they contribute in a separate and specific manner to the same action and to the same performance. But if performers fail to reach the anticipated effects, and do not perform well, it is necessary to look closer to the covert regulation processes, to distinguish between deficits in basic regulation and deficits in process regulation. In some cases, one may suffer deficits in both regulation types. To improve process regulation one must use training procedures in general (e.g., imagery training), which are determined at the level of mental representation (see Fig. 1). To improve basic regulation one must use specific self-regulation trainings (e.g., stress-regulation trainings), which are working at the level of mental control. Here, one must develop strategies to control emotions in an appropriate manner.

Physiological and cognitive consequences of emotions on performance

Rooted on the premise that negative emotions have very specific autonomic changes, and positive emotions are characterized by a relative lack of autonomic reactivity, Fredrickson et al. (2000) described positive emotion action tendencies as nonspecific. Based on the different action tendencies of both types of emotions, it was hypothesized that positive emotions undo the cardiovascular after-effects of negative emotions. Because negative emotions narrow individuals' thought–action repertoires while positive emotions broaden them (Fredrickson, 1998), the latter might serve to correct the effects of the former. Therefore, if negative emotions increase cardiovascular activity, preparing the body for quick and vigorous action, positive emotions facilitate recovery of cardiovascular function. Indeed, Fredrickson et al. (2000) found that after anxiety-induced cardiovascular reactivity, contentment- and amusement-eliciting films produced faster cardiovascular recovery than neutral or sad films. When these films were viewed after an emotionally neutral film, positive emotional states were not significantly different than neutral states in maintaining the physiological activation. That is, the reduced autonomic reactivity occurred only after initial heightened physiological activation due to negative emotions. In addition, cardiovascular activation after negative emotions was found to last longer than after positive emotions, regardless of reactivity (i.e., initial activation). The prolonged activation observed after negative emotions was speculated to be the product of ruminative thoughts (Brosschot and Thayer, 2003).

The findings of prolonged activation and faster recovery of cardiovascular activity associated with positive emotions (Fredrickson et al., 2000) have relevant implications as they support the need for intervening, in clinical contexts, with both types of emotions. When excessive activation is associated with an unpleasant hedonic tone, cognitive restructuring may facilitate positive emotions and rapid recovery. This has a significant impact on performance, not only because excessive arousal may reduce information processing (Easterbrook, 1959), but also because athletes may use this physiologic activation as a cue to build their self-efficacy (Bandura, 1997).

Arousal is thought to have a curvilinear relationship with performance (Easterbrook, 1959). Experienced at an optimal level, arousal maintains an optimal attention focus that consists of allocation of sufficient attention resources to task-relevant cues, while filtering or ignoring task-irrelevant cues. Departures from those optimal states results in performance impairment by altering performer's attention focus, resulting in the inability to pay attention to relevant cues (i.e., overarousal induces excessive narrowing of attention focus) or inability to ignore irrelevant cues (i.e., underarousal induces an excessively broad attention focus).

In a dual-task autoracing simulation, drivers who were highly anxious experienced an altered ability to acquire peripheral information at the perceptual level (Janelle et al., 1999). At higher levels of anxiety, the identification of peripheral lights became slower and less accurate, and significant performance decrements occurred in central and peripheral tasks. A variety of negative and positive emotions are also associated with increased activation; making it unclear whether it is the arousal or the valence of the emotions responsible for alterations on information processing. It is likely that arousal may interfere differently with cognitive processes, and consequently with motor performance when associated with different hedonic tones.

Variation of psychological and physiologic activation due to the stress response has an effect on the width of attention field, level of distractibility, amount of investment in controlled processing, and efficiency of attention processing (Janelle, 2002). These processes depend largely on the emotional experience (Mellalieu, 2003). In line with the research on specific psychological and physiologic appraisal responses (e.g., Tomaka et al., 1997), it would be important to determine how the cognitive changes are influenced by different appraisals of the stressful situation (i.e., threatening vs. challenging situation).

Extensive research studying marksmen has focused on the impact of arousal on sustained attention and performance; mainly distinguishing arousal, a nonspecific concept, from vigilance, a "preparatory attention state associated with anticipated cognitive/perceptual or behavioral activity independent of arousal" (Tremayne and Barry, 2001). Vigilance has a marked behavioral connotation involving "sustained readiness to detect and respond to environmental changes: it is an active, performance-related process involving perception of stimuli and information processing" (Collet et al., 1996, p. 954).

An increased activation level, measured by electrodermal activity, has been found to limit performance in sharp (rifle) shooters (Caterini et al., 1995). Caterini et al. described good shooting performances as characterized by low amplitude and increased duration of response of physiologic adaptations representing better mastery of emotional reactivity and greater concentration time, respectively. In contrast, systematic decrease in activation prior to the shot did not discriminate between best and worst shots (Tremayne and Barry, 2001). However, experts developed a marked state of attention focusing in the 10–15 s prior to shot initiation. If this state is enhanced and extended to the shot, better performance occurs. Worse shots take place when there is less vigilance, and when shot-initiation is delayed until reduced maximum vigilance occurs.

Tremayne and Barry (2001) compared physiological patterns of best versus worst shots. In expert pistol shooters, lower pre-shot electrodermal levels and longer and more systematic pre-shot cardiac deceleration characterized best shots compared with worst shots. Comparing with novices, experts underwent a slow reduction in skin conductance and heart rate levels before the shot, and "rebound" increase immediately after the shot, which were not observed in novices.

The concept presented in Table 1 and Fig. 1 allows studying the cognitive patterns related to different tasks, stress sources (pressure information), and related emotions. The patterns concerning the "pressure" information representation and the relationship between "pressure" information and emotions are the reference point for appraisal in real situations. These links allow predicting emotional states, and modifying the reference points (i.e., benchmarks) of the appraisal systems, and measure the link between the representation of threatening situations and the representation of action organization in these situations. The goal is to strengthen the productive links between the representation of the situation and the representation of the action (Nitsch, 2004; Schack and Hackfort, 2007). In the motor performance domain, one must capture the mental representations of situational components, such as the representation of the team structure, or the representation of the environmental conditions of a specific competition. Furthermore, one must know what type of activities performers have represented to cope with situations that vary in many respects. Finally, it is essential to measure the types of event-representations and action representations, and look for the fit between these two representation types. A misfit between these representations may lead to emotions of stress and anxiety. From this perspective, looking at the emotional process itself is insufficient, without measuring the cognitive benchmarks (especially event and action representations) of the appraisal process, which constitutes the development of emotions in action (see Fig. 1).

Cerebral cortical activity during skilled visuo-motor performance — a model to assess the impact of emotions during performance

The superior performer shows relaxation in non-essential areas of the brain and minimizes communication between the thinking and the motor regions — in essence, they become instinctive and efficient. Figure 2 illustrates brain electrical activity maps that reveal higher cortical activity (i.e., electroencephalogram (EEG) gamma power) in novice shooters compared to expert shooters. The gamma power is positively related to cerebral cortical activation and the higher levels exhibited by the novices implies a "noisy" or less stable platform from which to initiate motor commands to the skeletal muscles. Such a state would result in greater performance variability and diminished

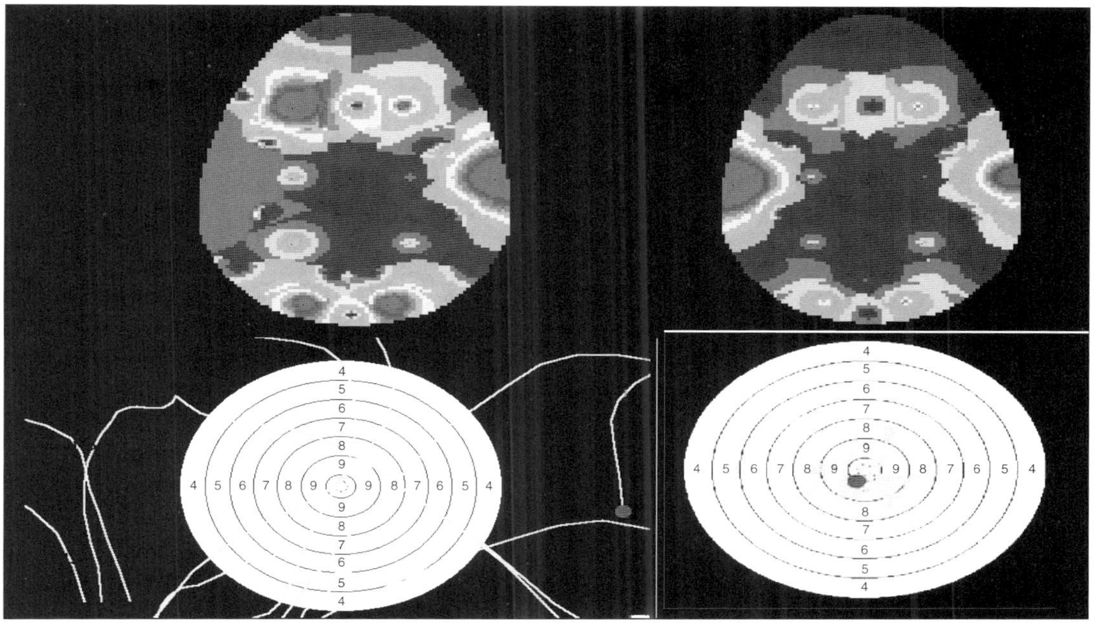

Novice Expert

Fig. 2. Cortical activity in novice and expert marksman. (See Color Plate 14.2 in color plate section.)

accuracy, which is typically the case with the novice. The lower levels of gamma power in the expert reduces such neuromotor noise, thus enabling greater consistency of performance, which is a hallmark of expert performance. In addition, the appearance of such cortical "noise" in the expert under conditions of psychological stress can result in a reversion of performance level to that of the novice (see Hatfield et al., 2004 for extensive review).

As such, the "busy" or noisy mind of the novices is associated with increased variability of the aiming trajectory as compared to that of the experts. More specifically, EEG studies of expert marksmen have revealed less activation in the left temporal region (T3), relative to that observed in the right homologous region, during the aiming period leading up to the trigger pull (Hatfield et al., 1984) implying an adaptive state of regional relaxation whereby potential interference from left-hemispheric verbal–analytic activity with automated motor processes is effectively managed. This finding has particular significance relative to earlier discussion in this paper in which negative moods were posited to elicit analytic step-by-step strategies. Accordingly, such strategic thinking is associated with the left hemisphere, and the study results imply both a lack of such an analytical cognitive approach, as well as a diminution of negative affect, during expert motor performance. Similarly, Haufler et al. (2000) observed less activation in the frontal, central, temporal, parietal, and occipital regions in expert shooters relative to novices. The skill difference was of greatest magnitude in the left temporal region, and revealed a similar pattern of temporal asymmetry in the experts to that observed earlier by Hatfield et al. (1984). The experts accomplished the task in a more efficient manner as indicated by the EEG. No differences in cortical activation were noted between the groups when they performed cognitive tasks with which the participants were equally familiar, suggesting that the task-specific EEG differences were obtained as a result of practice and skill level. It could be theorized that those who perform best under pressure are able to maintain cortical relaxation during challenge, while those who fail to maintain such a state are more likely to "choke."

The relationship between expertise, cortical activation, and order formation in long-term memory is considered a major challenge. To learn more about these interactions and networks, one must combine the neurophysiological methods, like EEG-coherence measures (Deeny et al., 2003), and methods for measuring the structure of mental representation (Schack, 1999). In order to learn more about such relationships, a pilot study was conducted by Schack (2003). Twelve experts and 12 novices performing rotational movements in the field of gymnastics were examined on representations of somersault flights. To examine cortical activity for these movement representations, a special measurement of the neuronal activity in the brain was employed. The assumption was that information recall in the memory is accompanied by space and time defined changes in excitability in neuronal network structures; this is the basis of the neurophysiologic measurement. The EEG-coherence method (Schack, 1999) was used for this purpose. Here activity of cortical areas is revealed by a sequence of coherence maps, which remain stable over time.

The analysis of the structural dimension analysis of motor memory (SDA-M) data pointed to differences between the mental representations according to the level of expertise. It was shown that the cognitive structure of experts is close to the biomechanical structure of the actual movement. The representations of experts were highly differentiated, and were more strongly function oriented than the novices' representations using a special invariance measure. The simultaneous neurophysiologic findings indicated that experts' cognitive activation of movement representations is accompanied by activation of several cortical areas. These patterns of activation provide a glimpse of the network of neurophysiologic areas with a high mental economy and a high stability. It seems that there are two different functions of activation concerning the time-related steering of movements and concerning space-oriented aspects of movement representations. A failure to obtain a stable activation of separate areas in movement representation was evident.

Another interesting question has to do with a systematic study of differences in cortical activation in perception of object and action concepts. Some studies were carried out to measure the cortical activation in perception of object concepts (Krause, 2000). Findings revealed the same kind of activation in the brain while performing and imagining actions (Jeannerod, 2004). Most of these studies were conducted through simple hand or finger movements. It would be challenging to measure the corresponding cortical activation in the brain using basic action concepts (BACS; e.g., the task-related elements, which constitute together the metal representations structure) of complex actions in experimental settings, and to find out about possible differences concerning the involved areas in object and action concepts.

Although evidence consisted of stationary self-paced target shooting, Kerick et al. (2004) recently extended the study of brain processes during skilled marksmanship to reactive shooting scenarios in which soldiers and marines had to react quickly to "pop-up" targets and discriminate between friend and foe stimuli. That is, they had to suppress firing on the friendly targets, and engage the enemy targets in an attempt to more closely approximate the kinds of challenges faced by soldiers in the field. The study attempted to achieve a degree of ecological validity for warfare conditions and, importantly, revealed the classic left temporal relaxation effect during the aiming period in the highly skilled group of participants. In addition, the investigators noted a decrease in alpha power as they progressively challenged the study participants with cognitive load (i.e., challenged them with increasing attention demands and decision making during shooting). Such a negative relationship between demand and alpha power provides a form of concurrent validation for the notion reported above that higher alpha levels are associated with cortical relaxation. However, studies on the relationship between physiological arousal, emotions, stages of brain activity, mental representations, and performance quality must be designed to investigate how mental frame tools can affect or alter performance systematically in case of physiological activation. Additional studies must focus on the question of how mental frames influence the quality of emotions, and to what extent the quality of emotions on performance (see Fig. 1). There is a need for experimental research to find out how strong performance is influenced by the type and quality of emotions, and to learn about the functional granularity of cognition–emotion units in the production of performance.

Management of brain processes to enhance performance

Affective neuroscience — brain processes during emotion. Based on fundamental neuroscience concepts provided by Bear et al. (2001), Hatfield et al. (2004) recently articulated a model with which they speculated on the neural structures involved in a system or circuit, which mediate the psychological and physiological responses to stress and its impact on motor performance. Generally, the stress response is orchestrated by the limbic system, but the central components of this functional circuit are the amygdalae, small almond-shaped structures located bilaterally, and anterior to the hippocampi on the inferior and medial aspect of the temporal lobes. Multiple sensory pathways converge in the basal lateral nuclei of the amygdalae so that environmental events are immediately processed (Pare et al., 2004). Depending on the valence of the stimuli, the lateral nuclei then communicate with the central nucleus in each amygdala and subsequent connections travel to critical forebrain, brainstem, autonomic, and endocrine structures that mediate the expression of emotion. Specifically, there are interconnections from the central nuclei to the: (1) hypothalamus, which results in sympathetic arousal and stimulation of stress hormones via the hypothalamic–pituitary–adrenal (HPA) axis; (2) the periaqueductal gray, which results in motor responses; and (3) the cingulate cortex, which results in additional cortico-cortical communication with neocortical association regions such as the temporo-parietal regions. Additionally, interconnections to pontine nuclei in the reticular formation result in an increase in overall arousal. In this manner orchestrated sequelae occur in response to a stressful environment, which collectively, can change the performer's mental and physical state in a profound manner. For example, heart rate and cortisol levels rise, as does muscle tension, and the performer may concomitantly experience excessive self-talk and "too much thinking" such that his/her attention is compromised and the execution of normally automated psychomotor skills, such as marksmanship, become explicitly managed — timing and coordination are then altered and likely reduced in quality while attention shrinks. In support of the "overthinking" hypothesis, Hung et al. (2005) provided psychophysiological evidence of increased networking between the left temporal region and the motor planning regions of the brain, by assessment of T3-Fz alpha EEG coherence levels, when study participants were asked to perform a dart-throwing task under the pressure of social evaluation. Relative to a nonstress control condition the increased "traffic" in the brain was accompanied by heightened reports of state anxiety and reductions in self-reported confidence levels. As expected, accuracy of performance or visuo-motor coordination was also reduced. However, how these changes were associated with mental representational changes was not studied, but can now add to bridging the understanding of covert and overt behaviors.

In light of the mental and physical change alterations that accrue, the activation of the amygdalae serves as a pivotal event in the manifestation of stress, and the control of activity in the amygdalae would exact a powerful influence on the performer's mental and physical state. Beyond the structures and processes outlined by Bear et al. (2001), Hatfield et al. (2006) also argued that a critical component of the circuitry underlying the neurobiology of fear is the executive control over limbic function and subcortical emotional circuits, which is housed anatomically in the frontal regions of the forebrain. Importantly, the anterior cortical regions have extensive anatomical connections with several subcortical limbic structures implicated in emotional behavior, particularly the amygdala (Davidson, 2002, 2004).

Davidson and colleagues have generated a significant body of literature that clearly shows a positive association between left frontal activation and positive affect while relative right activation is associated with negative affect (Tomarken et al., 1992; Davidson, 1998). Although the lateralization of frontal activation is robustly related to the valence of emotion, recent evidence points to a more fundamental association such that left frontal activation mediates approach-oriented behavior while right frontal activation is associated with avoidance or withdrawal-oriented behavior (Davidson, 2004). For example, left frontal activation is manifested during hostile behavior, which is certainly not a positive affective state, but most definitely involves approach toward an intended target. Whether positive in nature, approach-oriented, or a combination of the two dimensions, it would appear that such a neurobiological state would be highly adaptive for the performer who must control his/her emotional level while actively engaged with challenging tasks while under great pressure. Because EEG alpha power is inversely related to activation (i.e., relaxation), R minus L alpha power (i.e., Log right frontal alpha power – Log left frontal alpha power), when positive, implies greater relaxation in the right region or, in other words, a state of left frontal activation. Hence, *positive numbers for this metric imply left activation and executive control over emotion structures and processes*. This state involves a coherent utilization of mental representations presented in Table 1 (level III), which correspond perfectly to self-regulation (level IV), and allows smooth motor coordination, information processing, decision making, and awareness to cues, which trigger changes in both cognitive and motor systems. Conversely, *a negative value implies greater relaxation in the left region and a lack of executive control over limbic circuits*, thus resulting in collapse in both III and IV levels (see Table 1), and probably also in level II and I. Therefore, this EEG metric provides an opportune target for neurofeedback training to enable a heightened level of executive control over emotional response and task engagement during stressful and challenging tasks.

Importantly, there is evidence that this circuit (as quantified by this metric) can be controlled through augmented feedback training involving learned control over the activation of the frontal region of the brain. The key to control this disruptive chain of events, which would be especially important for those prone to anxiety, is to attain command of the frontal-limbic circuit and maintain appropriate mental representation functioning. Recent neurofeedback studies (Allen et al., 2001) provide strong evidence that frontal control over responsivity to fear-eliciting stimuli can be learned with neurofeedback training. Fortunately, a convenient frontal asymmetry metric can index relative frontal activation such that positive scores indicate relative left activation while negative scores indicate relative right activation. *Such a metric offers a pivotal candidate or powerful target for neurofeedback training to achieve control over a full cascade of mental and physical events during intense emotional challenge.* The goal of such an intervention would be for the performer to acquire systematic control over the emotional brain circuit, and to elicit it automatically so as to facilitate learning and maintain performance under the stress of competition.

Finally, the *principle of specificity* mandates that there is a need to assess the efficacy of neurofeedback within the context of military/dynamic challenge settings. That is, training must be similar to field maneuvers. *Thus, how can we get expert shooters to progress faster? We can improve the essence of attention and emotion regulation by learning control over critical brain processes.*

Pressure-induced performance failure

According to the conceptual framework presented here, the mental representational schema must be distorted in a way that task-irrelevant cues will be associated with relevant cues, and interfere with motor control. This interference can be detected by both distorted mental representations and "noisy" activation of the cortical system – allowing us to understand the underlying mechanism of choking under pressure.

For highly motoric tasks, according to explicit monitoring theory, the mechanisms underlying pressure-induced performance failure result from pressure shifting attention to the details of skill execution (Baumeister, 1984). This attention-to-skill execution disrupts the automaticity of well-learned skills, resulting in performance failure. Such explicit monitoring may be associated with excessive or unnecessary cortico-cortical communication, thus introducing "noise" into the motor planning regions of the brain resulting in degraded quality of performance. The foundations for the explicit monitoring theory are anchored in both self-awareness and skill-acquisition literature (Hardy et al., 1996; Lewis and Linder, 1997). The self-awareness literature (e.g., Baumeister and Showers, 1986) supports the notion that situational factors such as audience presence, ego-relevance, competition, and reward/punishment contingency can induce inward shifts of attentional focus. The argument linking pressure to self-focus draws on the idea that under pressure, the importance of the task becomes more salient. In order to ensure correctness of execution, more attention is given to monitoring the processes of performance (Baumeister, 1984; Lewis and Linder, 1997). It is this monitoring of skill execution that ultimately leads to performance failure in well-learned skills. Baumeister (1984) suggested that the increased self-focus caused by pressure leads to a disruption in the automatic nature of skill execution resulting in performance breakdown. A number of studies have found support indicating that increased attention to the details of a well-learned movement leads to performance breakdown (e.g., Beilock and Carr, 2001; Beilock et al., 2002, 2004). However, taking into consideration performers' skill level, and measuring cortical activity with mental representation simultaneously will allow one to conclude whether performance changes or maintenance (in some cases facilitation) is due to shifting attention inwards and onto task details or to external interfering emotional–cognitive elements.

The specific mechanism underlying breakdown in automaticity is offered by Masters' (1992) *reinvestment hypothesis*. Masters hypothesized that under stress, increased self-focus causes expert performers to reinvest declarative or explicit knowledge learned during earlier stages of skill acquisition (Gray, 2004; Mullen et al., 2005). This reinvestment of explicit knowledge leads to the automatic control structures that normally run uninterrupted to be dechunked, resulting in smaller sequences of independent units, which must be run and activated separately (Beilock and Carr, 2001). Once the control structure has been broken down, the process of activating and running each independent unit not only slows the performance, but also increases the likelihood of errors at each transition between the units (Beilock and Carr, 2001; Beilock and Gray, 2007). The breakdown in the automatic control structure results in the return to novice control strategies seen early in skill acquisition. If this is the underlying process that accounts for coordination breakdown, then mental representations must show it in the form of structure change. The new method (see Fig. 1 and Table 1; Schack, 2004a, b; Schack and Mechsner, 2006) must be implemented to show how mental representations change under pressure in relation to skill level and task characteristics; evidence that does not exist today.

Summary

A conceptual framework, which offers an integrative approach to study the underlying mechanisms of mental, emotional, and motor operations under neutral and pressure condition was developed. The structural components of human performance, such as emotional processes, cognitive processes and structures, motor processes, and the neurophysiological basis of these structural components (i.e., activation of cortical areas) have been integrated into a holistic framework that enables a better understanding of human performance, and allows generating applications that share scientific validity.

The conceptual framework consists of four levels, each with a designated function of mental control, mental representations, sensorimotor representations, and sensorimotor control. Once performance suffers under pressure, all structures remain stable, but the mental control level alters as it cannot meet the environmental or inner requirements of the task's demands. Mental control (self-regulation) breaks down because the performer lacks the sufficient strategies required for coping under external and/or stressful factors. Within this conceptual framework, motions are functionally adaptive and direct behavior. Positive and negative emotions can have a facilitative or debilitative effect on cognitive performance, depending on the nature of the task.

The model places an important role to the process of situational appraisal and evoking emotions. At this stage decision making and intention to reach specific action and action plans are created, and mental-control processing runs to a module, which is responsible for action execution. This module is linked to the level of sensorimotor control (see Fig. 1), and includes all motor components, which are necessary for production of goal-directed action effects. Once the action effects are not congruent to the intended action effects, the appraisal system "reads" an insufficient action, and evokes negative emotions. In some cases the control must consist of action strategies, such as control of attention, control of emotion, or motivational control in order to stabilize the motor action. Once the performer lacks such strategies, he/she may choke. Thus, one important link between emotion and information storage is caused by the representation of emotionally induced action effects in long-term memory. From this point of view emotions are a part of information storage in general.

The stronger the activation of particular emotional nodes are, the greater the mood-congruent effects. When emotions are strongly activated, emotion-congruent constructs (e.g., concepts, words, themes, rules of inference) become primed and available for use bringing into readiness certain perceptual categories, themes, or ways of interpreting the world congruent with current emotional states. Consequently, when athletes imagine positive (performance supporting) emotions while preparing for competition, they activate relevant knowledge that lead to emotion-directed forms of imagery, which focus their attention. Failing to perform well under pressure may be linked to both emotion state and the failure to utilize appropriate coping strategies.

Emotions operate as somatic markers by helping to consider or eliminate certain options, and increasing the accuracy of decision making. The model proposed in Fig. 1 illustrates the interaction between emotion, judgment, and decision making. The appraisal of a particular situation is based on different memory traces. In turn, emotions influence the judgments about situations both directly and indirectly. The direct influence is based on the development of different types of motivation, which result in appraisal-based emotions. The indirect influence has to do with the informational quality of emotions, and is a reference point for decision making and judgments. Thus, one may decide more holistically, or nonanalytically, in case of positive mood. From this point of view, the direct influence of emotions has to do with the content of decision making, especially the number and quality of motivation, while the indirect influence of emotions is defined by the type of decision making (holistic vs. analytic mode of decision making). Positive mood initiates a holistic mode of decision making, while negative mood triggers a more analytic mode in decision making and mental control.

As one can notice in the model (see Table 1), basic regulations are required to process information and for action control. They take place at the lower sensorimotor levels, and have to do with the use of basic routines and sensorimotor representation structures. At the level of mental control, one generates valid plans and controls attention and emotions in case of unexpected external events in the action organization. Basic regulation must generate and implement such tools like routines at the lower level, and implement mental control strategies at the mental control level.

The model consists of the assumption that the interruption of activated performance plans and the increasing inability to subordinate the action performance to an action program are emotionally negatively attributed. Particularly, through the collapse of solid connections within the system of mental control the inner speech loses the function of an instrument of action control (see Schack, 1997). The result is a functional damage of mental control: a mental control deficit due to inappropriate action plan, the deficient availability of action control strategies or a collapse of the control anticipation.

The concept presented in Fig. 1 allows studying the cognitive patterns related to different tasks, stress sources (pressure-information), and related emotions. The patterns concerning the "pressure" information representation and the relationship between "pressure" information and emotions are the reference point for appraisal in real situations. These links allow predicting emotional states, and modifying the reference points (i.e., benchmarks) of the appraisal systems, and measure the link between the representation of threatening situations and the representation of action organization in these situations. The goal is to strengthen the productive links between the representation of the situation and the representation of the action (Nitsch, 2004; Schack and Hackfort, 2007). In the sport domain, one must capture the mental representations of situational components, such as the representation of the team structure, or the representation of the environmental conditions of a specific competition. Furthermore, one must know what type of activities athletes have represented to cope with situations that vary in many respects. Finally, it is essential to measure the types of event representations and action representations, and look for the fit between these two representation types. A misfit between these representations may lead to emotions of stress and anxiety. From this perspective, looking at the emotional process itself is insufficient, without measuring the cognitive benchmarks (especially event and action representations) of the appraisal process, which constitutes the development of emotions in action.

For understanding the components of performance and their interaction, it is necessary to find out the relationship between skill level, cortical activation, and order formation in long-term memory. To learn more about these interactions and networks, one must combine the neurophysiological methods, like EEG-coherence measures (Schack, 1999; Deeny et al., 2003), and methods for measuring the structure of mental representation. The mental representations of experts were highly differentiated from those of novices, and were more strongly function oriented than the novices' representations. The simultaneous neurophysiological findings indicated that experts' cognitive activation of movement representations is accompanied by activation of several cortical areas. These patterns of activation provide a glimpse of the network of neurophysiological areas with a high mental economy and a high stability. It seems that there are two different functions of activation concerning the time-related steering of movements and concerning space-oriented aspects of movement representations. A failure to obtain a stable activation of separate areas in movement representation was evident. According to the conceptual framework presented here, the mental representational schema must be distorted in a way that task-irrelevant cues will be associated with relevant cues, and interfere with motor control. This interference can be detected by both distorted mental representations and "noisy" activation of the cortical system allowing the detection of the underlying mechanism of choking under pressure. The implementation of a research agenda, which takes into account all the overt and covert behaviors simultaneously, can clarify the functionality of the model presented here.

References

Allen, J., Harmon-Jones, E., & Cavender, J. (2001). Manipulation of frontal EEG asymmetry through biofeedback alters self-reported emotional responses and facial EMG. *Psychophysiology*, *38*, 685–693.

Bandura, A. (1991). Social cognitive theory of self-regulation. *Organizational Behavior and Human Decision Processes*, *50*, 248–287.

Bandura, A. (1997). *Self-efficacy: the exercise of control*. New York: W. H. Freeman Company.

Barry, E. S., Naus, M. J., & Rehm, L. P. (2004). Depression and implicit memory: understanding mood congruent bias. *Cognitive Therapy and Research*, *28*(3), 387–414.

Baumeister, R. F. (1984). Choking under pressure: self-consciousness and paradoxical effects of incentives on skillful performance. *Journal of Personality and Social Psychology*, *46*, 610–620.

Baumeister, R. F., & Showers, C. J. (1986). A review of paradoxical performance effects: choking under pressure in sports and mental tests. *European Journal of Social Psychology*, *16*, 361–383.

Bear, M. F., Connors, B., & Paradiso, M. A. (2001). *Neuroscience: exploring the brain*. Philadelphia, PA: Lippincott Williams & Wilkins.

Beilock, S. L., Bertenthal, B. I., McCoy, A. M., & Carr, T. H. (2004). Haste does not always make waste: expertise, direction of attention, and speed versus accuracy in performing sensorimotor skills. *Psychonomic Bulletin & Review*, *11*, 373–379.

Beilock, S. L., & Carr, T. H. (2001). On the fragility of skilled performance: what governs choking under pressure? *Journal of Experimental Psychology — General*, *130*, 701–725.

Beilock, S. L., & Gray, R. (2007). Why do athletes choke under pressure? In G. Tenenbaum & R. C. Eklund (Eds.), *Handbook of sport psychology* (3rd ed., pp. 425–444). New York: Wiley.

Beilock, S. L., Wierenga, S. A., & Carr, T. H. (2002). Expertise, attention, and memory in sensorimotor skill execution: impact of novel task constraints on dual-task performance and episodic memory. *Quarterly Journal of Experimental Psychology*, *55*, 1211–1240.

Bernstein, N. A. (1988). *Bewegungsphysiologie [Movement physiology]* (2nd ed.). Leipzig, Germany: Barth.

Bower, G. H. (1981). Mood and memory. *American Psychologist*, *36*, 129–148.

Bower, G. H. (1991). Mood congruity of social judgments. In J. P. Forgas (Ed.), *Emotion and social judgments* (pp. 31–53). New York: Pergamon Press.

Brosschot, J. F., & Thayer, J. F. (2003). Heart rate response is longer after negative emotions than after positive emotions. *International Journal of Psychophysiology*, *50*, 181–187.

Carver, C. S., & Scheier, M. F. (1988). A control-process perspective on anxiety. *Anxiety Research*, *1*, 17–22.

Carver, C. S., & Scheier, M. F. (1991). *A control-process perspective on anxiety*. Cambridge: Cambridge University Press.

Caterini, R., Delhomme, G., Deschaumes-Molinaro, C., Dittmar, A., Economides, S., & Vernet-Maury, E. (1995). Increased activation as a limiting factor of performance in sharp shooters. *Neuropsychologia*, *33*, 385–390.

Collet, C., Roure, R., Rada, H., Dittmar, A., & Vernet-Maury, A. (1996). Relationships between performance and skin resistance evolution involving various motor skills. *Physiology & Behavior*, *59*, 953–963.

Coombes, S. A., Cauraugh, J. H., & Janelle, C. M. (2006). Emotions and movement: Activation of defensive circuitry alters the magnitude of sustained contraction. *Neuroscience Letters*, *396*, 192–196.

Damásio, A. (1994). *Descarte's error: emotion, reason and human brain*. New York: GP Putnam.

Davidson, R. J. (1998). Affective style and affective disorders: perspectives from affective neuroscience. *Cognition & Emotion*, *12*, 307–320.

Davidson, R. J. (2002). The neural circuitry of emotion and affective style: prefrontal cortex and amygdale contributions. *Social Science Information*, *40*, 11–37.

Davidson, R. J. (2004). Affective style: causes and consequences. In J. T. Cacioppo & G. G. Berntson (Eds.), *Essays in social neuroscience* (pp. 77–91). Cambridge: MIT Press.

Deeny, S. P., Hillman, C. H., Janelle, C. M., & Hatfield, B. D. (2003). Cortico-cortical communication and superior performance in skilled marksmen: an EEG coherence analysis. *Journal of Sport & Exercise Psychology*, *25*, 188–204.

Easterbrook, J. A. (1959). The effect of emotion on cue utilization and the organization of behavior. *Psychological Review*, *66*, 183–201.

Endler, N. S., & Parker, J. D. (1990). Multidimensional assessment of coping: a critical evaluation. *Journal of Personality and Social Psychology*, *58*, 844–854.

Folkman, S., & Lazarus, R. S. (1990). Coping and emotion. In N. L. Stein, B. Leventhal, & T. Trabasso (Eds.), *Psychological and biological approaches to emotion* (pp. 313–332). Hillsdale, NJ: Lawrence Erlbaum Associates.

Forgas, J. P. (1991). Affect and social judgments: an introductory review. In J. P. Forgas (Ed.), *Emotion and social judgments* (pp. 3–30). New York: Pergamon Press.

Fredrickson, B. (1998). What good are positive emotions? *Review of General Psychology*, *3*, 300–319.

Fredrickson, B. L., Mancuso, R. A., Branigan, C., & Tugade, M. M. (2000). The undoing effects of positive emotions. *Motivation and Emotion*, *24*, 237–258.

Gray, R. (2004). Attending to the execution of a complex sensorimotor skill: expertise differences, choking and slumps. *Journal of Experimental Psychology — Applied*, *10*, 42–54.

Hardy, L., Mullen, R., & Jones, G. (1996). Knowledge and conscious control of motor actions under stress. *British Journal of Psychology*, *87*, 621–636.

Hatfield, B. D., Haufler, A. J., Hung, T., & Spalding, T. W. (2004). Electroencephalographic (EEG) studies of skilled psychomotor performance. *Journal of Clinical Neurophysiology*, *21*, 144–156.

Hatfield, B. D., Haufler, A. J., & Spalding, T. W. (2006). A cognitive neuroscience perspective on sport performance. In E. O. Acevedo & P. Ekkekakis (Eds.), *Psychobiology of physical activity* (pp. 221–240). Champaign, IL: Human Kinetics.

Hatfield, B. D., Landers, D. M., & Ray, W. J. (1984). Cognitive processes during a self-paced motor performance: an electroencephalographic profile of skilled marksman. *Journal of Sport Psychology*, *6*, 42–59.

Haufler, A. J., Spalding, T. W., Santa Maria, D. L., & Hatfield, B. D. (2000). Neuro-cognitive activity during self-paced visuospatial task: comparative EEG profiles in marksmen and novice shooters. *Biological Psychology*, *53*, 131–160.

Hung, T. M., Lin, J. H., Lo, L. C., Kao, J. F., Hung, C. L., Chen, Y. J., & Lai, Z. (2005). Effects of anxiety on EEG coherence during dart throw. In T. Morris, P. Terry, S. Gordon, S. Hanrahan, L. Ievleva, G. Kolt, & P. Tremayne (Eds.), *Promoting health and performance for life: proceedings of the ISSP 11th world congress of sport psychology [CD-ROM]*. Sydney: International Society of Sport Psychology.

Janelle, C. M. (2002). Modification of visual attention parameters under conditions of heightened anxiety and arousal. *Journal of Sports Science, 20*, 237–251.

Janelle, C. M., Singer, R. N., & Williams, M. A. (1999). External distraction and attentional narrowing: visual search evidence. *Journal of Sport & Exercise Psychology, 21*, 70–91.

Jeannerod, M. (2004). Actions from within. *International Journal of Sport and Exercise Psychology, 2*, 376–402.

Jones, J. G., & Hanton, S. (1996). Interpretation of competitive anxiety symptoms and goal attainment expectancies. *Journal of Sport & Exercise Psychology, 18*, 144–157.

Jones, J. G., Hanton, S., & Swain, A. (1994). Intensity and interpretation of anxiety symptoms in elite and non-elite sport performers. *Personality and Individual Differences, 17*, 657–663.

Jones, J. G., & Swain, A. (1995). Predispositions to experience debilitative and facilitative anxiety in elite and non-elite performers. *The Sport Psychologist, 9*, 201–211.

Jones, J. G., Swain, A., & Hardy, L. (1993). Intensity and direction dimensions of competitive state anxiety and relationships with performance. *Journal of Sports Science, 11*, 1–15.

Kerick, S. E., Douglass, L. W., & Hatfield, B. D. (2004). Cerebral cortical adaptations associated with visuomotor practice. *Medicine and Science in Sports and Exercise, 36*, 118–129.

Krause, W. (2000). *Denken und Gedächtnis aus naturwissenschaftlicher Sicht [Thinking and memory from a scientific perspective]*. Göttingen, Germany: Hogrefe.

Lang, P. J., Bradley, M. M., & Cuthbert, B. N. (1997). Motivated attention: affect, activation, and action. In P. J. Lang, R. F. Simons, & M. T. Balaban (Eds.), *Attention and orientating: sensory and motivational processes* (pp. 97–135). Mahwah, NJ: Lawrence Erlbaum Associates.

Lazarus, R. S. (1999). *Stress and emotion: a new synthesis*. London: Free Association Books.

Lazarus, R. S., & Folkman, S. (1984). *Stress, appraisal, and coping*. New York: Springer.

Lewis, B. P., & Linder, D. E. (1997). Thinking about choking? Attentional processes and paradoxical performance. *Personality and Social Psychology Bulletin, 23*, 937–944.

Mandler, G. (1979). Thought processes, consciousness, and stress. In V. Hamilton & D. A. Warburton (Eds.), *Human stress and cognition: an information processing approach*. London: Wiley.

Mandler, G. (1985). Consciousness, imagery, and emotion — with special reference to autonomic imagery. *Journal of Mental Imagery, 8*, 87–94.

Masters, R. S. W. (1992). Knowledge, knerves and know-how: the role of explicit versus implicit knowledge in the breakdown of a complex motor skill under pressure. *British Journal of Psychology, 83*, 343–358.

Mellalieu, S. D. (2003). Mood matters: but how much? A comment on Lane and Terry (2000). *Journal of Applied Sport Psychology, 15*, 99–114.

Mullen, R., Hardy, L., & Tattersall, A. (2005). The effect of anxiety on motor performance: a test of the conscious processing hypothesis. *Journal of Sport & Exercise Psychology, 27*, 212–225.

Nitsch, J. R. (2004). The action-theoretical perspective: a frame of reference for research and application in sport psychology. *Zeitschrift für Sportpsychologie, 11*, 10–23.

Pare, D., Quirk, G. J., & Ledoux, J. E. (2004). New vistas on amygdale networks in conditioned fear. *Journal of Neurophysiology, 92*, 1–9.

Plutchik, R. (2003). *Emotions and life: perspectives of psychology, biology and evolution*. Washington, DC: American Psychological Association.

Prinz, J. J. (2005). Emotions, embodiment, and awareness. In L. F. Barrett, P. M. Niedenthal, & P. Winkeilman (Eds.), *Emotion and consciousness* (pp. 363–383). New York: Guilford Press.

Rosenberg, E. (1998). Levels of analysis and the organisation of affect. *Review of General Psychology, 3*, 247–270.

Schack, B. (1999). Dynamic topographic spectral analysis of cognitive processes. In C. Uhl (Ed.), *Analysis of neurophysiological brain functioning*. New York: Academic Press.

Schack, T. (1997). *Ängstliche Schüler im Sport — Interventionsverfahren zur Entwicklung der Handlungskontrolle [Anxious students in sports — intervention procedures for the development of action control]*. Schorndorf: Hofmann.

Schack, T. (2003). Cognitive representation of complex movements in motor memory — a study in ski alpine freestyle. In: E. Müller, H. Schwameder, G. Zallinger, V. Fastenbauer (Eds.), *Proceedings of 8th annual congress of the European college of sport science*, (p. 198), July 9–12. Salzburg: European College of Sport Science.

Schack, T. (2004a). Relation of knowledge and performance in action. *Journal of Knowledge Management, 4*, 38–53.

Schack, T. (2004b). The cognitive architecture of complex movement. *International Journal of Sport and Exercise Psychology, 2*, 403–438.

Schack, T., & Hackfort, D. (2007). An action theory approach to applied sport psychology. In G. Tenenbaum & R. C. Eklund (Eds.), *Handbook of sport psychology* (3rd ed., pp. 332–351). New York: Wiley.

Schack, T., & Mechsner, F. (2006). Representation of motor skills in human long-term memory. *Neuroscience Letters, 391*, 77–81.

Schwarz, N., & Bless, H. (1991). Happy and mindless, but sad and smart? The impact of affective states on analytical reasoning. In J. P. Forgas (Ed.), *Emotion and social judgments* (pp. 55–71). New York: Pergamon Press.

Tenenbaum, G. (2003). Expert athletes: an integrated approach to decision making. In J. L. Starkes & K. A. Ericsson (Eds.), *Expert performance in sports* (pp. 191–218). Champaign, IL: Human Kinetics.

Tomaka, J., Blascovich, J., Kibler, J., & Ernst, J. M. (1997). Cognitive and physiological antecedents of threat and challenge appraisals. *Journal of Personality and Social Psychology*, *73*, 63–72.

Tomarken, A. J., Davidson, R. J., Wheeler, R. E., & Doss, R. C. (1992). Individual differences in anterior brain asymmetry and fundamental dimensions of emotion. *Journal of Personality and Social Psychology*, *62*, 676–687.

Tremayne, P., & Barry, R. J. (2001). Elite pistol shooters: physiological patterning of best vs. worst shots. *International Journal of Psychophysiology*, *41*, 19–29.

M. Raab et al. (Eds.)
Progress in Brain Research, Vol. 174
ISSN 0079-6123

CHAPTER 15

Visual cues influence motor coordination: behavioral results and potential neural mechanisms mediating perception–action coupling and response selection

Nicole Wenderoth[1,*] and Matthias Weigelt[2]

[1]Motor Control Laboratory, Department of Biomedical Kinesiology, KU Leuven, Belgium
[2]Neurocognition and Action Research Group and Center of Excellence "Cognitive Interaction Technology" (CITEC), Bielefeld University, Bielefeld, Germany

Abstract: Here we review behavioral and brain imaging results on stimulus–response selection in the context of bimanual movements, which is a prototypical paradigm frequently used to investigate the coordination of complex motor behavior. We propose that stimulus–response selection is constrained at the motor, perceptual, and cognitive levels, with the relative importance of each dependent on the task context. Motor constraints seem to dominate when response selection requirements are low, whereas perceptual and cognitive constraints become increasingly important when the appropriate movement has to be associated with a visual cue. We argue that certain cue features determine how task goals are conceptualized, which influences how a particular motor response is selected and implemented by the nervous system.

Keywords: bimanual control; task switching; parieto-premotor networks; stimulus–response mapping; motor binding; action representation; action goals

Introduction

Decision making in sports and many other situations is often linked to selecting the appropriate action in response to a set of environmental stimuli (e.g., when trying to surpass the defensive block of an opponent in volleyball, when anticipating the trajectory of a curve ball in baseball, or when following street signs while riding a bicycle). Even though this type of stimulus–response selection appears to be mainly determined by the effectiveness of several higher-order cognitive functions (see Chapter 11: Crossmodal interaction in speeded responses: time window of integration model; Chapter 12: Embodied cognition of movement decisions: a computational modeling approach), it is additionally constrained by other factors, such as the motor repertoire of the actor, preferred pathways for perception–action coupling, and the conceptualization of the task at a more abstract level. The ease with which a particular movement is selected from a number of alternatives often depends on the compatibility between the stimulus triggering the action and the characteristics of the appropriate movement.

*Corresponding author.
Tel.: +32-16-329-157; Fax: +32-16-329-197;
E-mail: Nicole.wenderoth@faber.kuleuven.be

DOI: 10.1016/S0079-6123(09)01315-6

For example, when observing someone else performing a particular gesture (i.e., someone lifting his/her index finger), people find it easier to initiate the same manual gesture compared to one that is different (Brass et al., 2000, 2001). In a similar vein, people react faster with a compatible foot response when observing a soccer kick and a compatible hand response when watching someone type on a computer keyboard, than when they have to select the incompatible action, respectively (Bach and Tipper, 2006; Bach et al., 2007). These studies support the notion that the organization of coordinated behavior is further influenced by the (online) association between motor activity and visual information available in the context in which actions are taking place.

In this chapter, we will discuss how actions in response to visual cues are constrained within the motor, visuo-perceptual, and cognitive domains, as well as in their interactions. On this basis we argue that decisions are not purely based on strategic plans, but are made within the constraints imposed by the neural architecture of the sensorimotor system, favoring the selection of some behavioral responses over others.

The button-pressing paradigms commonly used in previous work addressing this question required overly simply motor responses. Most everyday movements, however, require the accurate control and coordination of many different muscles, joints, and limbs, such that all components are implemented in accordance with a common movement plan. Bimanual movement tasks have proven useful in investigating more complex coordinated behavior in the laboratory environment. Some bimanual movements can be performed easily (e.g., applauding with both hands), while others are more difficult to control (e.g., patting the head with one hand, while rubbing the stomach with the other). Analogous to these two examples, certain bimanual coordination patterns are naturally preferred (indicated by stable and accurate performance) over others (nonpreferred, requiring considerable effort when executed). Importantly, the effectiveness of bimanual actions is not only affected by preferred versus nonpreferred motor patterns, but can also be substantially influenced by visual information and the conceptualization of the task at hand. For these reasons, we will focus on bimanual coordination paradigms in our discussion of the interplay between the motor domain and the perceptual/cognitive domain for selecting motor actions in the context of visual information.

Research on bimanual actions has generated two competing theories of coordination. In the first section we will outline the viewpoint of the "actionist," who argues that bimanual coordination is mainly constrained at the motor level. In the second section we will outline the viewpoint of the "cognitionist," who argues that bimanual coordination is strongly influenced by perceptual and cognitive constraints. Please note that in these two first sections previous findings will be summarized from the two extreme viewpoints of a pure "actionist" and a pure "cognitionist" approach to emphasize the differences between these notions. In the third section, we will outline what these two viewpoints actually have in common and argue that they are complementary, rather than incompatible. In the forth section, we describe potential neural mechanisms mediating some of the behavioral results that have been reported in the literature focusing on bimanual coordination. Finally, in the fifth section we will relate these results to the general issue of stimulus–response selection and further discuss goal coding and task conceptualization.

Bimanual movements from the perspective of the pure "actionist"

This section will review the "actionist's" view, which assumes that symmetrical bimanual movements are the preferred mode of coordination because of the organization of the human neuromuscular and skeletal system.

Much research supporting this assumption was strongly influenced by the use of rhythmical bimanual paradigms. Here, one of the most robust findings is that mirror symmetrical bimanual movements (in-phase) are executed with higher accuracy and stability than parallel movements (antiphase) (for reviews see Kelso, 1995; Mechsner et al., 2001; Swinnen, 2002; Swinnen and Wenderoth, 2004).

At a theoretical level, this behavior was described as a dynamical, self-organizing system, consisting of two mutually coupled oscillators (Yamanishi et al., 1980; Kelso, 1984). In the context of this model, each coordination pattern represents an attractor with the symmetrical pattern being more stable than the parallel pattern. While this difference in stability is only subtle for simple bimanual movements, it becomes particularly prominent when the system is challenged. For example, when movements have to be performed at higher frequencies or in parallel with a secondary task (Kelso, 1984; Peper et al., 1995; Temprado et al., 2001; Monno et al., 2002), large stability fluctuations in the antiphase pattern can be readily observed under these situations. At some critical value, these stability fluctuations lead to an involuntary transition to the more stable in-phase pattern. Similar effects of the stability mode on the control of bimanual movements have been reported for voluntary switches between different types of coordination patterns. Planned switches are performed faster from the antiphase pattern (asymmetrical or parallel) to the in-phase pattern (symmetrical), than vice versa (Scholz and Kelso, 1990; Byblow et al., 2000).

The increased stability of symmetrical movements also generalizes to bimanual tasks of higher spatial complexity. For example, when the left and right arm are moved rhythmically along different trajectories in space, performance is best when the trajectories are symmetrical with respect to the body midline (e.g., when both arms perform vertical up and down movements), whereas performance decreases and interference arises for nonsymmetrical trajectories (e.g., one arm moves up and down and the other left and right) (Swinnen et al., 1997, 1998a, 2002; Wenderoth et al., 2004, 2005, 2006). Similarly, bimanual performance is most accurate when both hands produce similar rather than different movement amplitudes or forces (Wenderoth et al., 2005). Differences between symmetrical and nonsymmetrical movements do not only emerge during movement execution, but also influence bimanual reactions to visual cues. Byblow et al. (2000) used a paradigm in which participants had to respond to visual cues by switching from one bimanual pattern to another. Switches were faster and less prone to errors when participants switched to symmetrical patterns (preferred coordination mode) in comparison to nonsymmetrical patterns (nonpreferred mode). Such differences between symmetrical and asymmetrical actions were further confirmed for discrete bimanual movements and it was argued that asymmetrical movements are prone to motor interference, which arises because of crosstalk at the motor programming and/or execution level (Heuer, 1993; Heuer et al., 1998, 2001).

In addition to the behavioral work, experiments exploiting neurophysiological methods have suggested that the advantage of symmetrical bimanual actions arises most likely from a spontaneous coupling tendency between homologous muscles (Serrien et al., 2002, 2003; Donchin and Cardoso De Oliveira, 2004). This coupling tendency is assumed to emerge via transcallosal connections between motor areas of the left and right hemisphere (Franz et al., 1996; Eliassen et al., 2000; Carson, 2005). As a consequence of such neural hard-wiring (in the form of transcallosal connections), intermanual coupling during the simultaneous coordination of two-handed actions occurs whenever similar motor signals are exchanged via the corpus callossum, whereas intermanual interference arises for different motor signals (Cardoso De Oliveira et al., 2001; Cardoso De Oliveira, 2002). From this perspective, limitations in the control of asymmetrical bimanual movements can be attributed to interference between different motor signals.

In summary, the results of the experiments described above have lead to the "actionist" viewpoint, which assumes that the symmetry advantage for rhythmical bimanual actions arises (in large part) at the motor level, and is a consequence of the organization of the neuromuscular and skeletal systems. In the following sections this will be referred to as the "motor outflow hypothesis."

Bimanual movements from the perspective of the pure "cognitionist"

Contrary to the "actionist's" perspective favoring the motor outflow hypothesis, an alternative view was put forward, arguing that the difference between symmetrical and nonsymmetrical bimanual actions depends mainly on perceptual instead of muscular symmetry. In an influential paper, Mechsner et al. (2001) demonstrated that even extremely difficult bimanual actions, such as cycling movements with a 3 to 4 ratio between the hands, can be performed rather easily when they are symbolized by symmetrical sensory feedback (e.g., visual or kinesthetic). Thus, from the view of the "cognitionist", bimanual movements are stabilized at the perceptual and cognitive level, where they are guided by the anticipation of sensory feedback, with the anticipation of similar or dissimilar perceptual feedback acting as the major constraint. We will refer to this view as the "perceptual symmetry hypothesis" in the remainder of the chapter.

In a related series of studies it was further argued that interference during asymmetrical bimanual movements results from the central processes involved "upstream" from the primary motor areas (Diedrichsen et al., 2001; Mechsner et al., 2001; Diedrichsen et al., 2003; Hazeltine et al., 2003; Kunde and Weigelt, 2005; Weigelt et al., 2006; Albert et al., 2007; Weigelt, 2007; Weigelt et al., 2007; Kunde et al., 2009). The majority of these studies used bimanual reaction time (RT) paradigms in which discrete bimanual actions were cued by a visual stimulus. In those studies, RT mainly reflected different stimulus characteristics, whereas the actual movement organization requiring either symmetrical or nonsymmetrical movements was subordinate. For example, Diedrichsen et al. (2001) used a bimanual RT paradigm and asked participants to initiate left and right hand actions in parallel, such that each hand moved either forward or sideward. The cues provided either symbolic information (i.e., "F" for forward and "S" for sideward, such that symmetrical movements were cued by "SS" or "FF" and nonsymmetrical movements by "SF" or "FS") or direct information by flashing light-emitting diodes (LEDs) at the end position of each hand. The results for symbolic cues showed a faster initiation of the bimanual responses (smaller RTs) when the bimanual movements were symmetrical, as compared to when they were asymmetrical. This symmetry advantage, however, was abolished when the movement targets were cued directly by flashing LEDs at the end position of each hand. Consequently, it was argued that interference during bimanual movement control arises rather at the cognitive level than at the motor outflow level (Diedrichsen et al., 2001; Albert et al., 2007). Two potential central processes associated with the increased RT costs for nonsymmetrical bimanual movements are stimulus identification and response selection (Diedrichsen et al., 2003; Hazeltine et al., 2003; Mechsner, 2004).

Recently, Weigelt et al. (2007) replicated the results of Diedrichsen et al. (2001, 2003), using symbolic instead of direct cues. In their study, participants were asked to perform bimanual actions to either close or far targets, conducting either symmetrical (both hands move to close, or far targets) or asymmetrical (one hand moves to a close target and the other to a far target) movement patterns. Movements were always cued by symbols (i.e., circles and/or crosses), with the symbols for the left and right hand being either congruent (identical symbols for both hands) or incongruent (different symbols for both hands). Interestingly, RT was faster when the cues showed congruent compared to incongruent symbolic stimuli, and this finding was independent of whether participants performed symmetrical or asymmetrical bimanual movements.

For bimanual object manipulation, Kunde and Weigelt (2005), Weigelt et al. (2006) and Kunde et al. (2009) reported that not only congruent versus incongruent visual stimuli, but more importantly the anticipation of congruent versus incongruent action goal states, influences the organization of action. They used two bars, each painted black at one end and white at the other. The bars were positioned horizontally and participants were instructed to rotate the bars by 90° such that either congruent task goals (both bars are positioned with the black or white end at the top) or incongruent task goals (one bar is positioned with the black end at the top, the other with the white end at the top) were achieved. Bimanual RT was faster for congruent than for incongruent task goals, irrespective of whether participants executed symmetrical or asymmetrical object rotations.

In summary, these lines of research have lead to an expansion of the "perceptual symmetry hypothesis," suggesting that bimanual performance is mostly dependent upon how the action is represented at the cognitive level. According to this extended view of the pure "cognitionist," bimanual actions are constrained by the manner in which abstract task goals are conceptualized. In this view, the conceptualization and maintenance of goal and target representations influences the ease with which bimanual movements are coordinated, and as a consequence, how fast bimanual actions can be initiated. Therefore, keeping an active representation of congruent goal and target states for the two hands is easier than representing different states. This will be referred to as the "task goal hypothesis" in the remainder of the chapter.

Is the perspective of the "actionist" really contradictory to that of the "cognitionist"?

One important aspect in discussing these contrasting points of view is that they have been formed on the basis of results obtained with quite different tasks. Even though all tasks were bimanual, an important difference is that many results supporting the motor outflow hypothesis were revealed with rhythmical paradigms (but see Heuer, 1993; Heuer et al., 1998, 2001; Heuer and Klein, 2006) and by measuring movement-related parameters (such as relative phase accuracy and stability), whereas nearly all experiments supporting the task goal hypothesis used discrete bimanual movements and measured processing costs, as indicated by RT. To acknowledge this difference is important, because recent functional neuroimaging results suggest that rhythmic movements are not controlled by the same neural process as discrete movements (Schaal et al., 2004). Therefore, results from one movement category may not generalize to the other.

To clarify whether this fundamental methodological difference has contributed to the apparent inconsistency of previous findings, we used a rhythmical, bimanual circle drawing paradigm, in which visual cues indicated that participants had to switch from symmetrical to asymmetrical patterns and vice versa. In addition, we investigated the influence of cues symbolizing either the new bimanual pattern either as an integrated entity (integrated cues) or as two dissociated components (dissociated cues). For example, integrated cues displayed "IN" for inward cycling, whereas dissociated cues symbolized the same pattern by "R L", that is, rightward cycling for the left hand and leftward cycling for the right hand. The main outcome variable was RT, which was measured as the delay between the appearance of the stimulus and the reversal of cycling direction. This experiment revealed three interesting results.

First, when participants could plan their switch from one pattern to another with ample time before it was initiated, RT was significantly faster when participants switched towards a symmetrical pattern, as compared to when they switched towards an asymmetrical pattern. This indicates a clear advantage of symmetrical bimanual movements under conditions in which central processing requirements in the form of stimulus–response mapping are low.

Second, when participants had to initiate the bimanual pattern immediately upon the appearance of a visual cue, RT was shorter when the left- and right-hand actions were cued as an integrated entity, as compared to when the action of each hand was symbolized separately. Importantly, higher RT costs arose because dissociated cues were processed serially.

Third, for the most direct cues, there was no difference between switches towards symmetrical or asymmetrical patterns. The latter two results are in strong agreement with the findings of Diedrichsen et al. (2001, 2003) and indicate that cue effects can also be observed for rhythmical movements when central processing costs are high.

These results therefore show that the "task goal hypothesis" holds also for rhythmical bimanual movements when central processing requirements are high. When central processing costs are low, however, the data provided support for the "motor outflow hypothesis." They further indicate that the coordination of two-handed actions is affected at multiple levels, and the extent to which motor and/or perceptual-cognitive constraints dominate bimanual performance is task dependent.

A similar conclusion can be drawn from a number of studies using continuous bimanual tasks, demonstrating that poor performance in the coordination of nonsymmetrical bimanual movements can be substantially improved when the resulting trajectories are visually integrated into a simpler "gestalt"(Swinnen et al., 1998b; Franz et al., 2001; Mechsner et al., 2001; Mechsner, 2004). Thus, integrating the action of each hand into a common movement plan appears to be crucial for the successful performance in complex bimanual coordination tasks. A successful integration might also occur at the action level for relatively simple symmetrical bimanual movements, while tasks with more complex spatio-temporal patterns might require conceptualization at the perceptual and/or cognitive level.

In summary, neither the extreme "actionist's" nor the extreme "cognitionist's" perspective is sufficient to explain all facets of coordinated behavior, which has resulted in the development of an intermediate position (Kelso et al., 2001; Ivry et al., 2004; Swinnen and Wenderoth, 2004; Oliveira and Ivry, 2008). Additionally, one might wonder whether the dichotomy between action constraint and perception/cognitive constraints of bimanual movements is in fact of an academic nature only. It has been well established that so-called "internal models" play an important role in motor planning and execution (Kawato, 1999). Internal models describe the relationship between motor actions and their sensory consequences. They are at the heart of current conceptions of motor control, for example, optimal feedback control theory (Todorov, 2004; Shadmehr and Krakauer, 2008), and they might play an important role in stimulus–response selection, as proposed by the modular selection and identification for control (MOSAIC) model (Haruno et al., 2001). Similar ideas were formulated previously in the form of the ideomotor principle or the event coding theory (Hommel et al., 2001). Furthermore, it has been argued recently that internal models contribute to action control even in the absence of online sensory feedback (e.g., during rapid eye movements; Chen-Harris et al., 2008). Thus, in contemporary concepts of motor behavior there is an inseparable link between motor action and sensory consequences, or, in other words, between the action, perception, and cognitive levels of motor control.

How does the brain link sensory information, goals, and motor activity?

The results described above indicate clearly that sensory cues have a dramatic impact on the general organization of coordinated behavior, and on bimanual tasks in particular. It has been argued that different types of sensory cues can result in radically different representations of action goals (Oliveira and Ivry, 2008), which in turn has an impact on response selection in bimanual performance. The idea that the conceptualization and representation of action goals influences motor behavior is appealing, because it provides for a broad framework that can be applied to a large variety of tasks and contexts. This raises the question of how action representations are implemented in the brain and how sensory information, goals, and motor activations are linked with them (see also Chapter 18: Motor imagery and its implications for understanding the motor system; Chapter 23: Juggling with the brain — thought and action in the human motor system).

To address this, we may turn to the direct cueing effect demonstrated by Diedrichsen et al. (2001, 2003). This effect might have a neural correlate in the form of the dorsal stream (Hazeltine et al., 2003). In the direct cueing condition, as exploited by these authors (Diedrichsen et al., 2001), bimanual reaching movements were performed to visual targets in space, indicating the endpoint position for each hand. In this regard, it was demonstrated previously that reaching movements driven by visual stimuli are controlled by the dorsal stream (an extrastriatal-posterior parietal-dorsal premotor circuit involved in representing visuo-motor mappings), which acts in an automatic manner without conscious awareness by the participant (Prablanc and Martin, 1992; Schindler et al., 2004; Fang and He, 2005). This specific feature of dorsal stream control has traditionally been tested by reaching paradigms in which the target position was unexpectedly changed at or around movement onset, resulting in online adjustments to hit the visual target (Pisella et al., 2000). A bimanual version of this task revealed that adjustments of the two hands were controlled nearly independently (Diedrichsen et al., 2004). This indicates that both hands can be controlled in parallel within the dorsal stream pathways of the left- and right-brain hemisphere. Interestingly, such automatic movement guidance occurs only when the endpoint position is indicated by a visual target but not when it is symbolized by a color cue (Pisella et al., 2000). Thus, the involvement of automatic dorsal stream processing might represent a possible neurophysiological mechanism underlying the RT benefits when movement endpoints are cued directly. While this neurophysiological explanation is plausible and in agreement with previous research, there is currently only limited support from functional neuroimaging results. Diedrichsen et al. (2006) went on to investigate which areas respond to direct versus symbolic cues in an imaging study, but did not demonstrate that the former evoked more dorsal stream activity than the latter.

By contrast, symbolic cues evoked activity in a left lateralized parieto-premotor network, which is known to be activated during stimulus–response mapping using arbitrary visuomotor associations (Schluter et al., 2001; Thoenissen et al., 2002; Rushworth et al., 2003; Diedrichsen et al., 2006). It was speculated that the left hemisphere might be specialized in representing abstract action goals and/or the translation of symbolic stimuli into actions (Ivry et al., 2004). Consequently, when movements were cued symbolically using dissociated cues, processing costs might increase because one specialized brain region has to compute two separate stimulus–response mappings serially, that is one for each hand (Ivry et al., 2004; Oliveira and Ivry, 2008). This limitation of computational resources, however, is most likely overcome when integrated cues are used, such that only one common response needs to be determined. Additionally, integrated cues might promote conceptual binding, such that the movements of each hand are easily combined into a common movement plan prior to execution.

Binding has also been studied in the vision domain. Within the visual system, it is well known that certain feature properties promote perceptual binding. For example, presenting visual stimuli in the same form or color activates grouping principles, such that they are perceived as one entity or gestalt (Wagemans, 1995, 1997). In addition, it is harder to generate and maintain visual representations encompassing several different features (Kosslyn et al., 1988). Such grouping also exists in the primary motor cortex to the extent that functional assemblies are formed (Jackson et al., 2003) and, specifically, that binding between the left and right hemisphere can be observed during bimanual movements. This was shown by analyzing the correlation of local field potentials simultaneously measured over the left and right primary motor cortex, which were found to be larger when bimanual movements were initiated compared to unimanual movements and larger again when symmetrical bimanual actions were compared to asymmetrical actions (Cardoso De Oliveira et al., 2001). Currently, we can only speculate whether binding in the visual and motor domain is promoted by similar mechanisms and whether these mechanisms can interact. Based on the results presented above, it might be possible that visual cues can facilitate binding processes in the motor domain, but further research is needed to either confirm or reject this hypothesis.

In summary, the previous paragraph described potential neurophysiological mechanisms that might explain how visual cues affect the control of bimanual actions. Many of the results revealed by bimanual RT paradigms can be explained by the differential involvement of dorsal stream areas on the one hand, and a parieto-premotor network associated with arbitrary stimulus–response mapping on the other hand. In particular, direct cues might be processed by the dorsal stream in both hemispheres, which appear to calculate sensorimotor transformations for the left and right hands in parallel, whereas dissociated symbolic cues seem to require specialized parieto-premotor networks of the left hemisphere, such that the stimulus–response mapping of the left and right hand is determined serially.

What else can we learn from bimanual paradigms about stimulus–response selection?

In the previous sections, we reviewed research on bimanual coordination, which served as a well-studied prototype for complex coordinated behavior. Additionally, we used RT as an indication of the processing costs related to response selection and implementation. We argued that some movement patterns are intrinsically more stable than others, which is most likely caused by the neural architecture of the motor system. Not surprisingly, preferred patterns of the movement repertoire are initiated faster than nonpreferred patterns, but only when response selection requirements are low. For example, when starting a 100-m sprint the appropriate response is already selected and prepared long before the "go" signal. Under these circumstances, RT will be shortest for preferred and well-trained movement patterns.

By contrast, motor preferences seem to be less important when response selection is involved and RT additionally reflects the mapping between the appropriate movement and a visual stimulus. In this situation, the properties of the cues have an enormous influence on response selection. For example, the position of defenders on a soccer field will influence the selection of the appropriate kicking action when we want to pass the ball to another team member. Apparently, direct targets in space activate dorsal visual stream control, which is a fast and highly efficient pathway that allows controlling different body parts in parallel. Using this direct neural coupling of perception and action might be an efficient way to teach complex multisegment movements.

Other neural pathways are activated when symbolic cues are used, for example, when one player shouts an instruction where to smash during a beach volleyball rally. These pathways perform more demanding computations, as reflected by longer RTs, and will benefit from an integrated conceptualization of the task. Thus, response selection processes will be faster when the symbolic cue provides integrated information concerning the action goal (e.g., smash to position I) than when it consists of dissociated information (e.g., smash left and long). In summary, the above research indicates that stimulus–response selection is strongly task dependent and that the coupling of perception and action is largely determined by the specific organization of the neural architecture of the human sensorimotor system.

Acknowledgements

Support for this work was provided through a grant from the Flanders Fund for Scientific Research (Projects G.0577.06, G.0593.08)) and from the Research Fund of K.U. Leuven, Belgium (CREA /07/037).

References

Albert, N. B., Weigelt, M., Hazeltine, E., & Ivry, R. B. (2007). Target selection during bimanual reaching to direct cues is unaffected by the perceptual similarity of the targets. *Journal of Experimental Psychology — Human Perception and Performance*, *33*, 1107–1116.

Bach, P., Peatfield, N. A., & Tipper, S. P. (2007). Focusing on body sites: The role of spatial attention in action perception. *Experimental Brain Research*, *178*, 509–517.

Bach, P., & Tipper, S. P. (2006). Bend it like Beckham: Embodying the motor skills of famous athletes. *Quarterly Journal of Experimental Psychology*, *59*, 2033–2039.

Brass, M., Bekkering, H., & Prinz, W. (2001). Movement observation affects movement execution in a simple response task. *Acta Psychologica*, *106*, 3–22.

Brass, M., Bekkering, H., Wohlschlager, A., & Prinz, W. (2000). Compatibility between observed and executed finger movements: Comparing symbolic, spatial, and imitative cues. *Brain and Cognition*, *44*, 124–143.

Byblow, W. D., Lewis, G. N., Stinear, J. W., Austin, N. J., & Lynch, M. (2000). The subdominant hand increases in the efficacy of voluntary alterations in bimanual coordination. *Experimental Brain Research*, *131*, 366–374.

Cardoso De Oliveira, S. (2002). The neuronal basis of bimanual coordination: recent neurophysiological evidence and functional models. *Acta Psychologica, 110*, 139–159.

Cardoso De Oliveira, S., Gribova, A., Donchin, O., Bergman, H., & Vaadia, E. (2001). Neural interactions between motor cortical hemispheres during bimanual and unimanual arm movements. *European Journal of Neuroscience, 14*, 1881–1896.

Carson, R. G. (2005). Neural pathways mediating bilateral interactions between the upper limbs. *Brain Research Reviews, 49*, 641–662.

Chen-Harris, H., Joiner, W. M., Ethier, V., Zee, D. S., & Shadmehr, R. (2008). Adaptive control of saccades via internal feedback. *Journal of Neuroscience, 28*, 2804–2813.

Diedrichsen, J., Grafton, S., Albert, N., Hazeltine, E., & Ivry, R. B. (2006). Goal-selection and movement-related conflict during bimanual reaching movements. *Cerebral Cortex, 16*, 1729–1738.

Diedrichsen, J., Hazeltine, E., Kennerley, S. W., & Ivry, R. B. (2001). Moving to directly cued locations abolishes spatial interference during bimanual actions. *Psychological Science, 12*, 493–498.

Diedrichsen, J., Ivry, R. B., Hazeltine, E., Kennerley, S., & Cohen, A. (2003). Bimanual interference associated with the selection of target locations. *Journal of Experimental Psychology — Human Perception and Performance, 29*, 64–77.

Diedrichsen, J., Nambisan, R., Kennerley, S. W., & Ivry, R. B. (2004). Independent online control of the two hands during bimanual reaching. *European Journal of Neuroscience, 19*, 1643–1652.

Donchin, O., & Cardoso De Oliveira, S. (2004). Electrophysiological approaches to bimanual coordination in primates. In S. P. Swinnen & J. Duysens (Eds.), *Neuro-behavioral determinants of interlimb coordination* (pp. 131–154). Boston, MA: Kluwer Academic Publishing.

Eliassen, J. C., Baynes, K., & Gazzaniga, M. S. (2000). Anterior and posterior callosal contributions to simultaneous bimanual movements of the hands and fingers. *Brain, 123*(12), 2501–2511.

Fang, F., & He, S. (2005). Cortical responses to invisible objects in the human dorsal and ventral pathways. *Nature Neuroscience, 8*, 1380–1385.

Franz, E. A., Eliassen, J. C., Ivry, R. B., & Gazzaniga, M. S. (1996). Dissociation of spatial and temporal coupling in the bimanual movements of callosotomy patients. *Psychological Science, 7*, 306–310.

Franz, E. A., Zelaznik, H. N., Swinnen, S. P., & Walter, C. B. (2001). Spatial conceptual influences on the coordination of bimanual actions: when a dual task becomes a single task. *Journal of Motor Behavior, 33*, 103–112.

Haruno, M., Wolpert, D. M., & Kawato, M. (2001). Mosaic model for sensorimotor learning and control. *Neural Computation, 13*, 2201–2220.

Hazeltine, E., Diedrichsen, J., Kennerley, S. W., & Ivry, R. B. (2003). Bimanual cross-talk during reaching movements is primarily related to response selection, not the specification of motor parameters. *Psychological Research, 67*, 56–70.

Heuer, H. (1993). Structural constraints on bimanual movements. *Psychological Research, 55*, 83–98.

Heuer, H., & Klein, W. (2006). The influence of movement cues on intermanual interactions. *Psychological Research, 70*, 229–244.

Heuer, H., Kleinsorge, T., Spijkers, W., & Steglich, W. (2001). Static and phasic cross-talk effects in discrete bimanual reversal movements. *Journal of Motor Behavior, 33*, 67–85.

Heuer, H., Spijkers, W., Kleinsorge, T., van der, L. H., & Steglich, C. (1998). The time course of cross-talk during the simultaneous specification of bimanual movement amplitudes. *Experimental Brain Research, 118*, 381–392.

Hommel, B., Musseler, J., Aschersleben, G., & Prinz, W. (2001). The Theory of Event Coding (TEC): a framework for perception and action planning. *Behavioral and Brain Sciences, 24*, 849–878.

Ivry, R. B., Diedrichsen, J., Spencer, R. M., Hazeltine, E., & Semjen, A. (2004). A cognitive neuroscience perspective on bimanual coordination and interference. In S. P. Swinnen & J. Duysens (Eds.), *Neuro-behavioral determinants of interlimb coordination* (pp. 259–296). Boston, MA: Kluwer Academic Publishing.

Jackson, A., Gee, V. J., Baker, S. N., & Lemon, R. N. (2003). Synchrony between neurons with similar muscle fields in monkey motor cortex. *Neuron, 38*, 115–125.

Kawato, M. (1999). Internal models for motor control and trajectory planning. *Current Opinions in Neurobiology, 9*, 718–727.

Kelso J. A., Fink, P. W., DeLaplain, C. R., & Carson, R. G. (2001). Haptic information stabilizes and destabilizes coordination dynamics. *Proceedings of the Royal Society of London Series B — Biological Sciences, 268*, 1207–1213.

Kelso, J. A. S. (1984). Phase transitions and critical behavior in human bimanual coordination. *American Journal of Physiology — Regulatory Integrative and Comparative Physiology, 15*, R1000–R1004.

Kelso, J. A. S. (1995). *Dynamic patterns: the self-organization of brain and behavior*. Cambridge, MA: MIT Press.

Kosslyn, S. M., Cave, C. B., Provost, D. A., & von Gierke, S. M. (1988). Sequential processes in image generation. *Cognitive Psychology, 20*, 319–343.

Kunde, W., Krauss, H., & Weigelt, M. (2009). Goal congruency without stimulus congruency in bimanual coordination. *Psychological Research, 73*, 34–42.

Kunde, W., & Weigelt, M. (2005). Goal congruency in bimanual object manipulation. *Journal of Experimental Psychology — Human Perception and Performance, 31*, 145–156.

Mechsner, F. (2004). A psychological approach to human voluntary movements. *Journal of Motor Behavior, 36*, 355–370.

Mechsner, F., Kerzel, D., Knoblich, G., & Prinz, W. (2001). Perceptual basis of bimanual coordination. *Nature, 414*, 69–73.

Monno, A., Temprado, J. J., Zanone, P. G., & Laurent, M. (2002). The interplay of attention and bimanual coordination dynamics. *Acta Psychologica, 110*, 187–211.

Oliveira, F. T. P., & Ivry, R. B. (2008). The representation of action: insights from bimanual coordination. *Current Directions in Psychological Science*, *17*, 130–135.

Peper, C. E., Beek, P. J., & van Wieringen, P. C. (1995). Frequency-induced phase transitions in bimanual tapping. *Biological Cybernetics*, *73*, 301–309.

Pisella, L., Grea, H., Tilikete, C., Vighetto, A., Desmurget, M., Rode, G., Boisson, D., & Rossetti, Y. (2000). An 'automatic pilot' for the hand in human posterior parietal cortex: toward reinterpreting optic ataxia. *Nature Neuroscience*, *3*, 729–736.

Prablanc, C., & Martin, O. (1992). Automatic control during hand reaching at undetected two-dimensional target displacements. *Journal of Neurophysiology*, *67*, 455–469.

Rushworth, M. F., Johansen-Berg, H., Gobel, S. M., & Devlin, J. T. (2003). The left parietal and premotor cortices: motor attention and selection. *NeuroImage*, *20*(Suppl. 1), 89–100.

Schaal, S., Sternad, D., Osu, R., & Kawato, M. (2004). Rhythmic arm movement is not discrete. *Nature Neuroscience*, *7*, 1136–1143.

Schindler, I., Rice, N. J., McIntosh, R. D., Rossetti, Y., Vighetto, A., & Milner, A. D. (2004). Automatic avoidance of obstacles is a dorsal stream function: evidence from optic ataxia. *Nature Neuroscience*, *7*, 779–784.

Schluter, N. D., Krams, M., Rushworth, M. F., & Passingham, R. E. (2001). Cerebral dominance for action in the human brain: the selection of actions. *Neuropsychologia*, *39*, 105–113.

Scholz, J. P., & Kelso, J. A. S. (1990). Intentional switching between patterns of bimanual coordination depends on the intrinsic dynamics of the patterns. *Journal of Motor Behavior*, *22*, 98–124.

Serrien, D. J., Cassidy, M. J., & Brown, P. (2003). The importance of the dominant hemisphere in the organization of bimanual movements. *Human Brain Mapping*, *18*, 296–305.

Serrien, D. J., Strens, L. H., Oliviero, A., & Brown, P. (2002). Repetitive transcranial magnetic stimulation of the supplementary motor area (SMA) degrades bimanual movement control in humans. *Neuroscience Letters*, *328*, 89–92.

Shadmehr, R., & Krakauer, J. W. (2008). A computational neuroanatomy for motor control. *Experimental Brain Research*, *185*, 359–381.

Swinnen, S. P. (2002). Intermanual coordination: from behavioural principles to neural-network interactions. *Nature reviews. Neuroscience*, *3*, 348–359.

Swinnen, S. P., Dounskaia, N., & Duysens, J. (2002). Patterns of bimanual interference reveal movement encoding within a radial egocentric reference frame. *Journal of Cognitive Neuroscience*, *14*, 463–471.

Swinnen, S. P., Jardin, K., Meulenbroek, R., Dounskaia, N., & Hofkens-Van Den Brandt, M. (1997). Egocentric and allocentric constraints in the expression of patterns of interlimb coordination. *Journal of Cognitive Neuroscience*, *9*, 348–377.

Swinnen, S. P., Jardin, K., Verschueren, S., Meulenbroek, R., Franz, L., Dounskaia, N., & Walter, C. B. (1998a). Exploring interlimb constraints during bimanual graphic performance: effects of muscle grouping and direction. *Behavioral Brain Research*, *90*, 79–87.

Swinnen, S. P., Verschueren, S., Bogaerts, H., Dounskaia, N., Lee, T. D., Stelmach, G. E., & Serrien, D. J. (1998b). Age-related deficits in motor learning and differences in feedback processing during the production of a bimanual coordination pattern. *Cognitive Neuropsychology*, *15*, 439–466.

Swinnen, S. P., & Wenderoth, N. (2004). Two hands, one brain: cognitive neuroscience of bimanual skill. *Trends in Cognitive Sciences*, *8*, 18–25.

Temprado, J. J., Chardenon, A., & Laurent, M. (2001). Interplay of biomechanical and neuromuscular constraints on pattern stability and attentional demands in a bimanual coordination task in human subjects. *Neuroscience Letters*, *303*, 127–131.

Thoenissen, D., Zilles, K., & Toni, I. (2002). Differential involvement of parietal and precentral regions in movement preparation and motor intention. *Journal of Neuroscience*, *22*, 9024–9034.

Todorov, E. (2004). Optimality principles in sensorimotor control. *Nature Neuroscience*, *7*, 907–915.

Wagemans, J. (1995). Detection of visual symmetries. *Spatial Vision*, *9*, 9–32.

Wagemans, J. (1997). Characteristics and models of human symmetry detection. *Trends in Cognitive Sciences*, *1*, 346–352.

Weigelt, M. (2007). Re-examining structural constraints on the initiation of bimanual movements: the role of starting locations, movement amplitudes, and target locations. *Human Movement Science*, *26*, 212–225.

Weigelt, M., Kunde, W., & Prinz, W. (2006). End-state comfort in bimanual object manipulation. *Experimental Psychology*, *53*, 143–148.

Weigelt, M., Rieger, M., Mechsner, F., & Prinz, W. (2007). Target-related coupling in bimanual reaching movements. *Psychological Research*, *71*, 438–447.

Wenderoth, N., Debaere, F., Sunaert, S., van Hecke, P., & Swinnen, S. P. (2004). Parieto-premotor areas mediate directional interference during bimanual movements. *Cerebral Cortex*, *14*, 1153–1163.

Wenderoth, N., Debaere, F., Sunaert, S., & Swinnen, S. P. (2005). Spatial interference during bimanual coordination: differential brain networks associated with control of movement amplitude and direction. *Human Brain Mapping*, *26*, 286–300.

Wenderoth, N., Toni, I., Bedeleem, S., Debaere, F., & Swinnen, S. P. (2006). Information processing in human parieto-frontal circuits during goal-directed bimanual movements. *NeuroImage*, *31*, 264–278.

Yamanishi, J., Kawato, M., & Suzuki, R. (1980). Two coupled oscillators as a model for the coordinated finger tapping by both hands. *Biological Cybernetics*, *37*, 219–225.

M. Raab et al. (Eds.)
Progress in Brain Research, Vol. 174
ISSN 0079-6123

CHAPTER 16

How do motoric realities shape, and become shaped by, the way people evaluate and select potential courses of action? Toward a unitary framework of embodied decision making

Daniel A. DeCaro[1,*], Michael Bar-Eli[2], Juliet A. Conlin[3], Adele Diederich[4], Joseph G. Johnson[1] and Henning Plessner[5]

[1]Department of Psychology, Miami University, Oxford, OH, USA
[2]Department of Business Administration, Guildford Glazer School of Business and Management, Ben-Gurion University of the Negev, Beer Sheva, Israel
[3]Center for Adaptive Behavior and Cognition, MPI for Human Development, Berlin, Germany
[4]School of Humanities and Social Sciences, Jacobs University, Bremen, Germany
[5]Institute of Psychology 1, University of Leipzig, Leipzig, Germany

Abstract: Until recently, the constraints imposed on decision makers by the human physical condition — situated both as a physical agent and within physical space — have played only an incidental, if not entirely inconsequential, role in conceptualizations of human decision making. The act of deciding has been positioned as the quintessence of traditional decision theory, while actual enactment of the decided action within physical space by a corporal actor, with all that this entails, has been regarded as the obvious and, therefore, scientifically uninteresting result of having made up one's mind (cf. Bagozzi et al., 2003). However, recent discoveries made in the area of embodied cognition regarding the involvement of fundamentally motoric representations in long-presumed "cognitive" systems (Wilson, 2002) potentially turned conventional wisdom upside-down. In this chapter, we go beyond prominent theories of action selection and decision making to rethink the link between mind and body as it pertains to the relatively novel frontier of embodied decision making. In particular, we reconceptualize what it means to evaluate one's options in light of recent advancements in embodied cognition, motor control, and dynamic decision making. In the process, we provide a much needed account of the primary theoretical issues that any good account would seem to be impelled to address. Perhaps the greatest contribution provided by the present chapter is an organizing framework that we hope will guide future research to the eventual answer to what it means to be an embodied decision maker.

Keywords: embodied cognition; decision making; motor system

Introduction

The premise of this chapter is a conceptualization of decision making as a cognitive process that is intimately rooted in real action and also the

*Corresponding author.
Tel.: +49-513-529-2400; Fax: +49-513-529-2420;
E-mail: d_decaro@yahoo.com

DOI: 10.1016/S0079-6123(09)01316-8

imposition of a challenge to consider action as principally embodied, which is to say, as a manifestation, or replication, of the physical environment within the decision maker (Wilson, 2002). An embodied cognition perspective of action, and in particular decision making, can be thought of as an at least three-part process of recognition and construction of the choice task, selection of a particular course of action from among a set of choices, and completion of the selected action, all of which are intimately shaped by the decision maker's physical, physiological, and environmental realities (see Chapter 12: Embodied cognition of movement decisions: a computational modeling approach). In this chapter, we address the second of these processes — *How do individuals' motoric realities shape, and become shaped by, the way they evaluate and select potential courses of action?*

The principle contribution of the present chapter is an explicit formalization of the research question itself, as informed by the authors' varied perspectives and respective areas of expertise. As will be seen, this feat is far from trivial and sorely needed. The chapter also outlines essential elements of any good — comprehensive and theoretically satisfying — response to the question, which is then used to generate novel predictions with which to guide future investigation. We retain the working group's substantive points of collective agreement and unique points of constructive disagreement throughout in order to do justice to the complexity and sophistication present in both the focal question of this chapter and to the diversity of expertise brought to bear on the matter. Our hope in writing this chapter is that it will prove an essential reading in the eventual answer to the myriad questions posed by the topic of embodied decision making.

Conceptualizing the research question and the essential qualities of a good response

Within the opening remarks of the working group's discussion at the "*Mind and motion: the bidirectional link between thought and action*" conference (Bielefeld, Germany) it was immediately clear that there were differences of perspective that were informative in themselves and, therefore, warranted full attention. Paramount among the contentions was a fundamental disagreement of how the research question should be framed and, therefore, what should be addressed in formulating an adequate account of embodied decision making. In testament of the necessity of these contentions, the majority of the discussion over the 3-day meeting was captured by them, resulting in a much-needed explicit exposition of the problem, the contributions and limitations of each perspective as a singular response to the problem, and a nuanced conceptualization of the essential qualities of any good — theoretically satisfying and complete — response. We present the product of this discussion as an unprecedented formalization of the problem itself and as a roadmap to the investigative framework to come.

Initial perspectives — their contributions and limitations

The original question posed to the working group owes is framing to cognitive science's popular conceptualization of cognitive and behavioral processes as separable components of a modular system (cf. Newell et al., 1958; Fodor, 1983). This framing naturally begs the question of how these separable systems interface: *how (if at all) do decision makers' physical and motoric realities play an interdependent role in the evaluation and selection of potential courses of action?* The conceptualization of the motoric components and cognitive components of processing as compartmental is also evident in the fact that this question was abstracted from its sister question of how choice options and their consequences are perceived and represented (see the Preface). The distinction warrants attention because such compartmentalization would not have been entertained by, for example, traditions favorable to embodied cognition (cf. Wilson, 2002; Faubel and Schöner, 2008). Notwithstanding, there are at least three distinct perspectives on the problem: (1) a sequential information-sampling model of preferential choice (cf. Busemeyer and Townsend, 1993; Roe et al., 2001; Chapter 12: Embodied cognition of movement decisions: a computational modeling approach) favorable to the information-processing assumptions of Newell and Simon (Newell et al., 1958); (2) a dynamic-systems-thinking perspective (Faubel and Schöner, 2008), which is conceptually amendable to both embodiment and the notion of ecological rationality espoused by the fast-and-frugal heuristics paradigm (cf. Gigerenzer and Goldstein, 1996; Chapter 10: Getting around: making fast and frugal navigation decisions); and (3) a perceptual categorization perspective stemming from the lens model (Hammond et al., 1975) of social judgment (Chapter 13: A multiple-cue learning approach as the basis for understanding and improving soccer referees' decision making).

According to sequential-sampling choice models of, for example, decision field theory (Busemeyer and Townsend, 1993), action selection is just a particular instance of preference-driven choice. Higher-order cognitive processes, such as selective attention, play an important role in determining what course of action will be selected by controlling what aspects of the information-rich environment are processed during action deliberation (Chapter 12: Embodied cognition of movement decisions: a computational modeling approach). Preference for any given course of action accumulates over time through selective attention to specific aspects of the decision situation. The process of information/preference accumulation can be quite heady, occurring consciously and deliberatively as in selection of a new car or home (Busemeyer and Townsend, 1993), or accumulation can be rapid, online, and driven by lower-order perceptual mechanisms (e.g., Link and Heath, 1975; Townsend and Ashby, 1983) as seems to occur in sports. Thus, sequential sampling accounts of decision making need not be instantiated at the level of higher-order cognitive processes or even for preferential choice in specific. Sequential sampling refers only to a generic information-processing constellation that is broadly characterized by information updating by receptive neuronal systems, accumulation of information until some threshold is met, and a subsequent system-specific output: each of these component processes can be automatic and subconscious or open to conscious control, depending on the particular model and circumstances (see Thomas et al., 2008 for review). Thus, while amenable to automatic updating, the particular position espoused by Johnson (Chapter 12: Embodied cognition of movement decisions: a computational modeling approach) in this volume (i.e., decision field theory) is taken at the level of deliberative preference accumulation. This perspective affords the advantage of a focused and detailed account of the decision event — option evaluation and selection.

By this view, motoric and physiological constraints, such as motor limitations of the decision maker (e.g., fatigue, physical obstacles) and embodied cognitions, can be considered modular and somewhat ancillary additions to a relatively separable decision event. Indeed, the theory itself has evolved first as a theory of mental-bound deliberation (Busemeyer and Townsend, 1993) and second as a physically active organism (Chapter 12: Embodied cognition of movement decisions: a computational modeling approach). Thus, researchers working within this perspective now conceptualize their own goal as one of identifying ways in which potentially separable cognitive and motoric systems impact one another during option evaluation, selection, and execution (cf. Chapter 11: Crossmodal interaction in speeded responses: time window of integration model; Chapter 15: Visual cues influence motor coordination: behavioral results and potential neural mechanisms mediating perception–action coupling and response selection). For instance, a midfielder faced with the decision of how to advance the next play in a ball sport may be subject to fatigue-induced preference vacillations, whereby the perceived value of more physically strenuous options is greatly reduced such that these options are rejected over less strenuous options; physical realities encountered during action execution may result in the midfielder actually completing an action other than his/her mentally chosen and intended action. The common theme among these examples is a conceptualization of motoric and decision systems as functionally distinct, yet interacting.

This more deliberative instantiation of sequential-sampling models of decision making may be contrasted with a strict dynamic systems thinking (DST) conceptualization of the problem, which disagrees fundamentally with any characterization that implies substantial separation of sensory-motor and "decision-making" processes (Faubel and Schöner, 2008). According to DST, there is an intimate link between an organism's sensory-motor experience and its behavioral responses, such that action selection ("decision making") arises naturally from the online interplay between an organism's physiological construction and the environment. Decision-making faculties are instantiated as embodied cognitions, wherein aspects of the physical environment are replicated within the neurophysiological structures of the organism and serve important functional roles. It is unclear how the information-processing perspective advocated by popular sequential-sampling models of deliberative choice would incorporate such an intimate integration of motoric and physical realities into the evaluative process, which precedes action selection (but see Link and Heath, 1975 for a potential bridge through rudimentary standard-target discriminatory mechanisms). DST also strongly resists attempts to isolate a "decision event" as a unit of functional operation. Instead, behavioral patterns are regarded as continuous sensory-perception–action sequences, consisting of multiple and interdependent action selection events. Thus, the goal of theorists working in this tradition is to describe the holistic sequence of action selection accurately and to uncover the lawful rules governing such a dynamic system. For instance, in the analysis of a male housefly actively tracking a potential mate through the air (Reichardt and Poggio, 1976), dynamic-systems theorists attempt to identify emergent patterns of flight as reducible to lawful sequences of activation between the male fly's retinal mapping of physical space and its motor system (Wilimzig et al., 2006). The analysis goes so far as to formalize the entire action sequence, beginning with the "decision" to visually orient toward the stimulus (female fly) or to maintain current visual fixation, to the momentary online "decisions" governing flight navigation during actual mate tracking (Wilimzig et al., 2006). The holistic analysis of action sequences by DST, coupled with its special attention to online, sensory-motor decision situations, provides a unique contribution *vis-à-vis* traditions that abstract decision making from place and time, such as in decision field theory and classical expected utility formulations of choice like subjectively weighted expected utility (see Schoemaker, 1982 for an overview of utility theories).

However, the DST approach encounters theoretical challenges of its own as a singular solution to the problem. The fact that DST analyses of decision making are currently limited to action selections analogous to saccadic eye movement and mate tracking by lower organisms, makes it difficult to extrapolate to decision-making situations that presumably involve higher-order cognitions, such as semantic and biographical memory, and which are far removed from the immediate sensory world (cf. Wilson, 2002), such as my deciding what to do when I return to the United States of America, while attending the Center for Interdisciplinary Research (ZiF) conference an ocean away in Germany (but see McGrath et al., 2000 for an extrapolation to social–psychological principles of group dynamics). At present, the information-processing perspective of decision field theory, as well as the perceptual categorization perspective (explained last), seem better suited to deal with such cognitive decision tasks. Moreover, the DST perspective appears at odds with the widely accepted observation that people spontaneously self-generate creative decision behaviors that are not quite so readily explained as arising from identical sensory-motor information from identical contexts.

One function of decision making may be to serve a self-determining role for the decision maker, whereby the decision maker decides between alterative actions for the specific purpose of defying or altering the dynamic system it momentarily finds itself acting within. Bari-Eli's report of expert athletes' socially rational tendency to optimize socially and personally defined utility functions instead of objective game-defined functions during high-stakes penalty kicks (Chapter 9: (Ir)rationality in action: do soccer players and goalkeepers fail to learn how to best perform during a penalty kick?) is a case in point. Social rationality highlights the necessity to consider boundary conditions wherein volitional control seems a natural element of the decision situation (deliberative preference-driven choice; e.g., consumer choice) and where it likely does not (perceptually driven motoric or proceduralized action sequences; e.g., choice during sports). This critique raises the concern that with regard to human decision making, DST advocates a form of deterministic, nonvolitional decision making that is most defensible when restricted to specific types of decision tasks that are outside of conscious awareness because they exist at lower levels of perceptual processing, are mostly reliant on subconscious processing, and/or are highly routinized.

The type of cognitive embodiment and ecologically rational cognitive system advocated by Simon (1956) and more recently by Gigerenzer (2000) in his adaptive tool box metaphor may offer a healthy compromise to the perceptually driven system posited by DST and the cognitively driven system posited by decision field theory. As Conlin relates elsewhere in this volume, (Chapter 10: Getting around: making fast and frugal navigation decisions) people may be able to solve the same problems in different ways. Individuals can pull from their adaptive tool boxes two general categories of information-processing strategies — algorithms and heuristics. Algorithmic strategies are approaches to a problem that require explicit and elaborate, and therefore cognitively taxing, mental operations. The classic example of an algorithmic decision strategy is the weighted-additive rule, in which the decision maker assesses the value of every potential course of action and its expectancy of success, multiplies these across every potential course of action to determine an overall "goodness" of fit for the decision situation, and then selects the action with the highest "goodness of fit" or expected utility (Svenson, 1979). In contrast, heuristics are relatively less cognitively taxing strategies that exploit the information inherent in the environment and in the person's physiological makeup to simplify the task at hand yet yield outcomes that approach or match those of algorithmic strategies (cf. Simon, 1956; Gigerenzer and Goldstein, 1996). For instance, the algorithmically complex act of tracking and catching a ball in a ball sport may be reducible to the simplifying strategy of moving around the field so as to keep constant the angle formed by the ball's trajectory through the sky relative to the horizon (McLeod and Dienes, 1996; cf. Chapter 10: Getting around: making fast and frugal navigation decisions). By this view, human cognition is not strictly reduced to deterministic embodied mechanisms, but inherently possesses the ability to rely on such a capacity if given the opportunity. Such a perspective leaves room for the core characteristics of both decision field theory and dynamic-systems-thinking, while sidestepping many of the more contentious elements of DST's embodiment position.

A third initial perspective on the question arises from the lens model (Hammond et al., 1975) conceptualization of decision making as a perceptual categorization task (Chapter 13: A multiple-cue learning approach as the basis for understanding and improving soccer referees' decision making; cf. Chapter 4: Perceiving and moving in sports and other high-pressure contexts). In contrast to both the sequential-sampling and DST approaches, decisions may not necessarily involve evaluation or arbitration of potential courses of action. Instead, experientially developed associative links between appropriate actions and particular environments may lead to associative memory-based retrieval of a relevant course of action (cf. Klein, 1999). Rather than ask how people make decisions per se, this framework asks theorists to consider how a physically grounded and constrained entity categorizes decision situations and recognizes appropriate courses of action. The decision maker's challenge is to acquire a veridical environment–action associative mapping whereby a particular action becomes selected for or highly accessible based on limited or imperfect information, such as it is. Consider a referee's task of deciding how to call an ambiguous sports play (Chapter 13: A multiple-cue learning approach as the basis for understanding and improving soccer referees' decision making). According to decision field theory, the referee's decision will depend principally on his/her attention to different aspects of the play (i.e., his/her evaluation); over time a preference will develop for one call over others, resulting in a decision. In contrast, the lens model construes the call task as fundamentally grounded in perceptual categorization; the referee's task is to categorize the play accurately. With the correct categorization, the appropriate call will be obvious. Thus, this perceptual-categorization perspective is able to solve the problem without making an (explicit) appeal to the notions of option evaluation and "decision making."

One potential concern with this approach is the extent to which a discrete distinction can be made between categorization and evaluation-driven choice. For instance, the information-processing approach advocated by decision field theory is not at odds with the perceptual-categorization approach. The same descriptive outcome can be achieved by a sequential-sampling model calibrated with a low threshold decision rule that thereby permits rapid option evaluation and selection within the typical time course at which a categorization-based explanation is often offered as a preferred account (Lee and Cummins, 2004). In this sense, perceptual-categorization-like strategies could be considered a special case of a decision system operating on a sequential-sampling mechanism (but see Beach, 1998 for a counter argument).

With so many relevant perspectives on this question of embodied decision making, the challenge is to determine if these conceptualizations can be reconciled or whether one must posit separate systems (and theories) of embodied decision making. We do not entertain separate theories of embodied decision making here (for that see the other chapters in this section of the present volume). Instead, we submit that reconciliation is a necessity borne out in this case by each perspective's shortcomings as a sole response to the question of how motor and evaluative faculties codetermine choices. For instance, a sequential-sampling model of deliberative choice based on Newell and Simon's (1958) principles of information processing seems to place too little emphasis on the situated and embodied nature of some types of decision tasks (but see Chapter 12: Embodied cognition of movement decisions: a computational modeling approach for a novel attempt). However, a purely dynamic-systems-thinking approach may emphasize situated decision making too much; a strict embodiment perspective elegantly accounts for sensory-driven action selection but risks painting an incomplete picture of human decision making if it leaves too little room for the type of heady, contextually abstracted decisions for which decision theories such as decision field theory and subjectively weighted utility are specialized. Thus, perhaps we can learn the qualities of a satisfactory response to the question and the basis for a unitary theory of embodied decision making by reflecting on what can be learned from these shortcomings.

Qualities of a good response

We have challenged the reader to consider the strengths and weaknesses of each perspective on the question of embodied decision making. Beyond yielding an explicit formalization of the research question, the exercise has revealed that the viability of a single theoretical perspective hinges on at least three major critical factors. The first factor concerns the level of abstraction at which the focal decision task exists – whether the task is primarily sensory-motor driven, such as frequently observed in sports (e.g., chosen goal location of a penalty kick) or is mostly abstracted from the immediate sensory-motor context. The difference is probably a matter of location on a continuum rather than membership in discrete categories (Hogarth, 2005). The distinguishing point between the two task types appears to depend on the extent to which the goal of the action selection is to execute an actual physical act in itself (e.g., penalty kick) or is to establish an instrumental preference structure (preference-based comparison of options) that will lead to subsequent nonfocal and nearly inconsequential physical acts. For instance, in the latter case, the act of actually driving home one's new car is far less focal to the goal of getting a new car than that of first establishing a psychological preference from among the available cars. The sensory-motor environment is so far removed from the focal task that the car selection could conceivably be completed in a sensory-motor vacuum (but see our later embodied cognition critiques). In any case, it is difficult to deny that there are inherent epistemic differences between a sports task on the one hand and an abstract deliberative preference-driven task on the other. These epistemic distinctions likely warrant substantive information-processing distinctions, which any complete framework of embodied decision making must address.

However, the distinction between sensory-motor and nonsensory-motor tasks is confounded with the reality that "decisions" made regarding physical acts as a goal, such as determining which trajectory to take while tracking a ball to catch it, typically are extremely time-pressured and defined on fleeting, dynamic information (cf. Klein, 1999); otherwise one might find ball sports being played like chess. However, even chess can be played something like ball sports in the sense that expert chess players transform initially totally "cognitive" evaluation-based decisions, as in decision field theory, into routinized perceptual-categorization tasks (Chase and Simon, 1973), as in the lens model. Hence, in differentiating between sensory-motor tasks and deliberative preferential choice tasks, one seems required to acknowledge the need to account for the role of task novelty and receptivity to proceduralization in any satisfactory framework of embodied decision making (cf. Smith and DeCoster, 2000).

A final, and highly related, boundary condition to contend with is the extent to which deterministic information-processing mechanisms underpin performance, given a particular decision situation. Online, sensory-motor driven activities are more likely to be executed by automatic processes that are relatively deterministic in nature, having little room for higher-order volitional control or spontaneous creativity (Posner and Snyder, 1975; Schneider and Shiffrin, 1977). Thus, performance on sports decision tasks, such as "deciding" what swing to use while hitting a golf ball, is optimal when highly proceduralized and outside conscious executive control (DeCaro et al., 2008). In contrast, tasks that pose unique informational challenges that are not easily subjected to rote procedures, such as hypothesis testing and choosing a new car (for novices), are better performed within the window of deliberate, conscious control (DeCaro et al., 2008). To the extent that human decision makers confront both types of tasks in life, they will need to be equipped with mechanisms capable of achieving optimal performance in both environments. So too will an adequate theory of embodied decision making need to leave room for both deterministic and volitional processing mechanisms in order to speculate about the likely nature of involvement by embodied cognitions for each type of processing.

Readers familiar with dual-systems accounts of information processing in learning (e.g., Smith and DeCoster, 2000), reasoning (Sloman, 1996; Hogarth, 2005), and judgment and decision making (Stanovich and West, 2000; Kahneman, 2003) will undoubtedly notice the parallels between such accounts and the major factors that we have discussed. Indeed, dual-systems accounts arose from attempts to make sense of these very same factors in myriad other domains of inquiry (see Sloman, 1996; Smith and DeCoster, 2000 for reviews). While specific formulations abound, the common thread among different dual-systems theories is the belief that there are two systems of thought that are specialized for particular processing demands and which exhibit a core constellation of processing characteristics.

System 1 processing is regarded as a rapid, highly intuitive or "gut"-driven (i.e., instinctual) mode of thinking that tends to operate largely outside of conscious awareness. Associative-learning mechanisms appear to play a central role in system 1 thinking, with appropriate responses being linked to innate or over-learned environmental representations that then map onto the given task via analogical principles. Thus, system 1 thinking is highly contextualized, highly situated, and data-driven (cf. Wilson, 2002). Because it primarily operates outside of conscious awareness, this form of thinking does not strongly tax higher-level cognitive resources, making it an ideal candidate for task situations characterized by time pressure, high stakes/stress, emotive content, and for which the actor is highly practiced (see Sloman, 1996; Smith and DeCoster, 2000; Kahneman, 2003 for reviews).

In contrast, system 2 thinking is slower and more deliberate and is characterized by a sequential (rather than parallel) mode of information processing. One hallmark of system 2 thinking is deductive, hypothesis-testing-like thinking, whereby individuals go through some explicit rule-based procedure to arrive at a decision or other product (but see Sloman, 1996). As such, system 2 thinking is regarded as relatively cognitively taxing, requiring much working memory and explicit attentional control. While system 2 modes of thought do not necessarily hold a clear-cut accuracy advantage over system 1 thinking in nonpressurized situations (cf. Stanovich and West, 2000), they are regarded as too time and resource consuming to support typical performance in pressurized situations. Thus, system 2 thinking is thought to be best suited for situations in which the immediate environmental context is of diminished import, novel decisions, and when the actor has much time to cogitate (see Sloman, 1996; Smith and DeCoster, 2000; Kahneman, 2003 for reviews).

The parallels of our previous embodied decision-making account and dual-systems theories notwithstanding, we must stress that we do not intend to present evidence for or against dual-systems accounts of information processing. Neither do we wish to rehash or provide yet another formulation of dual-systems thinking. Instead, we simply want to bring to theorist's attention that any comprehensive account of embodied decision making would seem to be compelled to contend with the very same extenuating situational and personal factors that dual-systems theories of information processing have had to address, namely differential task demands imposed by the skill level of the actor (expert vs. novice), extreme time pressure, richly contextualized versus abstracted decision situations (e.g., hitting a ball vs. buying a car), deterministic versus volitional processing, and so on. This is not to say that a dual-systems perspective is not welcomed. For instance, theorists may benefit from conceptualizing embodied decision making as a question of how embodiment may be differentially involved in system 1 versus system 2 modes of thinking. However, for the present purposes, we find it more focal to couch the question in terms of task demands *vis-à-vis* the characteristics of the person.

Thus, in surveying the various pros and cons of singular accounts of embodied decision making, we hope to have convinced the reader that any good — complete and theoretically appealing — framework of embodied decision making must come to terms with the following potentially essential qualities of a human embodied decision maker, as they may strongly qualify the ways in which the mind and body interact during decision making.

A lower-order, sensory-driven component in which action selection arises naturally from the decision maker's physiology and sensory-motor experience — including ecologically rational mechanisms whereby the environment is functionally replicated within, and able to be exploited by, the decision maker.

An associative, recognition-based component that is amendable to learning and capable of rapidly categorizing a decision situation so that the appropriate behavioral response can be retrieved from memory and executed.

A higher-order cognitive component that affords contextually removed deliberation and (potentially conscious) evaluation of alternative actions, especially when the decision maker is confronted by novel decision situations.

But this proposition is not so much a ready-made solution to the question of how (whether) mind and body interact during decision making as it is a coherent framework to guide future investigation.

A unitary framework for embodied decision-making research

At the outset of the ZiF consortium on the link between mind and body in action representation, selection, and execution, it was immediately clear that a general framework for how to proceed was sorely needed. What was most lacking was a clear definition of the problem — a task we have tried to achieve in the majority of this chapter. Having clearly framed the problem, we now turn to a research framework capable of accounting for each of the key components identified as potentially essential to human embodied decision making. Our hope is that the framework presented here, if not entirely veridical, will, as the first of its kind, challenge researchers in this domain to ask demanding questions and find solutions. Thus, the framework is presented mostly as a tool for generating hypotheses and as a guide to future attempts at solving the problem of embodied decision making.

A unitary, hierarchical-processing framework of embodied decision making

We have distinguished between two types of decision tasks — sensory-motor and deliberative preferential-choice — in order to raise the possibility that mind and body may interact differently depending on the type of decision task and its underlying information-processing system. It is also useful to clarify the stream of processing in which the tasks are embedded. Doing so will ease the application of the forthcoming framework to decision tasks in which the type of decision is strictly time-locked, such as in ball sports in which individuals make several different types of inter-related decisions within a matter of milliseconds (e.g., tracking the ball through the field), seconds (rapid action selection — to lob vs. rocket a pass), and minutes (e.g., deciding between novel plays) to fulfill a single goal of, for example, scoring the winning point.

It is possible to think of decision making as a hierarchy of interconnected dynamic processes, differing in terms of information-processing complexity (lower vs. higher order) and content or function — that is, detection, categorization, evaluation, deliberation, and so on (cf. Hogarth, 2005; but see Heekeren et al., 2008 for an alternative, heterarchic formulation). As de Oliveira et al. relate elsewhere in this volume (Chapter 8: The bidirectional links between decision making, perception, and action), qualitatively different "decision tasks" may be linked together in a parallel and continuous, hierarchical chain of active–reactive, perception–action events, with the type of decision task that must be completed at a particular point in the chain being dependent upon its hierarchical location in the stream of processing. For instance, in physical sports, individuals are engaged in active movement throughout a playing area; self-movement continually changes one's environmental array, thereby changing the types of decisions individuals must make throughout the course of play (cf. Chapter 11: Crossmodal interaction in speeded responses: time window of integration model). As with dual-systems accounts of information processing, the streams of processing could occur quite independently and essentially compete for behavioral expression, could be deeply integrated and interacting, could be sequential or parallel, or any combination in between (cf. Maddox and Ashby, 2004; Kahneman, 2003; Heekeren et al., 2008; Thomas et al., 2008). For the present purposes, we feel that it would be tangential to the focal task of this chapter, which is to speculate about the form of involvement of embodied cognitions in action selection, to distinguish between possible instantiations of the processing structure. Hence, we will not entertain a discussion of the various modes of information processing here. However, we will distinguish between *content* of processing, regardless of its temporal sequencing in the chain of processing (parallel vs. sequential), because even though a particular actor's situation — for example, being in a sports situation — may place the actor in a strict temporal sequence of processing, the content of the processes themselves are separable and superordinate units of inquiry as far as our analysis of embodied cognitions is concerned. Hence, in the explanation that follows, we will discuss various ways in which embodiment may be involved in decision making as if it is locked within a temporal sequence of processing merely for the sake of simplifying an otherwise complex discussion of embodiment's involvement within a unitary, hierarchical information-processing system that involves both temporal- and content-based specialization of processing. However, we must stress that as far as we are concerned, the defining characteristic for positing the involvement of embodied mechanisms is a matter of content of processing, not temporal sequencing, per se.

Table 1 details each content level of the hierarchy and corresponding hypotheses regarding the involvement of embodied cognitions (discussed later). With the onset of a stimulus, the first "decision task" may be strongly sensory-perception driven, such as online saccadic orientation and disengagement effects, which the DST perspective has been interested in studying (e.g., Wilimzig et al., 2006). Rapid, proceduralized, and primarily sensory-motor driven tasks would also exist at this level of the hierarchy. For example, a sizable portion of the "decisions" made in sports that are reflexive in nature would fall under this heading (e.g., a baseball player "selecting" the most appropriate swing when at bat).

Higher in this hierarchy of processing, the "decision task" may be less completely governed by direct sensory-perception inputs and may actually constitute the (learned) identification and categorization of a decision stimulus and a context-dependent appropriate response — hence, perceptual-categorization, as in the lens model. Klein and associates (e.g., Klein, 1999), working within the naturalistic decision-making paradigm, have documented decision making at this level of processing in great detail (see Lipshitz et al., 2001 for review). Action selection by expert firefighters, doctors, and individuals in military combat settings represent prime examples of "decision making" at the perceptual-categorization level of processing; many sports activities would also seem to fall within this category due to similar processing constraints (e.g., "deciding" whether a pitch is hittable).

Still higher in the hierarchy, the "decision task" may be quite remote from direct physical and sensory constraints, such that the action selection task involves contextually removed, deliberative evaluation and is substantially preference-driven. The latter distinction depends on the novelty of the task; procedural learning could transform the initially deliberative decision task into a recognition procedure executed at the level of perceptual categorization (Chase and Simon, 1973; Klein, 1999). The situation in which a coach decides during timeout what pitcher to substitute for the next play by deliberating the pros and cons of various pitchers *vis-à-vis* opponents and other conditions of the game is one example of preference-driven choice.

The apparent differences among the various perspectives that motivated much of the preceding discussion may be relatively superficial when judged from the standpoint of a hierarchical decision system such as this. For instance, the earlier disconnect between DST's conceptualization of decision making as being largely sensory-motor and environment based and decision field theory's conceptualization of decision making as being largely separable from sensory-motor information may simply be due to a failure to specify the level of processing at which one's theoretical treatment is targeted. A similar argument can be made regarding the lens model's critique of

Table 1. Hypotheses — a unitary, hierarchical-processing framework for studying embodied decision making

Type of information processing	Hypothesized nature of involvement by embodied mechanisms
Deliberative preference accumulation (e.g., decision field theory)	*Body, mind are modular and interacting* Body → Mind Proprioceptive-dependent evaluation Mind → Body ?
Perceptual categorization (e.g., lens model, recognition-primed decision making)	*Body, mind, action one via learning* Body → Mind Proprioceptive-learning; evaluation Mind → Body ?
Sensory-perception driven (e.g., dynamic systems thinking)	*Body, mind, and action are one* Action selection direct result of information processing by systems that replicate the environment within actor

decision field theory as underemphasizing the special status of perceptual-categorization decision tasks. Thus, a hierarchical processing interpretation seems to point naturally toward a unitary theory of embodied decision making. By this view, the bidirectional pathways of influence held between mind and body can be conceptualized as playing out among and between these three levels of (1) sensory-motor performance, (2) perceptual categorization, and (3) cognitive deliberation. The framework preserves the strengths of each perspective initially used to address the problem of embodied decision making and obviates the majority of their shortcomings. Moreover, the framework permits — theoretically justifies — the capacity of human decision makers to conduct online deterministic processing that exploits the sensory-motor context, perceptual categorization, and offline volitional processing. Above all, this interpretation puts a new light on the question originally posed to the working group, asking now how motor and physical realities impact decision making as it exists at and between each level of the hierarchical system.

Hypotheses — what is embodiment for each level of processing?

Perhaps the most obvious of hypotheses generated by this framework is that which most directly targets the framework itself. Researchers may ask themselves whether motoric influences on decision making are the same across each level of the hierarchy, what types of motoric influence will be present, and the conditions qualifying the presence of various forms of influence. As discussed here, researchers may find that the nature of embodiment's role in decision making depends principally on the characteristics of the focal decision task (see Table 1 for an overview).

Motoric systems, in particular proprioceptive and situated information, likely nearly completely dominate processing at the sensory-motor level of the decision-making hierarchy, such that sensory-motor performance is essentially the consequence of activated embodied cognitions, or reflections of the environment in the organism (see Wilson, 2002 for review). However, proprioceptive and situated information may play only a specialized role, if any, in the evaluative mechanism underlying deliberative preferential choice (cf. Hogarth, 2005; Kahneman, 2003). The overwhelming majority of preferential choice models posit some form of evaluative mechanism responsible for assessing the value or utility of potential actions and their consequences (Schoemaker, 1982; Edwards, 1992). To date, numerous studies have reported the potential for evaluations rendered on a positive–negative psychological scale (e.g., like–dislike, prefer–reject) to be systematically impacted by proprioceptive information (Förster and Strack, 1998; Niedenthal, 2007). For instance, Wells and Petty (1980) found that persuasive arguments presented via headphones during a contrived headphone-testing session were more likely to be accepted by individuals told to assess the comfort of the headphones by nodding up-and-down than by individuals instructed to shake their heads left-to-right. This finding was interpreted as evidence that nodding and shaking the head have acquired motivational significance in themselves — approach and avoidance, respectively — as a result of their frequent pairing in Western culture as gestures that signify agreement or acceptance (nodding) and disagreement/rejection (shaking). A similar interpretation was taken by Cacioppo et al. (1993), but for the effect of flexion (approach) and extension (avoidance) arm positions on judgments of likeability for innocuous Chinese ideographs. Finally, Strack et al. (1988) have found that individuals holding the end of a pen between their teeth (simulating an open-mouth smile) judge humorous cartoons as more funny than do individuals holding the pen between their lips (simulating a pucker). Thus, the evaluative component of deliberative preferential choice tasks may be systematically influenced by (presumably) incidental proprioceptive information to the extent that the decision task involves selection among options whose characteristics can be defined and evaluated on a good–bad psychometric scale (but see Raab and Green, 2005 for a counter argument).

Proprioceptive information may play a more central role in perceptual-categorization-based decision making, but in an additional form than transformations on underlying evaluative scales. While the performance of expert chess players convincingly demonstrates that perceptual-categorization tasks are not restricted to situations involving strenuous focal motor activity, time pressure, or chaotic/dynamic sensory information, a sizable proportion of real-world perceptual-categorization tasks do fall within these bounds — for example, decisions in sports, combat, firefighting, and driving (Klein, 1999; Lipshitz et al., 2001). In these latter situations, proprioceptive and sensory-motor information may be an essential content included in the memory stores that allow individuals to engage in the perceptual-categorization decision making (Chapter 13: A multiple-cue learning approach as the basis for understanding and improving soccer referees' decision making). All things being equal, proprioceptive information may be one of many cues to task-appropriate responding that is exploited during decision making by perceptual-categorization. For example, expert firefighters appear to rely on proprioceptive information, such as the sponginess or squishiness of the floor beneath their feet, to assess the risk level of dynamically changing fires rapidly (and nonconsciously) (Klein, 1999). As Plessner has pointed out elsewhere in this volume (Chapter 13: A multiple-cue learning approach as the basis for understanding and improving soccer referees' decision), research on the utility of immersing decision makers in ecologically representative training environments may be an informative diagnostic tool for investigating the role of sensory-motor inputs in expert judgment.

Ultimately, it remains unclear how much proprioceptive information influences perceptual-categorization across the range of such tasks. However, proprioceptive information may play an increasingly larger role in determining decision performance under circumstances in which such information is the most stable or strong signal available within the task environment. For instance, ambiguous situations could pose a special obstacle to the categorization process, whereby an appropriate action response is not readily activated because available information external to the motor system is unclear. Consider the ambiguous decision task that a ball sport referee frequently faces while attempting to render a ruling on an ambiguous play. In the absence of stronger signals, the referee's categorization of the task, and his/her subsequent action path, may be open to influence from his/her incidental postures, such as having hands forward or holding a pencil between his/her teeth. Alternatively, the referee's incidental postures at the time of training could serve as retrieval cues (cf. Tulving and Thompson, 1973) for calls in ambiguous situations in which the postures happen to have been replicated, thereby impacting decision.

Another question revealed, though left unanswered, by the present treatment is the extent to which the interplay between motoric and cognitive systems of decision making is *bidirectional* (Table 1). Few theorists would object to characterizing action selection at the level of sensory-perception processing as being fundamentally embodied and bidirectional, especially if the focal task entailed physical movement as its ultimate goal (Wilson, 2002; Faubel and Schöner, 2008). We have discussed the potential involvement of embodied cognition in the evaluative and learning components of more higher-order forms of decision making — a body-to-mind linkage. However we are at present unable to speculate in any progressive way about how higher-order components of decision making might impact embodied systems — a mind-to-body linkage (see Table 1). It may be that the only substantive contribution of the mind to the body in this respect is in the form of already documented executive control functions, whereby higher-order cognitions reconcile lower-level conflicts in processing, such as with the resolution of opposing bimanual movements such as patting one's head while rubbing one's stomach (Chapter 15: Visual cues influence motor coordination: behavioral results and potential neural mechanisms mediating perception-action coupling and response selection), and also modulate learning, movement, perception, and so on (cf. Conway et al., 2005). For instance, Johnson (Chapter 12: Embodied cognition of movement decisions: a computational modeling approach) discusses the role of higher-order cognitions, in particular assessment of decision outcomes (feedback), in the learning of situation-appropriate responding. Future research will undoubtedly need to devote special attention to the question of bidirectionality.

Finally, in the spirit of presenting this framework in order to inspire embodied decision-making research, we encourage researchers who would use this framework to ask themselves what adaptive value, if any, these (presumed) embodied decision-making effects may have for the human decision maker. In some senses, these effects might seem like nuisances: what good could come of one's decisions being whimsically changed by incidental proprioceptive information? This is a question future researchers will have to wrestle with, but some insight could be gained from the earlier discussion. As we have claimed, the centrality of embodied cognitions in decision-making performance may hinge critically upon the type of task, with tasks reliant on lower-order, sensory-motor systems being the most receptive to such influence and deliberative preferential choice tasks being the least. Indeed, preferential choice, for example, might only be impacted under specific circumstances, such as in decision tasks explicitly involving good–bad evaluations and when the individual is confused by ambiguous information. These, and other boundary conditions, likely limit the extent to which embodied decision making will result in nonadaptive behaviors. However, this even seems to ask the question of why embodied decision making would be advantageous under these latter circumstances. At least in the case of environmental ambiguity, embodied cognitions could serve an adaptive function by providing much-needed veridical information regarding the appropriate actions to take in an otherwise debilitating circumstance. Regardless, the thought exercise should prove useful in generating fruitful hypotheses of embodied decision making, which is the ultimate scope of the unitary hierarchical framework of embodied decision making.

Conclusion

Cognitive psychology has only recently begun to appreciate the involvement of embodied cognitions in human thinking (Wilson, 2002), and the decision sciences have only recently begun to conceptualize decision making as a dynamic mental process (see Busemeyer and Johnson, 2004 for review). As a result, the consequences of the human physical predicament for decision making have tended to be overlooked (Bagozzio et al., 2003; Chapter 12: Embodied cognition of movement decisions: a computational modeling approach). In this chapter, we have tried to push current understanding to new heights by pondering the intersection of these two promising frontiers of research — what is embodied decision making?

While we do not have a definitive answer to this question, we are confident that the framework offered in this chapter brings theorists one step closer. In the beginning, our naïve understanding of the problem led us to pose the question of embodied decision making in such a way that would have neglected important advancements to be had from the diverse perspectives on action selection, motor control, and decision making offered by experts in each of these fields. By this point, we have a sophisticated formulation of the challenge stretched out before us and a framework to guide further work in this fascinating area of inquiry. Because human decision makers face a variety of decision situations they have a variety of decision-making processes at their disposal. Theorists must, therefore, ask themselves how embodied cognitions exist within the bounds of each of these decision processes, whether conceptualized as heuristics and algorithms or intuitive and deliberative modes of thought. Armed with this clearer point of view, theorists can begin to confront the deeper theoretical issues of how decision making looks in a world of thinking heads *and* bodies.

References

Bagozzi, R. P., Dholakia, U. M., & Basuroy, S. (2003). How effortful decisions get enacted: the motivating role of decision processes, desires, and anticipated emotions. *Journal of Behavioral Decision Making*, *16*, 273–295.

Beach, L. R. (1998). *Image theory*. Mahwah, NJ: Lawrence Erlbaum Associates.

Busemeyer, J. R., & Johnson, J. G. (2004). Computational models of decision making. In D. J. Koehler, & N. Harvey (Eds.), *Blackwell handbook of judgment and decision making* (pp. 133–154). Oxford: Blackwell.

Busemeyer, J. R., & Townsend, J. T. (1993). Decision field theory: a dynamic-cognitive approach to decision making in an uncertain environment. *Psychological Review*, *100*, 432–459.

Cacioppo, J. T., Priester, J. R., & Berntson, G. G. (1993). Rudimentary determinants of attitudes. II: arm flexion and extension have differential effects on attitudes. *Journal of Personality and Social Psychology*, *65*, 5–17.

Chase, W. G., & Simon, H. A. (1973). The mind's eye in chess. In W. G. Chase (Ed.), *Visual information processing* (pp. 215–281). New York: Academic Press.

Conway, A. R. A., Kane, M. J., Bunting, M. F., Hambrick, D. Z., Wilhelm, O., & Engle, R. W. (2005). Working memory span tasks: a methodological review and user's guide. *Psychonomic Bulletin & Review*, *12*, 769–786.

DeCaro, M. S., Thomas, R. D., & Beilock, S. L. (2008). Individual differences in category learning: sometimes less working memory capacity is better than more. *Cognition*, *107*, 284–294.

Edwards, W. (1992). *Utility theories: measurements and applications*. Boston: Kluwer.

Faubel, C., & Schöner, (2008). Learning to recognize objects on the fly: a neurally based dynamic field approach. *Neural Networks*, *21*(4), 562–576.

Fodor, J. A. (1983). *The modularity of mind*. Cambridge, MA: MIT Press.

Förster, J., & Strack, F. (1998). Motor actions in retrieval of valenced information: II. boundary conditions for motor congruence effects. *Perceptual and Motor Skills*, *86*, 1423–1426.

Gigerenzer, G. (2000). *Adaptive thinking: rationality in the real world*. Oxford: Oxford University Press.

Gigerenzer, G., & Goldstein, D. G. (1996). Reasoning the fast and frugal way: models of bounded rationality. *Psychological Review*, *103*, 650–669.

Hammond, K. R., Stewart, T. R., Brehmer, B., & Steinmann, D. (1975). Social judgment theory. In M. F. Kaplan, & S. Schwartz (Eds.), *Human judgment and decision processes* (pp. 271–312). New York: Academic Press.

Heekeren, H. R., Marrett, S., & Ungerleider, L. G. (2008). The neural systems that mediate human perceptual decision making. *Nature Reviews. Neuroscience*, *9*, 467–479.

Hogarth, R. M. (2005). Deciding analytically or trusting your intuition? The advantages and disadvantages of analytic and intuitive thought. In T. Betsch, & S. Haberstroh (Eds.), *The routines of decision making* (pp. 67–82). Mahwah, NJ: Lawrence Erlbaum Associates.

Kahneman, D. (2003). A perspective on judgment and choice — mapping bounded rationality. *American Psychologist*, *58*(9), 697–720.

Klein, G. (1999). *Sources of power: how people make decisions*. Cambridge, MA: MIT Press.

Lee, M. D., & Cummins, T. D. R. (2004). Evidence accumulation in decision making: unifying the 'take the best' and 'rational' models. *Psychonomic Bulletin & Review*, *11*, 343–352.

Link, S. W., & Heath, R. A. (1975). A sequential theory of psychological discrimination. *Psychometrika*, *40*(1), 77–105.

Lipshitz, R., Klein, G., Orasanu, J., & Salas, A. (2001). Focus article: taking stock of naturalistic decision making. *Journal of Behavioral Decision Making*, *14*, 331–352.

Maddox, W. T., & Ashby, F. G. (2004). Dissociating explicit and procedural-learning based systems of perceptual category learning. *Behavioral Processes*, *66*(3), 309–332.

McGrath, J. E., Arrow, H., & Berdahl, J. L. (2000). The study of groups: past, present, and future. *Personality and Social Psychology Bulletin*, *4*, 95–105.

McLeod, P., & Dienes, Z. (1996). Do fielders know where to go to catch the ball or only how to get there? *Journal of Experimental Psychology — Human Perception and Performance*, *22*, 531–543.

Newell, A., Shaw, J. C., & Simon, H. A. (1958). Elements of a theory of human problem solving. *Psychological Review*, *65*, 151–166.

Niedenthal, P. M. (2007). Embodying emotion. *Science*, *316*, 1002–1005.

Posner, M. I., & Snyder, C. R. R. (1975). Facilitation and inhibition in the processing of signals. In P. M. A. Rabbitt, & S. Dornic (Eds.), *Attention and performance V* (pp. 669–682). New York: Academic Press.

Raab, M., & Green, N. (2005). Motion as input: a functional explanation of movement effects on cognitive processes. *Perceptual and Motor Skills*, *100*, 333–348.

Reichardt, W., & Poggio, T. (1976). Visual control of orientation behaviour in the fly: I. a quantitative analysis. *Quarterly Reviews of Biophysics*, *9*, 311–375.

Roe, R. M., Busemeyer, J. R., & Townsend, J. T. (2001). Multialternative decision field theory: a dynamic connectionist model of decision making. *Psychological Review*, *108*, 370–392.

Schneider, W., & Shiffrin, R. M. (1977). Controlled and automatic human information processing: I. detection, search, and attention. *Psychological Review*, *84*, 1–66.

Schoemaker, P. J. H. (1982). The expected utility model: its variants, purposes, evidence, and limitations. *Journal of Economic Literature*, *20*, 529–563.

Simon, H. A. (1956). Rational choice and the structure of environments. *Psychological Review*, *63*, 129–138.

Sloman, S. A. (1996). The empirical case for two systems of reasoning. *Psychological Bulletin*, *119*(1), 3–22.

Smith, E. R., & DeCoster, J. (2000). Dual-process models in social and cognitive psychology: conceptual integration and links to underlying memory systems. *Personality and Social Psychology Review*, *4*(2), 108–131.

Stanovich, K. E., & West, R. F. (2000). Individual differences in reasoning: implications for the rationality debate? *Behavioral and Brain Sciences*, *23*(5), 645–726.

Strack, F., Martin, L. L., & Stepper, S. (1988). Inhibiting and facilitating conditions of the human smile: a nonobtrusive test of the facial feedback hypothesis. *Journal of Personality and Social Psychology*, *54*(5), 768–777.

Svenson, O. (1979). Process descriptions of decision making. *Organizational Behavior and Human Decision Processes*, *23*(1), 86–112.

Thomas, O., Johnson, J., Rieskamp, J., Allenby, G. M., Brazell, J. D., Diederich, A., Hutchinson, J. W., MacEachern, S., Ruan, S., & Townsend, J. (2008). Sequential sampling models of choice: some recent advances. *Mark Letters*, *19*, 255–267.

Townsend, J. T., & Ashby, F. G. (1983). *Stochastic modeling of elementary psychological processes*. Cambridge: Cambridge University Press.

Tulving, E., & Thompson, D. M. (1973). Encoding specificity and retrieval processes in episodic memory. *Psychological Review*, *80*, 352–373.

Wells, G. L., & Petty, R. E. (1980). The effects of overt head movement on persuasion: compatibility and incompatibility responses. *Basic and Applied Social Psychology*, *1*, 219–230.

Wilimzig, C., Schneider, S., & Schöner, G. (2006). The time course of saccadic decision making: dynamic field theory. *Neural Networks*, *19*, 1059–1074.

Wilson, M. (2002). Six views of embodied cognition. *Psychonomic Bulletin & Review*, *9*(4), 625–636.

SECTION III

How is a Cognitively Intended Course of Action Physically Implemented?

M. Raab et al. (Eds.)
Progress in Brain Research, Vol. 174
ISSN 0079-6123

CHAPTER 17

Perceptual decision making: a bidirectional link between mind and motion

Nikos Green* and Hauke R. Heekeren

Neurocognition of Decision Making Group, Max Planck Institute for Human Development, Berlin, Germany

Abstract: Perceptual decision making is the process by which information gathered from sensory systems is combined and used to influence our behavior. Importantly, however, the route from perception to action is not a one-way street, rather, perception and action interact continuously. In this chapter, we synthesize a bidirectional view on perceptual decision making by discussing behavioral, neuroscientific, and computational findings in the context of one embodiment framework. We conclude that there is evidence for the liaison of action and perception in simple decision-making tasks. This framework may also be extended to reward-based decision making in humans.

Keywords: perceptual decision making; bidirectional link; perception; action; embodiment

Introduction

Perceptual decision making is the process by which information gathered from sensory systems is combined and used to influence our behavior. Importantly, however, the route from perception to action is not a one-way street, rather, perception and action interact continuously. A comprehensive account of perceptual decision making has to include this bidirectional link between mind and motion. Moreover, such a framework spans different levels of description (behavior, neural mechanisms, and computational mechanisms) and enables us to understand better how this link is instantiated in terms of information processing, brain structures, and resulting behavior.

In this chapter, we will synthesize a bidirectional view on perceptual decision making by discussing behavioral, neuroscientific, and computational findings in the context of the embodiment framework (cf. Clark, 1997; Wilson, 2002). First, we will present examples of how the principles of embodiment apply to the behavioral level. Furthermore, we will delineate the bidirectional link on the neural level and discuss the computational and representational capacities these neural components can have in computational models.

Perceptual decision making: embodied cognition?

The basic tenet of embodiment states that an organism is coupled via a sensorimotor loop with the environment (Clark, 1997; Wilson, 2002). This coupling results in a fusion of cognition, action, and perception, thereby shaping invariants that constrain cognition. These influence the

*Corresponding author.
Tel.: +49-308-240-6620; Fax: +49-308-240-6616;
E-mail: green@mpib-berlin.mpg.de

DOI: 10.1016/S0079-6123(09)01317-X

organism's ability to act in real time and with appropriate behavior in the environment.

To resolve whether this framework applies to the domain of perceptual decision making, we will pursue an approach proposed by Clark (1997) who argues for three related explanatory and descriptive accounts when taking an embodied standpoint. These accounts suggest the explanation of (a) gross behavior; (b) the identification of neural components that give rise to that behavior and their interaction; as well as (c) an account of the computational and representational capacities of the components. Therefore we have to tackle questions such as: How can we explain gross behavior of the entire organism with an embodied approach? What are the neural systems mediating perceptual decisions? How do sensory and motor systems interact? And what are the representational mechanisms and computational abilities of these systems?

Hence, we will briefly discuss the behavioral account by looking at the range of its applicability in robotics and psychology, and explore the insights gained from applying an embodied framework. Next, we will discuss the neural components by examining results from neuroscientific perceptual decision-making studies in monkeys and humans. Finally, we will discuss some examples of models that specify computational and representational mechanisms by which these components function.

Gross behavior

The first account describes gross behaviors of well-functioning organisms in the environment that span brain, body, and world. Research in the domains of robotics and psychology has addressed the adaptive behavior of organisms in the environment. The artificial intelligence community, particularly in robotics, has influenced the framework of embodiment significantly. While dealing with engineering problems of agent-to-world interaction it was realized that a serial account of cognitive processing (sense–think–act) faces serious problems, such as limitations in real-time interaction and problems with the generalization of learned knowledge to novel problems. Thus, ideas from various domains (e.g., cybernetics) were used to develop solutions, which contributed to the concept of embodiment. Thinking about embodiment in this way helped scientists to develop and establish design principles for robots with simple but realistic adaptive behavior (Pfeier and Bongard, 2007). This approach was termed *behavior-based robotics* (Brooks, 1995) and makes use of a coupled link between action and perception to enable intelligent behavior.

Studies using simulated robot models and real robots show that a coupling of action and perception systems is beneficial when having to interact with the world. For example, Wyss and colleagues (2004) used a virtual robot model that integrates sensory and motor information in activity maps by mapping sensory information onto motor action using reinforcement learning. They then implemented this mechanism in a real robot, which was behaviorally flexible due to sensorimotor integration in a population-activity-based neural network. The robot was enabled to navigate sensibly in the environment, for example, it could optimally explore the given simple environment by making the right decisions of when to change its direction facing a bifurcation point. Their results indicate the beneficial use of sensorimotor integration for decision making. For example, this mechanism constrains via the motor system the action space of the robot and thereby the degrees of freedom in the decision space, reducing the complexity of the problem at hand.

Another robotics study, using an exploration task where the aim for the robot was to minimize collisions and increase the number of visited targets (Verschure et al., 2003), pointed out that the robots' decision-making processes can be optimized by using interacting action and perceptual learning mechanisms linked using a contextually constraining process (a simplified example of the idea to put the brain, body, and mind together).

These examples illustrate that using the bidirectional action–perception link as a design principle solves some major problems in robotics (Pfeier and Bongard, 2007). For example, real-time interaction or the generalization of learned skills to novel problems is possible with such a design, albeit to date it has only been demonstrated for a limited set of problems, such as navigation or movement coordination (for a detailed account on robotics and the bidirectional link see Chapter 19: The cognitive nature of action — functional links between cognitive psychology, movement science and robotics). But is it possible to apply the bidirectional link as an explanation for more complex behavior? This question can be addressed by certain results in psychology: experimental studies and theoretical explanations support an embodied framework for understanding more complex cognitive processes.

For example, Raab and Green (2005) found that continuous positive (approach) or negative (avoidance) hedonic motor signals instantiated in a participant by pressing a lever during a word generation task influence the way visually presented words are processed. In this experiment, the motor signal quality resulted in differences in reaction time (RT) and word-associations generated, leading to the conclusion that motor signals are integrated into the decision process, suggesting a bidirectional link. Furthermore, these basic interactions could be the basis for more complex cognitive processes (cf. Sebanz et al., 2006).

As reviewed in this section, psychology and robotics suggest that there is a bidirectional link between perception and action. As a next step, it is important to identify the neural systems mediating perceptual decision making and to account for how these components interact.

Neural components: monkey physiology and human neuroimaging

Perceptual decision making is not only influenced by the sensory information at hand, but also by such factors as attention, task difficulty, the prior probability of the occurrence of an event, the outcome of the decision, and the means by which a choice is enacted (for a review see Heekeren et al., 2008). Traditional psychological theories posit that the decision-making process consists of components that act in a hierarchical manner, with a serial progression from perception to action (e.g., Tversky and Kahneman, 1981; cf. Cisek, 2007). More recent neuroscientific findings, however, indicate that the components of this process happen in parallel. The neural architecture for perceptual decision making can be viewed as a system consisting of four distinct, but interacting, processing modules (cf. Heekeren et al., 2008). One of these systems accumulates and compares sensory evidence and another system represents decision variables. We focus on these two systems because they are most relevant for the neural implementation of the bidirectional link between perception and action.

In this section we first provide a short overview of findings from studies in monkeys using direct recordings of neural activity (single- or multiunit recordings). These pioneering studies are the first to provide evidence of the bidirectional interaction of action and perception in perceptual decision making. After considering these results we will discuss the role of the motor system in perceptual decision making in monkeys. This discussion will motivate questions about the inferences we can draw from these results. We will continue by presenting neuroscientific results from human studies, including findings from functional magnetic resonance imaging (fMRI) and electroencephalography (EEG) (cf. Philiastides and Heekeren, in press). Following the neuroscientific findings in humans we will discuss whether the human and monkey results correspond, and how they demonstrate a bidirectional link.

Perceptual decision making in monkeys

The most recent substantial neuroscientific findings in perceptual decision making indicating parallel processing of motor and sensory information have been made in primate research. In those studies, mostly done in the tactile and visual domains, a decision problem is often framed in a simple two-alternative choice paradigm in which a monkey has to discriminate visual (cf. Newsome et al., 1989) or tactile stimuli (cf. Romo et al., 2003), and is trained to answer via a saccade or a button press to indicate the correct choice. These studies include parametric variation of the difficulty of the stimulus or the answer, for example, by changing the stimulus coherence of a visual stimulus, by varying the time partly for decision making, or by introducing multiple alternatives. The aim is to identify brain processes and brain areas that represent sensory evidence, integrate sensory and motor signals, and compare action options during perceptual decision making because these are necessary functions to instantiate a bidirectional link between action and perception.

Representation of sensory evidence

Single-unit recording studies in monkeys have provided evidence for a close link between behavior and the activity of neuronal populations in sensory regions (such as the primary somatosensory cortex (Salinas et al., 2000) and the middle temporal area (area MT/V5) (Britten et al., 1992; Shadlen et al., 1996). Further, electrical microstimulation studies have provided causal evidence for this close link in both the somatosensory (Salinas et al., 2000) and the visual domain (Ditterich et al., 2003). Romo and colleagues (1998, 2000) recorded from single neurons in the primary somatosensory cortex (SI) while monkeys performed a vibrotactile task in which they decided which of two sequentially presented flutter stimuli had a higher frequency. Trial-to-trial fluctuations in the firing rate of SI neurons predicted the monkeys' choices (Salinas et al., 2000), and the average firing rate of SI neurons increased monotonically with increasing stimulus frequency. Furthermore, when the vibrotactile stimuli were replaced with direct electrical microstimulation of SI, the monkeys showed a behavioral pattern very similar to when under normal conditions (Romo et al., 1998, 2000). Using a similar approach, Newsome and colleagues (e.g., Newsome et al., 1989) recorded from neurons involved in visual motion processing in area MT, while monkeys performed a direction-of-motion discrimination task. Here, the monkeys decided whether a noisy field of dots was moving in one direction or the opposite direction (e.g., upward or downward) and indicated their choice with a quick eye movement to the target on the respective side. The activity of a relatively small number of directionally selective neurons closely matched the monkeys' behavior (Shadlen et al., 1996). Subsequent studies showed that electrical microstimulation of directionally selective neurons in area MT caused the monkey to choose the neurons' preferred direction more often. For example, when neurons tuned to rightward motion were stimulated, the monkey was more likely to make an eye movement to the target on the right (Ditterich et al., 2003). Also, microstimulation of these neurons quickened the decision in favor of the preferred direction and slowed the decision in the opposite direction (Ditterich et al., 2003). Thus, in both the visual and the somatosensory domain, microstimulation studies have provided causal evidence for a tight link between the representation of sensory evidence in sensory regions and perceptual decisions. As these studies have provided us with detailed knowledge about these representations, and also with causal evidence that these representations of sensory evidence in sensory regions are used to make perceptual decisions, we can now go one step further by examining how these representations are used to make a decision.

Integration of sensory evidence and formation of the decision variable

Another important concept that has emerged from neurophysiological studies in monkeys is that perceptual decisions are made, at least in part, by integrating the sensory evidence represented by sensory neurons (Gold and Shadlen, 2002; Romo and Salinas, 2003). These studies have shown that the neuronal activity in areas involved in decision making gradually increases and then remains elevated until a response is made. Importantly, the rate of increase in neural activity is slower during more difficult trials than during easier trials. During the vibrotactile frequency-discrimination task described above (cf. Romo et al., 2003), cells in regions of the monkey brain that are downstream (i.e., further along the processing chain) from the primary and secondary sensory areas, such as the prefrontal, medial premotor, and ventral premotor cortices, form a decision by computing the difference in the activities of populations of sensory neurons in the secondary somatosensory cortex (SII) that prefer high and low frequencies, respectively (Hernandez et al., 2002; Romo et al., 2003, 2004; de Lafuente and Romo, 2005). Similarly, during the direction-of-motion visual discrimination task, cells in regions downstream from area MT, such as the lateral intraparietal area (LIP), the frontal eye field (FEF), and the dorsolateral prefrontal cortex (DLPFC), form a decision by computing the difference in the activities of populations of neurons in area MT that code for opposite directions of motion (Kim and Shadlen, 1999). Thus in both sensory systems, a simple subtraction operation appears to explain the behavioral effect. Moreover, in another study in the visual domain by Cisek and Kalaska (2005), monkeys performed an instructed-delay task in which they had to respond to different target cues by moving a cursor through an out-of-sight handle. The population-based activity in the dorsal premotor cortex (PMd) predicts the monkey's responses and thus reflects the selection between various action options. Together these findings indicate that brain areas such as DLPFC — but also sensorimotor areas such as LIP and FEF — integrate sensory evidence on which actions (saccades, arm movements) are based. But how are motor signals linked to the perceptual evidence?

Role of the motor system

It is noteworthy that regions in the monkey brain that have been implicated both in representing decision variables and in performing the comparison operation — LIP, FEF, DLPFC, PMd — are the same areas that select, plan, and execute motor responses. In other words, the boundaries between sensory processing, decision-related activity, and motor-related processing are not as distinct as previously thought. For example, when monkeys must choose the direction in which a random-dot-motion stimulus is moving and indicate their decision with an eye movement, decision-related as well as saccade-related activity can be found in the FEF (Gold and Shadlen, 2003). Similarly, when monkeys perform the vibrotactile discrimination task, activity in medial and ventral premotor cortex reflects the temporal evolution of the decision-making process leading to action selection (Hernandez et al., 2002; Romo et al., 2004). Other neurophysiological studies have revealed that decision variables are represented in the superior colliculus (SC), a midbrain region involved in the generation of saccadic eye movements (Gold and Shadlen, 2000; Horwitz et al., 2004). Taking an embodied perspective on these results gives us an account of how the bidirectional link between mind and motion is established in the brain: through the representation of decision variables in motor structures that are based on the integration of sensory information represented in sensory brain regions.

It should be noted that in most of the monkey studies presented above, the monkeys were trained to indicate their perceptual decision with a particular action. In other words, the monkeys could treat the perceptual decision as a problem in the motor domain. Therefore a common critique is that the role of the motor system and its link to perception should be restricted to these kinds of simple sensory motor tasks, where the choice alternatives can be understood as specific problems in the motor domain. Seen in this light, it is not surprising that motor structures play a role in decision formation. It is not yet clear, however, how these structures contribute to decisions that are not linked to particular actions (Gold and Shadlen, 2003, 2007) or to more complex cognitive functions.

Perceptual decision making in humans

Is there evidence that the principles that have emerged from the neurophysiological work in monkeys also hold for the human brain? If so, then these basic neural mechanisms have been conserved during brain evolution. Here we review findings from recent neuroimaging studies that have investigated perceptual decision making in the domains of somatosensation and vision (for research in other sensory domains cf. Uchida et al., 2006, Heekeren et al., 2008). Similar to the studies in monkeys, the representations of sensory evidence can now also be measured and manipulated in the human brain and can be distinguished from representations of decision variables.

Inspired by the work of Romo and colleagues, recent fMRI studies have used vibrotactile frequency tasks to study somatosensory decision making in the human brain. In this task, individuals had to decide which of two successive vibratory stimuli had a higher frequency. Consistent with neurophysiological data in monkeys, the primary somatosensory cortex exhibited increased activity during the encoding phase (processing of the first stimulus) of tactile decision making (Preuschhof et al., 2006). Similarly, using a somatosensory discrimination task, in which participants had to compare the frequency of two successive electrical tactile stimuli, Pleger et al. (2006) found that tactile stimuli evoked activity in, among other regions, the somatosensory cortex.

The most direct evidence in support of the concept of representation of sensory evidence in the somatosensory domain comes from a transcranial magnetic stimulation (TMS) study which showed that stimulation of the primary somatosensory cortex lowered two-point discrimination thresholds of the right index finger and enlarged its neural representation as assessed with fMRI (Tegenthoff et al., 2005). Notably, this enlargement correlated with the individual TMS-induced perceptual improvement. Taken together, the results of the studies described above provide support for the idea that, similar to the findings in monkeys, the primary somatosensory cortex represents sensory evidence during tactile decision making.

Similar studies were conducted in the visual domain. For example, Heekeren et al. (2004) used fMRI and a face–house categorization task to investigate perceptual decision making in the visual domain. Previous neuroimaging studies had identified regions in the human ventral temporal cortex that are activated more by faces than by houses, and vice versa: the fusiform face area (FFA) and the parahippocampal place area (PPA), respectively (Haxby, 1994; Kanwisher et al, 1997; McCarthy et al., 1997; Epstein and Kanwisher, 1998; Ishai et al., 1999). The face–house task can thus be used to identify two brain regions, and to test whether they represent the sensory evidence relevant for the task. There was a greater response in face-selective regions to clearer images of faces ("easy" trials) than to degraded images of faces ("difficult" trials), whereas degraded houses showed a greater response than clearer houses in these face-selective areas. The opposite pattern was found in house-selective regions, namely, a greater response to clearer images of houses ("easy" trials) than to degraded images of houses ("difficult" trials), but a greater response to degraded than to clearer images of faces . These results support the concept that face- and house-selective regions represent the sensory evidence for the two respective categories.

How is this information integrated in sensorimotor areas in the human brain to make a decision, and where is the bidirectional link instantiated? Next, we review recent human neuroimaging studies that provide evidence for a comparison of accumulated sensory evidence as a mechanism for perceptual decision making.

Integration of sensory evidence and formation of the decision variable

The single-unit recording studies in monkeys have shown that neuronal activity in areas involved in decision making gradually increases and then remains elevated until a response is made. Importantly, the rate of increase in neural activity is slower during more difficult trials than during easier trials. Furthermore, these studies have shown that downstream cortical regions such as LIP and the DLPFC could form a decision by comparing the output of pools of selectively tuned sensory neurons and linking them to actions.

A recent fMRI study showed how the blood-oxygen-level-dependent (BOLD) signal can be used to examine the process of accumulation of sensory evidence (Ploran et al., 2007). Pictures were revealed gradually over the course of 12–20 s and participants signaled the time of recognition with a button press. In several occipital regions, the fMRI signal increased primarily as stimulus information increased, suggesting a role in lower-level sensory processing. There was a gradual buildup in fMRI signal peaking in correspondence with the time of recognition in inferior temporal, frontal, and parietal regions, suggesting that these regions accumulate sensory evidence. Heekeren et al. (2004) directly tested whether a comparison operation is also at work in the human brain using the face–house discrimination task described above. Specifically, based on the neurophysiological data in monkeys, Heekeren et al. proposed that higher-level decision areas should fulfill two criteria. First, they should show the greatest BOLD activity on trials in which the weight of evidence for a given perceptual category is greatest, namely, a greater fMRI signal during decisions about clear images of faces and houses ("easy trials") than during decisions about degraded images of these stimuli ("hard trials"). Second, their BOLD signals should correlate with the difference between the signals in brain areas selectively tuned to the different categories involved; that is, those in face- and house-responsive regions. Only one brain region fulfilled both criteria (Heekeren et al., 2004): the posterior portion of the left DLPFC uniquely responded more to clear relative to degraded stimuli, and the activity of this region correlated with the difference between the output signals of face- and house-responsive regions. Thus, when people make categorical decisions about face and house stimuli, this brain region appears to integrate the outputs from lower-level sensory regions and use a subtraction operation to compute perceptual decisions. Notably, activity in the left DLPFC also predicted behavioral performance in the categorization task (Heekeren et al., 2004). Hence, even for complex object categories, the comparison of the outputs of different pools of selectively tuned neurons appears to be a general mechanism by which the human brain computes perceptual decisions.

Is this mechanism involved in establishing the bidirectional link between perception and action?

The role of the motor system in humans

Neurophysiological studies in monkeys suggest that the brain regions involved in selecting and planning a certain action play an important role in forming decisions that lead to that action. To test whether the result and models from research in monkeys reveal basic principles that hold for humans, Heekeren et al. (2003) asked human observers to make direction-of-motion judgments about dynamic random-dot-motion stimuli and indicate their judgments with an eye movement to one of two visual targets. In each individual, the authors localized regions that are part of the oculomotor network, namely the FEF, and an eye-movement-related region in the intraparietal sulcus (IPS) that presumably corresponds to the LIP of monkeys (Sereno et al., 2001). Importantly, during the decision formation period (between the onset of visual motion and the cue to respond), the percent change in the BOLD signal in both the FEF and the IPS was highly correlated with the strength of the motion signal in the stimuli (Heekeren et al., 2003). These data are thus consistent with the single-unit studies in monkeys that reported that the FEF and LIP participate in the process of forming a perceptual decision and might be brain structures that support a bidirectional link between action and perception.

The results are also similar to those from a study of oculomotor decision making by Heinen et al. (2006), who had participants play "ocular baseball" while undergoing fMRI. In this game, participants had to decide whether or not the trajectory of a dot moving across a computer screen was likely to cross into a visible "strike" zone. If participants decided that the dot was likely to enter the strike zone, they had to make an eye movement; otherwise their eyes had to remain fixed on a point in the center of the screen. When the results of a task with identical motor behavior were compared to the "baseball" trials, decision-related signals were found in the superior parietal lobule, the FEF and the ventrolateral prefrontal cortex. In line with the monkey data, these results suggest that when a decision is associated with a specific action, the formation of the decision and the preparation of the behavioral response have a common neural substrate.

More recently, Heekeren et al. (2006) investigated whether decisions might be transformed into motor actions in the human brain independently of motor planning and execution — that is, at an abstract level. Individuals performed the direction-of-motion discrimination task and responded with either button presses or saccadic eye movements. Areas that represent decision variables at a more abstract level should show a greater response to high coherence (easy) relative to low coherence (difficult) trials, independently of the motor system that is used to express the decision. Heekeren et al. (2006) found four such areas: the left posterior DLPFC, the left posterior cingulate cortex, the left IPS, and the left fusiform/parahippocampal gyrus. Most importantly, the increase in BOLD activity in these regions was independent of the motor system that the participants used to express their decision. The results from this fMRI study are in line with the finding by Kim and Shadlen (1999) that, in monkeys, neural activity increases proportionally with the strength of the motion signal in the stimulus. However, the findings in humans suggest that the posterior DLPFC is an important component of a network that not only accumulates sensory evidence to compute a decision but also translates this evidence into an action independently of response modality, thereby challenging a strong reading of a bidirectional link between mind and motion.

Notably, to date neurophysiological studies in monkeys have not found neurons with an activity that reflects decisions independently of response modality. In fact, one could conclude from the neurophysiological studies in monkeys "to see and decide is, in effect, to plan a motor response" (Rorie and Newsome, 2005). By contrast, in humans, Heekeren et al. (2006) found regions of the cortex that responded independently of the motor effectors used. Based on these findings, we could speculate that humans might have evolved a more abstract decision-making network that allows a more flexible link between perception and action during decision making.

But how is action linked to perception in humans? Are there any general principles that we can infer both from human neuroimaging and monkey physiology? One possible answer to this question is that the motor system provides a temporal structure for cognition by the sensorimotoric coupling of the organism to the world (cf. Walsh, 2003; Thornton and Knoblich, 2006). This idea has been long discussed in philosophy and cognitive science and it has been tested in several fields of research (e.g., in the domain of robotics, see Chapter 19: The cognitive nature of action — functional links between cognitive psychology, movement science and robotics). It has been shown in humans that perception-based performance can be enhanced by practicing a learning problem in the motor domain. For example, blindfolded participants practice unfamiliar movement patterns with their arms, later performance on a visual task is enhanced, that is, participants are faster in recognizing full-body movements when (Casile and Giese, 2006). How can proprioceptive information for perceptual processes that have to rely upon spatio-temporal representations be added to a cognitive process such as decision making?

By presenting the neuroscientific evidence that shows how sensory evidence is represented and used to compute decision variables, we discussed where to look in the brain and how to use this information to bidirectionally link action and perception. However, there are more aspects to this link. As Trommershäuser (Chapter 21: Biases and optimality of sensory-motor and cognitive decisions) discusses, there are differences in how the cognitive and motor systems are able to represent probabilities and optimize behavior although the problem at hand may share the same formal structure. Thus it seems as if the bidirectional link is established differently between the perceptual and the motor brain structures for different problems. Perhaps the flexible link between sensory and motor structures, as found in humans, is instantiated in the brain by varying the informational roles the distinct systems (motor, sensory but also attention, reward, etc.) play during a cognitive process. Therefore, to complete our account described at the beginning of the chapter, we will next give examples of two models that assign computational roles and representational capacities to the neural subsystems already discussed.

Information-processing roles: computation and representation

Now that we have related gross behavior and brain systems in an embodied perspective of decision making, computational models can help us understand which capacities and properties are used to establish a bidirectional link.

An interesting approach is laid out in a model by Cisek (2006), in which the activity of different neural populations (that represent different brain areas) reflects the likelihood of actions to be selected. Importantly, these neural populations can represent sensory and motor information at the same time. Hence, each group of neurons can represent either several distinct properties of one action or several action options at once. The model simulates neural network activity of the motor system in simple reach-decision tasks and recreates behavior observed in experimental approaches in monkey physiology of a similar task (Cisek and Kalaska, 2005). In other words, this model implements the idea that the motor system and the sensory system inform each other when a decision has to be made. Importantly, with this model comes a unifying perspective on action and perception (cf. Cisek, 2007): brain functions are not grouped according to the different systems (i.e., perceptual, cognitive) they belong to, but are grouped by the process they support (decision making, attention, etc.). This framework uses the bidirectional link between mind and motion to conceptualize the process of decision making. Additionally it has been pointed out (cf. Bestmann, 2006) that this model can account for the correlated activity found in prefrontal as well as sensorimotor regions in experimental settings and thereby gives a reasonable account of how perceptual, cognitive, and motor information are fused in neural networks.

Other models have used different approaches to expose the inter-relation of motor and sensory systems. One notable experimental phenomenon in the above-mentioned two alternative choice tasks is the so-called speed accuracy tradeoff, a psychophysical effect that links RT and accuracy of decisions: when the response time decreases, the mean accuracy of the answer decreases and vice versa. Monkeys can treat the decision problem in this kind of task as a motor problem because they have to saccade toward one or the other target to indicate the correct choice. It explains the relation of RT to accuracy and conceptualizes the observed psychophysical results in the following way: as RT decreases, accuracy diminishes, and vice versa. In a biologically constrained neural network model, Lo and Wang (2006) explain how different brain regions such as the SC, sensory cortical areas, and the basal ganglia implement perceptual decision-making processes by changing their coupling strength and thus influence their interaction. The network is able to replicate the speed-accuracy tradeoff and neural activity as found in experimental studies presented earlier. Thus, this model presents another way that action and perception interact in the brain during perceptual decision making — by changing the coupling strength between action and perception systems. Using this mechanism, the brain is able to set the threshold that determines when enough perceptual evidence leads to an action.

The two models tackle the account on information-processing roles differently. Whereas the first explains the representational component when linking action and perception, the second puts forward a computational mechanism by which this link could be instantiated.

Synthesis

In this article we have focused on the bidirectional link between mind and motion in perceptual decision making, and we have applied Clark's (1997) embodiment framework to the bidirectional problem. The notion of embodiment is still disputed and some readings of it are quite radical and use novel — and in parts contradictory — interpretations of some of the concepts such as computation and representation. We have not tried to resolve these problems here, instead we use a specific framework (Clark, 1997) in a constructive way to link insights from different levels of analysis concerning the bidirectional link between mind and motion. We have briefly discussed which decision behavior can be obtained by linking perception and action in parallel, as exemplified by the adaptive behavior of robots, which have been constructed using an embodied perspective. Next, we have discussed which brain structures could subserve this behavior. This was the most emphasized part of our contribution and it shows that there are sensorimotor structures (i.e., LIP, FEF, SC) as well as prefrontal structures in humans (i.e., DLPFC) that play a role in integrating sensory and motor information. This has been shown by the ability to represent both types of information at the same time and in the same structure. Moreover, we have described the information-processing roles of these structures that could be important for the modulation of the bidirectional link.

In the present chapter we outlined how the embodied framework can be used to link cognitive theories, psychological accounts, neuroscience, and computational approaches to provide a unified conceptual view on perceptual decision making. This framework may also be extended to reward-based decision making in humans. As discussed above, studies in the domain of robotics have used reinforcement learning to map sensory information onto motor action (e.g., Wyss et al., 2004). At the neural level reinforcement learning has been linked to the brain reward system (O'Doherty, 2004; Daw et al., 2006; Montague et al., 2006; O'Doherty et al., 2007; Rushworth and Behrens, 2008). Surprisingly little research has, however, been done to explore the potential effects of reward on perceptual decision making, whether on sensory function or motor planning and action selection. One hypothesis is that in situations where rewards depend on decisions associated with different perceptual tasks, reward signals are propagated back to sensory systems in the form of a "teaching signal" where they can shape early sensory representations to optimize reward.

To date the only study that has addressed this issue is by Pleger et al. (2008). In this study a tactile discrimination task was used, where participants had to discriminate the relative frequency of two successive somatosensory stimuli applied to the same finger, while manipulating the reward rate received at the end of each trial. Not only did higher rewards improve performance but they also led to increased BOLD responses in the ventral striatum, a key component of the human reward system. More importantly however, these authors demonstrated that during reward delivery and in the absence of somatosensory stimulation, the S1 contralateral to the judged finger was reactivated and this reactivation was proportional to the amount of reward. Finally, they showed that reward magnitude on a particular trial influenced responses on the subsequent trial, with better behavioral performance and greater contralateral S1 BOLD responses for higher rewards.

These results clearly demonstrate that the systems involved in valuation interact with early sensory systems, however, it still remains elusive how these interactions are mediated.

As we can see from this example, our understanding of the bidirectional link between perception and action has to include additional contributors such as the valuation system. New human neuroimaging approaches will help to study this multitude of interacting systems (cf. Philiastides and Heekeren, in press) the present chapter highlights that it is important to relate the identified neural components to gross behavior and to give an account of their informational role.

References

Bestmann, S. (2006). A new unified framework for making and implementing decisions. *Journal of Neuroscience*, *26*, 13121–13122.

Britten, K. H., Shadlen, M. N., Newsome, W. T., & Movshon, J. A. (1992). The analysis of visual motion: a comparison of neuronal and psychophysical performance. *Journal of Neuroscience*, *12*, 4745–4765.

Brooks, R. (1995). Intelligence without reason. In L. Steels & R. Brooks (Eds.), *The artificial life route to artificial intelligence: building embodied, situated agents*. Hillsdale, NJ: Lawrence Erlbaum Associates.

Casile, A., & Giese, M. (2006). Nonvisual motor training influences biological motion perception. *Current Biology, 16*(1), 69–74.

Cisek, P. (2006). Integrated neural processes for defining potential actions and deciding between them: a computational model. *Journal of Neuroscience, 26*(38), 9761–9770.

Cisek, P. (2007). Cortical mechanisms of action selection: the affordance competition hypothesis. *Philosophical Transactions of the Royal Society of London Series B — Biological Sciences, 362*(1485), 1585–1599.

Cisek, P., & Kalaska, J. (2005). Neural correlates of reaching decisions in dorsal premotor cortex: specification of multiple direction choices and final selection of action. *Neuron, 45*(5), 801–814.

Clark, A. (1997). *Putting brain, body and world together again.* Cambridge, MA: MIT Press.

Daw, N., O'Doherty, J., Dayan, P., Seymour, B., & Dolan, R. (2006). Cortical substrates for exploratory decisions in humans. *Nature, 441*, 876–879.

de Lafuente, V., & Romo, R. (2005). Neuronal correlates of subjective sensory experience. *Nature Neuroscience, 8*, 1698–1703.

Ditterich, J., Mazurek, M. E., & Shadlen, M. N. (2003). Microstimulation of visual cortex affects the speed of perceptual decisions. *Nature Neuroscience, 6*, 891–898.

Epstein, R., & Kanwisher, N. (1998). A cortical representation of the local visual environment. *Nature, 392*, 598–601.

Gold, J., & Shadlen, M. (2000). Representation of a perceptual decision in developing oculomotor commands. *Nature, 404*(6776), 390–394.

Gold, J. I., & Shadlen, M. N. (2002). Banburismus and the brain: decoding the relationship between sensory stimuli, decisions, and reward. *Neuron, 36*, 299–308.

Gold, J. I., & Shadlen, M. N. (2003). The influence of behavioral context on the representation of a perceptual decision in developing oculomotor commands. *Journal of Neuroscience, 23*(2), 632–651.

Gold, J. I., & Shadlen, M. N. (2007). The neural basis of decision making. *Annual Review of Neuroscience, 30*, 535–574.

Haxby, J. (1994). The functional organization of human extrastriate cortex: a PET rCBF study of selective attention to faces and locations. *Nature Neuroscience, 14*, 6336–6353.

Heekeren, H. R., Marrett, S., Bandettini, P. A., & Ungerleider, L. G. (2003). Human fMRI evidence for representation of a perceptual decision in oculomotor areas. In *Proceedings of the 2003 Society for Neuroscience conference.* New Orleans, LA. Abstract 228.8.

Heekeren, H. R., Marrett, S., Bandettini, P. A., & Ungerleider, L. G. (2004). A general mechanism for perceptual decision-making in the human brain. *Nature, 431*(7010), 859–862.

Heekeren, H. R., Marrett, S., Ruff, D. A., Bandettini, P. A., & Ungerleider, L. G. (2006). Involvement of human left dorsolateral prefrontal cortex in perceptual decision making is independent of response modality. *Proceedings of the National Academy of Sciences of the United States of America, 103*(26), 10023–10028.

Heekeren, H. R., Marrett, S., & Ungerleider, L. G. (2008). The neural systems that mediate human perceptual decision making. *Nature Reviews. Neuroscience, 9*(6), 467–479.

Heinen, S. J., Rowland, J., Lee, B. T., & Wade, A. R. (2006). An oculomotor decision process revealed by functional magnetic resonance imaging. *Journal of Neuroscience, 26*, 13515–13522.

Hernandez, A., Zainos, A., & Romo, R. (2002). Temporal evolution of a decision-making process in medial premotor cortex. *Neuron, 33*(6), 959–972.

Horwitz, G. D., Batista, A. P., & Newsome, W. T. (2004). Representation of an abstract perceptual decision in macaque superior colliculus. *Journal of Neurophysiology, 91*, 2281–2296.

Ishai, A., Ungerleider, L., Martin, A., Schouten, J., & Haxby, J. (1999). Distributed representation of objects in the human ventral visual pathway. *Proceedings of the National Academy of Sciences of the United States of America, 96*, 9379–9384.

Kanwisher, N., McDermott, J., & Chun, M. (1997). The fusiform face area: a module in human extrastriate cortex specialized for face perception. *Journal of Neuroscience, 17*, 4302–4311.

Kim, J. N., & Shadlen, M. N. (1999). Neural correlates of a decision in the dorsolateral prefrontal cortex of the macaque. *Nature Neuroscience, 2*, 176–185.

Lo, C. C., & Wang, X. J. (2006). Cortico-basal ganglia circuit mechanism for a decision threshold in reaction time tasks. *Nature Neuroscience, 9*(7), 956–963.

McCarthy, G., Puce, A., Gore, J., & Allison, T. (1997). Face-specific processing in the human fusiform gyrus. *Journal of Cognitive Neuroscience, 9*, 605–611.

Montague, P. R., King-Casas, B., & Cohen, J. D. (2006). Imaging valuation models in human choice. *Annual Review of Neuroscience, 29*, 417–448.

Newsome, W. T., Britten, K. H., & Movshon, J. A. (1989). Neuronal correlates of a perceptual decision. *Nature, 341*, 52–54.

O'Doherty, J. (2004). Reward representations and reward-related learning in the human brain: insights from human neuroimaging. *Current Opinions in Neurobiology, 14*, 769–776.

O'Doherty, J., Hampton, A., & Kim, H. (2007). Model-based fMRI and its application to reward learning and decision making. *Annals of the New York Academy of Science, 1104*, 35–53.

Pfeier, R., & Bongard, J. (2007). *How the body shapes the way we think.* Cambridge, MA: MIT Press.

Philiastides, M. G., & Heekeren, H. R. (in press). Spatiotemporal characteristics of perceptual decision making in the human brain. In J.-C. Dreher, & L. Tremblay (Eds.), *The handbook of reward and decision making.* Amsterdam: Elsevier.

Pleger, B., Blankenburg, F., Ru, C., Driver, J., & Dolan, R. (2008). Reward facilitates tactile judgments and modulates hemodynamic responses in human primary somatosensory cortex. *Journal of Neuroscience, 28*, 8161–8168.

Pleger, B., Ruff, C. C., Blankenburg, F., Bestmann, S., Wiech, K., Stephan, K. E., Capilla, A., Friston, K. J., & Dolan, R. J. (2006). Neural coding of tactile decisions in the human prefrontal cortex. *Journal of Neuroscience*, *26*, 12596–12601.

Ploran, E., Nelson, S., Velanova, K., Donaldson, D., Petersen, S., & Wheeler, M. (2007). Evidence accumulation and the moment of recognition: dissociating perceptual recognition processes using fMRI. *Journal of Neuroscience*, *27*, 11912–11924.

Preuschhof, C., Heekeren, H. R., Taskin, B., Schubert, T., & Villringer, A. (2006). Neural correlates of vibrotactile working memory in the human brain. *Journal of Neuroscience*, *26*, 13231–13239.

Raab, M., & Green, N. (2005). Motion as input: a functional explanation of movement effects on cognitive processes. *Perceptual and Motor Skills*, *100*, 333–348.

Romo, R., Hernandez, A., & Zainos, A. (2004). Neuronal correlates of a perceptual decision in ventral premotor cortex. *Neuron*, *41*(1), 165–173.

Romo, R., Hernandez, A., Zainos, A., Brody, C. D., & Lemus, L. (2000). Sensing without touching: psychophysical performance based on cortical microstimulation. *Neuron*, *26*, 273–278.

Romo, R., Hernandez, A., Zainos, A., & Salinas, E. (1998). Somatosensory discrimination based on cortical microstimulation. *Nature*, *392*, 387–390.

Romo, R., Hernandez, A., Zainos, A., & Salinas, E. (2003). Correlated neuronal discharges that increase coding efficiency during perceptual discrimination. *Neuron*, *38*, 649–657.

Romo, R., & Salinas, E. (2003). Flutter discrimination: Neural codes, perception, memory and decision making. *Nature Reviews Neuroscience*, *4*, 203–218.

Rorie, A. E., & Newsome, W. T. (2005). A general mechanism for decision-making in the human brain? *Trends in Cognitive Sciences*, *9*, 41–43.

Rushworth, M., & Behrens, T. (2008). Choice, uncertainty and value in prefrontal and cingulate cortex. *Nature Neuroscience*, *11*, 389–397.

Salinas, E., Hernandez, A., Zainos, A., & Romo, R. (2000). Periodicity and firing rate as candidate neural codes for the frequency of vibrotactile stimuli. *Journal of Neuroscience*, *20*, 5503–5515.

Sebanz, N., Bekkering, H., & Knoblich, G. (2006). Joint action: bodies and minds moving together. *Trends in Cognitive Sciences*, *10*(2), 70–76.

Sereno, M. I., Pitzalis, S., & Martinez, A. (2001). Mapping of contralateral space in retinotopic coordinates by a parietal cortical area in humans. *Science*, *294*, 1350–1354.

Shadlen, M. N., Britten, K. H., Newsome, W. T., & Movshon, J. A. (1996). A computational analysis of the relationship between neuronal and behavioral responses to visual motion. *Trends in Cognitive Sciences*, *16*, 1486–1510.

Tegenthoff, M., Ragert, P., Pleger, B., Schwenkreis, P., Förster, A. F., Nicolas, V., & Dinse, H. R. (2005). Improvement of tactile discrimination performance and enlargement of cortical somatosensory maps after 5 Hz rTMS. *PloS Biology*, *3*, 2031–2040.

Thornton, I., & Knoblich, G. (2006). Action perception: seeing the world through a moving body. *Current Biology*, *16*(1), R27–R29.

Tversky, A., & Kahneman, D. (1981). The framing of decisions and the psychology of choice. *Science*, *211*, 453–458.

Uchida, N., Kepecs, A., & Mainen, Z. F. (2006). Seeing at a glance, smelling in a whiff: rapid forms of perceptual decision making. *Nature Reviews. Neuroscience*, *7*, 485–491.

Verschure, P., Voegtlin, T., & Douglas, R. (2003). Environmentally mediated synergy between perception and behaviour in mobile robots. *Nature*, *425*(6958), 620–624.

Walsh, V. (2003). Time: the back-door of perception. *Trends in Cognitive Sciences*, *7*(8), 335–338.

Wilson, M. (2002). Six views of embodied cognition. *Psychonomic Bulletin & Review*, *9*(4), 625–636.

Wyss, R., König, P., & Verschure, P. F. (2004). Involving the motor system in decision making. *Proceedings of the Biological sciences/The Royal Society*, *271*(3), 50–52.

M. Raab et al. (Eds.)
Progress in Brain Research, Vol. 174
ISSN 0079-6123

CHAPTER 18

Motor imagery and its implications for understanding the motor system

Jörn Munzert* and Karen Zentgraf

Institute for Sports Science, Justus-Liebig University, Giessen, Germany

Abstract: In the neurosciences, motor imagery (MIm) has not just been a topic of basic research. It has also attracted attention in applied research as a therapeutic tool. MIm is conceptualized as an internal simulation of motor acts that generates images on the basis of motor representations. Therefore, MIm is associated with neural activation of the cortical and subcortical motor system. The resulting concept of functional equivalence between MIm and execution opens a window to study the organization of motor processes and, more generally, to understand the neural plasticity of the motor system.

Keywords: motor imagery; mental simulation theory; primary motor cortex; verbal instruction; motor representation

Introduction

Research on motor imagery (MIm) and mental training (MT) has made substantial contributions to both sport and exercise psychology. Whereas findings on the impact of MT on athletes' skills and performance go back to early research conducted by Russian psychologists and their European followers (Puni, 1961; Pickenhain, 1976), new and promising results on MIm and MT have come from studies related to rehabilitation exercise (Page et al., 2001). Applied research has profited greatly from neuroscientific research on MIm. In the present review, we shall focus on work on MIm and MT that expands our knowledge about motor control and learning. More specifically, we shall seek to demonstrate that MIm offers a window for observing the neural processes underlying motor performance.

We argue that MIm reflects the powerful combination of perceptual processes, decisions, and action implementation (cf. Chapter 24: How are actions physically implemented?), because it encompasses perceptual qualities and specifically early stages of action implementation after a decision for a specific action has been made. We shall also show how action implementation relying on neural activation in primary motor cortex (M1) has strong connections with cognitive processes.

Defining motor imagery

First, we shall examine the methodological assumptions underlying neuroscientific research on a subjective state such as MIm. Most studies start with verbal instructions to imagine a certain action. They presuppose that participants follow

*Corresponding author.
Tel.: +49-641-9925220; Fax: +49-641-9925209;
E-mail: Joern.Munzert@sport.uni-giessen.de

DOI: 10.1016/S0079-6123(09)01318-1

this instruction obediently, and even the manipulation checks designed to assess whether participants really do follow this instruction are based on verbal self-reports and are, therefore, scarcely objective. In the next paragraph, we shall discuss how instructions influence MIm in a subtle manner.

All basic concepts of MIm have to take into account that the imagery process refers to some kind of subjective experience. A participant in an experiment is asked to imagine himself/herself performing a specific action and then to generate the perception-like sensations that would occur if he/she actually were to perform it. In other words, he/she follows a verbal instruction to generate a perception-like mental state. No direct behavioral output indicates whether he/she has done this successfully. However, he/she can signal the beginning and the end of the imagery process. Although this once again reflects his/her mental state, information on the time course of imagery has proved to be a valid manipulation check. It has been demonstrated repeatedly that the actual and the mental durations of an action correlate closely (Decety et al., 1989; Munzert, 2008; for an overview see Guillot and Collet, 2005). When this correlation can then be shown in a specific imagery task, it is taken as an indicator that the participant has followed the verbal instructions for MIm.

Imagery can be defined as a perception-like process in the absence of external stimuli (Kosslyn, 1994; Annett, 1995). In this definition, internal processes form the basis for images. Farah (1984) was the first to show that imagery cannot be described as a single process, but as one involving several components that can each be examined separately. First, retrieval of information from long-term memory is necessary because of the lack of external stimulus input. Information about a motor act has to be processed from long-term memory to working memory in which it forms the perception-like process. This stage of processing is called *image generation*. Second, information in working memory can be subject to different *transformations*. For example, persons can focus their attention on a specific modality such as the kinesthetic sensations of the movement, or they can change the speed of an imagined movement (Louis et al., 2008). Third, a necessary precondition for different imagery operations is that the image can be held in working memory. This is the *maintenance* phase of an image as emphasized by Kosslyn (1987, 1994). Normally, images are accessible to conscious operations within only a limited timeframe after which they fade and disappear. Maintaining an image requires the allocation of mental effort. A person can *inspect* images with the aim of detecting details, comparing specific aspects with former percepts, or verbally reporting imagined sensations. For example, reporting the start and the end of an image requires a verbal report of subjective states referring to this image. Whether these processes actually do occur depends on the imagery task at hand and, more specifically, on the instructions.

Figure 1 depicts the basic subprocesses of mental imagery and their inter-relation according to Farah's (1984) model. An additional path for image generation that is not derived from information stored in long-term memory is also implemented in the model. This describes a condition in which information in working memory is a result of an action just performed by the participant. This relies on *recognition* of sensory inputs. For example, a person may look at a picture and then rely on this information to imagine the depicted scene. For MIm, this corresponds to imagery of an action performed shortly before image generation. This procedure is common in sports when athletes analyze their motor performance after a trial using information maintained in working memory. Farah's (1984) model is based on extensive neuropsychological evidence from patients suffering from deficits in a specific subprocess of imagery as specified within the model. The model can also be applied to MIm where it offers a possibility of categorizing different approaches to assess individual differences in imagery abilities.

Any study on MT presupposes that participants have the competence to generate mental images. However, this competence may vary differentially. Applied sport psychology and applied neuropsychology have therefore developed tests

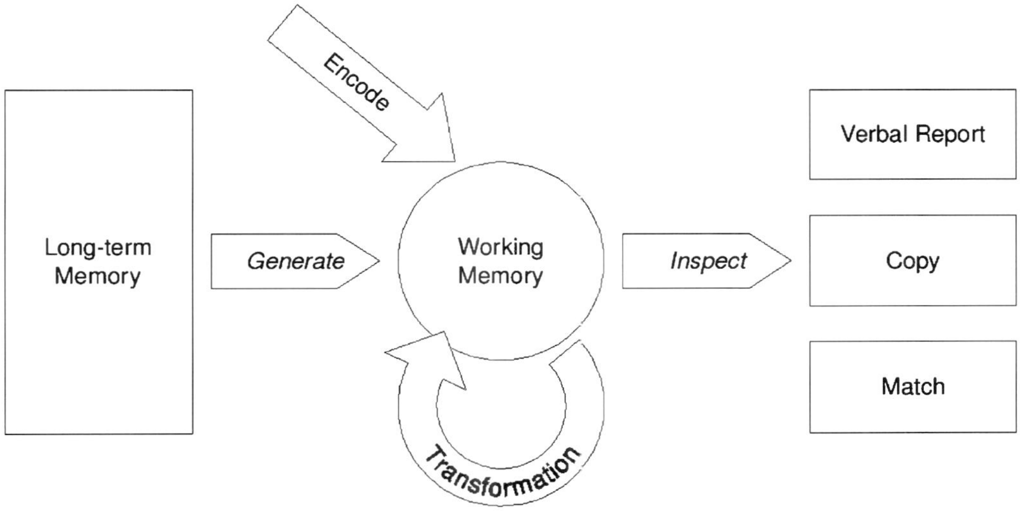

Fig. 1. Computational model of the components involved in motor imagery (adapted from Farah, 1984).

to assess this ability, and such assessments have been a major issue in motor imagery research (McAvinue and Robertson, 2008). As we have argued above, we want to emphasize that the assessment of a subjective state in imagery has to be instructed verbally. One adaptation of the Vividness of Movement Imagery Questionnaire (VMIQ-2; Roberts et al., 2008) asks participants to imagine specific actions in terms of different imagery modalities. The imagery task is presented verbally, and participants have to recall information about the action from long-term memory in order to generate the required MIm. The Movement Imagery Questionnaire (MIQ-R; Hall and Martin, 1997) uses a very different procedure: here, the participant first has to perform the specific action and then engage in MIm. In terms of Farah's model (cf. Fig. 1), a participant using the MIQ-R can rely on information present in working memory. Although it is also possible for them to use information from long-term memory, they do not have to do this. This differs from the VMIQ-2 procedure. Here, no actual performance precedes MIm, obliging participants to use information from long-term memory. The two procedures differ markedly in terms of the extent of centrally triggered top-down processes. The MIQ-R can be linked to the recognition path in Farah's model; the VMIQ-2, to the generation path. This difference shows how assumptions in Farah's model reflect basic processes of imagery and, more specifically, of MIm. Both tests also differ with respect to the underlying processes to be assessed. Whereas the VMIQ-2 asks participants explicitly to assess the vividness of the specific image, the MIQ-R directs the participant's focus onto the level of difficulty in generating it. Despite these differences, the two tests correlate significantly (Roberts et al., 2008).

Dynamic images of the motor act typically focus on kinesthetic or visual information. For the latter, one can distinguish between an external and an internal perspective. The external, or third-person, perspective is defined as corresponding to the view one obtains when watching a video recording. The internal, or first-person, perspective corresponds to the visual view of the actor during performance. This is similar to the view offered by a helmet camera. The VMIQ-2 assesses both visual perspective and kinesthetic images, whereas the MIQ-R differentiates only between the visual and the kinesthetic modality. The VMIQ-2 explicitly instructs participants to imagine themselves performing the action (in contrast to a former version of the VMIQ). This is not the case for the MIQ-R, in which it can be argued that the instruction to imagine oneself is given implicitly, because the specific action has to be performed before imagery. Nevertheless, imagining oneself and recalling the sensations induced by self-action is a very important issue in imagery research. Differences between the procedures implemented in the tests are again due to differences in the imagery instructions.

We shall use the term *motor imagery* (MIm) for imagery processes requiring participants to imagine their own actions without displaying any overt behavior. Hence, the agency of actions is a critical feature for MIm. MIm can include both visual perspectives (first- or third-person) and kinesthetic imagery depending on the level of experience and individual imagery preferences. The majority of neuroscientific studies in MIm apply the term in this sense, but focus on kinesthetic imagery.

Motor imagery as an internal simulation of action

One of the basic findings in imagery research is that visual imagery and perception (Farah, 1984; Kosslyn, 1994) as well as MIm and action execution (Jeannerod, 1994, 2001) share similar neural representations. However, both processes differ fundamentally with respect to the input they receive from the periphery. As outlined in Farah's model (cf. Fig. 1), imagery is based on input from memory. Visual imagery and MIm both refer to experiences gained in the past that can be recalled voluntarily. However, perception and motor execution basically rely on input from the external world, although motor execution may also have a strong memory component depending on which specific motor task is required. In the case of MIm, one may ask whether the central components in MIm and motor execution differ, or whether they rely on common structures. Jeannerod (2001) proposed a concept of functional equivalence between imagery and execution of actions based on this idea. In his view, motor images are based on representations that would normally trigger cortical and subcortical motor execution processes, but are inhibited at a certain stage of processing. The inhibition of excitatory motor signals traveling to the periphery is the essential prerequisite for the distinction between MIm and execution. However, an overlap of neural representations is not just found between motor execution and MIm. It is also found between execution and movement observation, execution and action planning, or execution and action verbalization. Jeannerod (2001) coined the term "S-states" for all processes that are viewed as covert stages of action and share common representations with execution. It can be argued that S-states simulate real actions to some degree. Obviously, this is conceivable for MIm, but the simulation concept can also be applied to the observation of movements (Zentgraf et al., 2005).

Support for a functional equivalence between execution and MIm also comes from research on internal models for motor control (Miall and Wolpert, 1996; Miall, 2003). Two internal models are considered to work in close interaction: given a desired goal for an action, the inverse model generates an appropriate motor command. An efference copy is fed into an internal predictive model that maps the motor command to the anticipated effect of the action. At the same time, the forward model allows for rapid adjustments during movement execution. It builds a template against which the incoming information can be compared. These computational models were originally designed to study motor performance. In the context of functional equivalence between S-states and motor execution, however, the predictive information can also be used to conceptualize actions performed by others (which might only be observed). In this sense, perceiving an action means to simulate it internally. A basic functional mechanism in simulation is for the observer to use his/her own body-related representations to predict and detect action intentions and effects as well as emotions or sensations experienced by others. In action observation, an external stream of visual signals updates the simulation process, whereas in MIm, a deliberate internal signal is needed. Although these kinds of simulation obviously differ, they share the same major motor fronto-parietal network (Munzert et al., 2008). All these aspects have been linked to a functional mechanism labeled "embodied simulation"

(Gallese, 2005). In this context, findings from imagery research can be viewed as empirical evidence supporting the paradigmatic shift to viewing action, perception, and bodily states as being tightly intertwined. This perspective has been labeled the embodiment approach and has been a source of much theoretical inspiration not only in psychology but also in the neurosciences.

Conceiving MIm as a simulation state entails two general hypotheses: first, common neural representations are predicted between MIm and motor execution, and, in addition, similar motor-relevant physiological responses should be seen for execution and MIm (cf. Jeannerod, 2001). Second, differences between MIm and motor execution should be found as well, reflecting the fact that MIm is the covert stage of an action, whereas execution also implies additional processes for the stage of overt behavior.

Motor areas of the cerebral cortex are subdivided into M1 and several premotor areas, including supplementary motor area (SMA), pre-supplementary motor area (pre-SMA), and ventral and dorsal parts of premotor cortex (PMC) (Roland and Zilles, 1996). Cortical motor areas are closely linked to the cerebellum and the basal ganglia creating feedback loops (Krakhauer and Ghez, 2000). Other nonmotor areas that are relevant for motor execution and possibly for MIm are the primary somatosensory cortex (S1) and areas within the parietal cortex, especially the inferior and superior parietal lobes. Whereas activation of PMC, pre-SMA, and SMA, as well as subcortical areas during motor imagery is well established (Jeannerod, 1994, 2001; Lotze and Halsband, 2006), the role of M1 for motor imagery is still under discussion.

The role of the primary motor cortex for motor imagery

It has often been argued that M1 represents the cortical level at which motor commands are processed and then transferred via the spinal cord to the muscle–joint system. As Jeannerod (2001) has argued from a position emphasizing the functional equivalence of motor imagery and motor execution, MIm demands an activation of the cortical motor system followed by an inhibition of motor commands in a way that will inhibit the production of overt behavior. This raises the question whether neural activation during motor imagery also implies activation of M1. On the one hand, early positron emission tomography (PET) studies (Roland et al., 1980; Decety et al., 1988) and a large number of subsequent studies failed to reveal M1 activation during MIm consistently. On the other hand, these findings have been challenged by a large number of studies revealing a significant M1 activation when participants imagine motor acts. Nonetheless, M1 activation is typically smaller compared with execution, and it is not always to be seen in all participants (cf. Lotze et al., 1999).

New data have shown that the inhibition of the efferent command during MIm is not complete. Repeated studies based on different methods have revealed that MIm may influence spinal excitability (Kiers et al., 1997; Li et al., 2004; for very small changes in H reflex see also Gandevia et al., 1997). During MIm, a significant, but subthreshold, signal may be generated in M1 that produces changes of excitability at the spinal level. One could conclude that a suprathreshold signal has to be suppressed, but that subthreshold signals may leave M1 and cause a change in excitability at the spinal level.

All these findings result in an interpretation that M1 also processes cognitive properties during motor control (Sanes and Donoghue, 2000; cf. Chapter 23: Juggling with the brain — thought and action in the human motor system). The concept of M1 as being the final level producing exclusively motor commands is not supported by data. For example, M1 also reveals anticipatory functions for the generation of movement sequences (Lu and Ashe, 2005). Another feature of M1 is that it receives input from the muscle spindles (Naito et al., 2002) and is therefore sensitive to sensory processes. This is in clear contrast to classical assumptions of M1 representing a core output level. Furthermore, it shows some sensitivity to attentional processes: the posterior part 4p is modulated by attention, whereas the anterior part 4a is not (Binkofski et al., 2002).

In general, the activation of M1 during MIm can be attributed to embodied cognitive processes. It has been argued that it is particularly the preparation for MIm that impacts significantly on M1 activation (Johnson et al., 2002; Zang et al., 2003). This is an important distinction that has rarely been considered in imagery research so far.

Empirical data supporting mental simulation theory

Mental simulation theory (MST) predicts (1) an activation of motor areas during processing of cognitive motor states (S-states such as action observation, MIm, and action verbalization); (2) a strong overlap of neural activations between different S-states; and (3) specific differences between neural activations of S-states due to the affordances of specific processing. Whereas predictions (2) and (3) require a comparison between different S-states, such a comparison is considered to be only helpful for prediction (1). As a result, it should be noted that, up to now, most of the evidence supporting MST is based on comparing results from different experiments but not from different conditions within one study (cf. Jeannerod, 2001).

Here, we shall report results from a functional magnetic resonance imaging (fMRI) study comparing neural activations in movement observation and MIm of gymnastic movements based on a within-subject design (Munzert et al., 2008). Participants watched videos of gymnastic movements under an instruction that they would subsequently have to imagine the observed movements. Compliance with the observational condition was investigated by a recall test of observed movements after the scanner session. Compliance with the mental imagery instruction (MIm) was checked by comparing observed and mental movement times. Neural activation was measured during action observation and the subsequent imagery condition. An analysis of areas activated in both conditions revealed an overlap of neural activation not only in core motor areas such as the M1, SMA, and PMC, but also in subcortical motor areas of the cerebellum and the basal ganglia. Additionally, activation of the inferior parietal sulcus and the inferior frontal gyrus was found for both S-states. Both of these areas have been discussed as showing "mirror qualities" (Rizzolatti and Craighero, 2004). The mirror system contains brain areas that are relevant for the observation of actions. Here, it could be shown that they are engaged not only in action observation but also in the S-state of MIm (cf. Fig. 2).

The results reveal strong support for predictions 1 and 2 in MST. We also found significant neural activation in M1 during MIm for 10 out of 10 participants. Again, this can be taken as strong support for Prediction 1. A contrast between neural activation in MIm and observation showed significant differences for the anterior cingulate cortex (ACC) and the insula. These effects can be interpreted in terms of specific processing needs in MIm. The ACC is seen as a source for motor attention (Rushworth et al., 2001), and this activation might be necessary to maintain an imagery process that is not receiving the continuous input available for observation. The insula is part of the secondary somatosensory cortex and provides a basis for the integration of different sensory processes originating in the visual world and in bodily senses. Again, this effect can be interpreted as a consequence of specific requirements of the imagery process. Both results are in accordance with prediction 3. The contrast between observation and imagery showed significant effects for occipital areas, the superior parietal lobe, the hippocampus, and parts of the cerebellum. All activations can be interpreted as a consequence of the observational task. This also supports prediction 3.

In a preceding study with the same stimulus material, we compared neural activation in two observational conditions (Zentgraf et al., 2005). Participants were asked to observe the same movement sequences either with the aim of subsequent mental simulation (observation for imitation: OI) or that of judging the accuracy of the movements (observation for evaluation: OE). In the first condition, participants were asked to observe movement sequences carefully followed by an internal simulation (MIm) of the same

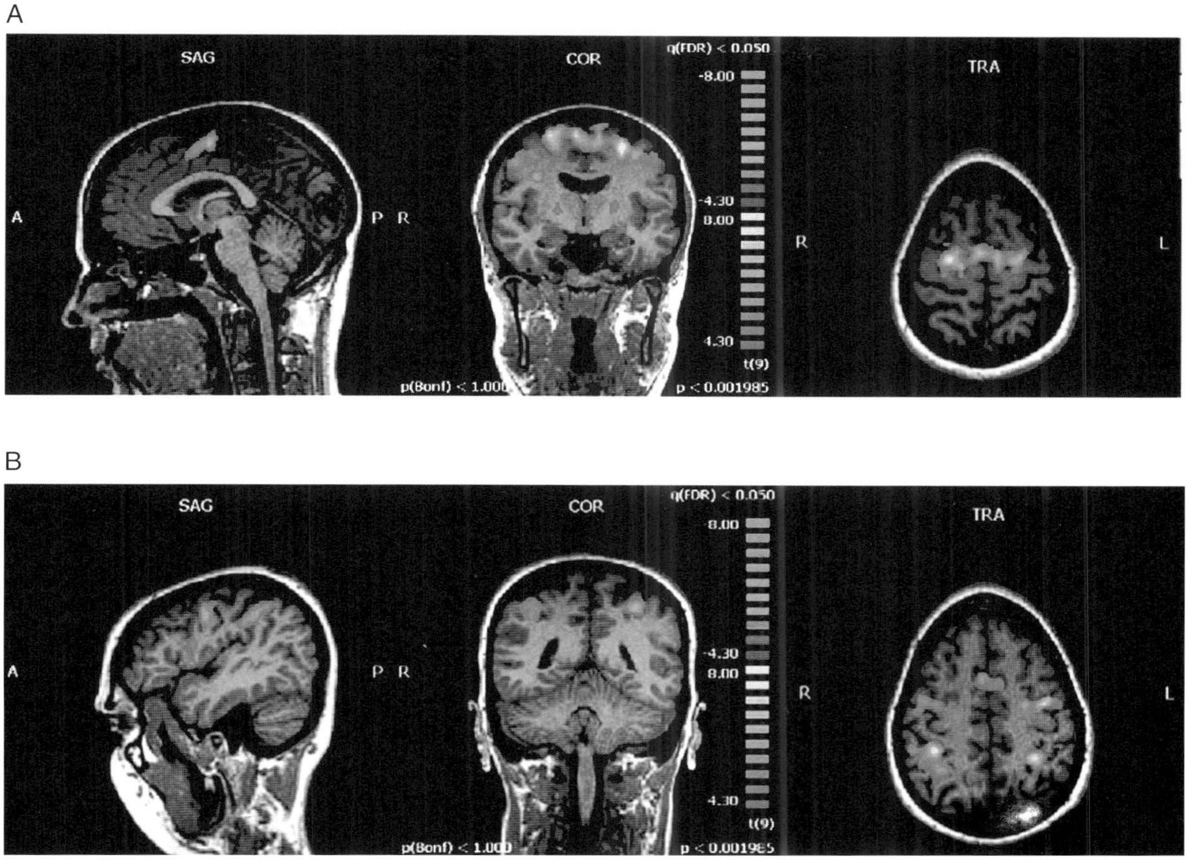

Fig. 2. Neural activations of cortical areas in movement observation and motor imagery (MIm). SAG, sagittal; COR, coronal; TRA, transversal. Reprinted from Munzert et al. (2008) with permission. (See Color Plate 18.2 in color plate section.)

sequence. In the OE condition, they had to assess movement accuracy. Action sequences contained repetitions of single parts of the sequence, and participants were asked to evaluate whether repetitions were congruent. Both observational conditions OI and OE relied on identical stimulus material for observation, whereas the subsequent task differed (motor imagery vs. evaluation of sequences). This allowed an analysis of the impact of the core instructional set on the same S-state of movement observation. More specifically, the study investigated whether cognitive processes influence neural activation in motor areas.

When both observational conditions were contrasted, the posterior part of the supplementary motor cortex (SMA proper) revealed stronger activation under the instruction for subsequent mental simulation (OI), whereas observation with the aim of judging movements (OE) led to stronger activation in the anterior part of the supplementary motor area (pre-SMA, see Fig. 3). This pattern of results fits in nicely with the different functions that have been attributed to SMA proper and pre-SMA (Picard and Strick, 1996, 2001). Pre-SMA has major connections with prefrontal areas. Therefore, a stronger activation during the more cognitive evaluation task supports the concept of mental simulation. In contrast, SMA proper is connected tightly to the M1. The present results show that an instruction that is closer to execution activates basic motor areas more strongly. This is notable, because

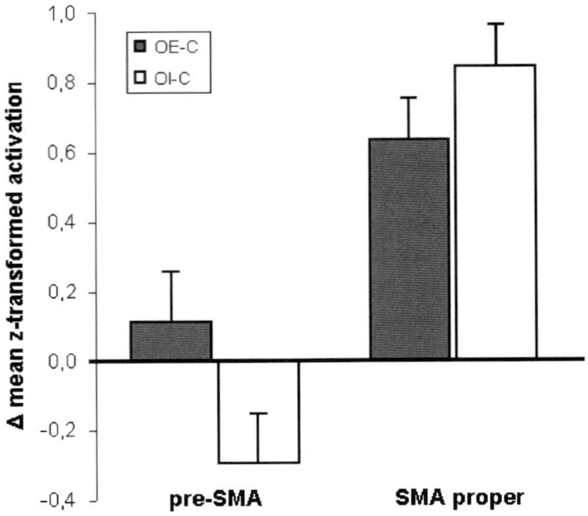

Fig. 3. Different activation patterns for pre-SMA and SMA proper as a function of movement observation instructions. OE-C, observation for evaluation-control; OI-C, observation for imitation-control. Reprinted from Zentgraf et al. (2005) with permission.

these different activations in motor areas were found for the observation of identical stimulus material in both conditions. Different goals of movement observation can therefore modulate activation of motor representations in a significant way. Different instructions also had a notable influence on neural activation in subcortical areas.

Neuroscientific studies confirm a profound and reproducible effect of MIm on neural activation of motor areas. Effects of MIm on motor learning are not restricted to laboratory tasks, but are also found for applications in physical therapy or in the sports setting. Training regimes applying MT for stroke patients have proved to be moderately effective (Page et al., 2001). In the context of power training, behavioral effects of MT have also been confirmed (Yue and Cole, 1992). Findings on MT effects in power training reveal that MIm actually does enhance power training without any active training. This constitutes a strong argument suggesting that central cortical processes subserve the enhancement of power.

MST provides convincing arguments in support of processes underlying MT that explain neural activation during MIm as well as behavioral outcomes of MT. The basic idea of MST refers to the activation of a neural motor network while imaging motor actions (Jeannerod, 2001). This includes activation of not only premotor and motor areas such as PMC, SMA, and M1 but also that of subcortical areas of the cerebellum (Lotze et al., 1999) and basal ganglia (Bonda et al., 1995). Whereas the activation of M1 has been confirmed in a large number of studies, the status of primary sensory areas during MIm remains unclear. For example, Munzert et al. (2008) found activation in S1 and in the insula during MIm (see also, Lotze et al., 1999; Solodkin et al., 2004; Sacco et al., 2006). Searching for further evidence of activation of sensory areas seems to be a promising goal for future research.

Studying activation in sensory areas as a consequence of MIm might help to create a theoretical association between MST and forward model concepts (Miall and Wolpert, 1996; Miall, 2003). Forward models contain an expectation about sensory consequences (Miall, 2003). They afford an activation of the motor network to create an image of future action consequences (see also Chapter 3: A cognitive movement scientist's view on the link between thought and action: insights from the "Badische Zimmer" metaphor). Therefore, it seems plausible to predict S1 activation during MIm. Nevertheless, this is an issue that is still under discussion.

It should also be considered that it is not only premotor and motor areas that are activated during MIm, but also prefrontal areas associated with attentional processes and working memory. It has been argued that a decline in imagery abilities in older adults is due to a more cognitive impairment of working memory capacities (Mulder et al., 2007). Such data demonstrate a close integration of cognitive and motor processes. Moreover, applications of MT in sports show that athletes can switch between imagery modalities quite easily despite having individual preferences. The ease of changing between imagery modalities depends on individual experience and the instructional setting. Neuroscientific studies have particularly failed to stress the latter point. This is astonishing in light of the significant differences found for neural activation associated with the instruction to perform motor versus visual imagery. For example, stronger effects of MIm on M1 activation (Solodkin et al., 2004) and M1 excitability (Fourkas et al., 2006; Stinear et al., 2006) have been reported repeatedly. A closer look at the experimental conditions in these studies reveals that imagery conditions differed only in terms of initial instructions, but not in terms of stimuli or tasks. In principle, it would seem difficult for any manipulation check to differentiate between imagery conditions and ascertain the compliance of participants. Although mental durations can be taken as an indirect measure to control imagery processes (Zentgraf et al., 2005; Munzert et al., 2008), no direct check of task compliance is available.

Taking these methodological constraints into consideration, one can ask how differences between instructions can cause significant changes in neural activation of cortical and subcortical areas. No other differences exist between MIm and visual imagery conditions than that of instructional content. We argue that different imagery instructions lead participants to focus attention on different aspects of motor control: visual imagery focuses on external aspects with special reference to the relation between the body and its environment, whereas MIm focuses on internal states of movement dynamics and force production. These differences can explain why MIm specifically activates motor areas. Attention is drawn to own movements and therefore to the agency of actions. This leads to activation of cortical motor areas and subcortical areas that are integrated in cerebello-thalamo-cortical loops.

Neural activations during MIm seem to support MST. MIm is seen to be a mental simulation of actions that relies on motor representations. However, it remains unclear why visual imagery of movements does not evoke similar activation. Is it possible that motor programs and kinesthetic representations are linked more closely than motor programs and visual imagery? Until now, no evidence exists that could provide a satisfactory answer to this question.

Conclusion

Neuroimaging data support the concept of a functional equivalence of MIm and motor execution (Jeannerod, 2001). Even though differences exist between both processes (e.g., MIm requires inhibition of motor processes and apparently lacks feedback information), one may ask how MIm can be used as a tool to analyze motor processes. As outlined, MIm is very similar to stages of action preparation. This points directly at the central aspect of the functional equivalence concept. Motor execution implies covert stages of motor planning and preparation. It has been argued that these processes can be studied through imagery research. Furthermore, MIm research has contributed to the paradigmatic shift in how motor system functioning is viewed. Data gained within the two last decades has shown multiple cognitive links within the motor system (Rizzolatti and Craighero, 2004), and this has encouraged embodiment approaches in psychological and neuroscientific research. Results supporting the concept of functional equivalence between MIm and motor execution likewise help us to understand that cognitive functions are embedded in body representations. Gallese (2005) has described this link as embodied simulation, showing that it is a prerequisite for understanding others' intentions and for acting in a social context.

One problem of imagery research in this context is that conscious processes are unavoidable. Up to now, whether this methodological issue limits research on MIm has not been studied comprehensively. The reliance on instructions seems to be a decisive issue in this field. First, instructions for MIm ask participants to place themselves in a mental state, and this demands mental effort and is thereby based on a voluntary decision. Second, several studies have shown that verbal instructions can modify neural activation in motor areas. From a traditional viewpoint, this seems surprising. How can this be explained? At this stage, we can go back to the basic concepts of ecological approaches introduced by Newell (1986). These view actions as being constrained by the person, the task, and the environment. Instructions are a tool to specify tasks. They circumscribe how a person has to image an action or how he/she should specifically observe an action. Taking these arguments as a central prerequisite of imagery research, future studies should consider imagery instructions more carefully. These may well confirm attentional focus as one of the basic subprocesses in S-states.

References

Annett, J. (1995). Motor imagery: perception or action? *Neuropsychologia*, *33*(11), 1395–1417.

Binkofski, F., Fink, G. R., Geyer, S., Buccino, G., Gruber, O., Shah, N. J., Taylor, J. G., Seitz, R. J., Zilles, K., & Freund, H.-J. (2002). Neural activity in human primary motor cortex areas 4a and 4p is modulated differentially by attention to action. *Journal of Neurophysiology*, *88*(1), 514–519.

Bonda, E., Petrides, M., Frey, S., & Evans, A. (1995). Neural correlates of mental transformations of the body-in-space. *Proceedings of the National Academy of Sciences, USA*, *92*(24), 11180–11184.

Decety, J., Jeannerod, M., & Prablanc, C. (1989). The timing of mentally represented actions. *Behavioural Brain Research*, *34*(1–2), 35–42.

Decety, J., Philippon, B., & Ingvar, D. H. (1988). RCBF landscapes during motor-performance and motor ideation of a graphic gesture. *European Archives of Psychiatry and Clinical Neuroscience*, *238*(1), 33–38.

Farah, M. J. (1984). The neurological basis of mental imagery: a componential analysis. *Cognition*, *18*(1–3), 245–272.

Fourkas, A. D., Ionta, S., & Aglioti, S. M. (2006). Influence of imagined posture and imagery modality on corticospinal excitability. *Behavioural Brain Research*, *168*(2), 190–196.

Gallese, V. (2005). Embodied simulation: from neurons to phenomenal experience. *Phenomenology and the Cognitive Sciences*, *4*(1), 23–48.

Gandevia, S. C., Wilson, L. R., Inglis, J. T., & Burke, D. (1997). Mental rehearsal of motor tasks recruits α-motoneurons but fails to recruit human fusiform neurons selectively. *Journal of Physiology*, *505*(1), 259–266.

Guillot, A., & Collet, C. (2005). Duration of mentally simulated movement: a review. *Journal of Motor Behavior*, *37*(1), 10–20.

Hall, C. R., & Martin, K. E. (1997). Measuring movement imagery abilities: a revision of the Movement Imagery Questionnaire. *Journal of Mental Imagery*, *21*(1–2), 143–154.

Jeannerod, M. (1994). The representing brain: neural correlates of motor intention and imagery. *Behavior and Brain Sciences*, *17*(2), 187–245.

Jeannerod, M. (2001). Neural simulation of action: a unifying mechanism for motor cognition. *Neuroimage*, *14*(1), S103–S109.

Johnson, S. H., Rotte, M., Grafton, S. T., Hinrichs, H., Gazzaniga, M. S., & Heinze, H. J. (2002). Selective activation of a parieto-frontal circuit during implicitly imagined prehension. *Neuroimage*, *17*(4), 1693–1704.

Kiers, L., Fernando, B., & Tomkins, D. (1997). Facilitatory effect of thinking about movement on magnetic motor-evoked potentials. *Electroencephalography and Clinical Neurophysiology*, *105*(4), 262–268.

Kosslyn, S. M. (1987). Seeing and imagining in the cerebral hemispheres: a computational approach. *Psychological Review*, *94*(2), 148–175.

Kosslyn, S. M. (1994). *Image and brain. The resolution of the imagery debate*. Cambridge: MIT Press.

Krakhauer, J., & Ghez, C. (2000). Voluntary movement. In E. R. Kandel, J. H. Schwarz, & T. M. Jesell (Eds.), *Principles of neural science* (pp. 756–781). New York: McGraw-Hill.

Li, S., Kamper, D. G., Stevens, J. A., & Rymer, W. Z. (2004). The effect of motor imagery on spinal segmental excitability. *Journal of Neuroscience*, *24*(43), 9674–9680.

Lotze, M., & Halsband, U. (2006). Motor imagery. *Journal of Physiology (Paris)*, *99*(4–6), 386–395.

Lotze, M., Montoya, P., Erb, M., Hülsmann, E., Flor, H., Klose, U., Birbaumer, N., & Grodd, W. (1999). Activation of cortical and cerebellar motor areas during executed and imagined hand movements: an fMRI study. *Journal of Cognitive Neuroscience*, *11*(5), 491–501.

Louis, M., Guillot, A., Maton, S., Doyon, J., & Collet, C. (2008). Effect of imagined movement speed on subsequent motor performance. *Journal of Motor Behavior*, *40*(2), 117–132.

Lu, X. F., & Ashe, J. (2005). Anticipatory activity in primary motor cortex codes memorized movement sequences. *Neuron*, *45*(6), 967–973.

McAvinue, L. P., & Robertson, I. H. (2008). Measuring motor ability: a review. *European Journal of Cognitive Psychology*, *20*(2), 232–251.

Miall, R. C. (2003). Connecting mirror neurons and forward models. *Neuroreport*, *14*(17), 2135–2137.

Miall, R. C., & Wolpert, D. M. (1996). Forward models for physiological motor control. *Neural Networks*, *9*, 1265–1279.

Mulder, T., Hochstenbach, J. B. H., van Heuvelen, M. J. G., & den Otter, A. R. (2007). Motor imagery: the relation between age and imagery capacity. *Human Movement Science*, *26*(2), 203–211.

Munzert, J. (2008). Does level of expertise influence imagined durations in open skills? Played versus imagined durations of badminton sequences. *International Journal of Sport and Exercise Psychology*, *6*(1), 24–38.

Munzert, J., Zentgraf, K., Stark, R., & Vaitl, D. (2008). Neural activation in cognitive motor processes: comparing motor imagery and observation of gymnastic movements. *Experimental Brain Research*, *188*(3), 437–444.

Naito, E., Kochiyama, T., Kitada, R., Nakamura, S., Matsumura, M., Yonekura, Y., & Sadato, N. (2002). Internally simulated movement sensations during motor imagery activate cortical motor areas and the cerebellum. *Journal of Neuroscience*, *22*(9), 3683–3691.

Newell, K. M. (1986). Constraints of the development of coordination. In M. G. Wade & H. T. A. Whiting (Eds.), *Motor development in children: aspects of coordination and control* (pp. 341–360). Dordrecht: Martinus Nijhoff.

Page, S. J., Levine, P., Sisto, S., & Johnston, M. V. (2001). A randomized efficacy and feasibility study of imagery in acute stroke. *Clinical Rehabilitation*, *15*(3), 233–240.

Picard, N., & Strick, P. (1996). Motor areas of the medial wall: a review of their location and functional activation. *Cerebral Cortex*, *6*(3), 342–353.

Picard, N., & Strick, P. (2001). Imaging the premotor areas. *Current Opinion in Neurobiology*, *11*(6), 663–672.

Pickenhain, L. (1976). Die Bedeutung innerer Rückkopplungskreise für den Lernvorgang (gezeigt am Beispiel des motorischen Lernens). [The relevance of internal feedback for motor learning]. *Zeitschrift für Psychologie*, *184*(4), 551–561.

Puni, A. Z. (1961). *Abriss der Sportpsychologie [Foundations of sport psychology]*. Berlin: Sportverlag.

Rizzolatti, G., & Craighero, L. (2004). The mirror neuron system. *Annual Review of Neuroscience*, *27*(1), 169–192.

Roberts, R., Callow, N., Hardy, L., Markland, D., & Bringer, J. (2008). Movement imagery ability: development and assessment of a revised version of the Vividness of Movement Imagery Questionnaire. *Journal of Sport & Exercise Psychology*, *30*(2), 200–221.

Roland, P. E., Larsen, B., Lassen, N. A., & Skinhoj, E. (1980). Supplementary motor area and other cortical areas in organization of voluntary movements in man. *Journal of Neurophysiology*, *43*(1), 118–136.

Roland, P. E., & Zilles, K. (1996). Functions and structures of the motor cortices in humans. *Current Opinion in Neurobiology*, *6*(6), 773–781.

Rushworth, M. F. S., Paus, T., & Sipila, P. K. (2001). Attention systems and the organization of the human parietal cortex. *Journal of Neuroscience*, *21*(14), 5262–5271.

Sacco, K., Cauda, F., Cerliani, L., Mate, D., Duca, S., & Geminiani, G. C. (2006). Motor imagery of walking following training in locomotor attention. The effect of "the tango lesson". *Neuroimage*, *32*(3), 1441–1449.

Sanes, J. N., & Donoghue, J. P. (2000). Plasticity and primary motor cortex. *Annual Review of Neuroscience*, *23*, 393–415.

Solodkin, A., Hlustik, P., Chen, E. E., & Small, L. S. (2004). Fine modulation in network activation during motor execution and motor imagery. *Cerebral Cortex*, *14*(11), 1246–1255.

Stinear, C. M., Byblow, W. D., Styvers, M., Levin, O., & Swinnen, S. P. (2006). Kinesthetic, but not visual, motor imagery modulates corticomotor excitability. *Experimental Brain Research*, *168*(1–2), 157–164.

Yue, G., & Cole, K. J. (1992). Strength increases from the motor program: comparison of training with maximal voluntary and imagined muscle contractions. *Journal of Neurophysiology*, *67*(5), 1114–1123.

Zang, Y., Jia, F., Weng, X., Li, E., Cui, S., Wang, Y., Hazeltine, E., & Ivry, R. (2003). Functional organization of the primary motor cortex characterized by event-related fMRI during movement preparation and execution. *Neuroscience Letters*, *337*(2), 69–72.

Zentgraf, K., Stark, R., Reiser, M., Künzell, S., Schienle, A., Kirsch, P., Walter, B., Vaitl, D., & Munzert, J. (2005). Differential activation of pre-SMA and SMA proper during action observation: effects of instructions. *Neuroimage*, *26*(3), 662–672.

M. Raab et al. (Eds.)
Progress in Brain Research, Vol. 174
ISSN 0079-6123

CHAPTER 19

The cognitive nature of action — functional links between cognitive psychology, movement science, and robotics

Thomas Schack[1,3,4,*] and Helge Ritter[2,3,4]

[1]Neurocognition and Action Research Group, Bielefeld University, Bielefeld, Germany
[2]Neuroinformatics Group, Bielefeld University, Bielefeld, Germany
[3]Center of Excellence "Cognitive Interaction Technology" (CITEC), Bielefeld University, Bielefeld, Germany
[4]Research Institute for Cognition and Robotics (CoR-Lab), Bielefeld University, Bielefeld, Germany

Abstract: This paper examines the cognitive architecture of human action, showing how it is organized over several levels and how it is built up. Basic action concepts (BACs) are identified as major building blocks on a representation level. These BACs are cognitive tools for mastering the functional demands of movement tasks. Results from different lines of research showed that not only the structure formation of mental representations in long-term memory but also chunk formation in working memory are built up on BACs and relate systematically to movement structures. It is concluded that such movement representations might provide the basis for action implementation and action control in skilled voluntary movements in the form of cognitive reference structures. To simulate action implementation we discuss challenges and issues that arise when we try to replicate complex movement abilities in robots. Among the key issues to be addressed is the question how structured representations can arise during skill acquisition and how the underlying processes can be understood sufficiently succinctly to replicate them on robot platforms. Working towards this goal, we translate our findings in studies of motor control in humans into models that can guide the implementation of cognitive robot architectures. Focusing on the issue of manual action control, we illustrate some results in the context of grasping with a five-fingered anthropomorphic robot hand.

Keywords: cognitive psychology; mental representation; robotics; computational models; manual intelligence

Introduction

Mental (action) representation plays a central role in the control and implementation of actions. In different fields of action, it makes it possible to select and combine effective sources of information. Regardless of whether a surgeon has to select the appropriate instrument for an operation; a mechanic, a suitable tool for repairing an engine; or a basketball player, which member of the team to pass the ball to, actors have to use their mental representation as a basis to identify possible and functionally relevant sensory inputs. Frequently, this identification has to be made under extreme time pressure. Hence, mental

*Corresponding author.
Tel.: +49-521-106-5127; Fax.: +49-521-106-6432;
E-mail: thomas.schack@uni-bielefeld.de

DOI: 10.1016/S0079-6123(09)01319-3

representation has to be available quickly and provide clear criteria for selecting relevant pieces of information. At the same time, mental representation forms the functional basis for a meaningful and, thereby, task-related reduction in the large number of potential behaviors available to us and our social and technological systems. Mental representation does not just facilitate information selection, but also more generally permits a target-related and purposeful adaptation of behavioral potentials to conditions in the environment. In other words, mental representation helps to shape interaction patterns in purposeful ways. This also includes storing the cognitive perceptual outcomes of learning processes as items in long-term memory (LTM) (see Chapter 14: A conceptual framework for studying emotions–cognitions–performance linkage under conditions that vary in perceived pressure; Chapter 22: Advances in coupling perception and action: the quiet eye as a bidirectional link between gaze, attention, and action).

To gain a better understanding of the functionality of representation and categorization in action, this chapter starts with a model addressing the cognitive architecture of action. It then considers relevant issues in research methodology and presents methods that can be used to assess action-relevant knowledge structures experimentally. Experimental studies based on these methods are used to show relations between cognitive representation and performance in human movements. In particular, the results of experiments on manual actions are selected for implementation in robotic platforms. The conclusion argues that experimental approaches in the field of cognitive psychology and in the motor control area benefit from insights gained from exploring the possibilities and limits of artificial control architectures for robot systems.

Cognitive building blocks and the architecture of human action

Ever since Wilhelm Wundt (1893) and William James (1890), there has been a variety of studies on mental processes in action control, and different theoretical models have been proposed. Particularly during the last two decades, it has become clear that perceptual-cognitive representations mediate between the anticipated effect of a movement and current activity (see Schack and Tenenbaum, 2004a, b for an overview). These have been mapped out step-by-step in recent studies.

The fact that something like a "model of the needed future" (Bernstein, 1967) plays a central role in the implementation of action can be seen clearly in a set of studies addressing the *end-state comfort effect* (e.g., Rosenbaum and Jorgensen, 1992; Rosenbaum et al., 2004; Weigelt et al., 2006; Rosenbaum et al., 2007). This research has shown that individuals are prepared to adopt uncomfortable positions with their hands or arms when initiating and executing object manipulations (movement constellations) as long as this leads to a comfortable position for the final state of the movement. For example, to pick up a pencil that is pointing upwards in a cup, one uses an awkward underhand grip ultimately to hold the pencil in a comfortable writing posture. Such observations show clearly that movements are planned, controlled, and performed with reference to the anticipated final position of the movement. Hence, they indicate the existence of a *mental model* (of the needed future) to which all control processes can be related.

In his now classic work *O postrojenii dvizenij* (On the construction of movement), Bernstein (1947) presents the most comprehensive compilation of descriptive and experimental data on the functional mediation of the building blocks of the movement system available at that time. His detailed model of the interplay between, for example, movement goals, motor representations, and perceptual feedback is composed of several interdependent levels in a hierarchy headed by an object-related action organization level. Alongside Bernstein's (1947) approach to the construction of action, there have been several formulations of the idea that movement control is constructed hierarchically. One set of studies has focused on a hierarchy of different levels of representation (e.g., see Saltzman, 1979; Keele, 1986; Rosenbaum, 1987; Perrig and Hofer, 1989).

Other studies, in contrast, have focused more strongly on the aspect of a hierarchic execution regulation (e.g., Rosenbaum, 1987; Greene, 1988; Keele et al., 1990; Hacker, 1998). In contrast, the model proposed here views the functional construction of actions (Schack, 2004b; Schack and Bar-Eli, 2007; Schack and Hackfort, 2007) on the basis of a reciprocal assignment of performance-oriented regulation levels and representational levels (see Table 1). These levels differ according to their central tasks on the regulation and representation levels. Each level is assumed to be functionally autonomous.

The level of sensorimotor control (I) is linked directly to the environment. In contrast to the level of mental control (IV), which, as explained below, is induced intentionally, the level of sensorimotor control (I) is induced perceptually. The level of sensorimotor control is built on functional units composed of perceptual effect representations, afferent feedback, and effectors. The essential invariant (set value) of such functional units is the representation of the movement effect within the framework of the action. The system is broadly autonomous. Automatisms emerge when this level possesses sufficient correction mechanisms to ensure the stable attainment of the intended effect.

The need for a level of sensorimotor representation (II) is obvious in this context. It can be assumed that this is where, among others, the modality-specific information representing the effect of the particular movement is stored. The relevant modalities change as a function of the level of expertise in the learning process and as a function of the concrete task.

The level of mental representations (level III) predominantly forms a cognitive workbench for level IV, which is the *mental control level* and has already been sketched for voluntary movement regulation and the coding or the anticipated outcome of movement. Level III is organized conceptually, and it is responsible for transferring the anticipated action outcome into a movement program that suffices to bring about the desired outcome. Because an action is "no chain of details, but a structure subdivided into details" (Bernstein, 1988, p. 27, translated), action organization has to possess a working model of this structure.

So-called basic action concepts (BACs) have been identified as major representation units for movement control (Schack, 2004a, b; Schack and Mechsner, 2006). BACs are based on the cognitive chunking of body postures and movement events concerning common functions in realization of action goals. They do not refer to behavior-related invariance properties of objects, as this is the case in object concepts, but to perception-linked invariance properties of movements. Their characteristic set of features results from the perceptive and functional properties of action effects: they tie together their functional and sensory features. These functional features are derived from action goals, and this link connects BACs to level IV. However, BACs also integrate sensory features of submovements, for example, through chunking (see Verwey et al., 2009). As a result, they also refer to the perceptual effects of movements. This connects BACs with level II. Finally, the connection between BACs and sensory effect representations permits the intentional manipulation of the cognitive framing conditions of sensorimotor coordination.

All together BACs can be viewed as the mental counterparts of functionally relevant elementary components or transitional states of complex movements. They are characterized by recognizable perceptual features. They can be described verbally as well as pictorially, and can often be labeled with a linguistic marker. "Turning the head" or "bending the knees" might be examples of such basic action concepts in the case of, say, a complex floor exercise. As mentioned above, each individual BAC is characterized by a set of closely interconnected sensory and functional features. For example, a BAC in tennis like "whole body stretch motion" is functionally related to providing energy to the ball, transforming tension into swing, stretching but remaining stable, etc. Afferent sensory features of the corresponding submovement that allow monitoring of the initial conditions are bended knees, tilted shoulder axis, and body weight on the left foot. Reafferent sensory features that allow monitoring of whether the functional demands of the submovements have been addressed successfully are muscles stretched and under tension, proprioceptive, and, finally, perhaps visual perception of the swinging arm and ball in view.

This leads us to ask how we can conceive the mental structures responsible for complex movements. Is it possible to confirm mutual overlaps between representation structures and movement structures in humans (similar categorization in representation and movement) and is it possible to implement such results on robot platforms to produce human-like movements? To answer such questions in the further development of the paper, we shall move along different lines of empirical research. Because of its high stability over time, we like to start with representation in LTM.

Structure formation in LTM

Simplification in the domain of cognitive operations and movement structures is accompanied by order formation. Such order formation in action knowledge reduces the cognitive effort required to activate relevant information. In general, cognitive structures improve when more problem-solving-related classifications (concepts) are formed.

In the present perspective, we have to solve *movement tasks* purposefully and step-by-step within the framework of a voluntary organization of movement. Therefore, it is of interest to learn about the task-related order formation of action knowledge.

Concrete research questions can be derived from the architecture model presented here. We shall focus on two of these: (1) is it possible to use experimental methods to confirm specific expertise-dependent (and, hence, learning-level-dependent) structures in LTM? (2) Is the structure of mental representations in LTM formed more by the spatiotemporal effect structure of the movement or by muscle and joint constraints?

The experimental approach has been documented in several contributions (Schack, 2004a, b; Schack and Mechsner, 2006; Hodges et al., 2007; Schack and Hackfort, 2007). To introduce our method for measuring motor (action) representation, we have chosen a special action, the tennis serve, because it seems well suited for an investigation of potential conceptual representational structures at different levels of expertise. In the tennis serve, many degrees of freedom in the musculoskeletal system have to be controlled, and performance quality is influenced considerably by training and expertise. However, it is a finite, recognizable, and thereby flexible action pattern whose overall structure is well defined by biomechanical demands (Schack and Mechsner, 2006).

We start by characterizing the task-adequate functional organization of the tennis serve and formulating a plausible and workable set of basic action concepts in collaboration with nonplayers, athletes of different levels, and coaches. A tennis serve consists of three distinct phases that each fulfills distinct functional and biomechanical demands. In a pre-activation phase, body and ball are brought into position, and tension energy is stored to prepare for the strike. The following BACs are identified for this phase: (1) ball throw, (2) forward movement of the pelvis, (3) bending the knees, and (4) bending the elbow. In the following strike phase, energy is conveyed to the ball. The following BACs are identified: (5) frontal upper body rotation, (6) racket acceleration, (7) whole body stretch motion, and (8) hitting point. In the final swing phase, the balance is maintained and the racket movement is decelerated after the strike. The following BACs are identified: (9) wrist flap, (10) forward bending of the body, and (11) racket follow-through.

Because the usual rating and sorting methods do not permit a psychometric analysis of the representational structure, we developed an experimental method for probing mental representation structures (Lander and Lange, 1996; Schack and Schack, 2005). It has now been modified (structural dimensional analysis-motoric, SDA-M, Schack, 2004b) for the analysis of action representation. The SDA-M method consists of four steps: in a first step, the method seeks to gain information on the distance between selected representation units (BACs). Because it can be assumed that the structure of movement representations can only be explicated to a limited extent, this is done with a special splitting technique. Second, the structured relations between the N concepts were obtained by compiling a distance matrix through the scaling procedure presented above and subjecting it to a hierarchic cluster analysis. Third, a factor analysis is applied to reveal the dimensions in this structured set of BACs. Fourth, the cluster solutions are tested for invariance within and between groups (for more details, see Hodges et al., 2007).

In the following example, we examined three groups of tennis players: the expert group consisted of 11 males (mean age: 24 ± 3.7 years) playing in upper German leagues and ranked between places 15 and 500 in the German men's rankings. The low-level group consisted of 11 males (mean age: 26 ± 4.8 years) playing in lower German leagues (district leagues) and not listed in the German men's rankings. The nonplayer group consisted of 11 males (mean age: 24 ± 6.7 years) who had virtually no experience of the game (maximum: 5 hours), and had never taken any tennis lessons.

In the first step, participants were familiarized with the above-mentioned BACs by looking at pictures with a verbal BAC name as a printed heading. These pictures were positioned in front of each participant throughout the experiment. In order to determine subjective distances between the BACs, the participants performed the following split procedure as the first step in the SDA-M. On a computer screen, one selected BAC was presented constantly as an "anchoring unit" in red writing. The rest of the BACs were presented in yellow writing as a randomly ordered list. The participant judged whether each of these additional yellow-colored BACs was "functional related" (associated) to the red anchor BAC "while performing the movement" or not. This produced two subsets that were submitted to the same procedure repeatedly until the experimenter decided to do no further splits. As every BAC was used as an anchoring unit, this procedure resulted in 11 decision trees per participant. In the second step of the SDA-M, we submitted the above-mentioned BACs to a hierarchical cluster analysis with the distances based on subjective distance judgments of all combinations of pairs of BACs obtained in the previous step. As a result, we obtained the individual partitioning of the BACs. In the third step, the dimensioning of the cluster solutions was performed using a factor analysis applied to a specific cluster-oriented rotation process. This resulted in a factor matrix classified by clusters (see Schack and Mechsner, 2006). Finally, in the fourth step of the SDA-M, a within-group and between-group comparison was used to test for significant differences between cluster solutions using an invariance measure λ_{crit}. Alpha was set at 5% in all significance tests.

Figure 1 presents dendrograms for the subjective distances of BACs based on the hierarchical cluster analysis of the means of experts and nonplayers. Experts (Fig. 1a) showed a cognitive structure close to the functional structure of the tennis serve. The three functional phases (i.e., pre-activation, strike, and final swing) could be identified as distinct tree-like structured clusters in the dendrograms. Experts seemed to group the BACs in their memory according to generic terms that conformed to the solution of special movement problems. An invariance analysis (step four of SDA-M) confirmed this interpretation. There was no significant difference between the cognitive BAC framework in experts and the biomechanical demand structure of the movement ($\lambda = 0.70$; $\lambda_{crit} = 0.68$). Results were rather different in nonplayers (Fig. 1b): BAC clusterings did not reflect the functionally and biomechanically necessary phases so well. BACs were less clearly grouped, with no close neighborhoods, and the partial clusters usually failed to attain significance. The difference between the cognitive BAC framework and the functionally demanded structure of the action even attained significance in nonplayers ($\lambda = 0.31$; $\lambda_{crit} = 0.68$). All together we summarize, that in high-level experts, the representational frameworks were organized in a distinctive hierarchical tree-like structure, were remarkably similar between individuals, and were well matched with the functional and biomechanical demands of the task. In comparison, action representations in low-level players and nonplayers were organized less hierarchically, were more variable between persons, and were less well matched with functional and biomechanical

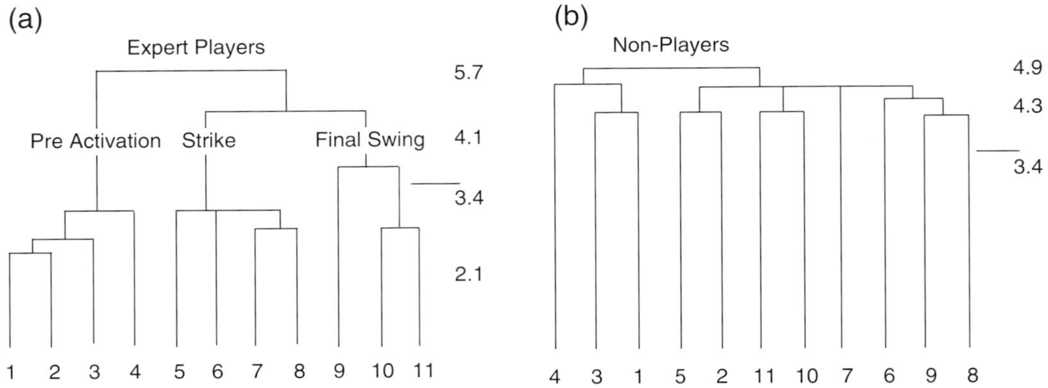

Fig. 1. Representation structures for two chosen expertise groups ((a) experts; (b) nonplayers) based on the hierarchical cluster analysis of basic action concepts (BACs) in the tennis serve. The horizontally aligned numbers denote the BACs (for the code, see text); the vertical numbers are the Euclidean distances. The lower the numbers, the lower the distances between BACs in long-term memory. For both groups, $n = 11$; $p = 0.05$; $d_{crit} = 3.46$ (Schack and Mechsner, 2006, reproduced with permission).

demands. This analysis is interesting to learn about the cognitive reference systems in action implementation. For instance in case of a non-functional representation we expect specific problems in action implementation.

Horizontal and vertical cooperation between cognitive systems and architecture levels

Results from two different lines of research addressing the mental representation level showed that not only the structure formation of representations in LTM but also chunk formation in working memory are built up on BACs and relate systematically to movement structures (Schack, 2004b). Experiments were designed to assess both the structure of mental representations in LTM (determined with the SDA-M) and chunking in working memory (determined with cognition and movement chronometry (CMC); Schack, 2004b). If the interaction assumption is true, identifying functional modules of the movement architecture with both groups of experiments should make it possible to match indications of structure in LTM with those in working memory. Results confirmed this, demonstrating that cognitive systems interact to produce complex movements. Our experiments have shown that both the order formation in LTM (Schack, 2004a, b; Schack and Mechsner, 2006; Schack and Bar-Eli, 2007; Schack and Hackfort, 2007) and the chunking in working memory (Schack, 2004b) build on the topological (spatio-temporal) structure of the movement. This provides experimental evidence that structures in movement and memory mutually overlap.

To get a closer understanding of the cognitive architecture of complex movements, it is not only interesting to know whether LTM and working memory cooperate horizontally on, for example, one level of mental representations. It is, for instance, also crucial to know whether there is a vertical cooperation between the level of mental representations and the level of sensorimotor control. One could ask whether biomechanically relevant features can be found in the structure of mental representations. Some of our studies have been designed to start a systematic search for paths between biomechanical aspects and mental movement representation (Schack, 2003; Heinen and Schack, 2004). This required us to develop methodological approaches to measure kinematic parameters *and* the structure of mental representations.

Experimental studies (Schack, 2003) showed that representational frameworks were organized in a hierarchical tree-like structure and revealed a good match with the biomechanical demands of the task. After measuring kinematic parameters, we investigated the relationship between the structure of motor representation and the kinematic parameters of different movements. Our studies (Heinen and Schack, 2003; Schack, 2003) have revealed significant correlations between kinematic parameters (time structure, angles according to the takeoff phase, tilt angle, angular velocities, etc.) of movement and the corresponding parts of mental representations. Hence, these results also suggest that there is a level in the organization of movement from which representations are translated directly into movement. According to this perspective, the representation structure can access all the topological properties that support the movement. This also means that no special translation mechanism is required between perception, representation, muscle control, and movement performance. All together, our experimental results support the hypothesis that voluntary movements are executed in a person–task–environment constellation and stored in memory directly through representations of their anticipated perceptual effects.

In the last years, the ongoing field of cognitive robotics offered new opportunities to study the construction and the functionality of cognitive representations at technical platforms. Research laboratories produced, for instance, impressive humanoid robots, robot musicians, dancing robots, robot arms, or brain–machine interfaces to study the organization and functioning of human action in more detail. Experimental approaches for investigating the action implementation in human actions benefit from insights gained from artificial control architectures for robot systems and vice versa (Pfeifer and Bongard, 2006). The research lines and the research areas created in the field of robotics aim to investigate systematically the principles needed to build artificial cognitive systems that can interact with a human in an intuitive way, including the acquisition of new skills by learning. Therefore, we have created a sustainable cooperation between research groups focusing on the more human aspects and research groups focusing on the more robot-like aspects of action generation and action control. At the Center of Excellence Cognitive Interaction Technology (CITEC, Bielefeld, Germany) and the Research Institute for Cognition and Robotics (CoR-Lab, Bielefeld, Germany), research cooperation is designed to study how the development of structured representation (action templates) proceeds in human skill acquisition and how it can be applied in robotics.

The challenge of cognitive robotics

The understanding of the neurocognitive architecture of actions based on experimental research is, on the one hand, an important step for applied fields, such as the mental preparation of athletes in high-performance sport (Schack et al., 2005; Schack and Bar-Eli, 2007; Schack and Hackfort, 2007) or decision making (see Chapter 20: Mental representations as an underlying mechanism for human performance). On the other hand, it is clearly linked to the ongoing field of *cognitive robotics* whose goal is to elevate the currently still rigid and rather narrow action repertoire of robots to a level where the robot can select and adjust its actions flexibly according to highly varying contexts, maintain a shared focus of attention with a human partner or instructor, and react to commands that are offered in a "natural" format, such as speech and demonstration.

Shaping the movements of advanced humanoid robots, or, more ambitiously, shaping their interaction with complex real-world environments, raises a substantial number of nontrivial research questions. Earlier approaches attempted to achieve closed-form solutions for robot actions on the basis of planning with synthetic world models that tried to capture as much detail as possible. These attempts have not lived up to their promises and made development any easier , for the insight that principles of biological motion control may have much to offer and may differ significantly from straightforward engineering approaches (Brooks, 1991; Hirose, 1993; Arkin, 1998). Historically, classical engineering approaches have been developed for systems where all relevant control parameters and system characteristics could be known and described in explicit models. However, when it comes to action in the real world, precise models are an unrealistic idealization and uncertainty about many parameters prevails, such as geometric shape, object weight, surface friction, or mechanical stiffness. For many common daily tasks (e.g., imagine dealing with a portion of spaghetti during lunch!) object models as they are currently in use in robotics with their focus on an explicit description of detailed material physics are simply unfeasible to parametrize properly.

Current engineering approaches also give rise to problems with regard to other important aspects of movement control. Traditionally favoring rigid mechanical parts for their "easier control," object contact becomes very difficult to handle since rigid objects when coming into contact very rapidly give rise to harmfully high contact forces. As a result, the best-developed area in robotics is the control and optimization of motion in the absence of any contact between robot arm and environment. However, when the robot has to do something useful, contacts become unavoidable in most cases and then either movements have to be performed excessively slowly or control strategies with cycle times in the sub-millisecond range need to be applied. Such rapid responses would be totally unfeasible for biological nervous systems — yet the capabilities if these "slow systems" for controlling contact movements are by a wide margin superior to current robotics approaches (Blickhan et al., 2007). Looking at how biological systems achieve their performance researchers have found that an essential part is the careful exploitation of elasticity: biological limbs are far from rigid and their elasticity properties are tightly combined with the underlying control architectures that then can work at much slower time scales. Some of these aspects have been addressed under the notion of "embodiment" (note: the term embodiment in computer science, robotics, and artificial intelligence (AI) emphasizes the role that the body and sensorimotor skills play in creating cognition and shaping the intelligent mind). This draws attention to the significant role that the body's physics can make when addressing the questions of: what needs to be computed for a movement and what can be obtained almost "for free" from suitably designed body physics? A deep understanding of these tradeoffs can enable what some researchers have very aptly termed "cheap design" (Pfeifer and Scheier, 1999): significant simplifications for the movement control architecture that are perhaps most impressively exemplified by the "passive walkers" — two-legged devices that can execute amazingly convincing downhill walking motion due to their body physics alone, without any need for sensors or joint actuations (McGeer, 1990).

Yet another important aspect is the role of world knowledge. Current robot control is largely focused on a very low level of abstraction that is very close to the sensors and actuators. In stark contrast, human actions are heavily informed by huge amounts of knowledge about the characteristics of the encountered objects, about goals, and about how to counteract the numerous disturbances and mishaps that usually occur during even moderately complex movements. It appears unlikely that this knowledge is laid down in the form of detailed geometrical or mechanical models leading to routes such as finite-element descriptions that would require a myriad of elements for describing situations such as the interaction between a fork and the spaghetti already mentioned. Instead, what is called for are much more "condensed" models that capture essential qualitative "topological" properties that can be robustly inferred and scaled into situations of realistic complexity. Useful concepts in this regard are projection techniques that map the high-dimensional interaction space of a robot manipulator onto a much lower-dimensional manifold which captures the essential degrees of freedom and which — due to its low dimensionality — makes planning and control much more feasible (Steffen et al., 2007).

Such manifolds can be seen as nonlinear generalizations of the concept of "principal components" — the familiar and widely used approach to create low-dimensional approximations of data distributions in some high-dimensional "feature space." At a yet higher level of abstraction manifolds as "nonlinear principal components" have to be generalized even further into what might be termed "principal interaction modes" — low-dimensional descriptions of the essential interaction patterns that can be connected with the current object.

While there is a large body of literature of how to deal with perceptual patterns and their analysis, comparably little attention has been focused on the representation and analysis of *interaction patterns*. Creating techniques for the systematic representation, analysis, and learning of interaction patterns constitutes a key task for advancing cognitive robotics and may turn out to be connected by ideas such as the concept of affordances.

While progress on this task is likely to benefit from many techniques developed in the more traditional field of pattern recognition and analysis, existing techniques are unlikely to be sufficient for the challenges ahead. Given the experiences briefly sketched above, any attempt to elevate the capabilities of technical systems into the domain of complexity observed in biological systems should carefully look into what can be known about working principles and strategies in such systems. This would allow benefit from the discoveries of biological evolution and augment our "Cartesian engineering" by approaches that may not fall along our natural axes of thinking but turn out to be useful in terms of performance. And, vice versa, some engineering approaches may turn out to have something to offer for describing or analyzing action control in biological systems, in the least by offering experimental platforms on which hypothesis about biological movement control can be put to more stringent tests than would be possible by theorizing or simulation alone (Adams et al., 2000).

Robot architectures

It is difficult to treat the above issues in isolation: control strategies, representations, and primitives chosen at the "lower levels" provide the basis on which the higher levels have to be erected and strongly shape the problems to be solved there. Moreover, due to the need also of top-down feedback influences, the converse is true as well. The ultimate challenge is the development of a suitable organizational architecture that supports the coordination of its functional parts while offering a sufficient amount of extensibility, for instance, to allow for abilities such as incremental addition of new functions or, more ambitiously, the acquisition of new skills through learning.

The case of human action control is a superb working example of such an architecture, realized under constraints (e.g., with respect to processing speed of its elements) that appear formidable from any engineering stand point. Ideas such as the previously sketched four-level organization (sensorimotor control (I), sensorimotor representation (II), mental representation (III), and mental control (IV) show that we may be at the verge of understanding some important structural aspects of this architecture with potentially far-reaching implications for the design of robot architectures that come closer to their natural counterpart. In contrast, our understanding of human action control is likewise highly fragmentary and partial at best. Advanced robots can provide a valuable platform to try out hypothesis about action organization "in the real" instead of being limited to realms of paper, pencil, and simulation only. Yet, research on robot architectures (Ritter et al., 2003) has, until now, proceeded relatively independently from cognitive architecture research in humans (Schack, 2004a, b). Reasons besides the notorious difficulty of truly interdisciplinary work are the, often very different, constraints posed by a technical robot platform as opposed to "human hardware." Only very recently, we begin to enjoy a broader availability of robot platforms whose body plan and movement capabilities at least offer coarse approximations to the human model. The perhaps most widely known example is the Honda ASIMO robot (Sakagami et al., 2002) which is regarded to be currently perhaps the farthest developed humanoid robot available today. The architecture of this robot is, however, proprietary to Honda and not open to the scientific public and the robot itself is currently not commercially available. Other major Japanese humanoid robot developments are the HRP-2, Kawada Industries, Inc. (Kaneko et al., 2004) that tries to replicate many of the features of the ASIMO robot, and the Sony QRIO humanoid (Geppert, 2004). The latter have been commercially available for some time and have become a research platform for a number of research laboratories. A very recent and entire open development is the iCub robot, Research group of Giulio Sandini, Italian Institute of Technology (IIT) (Vernon et al., 2007), which originated from a European research initiative to create an open platform for robot architecture research. In the near future, we are going to implement the measured representation of human tennis serve movements (see Fig. 1) on such robot platforms to produce human-like tennis serves.

Platforms such as these not only allow the testing of ideas about robot architectures and innovative control schemes but also create a very strong need for advances in this regard: even a "simple" humanoid robot easily comprises more than 30° of freedom, and more advanced designs confront us with the prospect of having to control in the order of 100° of freedom or more (which is still far less than the 600 or so muscles in the human body).

To cope with such levels of complexity requires principled approaches to structure robot architectures in ways that help to split the overall problem into more manageable chunks. An emerging structure that is close in spirit to the organization of human action control as described previously (see Table 1) is to partition the architecture into layers, each of which is focused on a different level of behavior organization. This shares the idea of a bottom-most sensorimotor control level providing low-level (level I; Table 1) control primitives (e.g., suitably parametrized feedback controllers). The issue of sensorimotor representations (see level II; Table 1) can be seen to be addressed at the next layer, where the sensorimotor control primitives of the first layer become packaged into "elementary behaviors," which then are used by a (usually top-most) planning layer to build task-directed action sequences. For instance, in the case of hand–arm control, the bottom-most sensorimotor layer would offer a set of controllers for individual joint motions such that prescribed low-level targets (hand position, stiffness against forces) are achieved. The second layer would coordinate the operation of several such controllers to realize behavioral primitives such as approaching an object, grasping, or inserting a held object in some target location. Finally, the third layer (level III; Table 1) would combine several such primitives in order to, for instance, fill a glass of water from a bottle. This still leaves room for a fourth layer (level IV in humans, Table 1) of "mental action control," which would have to host capabilities such as intentionality and a high-level situational understanding, issues for which current robotics offers only very tentative approaches so far.

The GRAVIS robot system that was developed in the context of the Collaborative Research Center 360: Situated Artificial Communicators in Bielefeld (Germany) may provide an illustration of some of the issues involved (Ritter et al., 2003; Fink et al., 2007). The goal of this system was to realize a feasible architecture for a robot system that combined several cognitive functions such that the robot became an "attentive communication partner" reacting to "natural" commands consisting of combined speech and manual gestures, such as pointing. The hardware consisted of a robot arm with a three-fingered hand moved in the viewing field of a binocular robot head. The software consisted of a three-layered system with basic sensorimotor, perceptual, and memory primitives at the lowest level. At the middle level, these primitives were connected into an "action graph" depicted in Fig. 2. This action graph represented the different ways how action primitives could be chained into useful subsequences. An additional top-level control system used this

Table 1. Levels of action organization (Schack, 2004b)

Code	Level	Main function	Subfunction	Means
IV	Mental control	Regulation	Volitional initiation control strategies	Symbols; strategies
III	Mental representation	Representation	Effect-oriented adjustment	Basic action concepts
II	Sensorimotor representation	Representation	Spatial-temporal adjustment	Perceptual effect representations
I	Sensorimotor control	Regulation	Automatization	Functional systems; basic reflexes

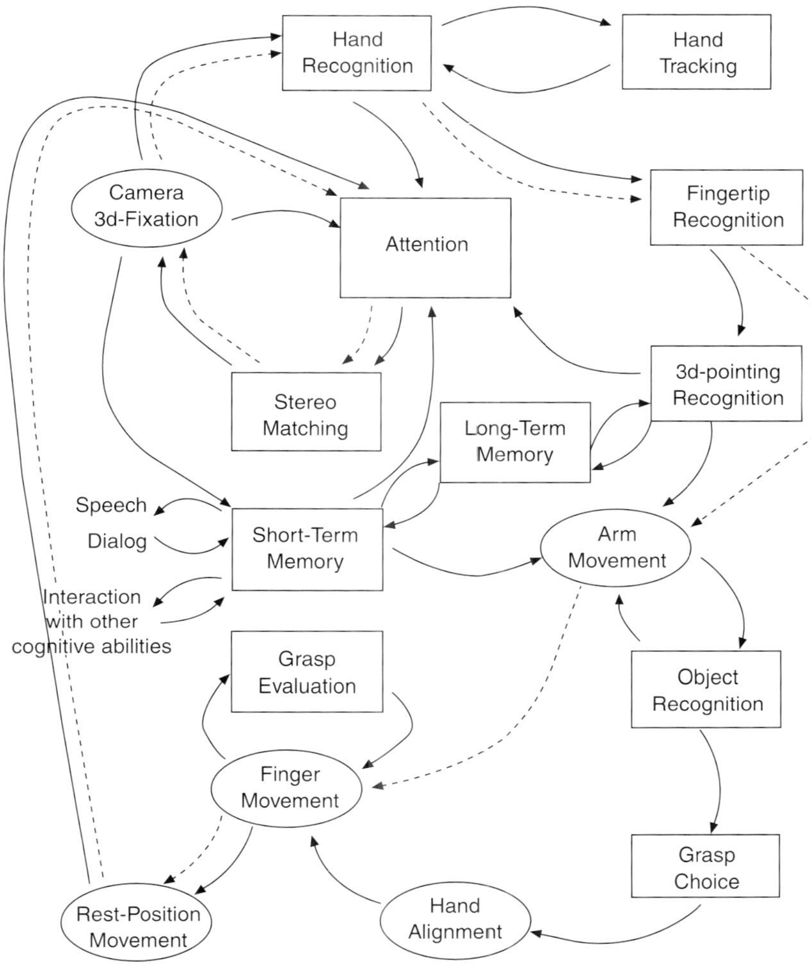

Fig. 2. Functional modules of the GRAVIS architecture. Memory modules (short-term memory; action representation in long-term memory) are integrated in the background. Altogether there are four module groups: speech processing/communication, attentive systems, memory, and motor control.

graph (together with a state memory not shown) to control the behavior of the overall system.

More recently, we have completed a major redesign of this system, extending the representation schemes of the middle and top-most layers in essential ways to improve the scalability of the system towards higher complexity (Ritter et al., 2007a). One major drive for this has been the transition to a more sophisticated robot hand (Shadow) that more closely mimics the kinematics of the human hand and that offers 20 degrees of freedom (DOF) to control. Additionally, we have now extended the original one-arm system to a bimanual system of such hands mounted on a pair of 7-DOF robot arms (Mitsubishi PA-10), leading to a system with 54 DOFs for the arms alone.

Working with this system made us aware that even subsets of our own motion intelligence may be sufficiently comprehensive to address and drive most of the research questions mentioned previously. One area that appears particularly rich and practically important at the same time is the understanding of *manual action*.

Manual action in robots and humans

We have argued elsewhere that a scientific understanding of manual action, or, more generally, *manual intelligence* might well play the role of a "Touch Stone" for cognitive systems research (Ritter et al., 2007a, b). This follows the idea that the development of cognition is linked closely with the capability of acting on one's environment and causing changes to it (Bernstein, 1967; Thelen and Smith, 1994; Klix, 1993). Therefore, we may expect that the need to control sophisticated manipulators, particularly in the form of arms and hands, have been a shaping factor for cognitive systems and the cognitive architecture in humans. Moreover, focusing on manual intelligence makes it inevitable to put *interaction* in the center of the research focus. This is complementary to classical AI, where intelligence was primarily equated with problem solving. As we now know, this focus on reasoning and logical operations caused a long deadlock and left out all the problems that have to be solved when actions are embedded in a physical world, under conditions of partial observability, high variability, and noise. From the perspective of robotics, it omitted precisely the "pre-rational" parts of intelligence (Cruse et al., 2000) that embodied robots require in the first place and as a basis for the more abstract intelligence functions to erect on.

In fact, in manual actions we find a most impressive integration of capabilities to shape physical interaction, comprising all levels ranging from micro to macro and even beyond: at the "micro" scale, we find the control of local finger contacts, involving different contact types and the exploitation of dynamic interaction patterns such as rolling and sliding. These local interactions become integrated into grasp patterns to constrain objects of widely varying shapes and firmness, or into haptic exploration behavior using controlled contact to identify objects and action affordances. Hand–eye coordination, bimanual coordination, and goal-directed sequences of manual actions introduce even more global levels of integration and give rise to the question how interaction patterns formulated originally at the level of physics can become connected with more abstract perspectives of action semantics, goal-directedness, and intentionality. On these higher levels, we find that hands also serve important roles in communication, thereby reaching even into the social sphere by contributing in an important way to the transfer of emotions and the experience of presence; qualities that have only more recently come into the focus of modern robotics.

Today, highly developed anthropoids, humans, and — to some extent — robots are able to perform manual actions. Manual actions allow for some sort of object-related organism–environment interaction and represent cognitive operations (classifications). Based on grasping movements, anthropoids are able to identify physically different objects as functionally equivalent tools for the solution of typical behavioral problems (such as the manipulation of a screw). Here, it has been possible to confirm the formation of concepts (terms without words; see Rensch, 1973). As current studies show, human grasping movements are cognitively represented on the basis of movement concepts (e.g., BACs) and build on effect-oriented target codes (in relation to space rather than to body; Schack, 2004b; Rosenbaum et al., 2007; Weigelt et al., 2009).

Hence, the perceptual-cognitive control of arm and hand movements has become a subject to be studied in the fields of psychology, biomechanics, cognitive motion sciences, and biocybernetics. To understand the biomechanical and cognitive background of manual action in more detail, we established research lines to determine movement primitives of manual action with the help of biomechanical and cognitive analyses. Because the production of manual actions is affected by a number of factors such as biomechanical constraints (Rosenbaum et al., 2001a, b; Weigelt et al., 2006), we have done some developmental studies on the relationship between biomechanical constraints like the end-state-comfort effect and the cognitive representation of grasping postures in children and adults (Weigelt et al., 2007; see also, Rosenbaum et al., 2007). Another group of studies have explored the link between semantic and motor memory especially in the case of grasping (Weigelt et al., 2009). In addition, a further line of studies focuses on the relationships between cognitive systems (like the representation of movement directions or executive systems) and sensorimotor adaptation in manual action (Lex et al., 2008). Cognitive robotics has a strong interest in questions regarding the segmentation of arm movements, the control of robot actuators via neural networks such as self-organizing maps (SOM; Barreto et al., 2003), and the combination of learning strategies for motion primitives in grasping movements (Steil et al., 2004).

Our recent research focuses in a first step on the question of how structures of sensory-motor representations are established and changed step-by-step in compliance with task constraints. It is of interest to learn about the relationship between the structure of mental representations and the performance in manual actions in humans and robots. Therefore, we studied how the development of structured representation (action templates) occurs in human skill acquisition, and how such research results can be applied in robotics. In a next step, we have to translate our findings from studies of human movement into sufficiently specific models that permit an implementation on a robotic system with a five-fingered dexterous hand. This research connection will be used in both directions: insights gained from the attempt to validate the hypothesis about action and representation structures in the robot learning scenario will be used to inform the design of experiments with human subjects. For instance, to learn about the "granularity" of cognitive building blocks in manual actions, we tried to gain a closer insight into the relationship between the structure of representations and the performance in manual actions, including situations in which actions result in error. Hence, an important linkage between cognitive architecture research in humans and robots deals with the cognitive benchmarks (hierarchical representations) of manual actions especially the representation of objects and grasping postures. Therefore, in a first step, we were interested in learning about the hierarchical representation of objects under different task constraints, and, complementary to this, to learn about the hierarchical representation of grasping movements (hierarchies of power and precision grips). The insights gained in these experiments will be implemented in robotic platforms (i.e., a seven DOF robot arm setup) and are clearly linked to such robot architectures as GRAVIS.

Grasping with human and with robot hands

The question of what is a good grasp and how such grasps may be formed is another shared focal point of researchers in robotics, neuroscience, and psychology (Shimoga, 1996; Rosenbaum et al., 2001a, b; Borst et al., 2005; Castiello, 2005). This has exposed fruitful interconnections between these disciplines: analytical approaches in robotics viewing grasp formation as a constrained selection of grasp points according to some optimization criterion (Borst et al., 2002, 2003) have found successful analogs in modeling aspects of human grasp selection (Smeets and Brenner, 1999). In the other direction, analysis and modeling of human reach-to-grasp behavior with respect to timing (Jeannerod, 1984) and the role of sensory guidance (Santello et al., 2002) has suggested low-dimensional "dynamic templates" for grasp behavior that are shaped by adjusting only a small number of parameters. Adopting such biologically motivated templates as behavioral primitives stimulated the realization of robust grasp behavior in robots (Hauck et al., 2000; Kragic and Christensen, 2003). With regard to the final grasping phase, these models replace the optimization-based grasp point selection by a dynamic finger closure process starting from a hand pre-shape and "wrapping" the fingers under tactile feedback around the object. This shares the idea of grasp generalization from prototypes (Pollard, 1996), but along a more behaviorally motivated route. A major issue then is the choice of a good hand pre-shape, which can be based on existing grasp taxonomies, such as power and precision grasps (Cutkosky and Wright, 1986). If this choice is carefully made, even as few as five different pre-shapes can enable the grasping of a wide range of different objects (Röthling et al., 2007), offering an approach to robust grasping in the absence of detailed object models. A more detailed study (Roethling, 2007), involving also measurements of human grasping, suggests further optimizations, such as the maximization of finger contact synchrony and thumb opposition. Finally, we only mention that grasping is connected with further nontrivial cognitive abilities, including the interplay of visual object recognition and nonvisual memory to predict object properties such as weight, firmness, and surface friction and the anticipation of the future state of the grasped object to constrain grasp choices properly (Weigelt et al., 2006), for instance, to minimize the need for regrasping.

Experimental research has taught us about the functional principles in human grasping (e.g., Jeannerod, 1984; Rosenbaum et al., 2001a, b, 2006). Rosenbaum and colleagues (2006) differentiate between grasping based on perception and grasping based on intention. Several phenomena have been described and investigated in grasping based on perception. For instance, when the human hand opens to reach for an object, the fingers open up much more than the thumb does (Rosenbaum et al., 2006, p. 10). However, the kinematics of the hand do not just depend on the object to be grasped. The way people reach for objects depends far more on what they plan to do with them. Such anticipatory effects were first observed by Leontjew and Zaporoshets (1960), and later by Marteniuk et al. (1987) who studied the speed of the hand as it either approached an object to be tossed or a different object to be screwed into a socket. Marteniuk and colleagues found that the speeds of hand approach differed under these two conditions. The hand moved more quickly when it approached the object to be tossed than when it approached the object to be screwed.

Preferring to examine how performance changes with just one object, Rosenbaum and colleagues (1990) showed that people chose different initial grips when reaching for the same rod depending on which end they planned to place on a disc on the table. Through this change in initial posture, participants avoided finishing their movements with awkward end postures (i.e., holding the rod with their thumb pointing down), even if this meant initially grasping the rod with an uncomfortable grip (i.e., an underhand grip). This tendency to avoid awkward postures at the final position of a movement was termed the *end-state comfort* effect (Rosenbaum et al., 1990).

Although these and many other principles are described and investigated for grasping with human hands, until now, they have been only partially implemented in robot hands. This is due to the technical limits of robot hands. Robot hand setups differ in terms of complexity (robot hand only vs. multimodal interaction capabilities), size, material, and number of DOF. There are differences not only in the size and structure of the hand, but also in the type and number of integrated actuators (modular vs. integrated systems). Examples for such hands are the Robonaut hand (Lovchik and Diftler, 1999), the Barrett hand (Townsend, 2000), or the Bielefeld Shadow hand (Shadow Robot Company, 2003; see Fig. 3). The number of DOF in these examples ranges between

(a) (b) (c) (d)

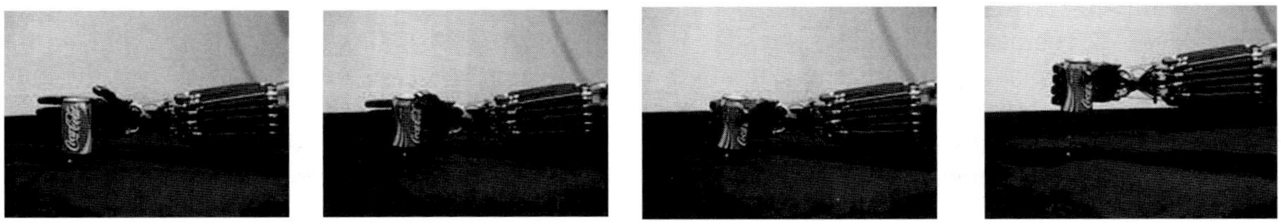

Fig. 3. An object is grasped with the Shadow hand using a grasp prototype. The target grasp is applied after the hand reached the grasp position. In humans, grasping starts immediately after the finger touches the object. (a) Pre-grasp position, (b) grasp position, (c) object grasped, (d) object lifted. Reproduced from Roethling (2007).

4 (Barrett hand) and 20 for the Bielefeld Shadow hand (Steil et al., 2004; Roethling, 2007).

To enable a flexible kind of grasping performance, we propose a taxonomy of grasping prototypes that has been carefully matched to our Shadow robot hand. This taxonomy of grasping postures is based on different research sources (e.g., Cutkosky and Wright, 1986; Cutkosky and Howe, 1990; Iberall, 1997). As we can see in Fig. 4, the differentiation between precision and power grasp is an important element of the grasping taxonomy by Cutkosky.

The taxonomy of Cutkosky is based on observations of grasping movements in a population of mechanics. However, up until now, it is unknown how such grasping postures are represented in human LTM. Therefore, we have done some explorative studies on the representation of grasping postures in human LTM in order to improve our recent taxonomy in our robot

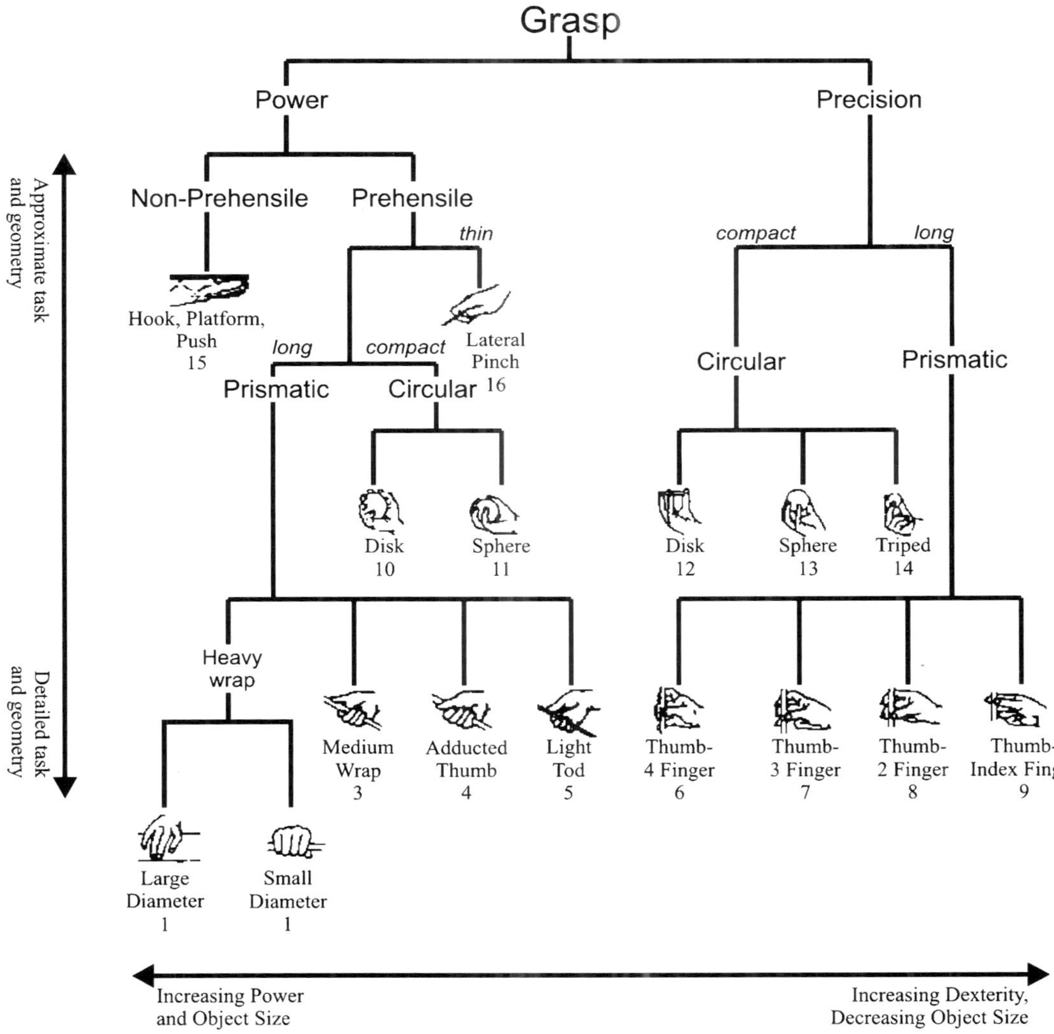

Fig. 4. A modified version of the taxonomy of grasps originally from Cutkosky and Howe (1990). This taxonomy is an important reference point for recent robot grasping taxonomies.

Shadow hand. We studied the representation of grasping postures in 18 subjects with the help of the SDA-M method. We defined the grasping postures of Cutkosky (Fig. 4) as relevant units (BACs) for our experimental procedure. We used not only verbal labels and pictures of these grasping postures but also real moves simulating the grasping of the pictorially shown objects.

As we can see in Fig. 5, the results of our experimental analysis of one student with many experiences as a mechanic show a clear similarity between the structure of mental representation of manufacturing grasps and the structure of the taxonomy following Cutkosky. Here we found a clear differentiation between power-oriented and precision-oriented grasping postures. Only grasping positions 15 and 16, which are located at the highest level of the Cutkosky taxonomy, are not integrated in the representation structure. However, we could argue that these grasping postures are also located at a separate (isolated) position in the Cutkosky taxonomy.

This type of integrated (interdisciplinary) research allows us to explore experimentally the interactions of action representation in memory (simulated with different kinds of artificial neural networks) and motor skills in the context of real-world tasks. Based on the present studies on the mental representation of grasping postures in humans and other experimental studies with the robot hand (Röthling, 2007), we will be able to refine robot hand grasping step-by-step.

Studies on the biomechanical functions and the cognitive representation of movement primitives (in the form of BACs) help us to differentiate further the taxonomy grasps with which the robot hand is equipped. For instance, when grasping objects with the robot hand, in some cases, one finger got stuck in front of one of the blades of a toy propeller (one of our grasping objects) rather

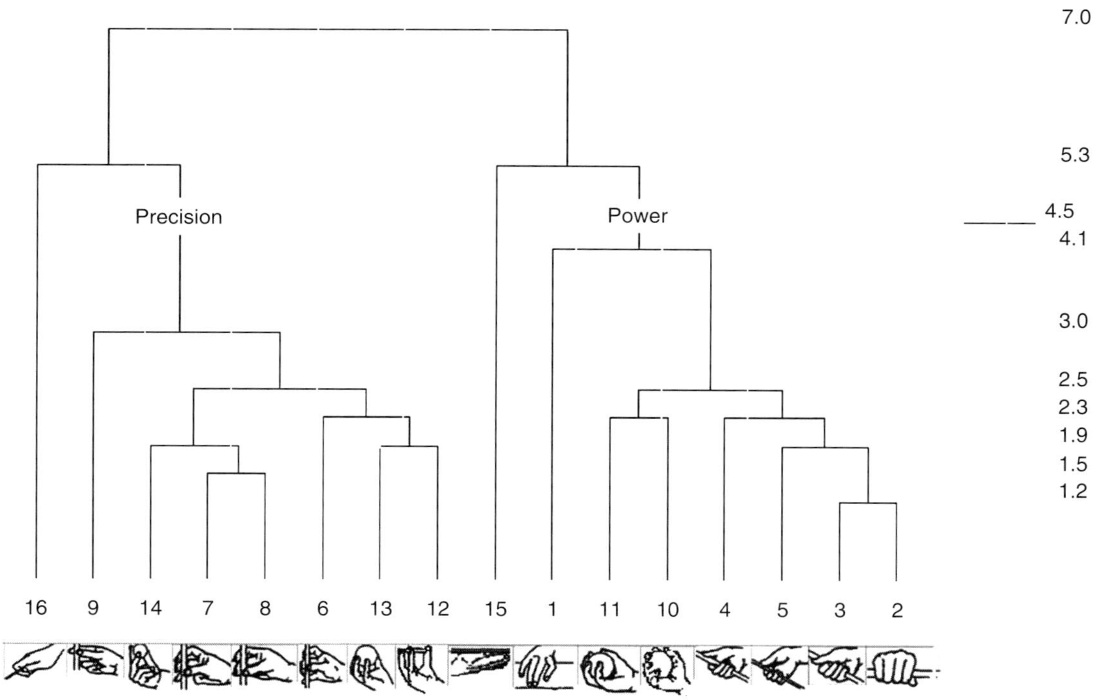

Fig. 5. An individual mechanic's representation structure (subject 7) of grasping postures ($d_{crit} = 4.3; = 5\%$). The lower the value of a diagonal connection between basic action concepts (BACs) (compare the value scale of Euclidian distances on the right), the lower the distance between concepts.

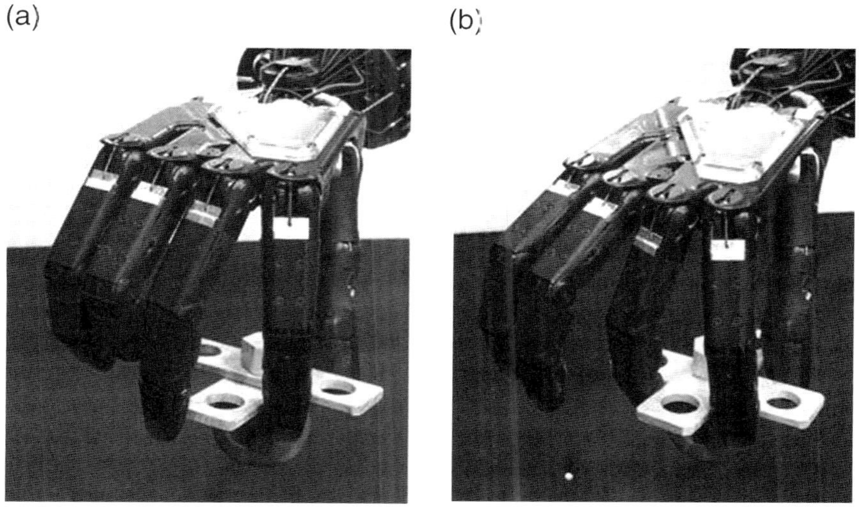

Fig. 6. Standard grasp (a) and a further developed three-finger special grasp of the Shadow hand grasping a toy propeller (b). While in (a), the middle finger touches the propeller at its blade, in (b), the fingers reach between the blades (see Roethling, 2007).

than successfully grasping the object between the blades (Röthling, 2007). Some of the reasons for this problem involve the represented grasping postures of the robot hand. Experimental studies on human grasping movements may well convey and lead to the implementation of a new and appropriate representation structure and a grasping posture assigned to it. This can be used to extend the robot hand's repertoire of grasping prototypes and, thereby, improve the robot's hand functionality (see Fig. 6).

Conclusions

As we can see in our research examples, experimental approaches for investigating the cognitive architecture of human actions benefit from insights gained from artificial control architectures for robot systems and vice versa. The lines of research presented here can help us not only to understand the cognitive background of human actions but also to improve human movements via, for instance, mental training. This research forms the basis for building artificial cognitive systems that can interact with a human in a more intuitive way and also acquire new skills by learning. Interestingly current robot technology has matured to the point at which it can approximate a reasonable spectrum of specialized perceptual, cognitive, and motor capabilities, allowing exploration of architectures for the integration of these functions into robot action control. This provides us the opportunity to fit models dealing with perception, representation, motor control, or decision making in human action together with architectures generated for robot actions. All together we are part of a process in which we are not only using our knowledge in computers but starting now to give our action skills to robots to learn more about us.

Acknowledgments

We wish to thank Christian Seegelke (Bielefeld University) for his work in the study of grasping postures in 18 subjects using the SDA-M method.

References

Adams, B., Breazeal, C., Brooks, R. A., & Scassellati, B. (2000). Humanoid robots: a new kind of tool. *Intelligent Systems and their Applications*, *15*, 25–31.

Arkin, R. C. (1998). *Behavior-based robotics*. Cambridge, MA: MIT Press.

Barreto, G., Araujo, A., & Ritter, H. (2003). Self-organizing feature maps for modeling and control of robotic manipulators. *Journal of Intelligent and Robotic Systems*, *36*, 407–450.

Bernstein, N. A. (1947). *O postrojenii dviženij (Ueber den Aufbau der Bewegungen) [On the construction of movement]*. Mozkva: Medgiz.

Bernstein, N. A. (1967). *The coordination and regulation of movements*. Oxford: Pergamon Press.

Bernstein, N. A. (1988). *Bewegungsphysiologie* (2.Aufl.). Leipzig: Barth.

Blickhan, R., Seyfarth, A., Geyer, H., Grimmer, S., Wagner, H., & Günther, M. (2007). Intelligence by mechanics. *Philosophical Transactions. Series A, Mathematical, Physical, and Engineering Sciences*, *365*, 199–220.

Borst, C., Fischer, M., & Hirzinger, G. (2002). Calculating hand configurations for precision and pinch grasps. *Proceedings of the 2002 IEEE/RSI/GI International Conference on Intelligent Robots and Systems*, 1553–1559.

Borst, C., Fischer, M., & Hirzinger, G. (2003). Grasping the dice by dicing the grasp. *Proceedings of the 2003 IEEE/RSJ International Conference on Intelligent Robots and Systems*, 3692–3697.

Borst, C., Fischer, M., & Hirzinger, G. (2005). Efficient and precise grasp planning for real world objects. In F. Barbagli, D. Prattichizzo, & K. Salisbury (Eds.), *Multi-point interaction with real and virtual objects* (pp. 91–111). Berlin: Springer.

Brooks, R. (1991). Intelligence without representation. *Artificial Intelligence*, *47*, 139–159.

Castiello, U. (2005). The neuroscience of grasping. *Nature Review Neuroscience*, *6*, 726–736.

CITEC Cognitive Interaction Technology Excellence Center: Homepage: http://www.cit-ec.org

CoR-Lab Research Institute for Cognition and Robotics: Homepage: http://www.cor-lab.de

Cruse, H., Dean, J., & Ritter, H. (2000). *Prerational intelligence — adaptive behavior and intelligent systems without symbols and logic* (Vol. 1–3). Dordrecht: Kluwer Academic Publishers.

Cutkosky, M. R., & Howe, R. D. (1990). Human grasp choice and robotic grasp analysis. In S. T. Venkatamaran & T. Iberall (Eds.), *Dextrous robot hands* (pp. 5–31). New York: Springer.

Cutkosky, M. R., & Wright, P. K. (1986). Modelling manufacturing grips and correlations with the design of robotic hands. *Proceedings of the 1986 IEEE International Conference on Robotics and Automation, San Francisco, CA*, 1533–1539.

Fink, G., Fritsch, J., Leßmann, N., Ritter, H., Sagerer, G., Steil, J., & Wachsmuth, I. (2007). Architectures of situated communicators: from perception to cognition to learning. In G. Rickheit & I. Wachsmuth (Eds.), *Situated communication* (pp. 357–376). Berlin: Mouton de Gruyter.

Geppert, L. (2004). Qrio, the robot that could. *IEEE Spectrum*, *41*, 34–37.

Greene, P. H. (1988). The organization of natural movement. *Journal of Motor Behavior*, *20*, 180–185.

Hacker, W. (1998). *Allgemeine Arbeitspsychologie [General industrial psychology]*. Bern: Huber.

Hauck, A., Passig, G., Schenk, T., Sorg, M., & Färber, G. (2000). On the performance of a biologically motivated visual control strategy for robotic hand–eye coordination. *Proceedings of the 2000 IEEE/RSJ International Conference*, *3*, 1626–1632.

Heinen, T., & Schack, T. (2003). Movement organization and kinematics in sports: in-group-coherences and inter-subject-differences. In F. Schmalhofer, R. M. Young, & G. Katz (Eds.), *Proceedings of the EuroCogSci 03* (p. 396). New York: Lawrence Erlbaum Associates.

Heinen, T., & Schack, T. (2004). Bewegungsgedächtnis und Bewegungsausführung – Optimierung von Rotationsbewegungen im Gerätturnen. In: H. Roscher (Hrsg.), *Lehren und Lernen im Turnen. Veröffentlichungsband zur Jahrestagung 2002 der DVS Kommission Turnen vom 23.-25. September 2002 in Kienbaum* (pp. S.85–S.96). Hamburg: Czwalina.

Hirose, S. (1993). *Biologically inspired robotics*. Oxford: Oxford University Press.

Hodges, N. J., Huys, R., & Starkes, J. L. (2007). Methodological review and evaluation of research in expert performance in sport. In G. Tenenbaum & R. C. Eklund (Eds.), *Handbook of sport psychology* (3rd ed., pp. 161–183). New York: Wiley.

Iberall, T. (1997). Human prehension and dexterous robot hand. *International Journal of Robotics Research*, *16*, 285–299.

James, W. (1890). *The principles of psychology* (Vol. 1). New York: Holt.

Jeannerod, M. (1984). The timing of natural prehension movements. *Journal of Motor Behavior*, *16*, 235–254.

Kaneko, K., Kanehiro, F., Kajita, S., Hirukawa, H., Kawasaki, T., Hirata, M., Akachi, T., & Isozumi, T. (2004). Humanoid robot HRP-2. *Proceedings of the 2004 IEEE International Conference on Robotics and Automation Conference*, 1083–1090.

Keele, S. W. (1986). Movement control in skilled motor performance. *Psychological Bulletin*, *70*, 387–403.

Keele, S. W., Cohen, A., & Ivry, R. (1990). Motor programs: concepts and issues. In M. Jeannerod (Ed.), *Attention and performance XIII* (pp. 77–110). Hillsdale, NJ: Lawrence Erlbaum Associates.

Klix, F. (1993). *Erwachendes Denken [Think awakening]*. Heidelberg: Spektrum.

Kragic, D., & Christensen, H. I. (2003). Biologically motivated visual servoing and grasping for real world tasks. *Proceedings of the IEEE/RSJ International Conference on Intelligent Robots and Systems*, *4*, 3417–3422.

Lander, H. J., & Lange, K. (1996). Untersuchung zur Struktur- und Dimensionsanalyse begrifflich-repräsentierten Wissens [Studies on the structure and dimension analysis conceptually represented knowledge]. *Zeitschrift für Psychologie*, *204*, 55–74.

Leontjew, A. N., & Zaporoshets, A. V. (1960). *Rehabilitation and hand function*. London: Pergamon.

Lex, H., Weigelt, M., & Schack, T. (2008). Mentale Repräsentationen von Bewegungsrichtungen und deren Einfluss auf sensomotorische Adaptationsleistungen [Mental representations of directions of movement and their influence on sensory adaptation benefits]. In G. Sudeck, A. Conzelmann, K. Lehnert, & E. Gerlach (Eds.), *Schriften der Deutschen Vereinigung für Sportwissenschaft: Vol. 176. Differentielle SportpsychologieSportwissenschaftliche Persönlichkeitsforschung [Writings of the German Association of Sports Science: Vol. 176. Differential Sports Psychology — Sports Personality Scientific Research]* (p. 94). Hamburg: Czwalina Verlag.

Lovchik, C. S., & Diftler, M. A. (1999). The robonaut hand: a dexterous robot hand for space. *Proceedings of the International Conference on Robotics and Automation (ICRA), 2*, 907–912. Detroit, Michigan: IEEE.

Marteniuk, R. G., MacKenzie, C. L., Jeannerod, M., Athenes, S., & Dugas, C. (1987). Constraints on human arm movement trajectories. *Canadian Journal of Psychology, 4*, 365–378.

McGeer, T. (1990). Passive dynamic walking. *International Journal of Robotics Research, 9*, 62–82.

Perrig, W. J., & Hofer, D. (1989). Sensory and conceptual representations in memory: motor images that cannot be imaged. *Psychological Research, 51*, 201–207.

Pfeifer, R., & Bongard, J. (2006). *How the body shapes the way we think — a new view of intelligence*. Cambridge, MA: MIT Press.

Pfeifer, R., & Scheier, C. (1999). *Understanding intelligence*. Cambridge, MA: MIT Press.

Pollard, N. S. (1996). Synthesizing grasps from generalized prototypes. *Proceedings of the International Conference on Robotics and Automation Conference, 3*, 2124–2130.

Rensch, B. (1973). *Handgebrauch und Verstaendigung bei Affen und Fruehmenschen [Hand and understanding early in monkeys and humans]*. Bern: Huber.

Ritter, H., Haschke, R., Röthling, F., & Steil, J. J. (2007b). Manual intelligence as a Rosetta Stone for robot cognition. *Proceedings of the 13th International Symposium of Robotics Research (ISRR) 2007*, 173–184.

Ritter, H., Haschke, R., & Steil, J. J. (2007a). A dual interaction perspective for robot cognition: grasping as a "Rosetta Stone". In: B. Hammer, & P. Hitzler (Eds), *Perspectives of neural-symbolic integration. Springer Series Studies in Computational Intelligence (SCI) 77*, 159–178.

Ritter, H., Steil, J. J., Noelker, C., Roethling, F., & McGuire, P. (2003). Neural architectures for robotic intelligence. *Reviews in the Neurosciences, 14*, 121–143.

Roethling, F. (2007). *Real robot hand grasping using simulation-based optimisation of portable strategies*. Unpublished doctoral dissertation, University of Bielefeld.

Rosenbaum, D. A. (1987). Successive approximations to a model of human motor programming. *Psychology of Learning and Motivation, 21*, 153–182.

Rosenbaum, D. A., Cohen, R. G., Jax, S. A., Van Der Wel, R., & Weiss, D. J. (2007). The problem of serial order in behavior: Lashley's legacy. *Human Movement Science, 26*, 525–554.

Rosenbaum, D. A., Cohen, R. G., Meulenbroek, R. G., & Vaughan, J. (2006). Plans for grasping objects. In M. Latash & F. Lestienne (Eds.), *Motor control and learning over the lifespan* (pp. 9–25). New York: Springer.

Rosenbaum, D. A., & Jorgensen, M. J. (1992). Planning macroscopic aspects of manual control. *Human Movement Science, 11*, 61–69.

Rosenbaum, D. A., Marchak, F., Barnes, H. J., Vaughan, J., Slotta, J., & Jorgensen, M. (1990). Constraints for action selection: overhand versus underhand grips. In M. Jeannerod (Ed.), *Attention and performance XIII: motor representation and control* (pp. 321–342). Hillsdale, NJ: Lawrence Erlbaum Associates.

Rosenbaum, D. A., Meulenbroek, R., & Vaughan, J. (2001a). Planning reaching and grasping movements: theoretical premises and practical implications. *Motor Control, 2*, 99–115.

Rosenbaum, D. A., Meulenbroek, R. G., Vaughan, J., & Jansen, C. (2001b). Posture-based motion planning: applications to grasping. *Psychological Review, 108*, 709–734.

Rosenbaum, D. A., Meulenbroek, R. G. J., & Vaughan, J. (2004). What is the point of motor planning? *International Journal of Sport and Exercise Psychology, 2*, 439–469.

Röthling, F. (2007). *Real robot hand grasping using simulation-based optimisation of portable strategies*. Unpublished dissertation, Faculty of Technology, Bielefeld University, Germany.

Röthling, F., Haschke, R., Steil, J. J., & Ritter, H. (2007). Platform portable anthropomorphic grasping with the Bielefeld 20 DOF shadow and 9 DOF TUM Hand. *Proceedings of the IEEE IROS Conference*, 2951–2956.

Sakagami, Y., Watanabe, R., Aoyama, C., Matsunaga, S., Higaki, N., & Fujimura, K. (2002). The intelligent ASIMO: system overview and integration. *IEEE/RSJ IROS Conference Proceedings, 3*, 2478–2483.

Saltzman, E. (1979). Levels of sensorimotor representation. *Journal of Mathematical Psychology, 20*, 91–163.

Santello, M., Flanders, M., & Soechting, J. F. (2002). Patterns of hand motion during grasping and the influence of sensory guidance. *Journal of Neuroscience, 22*, 1426–1435.

Schack, T. (2003). The relationship between motor representation and biomechanical parameters in complex movements — towards an integrative perspective of movement science. *European Journal of Sport Science, 2*, 1–13.

Schack, T. (2004a). Knowledge and performance in action. *Journal of Knowledge Management, 8*, 38–53.

Schack, T. (2004b). The cognitive architecture of complex movement. *International Journal of Sport and Exercise Psychology; Special Issue Part II: The Construction of Action — New Perspectives in Movement Science, 2*, 403–438.

Schack, T., & Bar-Eli, M. (2007). Psychological factors in technical preparation. In B. Blumenstein, R. Lidor, & G. Tenenbaum (Eds.), *Psychology of sport training* (pp. 62–103). Oxford: Meyer & Meyer Sport.

Schack, T., & Hackfort, D. (2007). An action theory approach to applied sport psychology. In G. Tenenbaum &

R. C. Prinz (Eds.), *Handbook of sport psychology* (3rd ed., pp. 332–351). New York: Wiley.

Schack, T., & Mechsner, F. (2006). Representation of motor skills in human long-term memory. *Neuroscience Letters, 391*, 77–81.

Schack, T., & Schack, E. (2005). In- and outgroup representation in a dynamic society: Hong Kong after 1997. *Asian Journal of Social Psychology, 8*, 123–137.

Schack, T., & Tenenbaum, G. (Eds.) (2004a). The construction of action — new perspectives in science. *Special Issue Part I: Perceptual and Cognitive Control. International Journal of Sport and Exercise Psychology*, 2.

Schack, T., & Tenenbaum, G. (Eds.) (2004b). The construction of action — new perspectives in science. *Special Issue Part II: Representation and Planning. International Journal of Sport and Exercise Psychology, 2.*

Schack, T., Whitmarsh, B., Pike, R., & Redden, C. (2005). Routines. In J. Taylor & G. Wilson (Eds.), *Applying sport psychology: four perspectives* (pp. 137–150). Champaign, IL: Human Kinetics.

Shadow Robot Company. (2003). *Design of a Dextrous hand for advanced CLAWAR applications*. CLAWAR Conference 2003, Catania, Italy. Available at http://www.shadow.org.uk/

Shimoga, K. B. (1996). Robot grasp synthesis algorithms: a survey. *International Journal of Robotics Research, 15*, 230–266.

Smeets, J. B., & Brenner, E. (1999). A new view on grasping. *Motor Control, 3*, 237–271.

Steffen, J. F., Haschke, R., & Ritter, H. (2007). Experience-based and tactile-driven dynamic grasp control. *IEEE IROS Conference proceedings*, pp. 2938–2943.

Steil, J., Roetling, F., Haschke, R., & Ritter, H. (2004). Situated robot learning for multi-modal instruction and imitation of grasping. *Robotics and Autonomous Systems, Special Issue on Robot Learning by Demonstration, 47*, 129–141.

Thelen, E., & Smith, L. B. (1994). *A dynamic systems approach to the development of cognition and action*. Cambridge, MA: MIT Press.

Townsend, W. T. (2000). MCB — industrial robot feature article — Barrett hand grasper. *Industrial Robot: An International Journal, 27*, 181–188.

Vernon, D., Metta, G., & Sandini, G. (2007). The iCub cognitive architecture: interactive development in a humanoid robot. *The 6th International IEEE Conference on Development and Learning (ICDL)*, 122–127.

Verwey, W. B., Abrahamse, E. L., & Jiménez, L. (2009). Segmentation of short keying sequences does not spontaneously transfer to other sequences. *Human Movement Science.*

Weigelt, M., Kunde, W., & Prinz, W. (2006). End-state comfort in bimanual object manipulation. *Experimental Psychology, 53*, 143–148.

Weigelt, M., Rosenbaum, D. A., Huelshorst, S., & Schack, T. (2009). Motor planning and memorization. Acta Psychologica (under review).

Weigelt, M., Wolfgram, M., & Schack, T. (2007). End-state Comfort Effekt und motorische Entwicklung. *Abstract-Band*, S. 23.

Wundt, W. (1893). *Grundzuege der Physiologischen Psychologie [Principles of Physiological Psychology]* (Vol. II). Leipzig: Engelmann.

M. Raab et al. (Eds.)
Progress in Brain Research, Vol. 174
ISSN 0079-6123

CHAPTER 20

Mental representations as an underlying mechanism for human performance

Gershon Tenenbaum* and William M. Land

Department of Educational Psychology and Learning Systems, Florida State University, Tallahassee, FL, USA

Abstract: This chapter presents a theoretical framework, which is supported by empirical evidence, where changes in human performance are accounted for by changes in mental representation structure (MRS). More specifically, the knowledge base, represented in the form of mental representations, controls the perceptual, cognitive, and motor systems when interacting with the environment. Once this interaction induces pressure, changes in the MRS lead to respective changes in the function of attention, anticipation, long-term working memory, the control system, and the motor systems. Such changes can be detected via both overt and covert behavior of the human system. This chapter presents the theoretical frameworks and accompanies them with graphical illustrations.

Keywords: representation; motor learning; variability; motor control; memory

Introduction

The structural components of human performance, such as emotional processes (i.e., feelings, mood), cognitive processes and structures (e.g., knowledge architecture, long-term working memory), motor processes (coordination, endurance), and the neurophysiological basis of these structural components (i.e., activation of cortical areas) have been studied independently. Our attempt is to integrate these structural components into a unified theoretical framework that enables a better understanding of human performance, which allows for generating applications that share scientific validity. Our working assumption is that every action made by humans is a consequence of response selection, whether intentional or unintentional. By definition, response selection indicates adaptive behavior based upon the capacity to solve problems. Cognitive processes and mental operations underlie this "behavioral effectiveness." The effectiveness of these processes consists of the richness and variety of perceptions processed at a given time; that is, the system's capacity to encode (store and represent) and access (retrieve) information relevant to the task being performed (Tenenbaum, 2003). Because tasks vary with respect to unique characteristics and requirements, it is assumed that the nature and integration of perceptual-cognitive components required for decision making and action execution are also unique, though they may share similar architecture when the tasks include common elements. From an information-processing perspective, motor

*Corresponding author.
Tel.: +1-850-644-8791; Fax: +1-850-644-8776;
E-mail: tenenbau@coe.fsu.edu

DOI: 10.1016/S0079-6123(09)01320-X

behaviors consist of encoding relevant environmental cues through the utilization of attentional strategies, processing information through an ongoing interaction between long-term and working memory, making action-related decisions, and executing actions while leaving room for refinements and modifications.

Under pressure, changes in each functional component may occur. These changes can affect the perceptual components, continuing with the cognitive components, and ending with the motor system. However, each of these components can be understood as a decision-making component, which has consequences for the final decision and action to be taken. To capture changes in the perceptual-cognitive–motor linkage under varied conditions of pressure and evoked emotions, we must use research paradigms that integrate the cognitive structure components and processes (cognitive appraisal), emotional system, and the self-regulation structure (i.e., emotional control, motivation control, attentional control, etc.). Both idiographic and group concepts are presented, allowing for detection of a collapse in the perceptual-cognitive linkage under altered emotional states, and their subsequent effects on the motor system.

Decision-making processes as a conceptual framework for studying actions

In most cases, the motor system responds to intentional cognitive processes, which are goal-directed, and thus rely on mental regulation. These mental regulations take into consideration person-related characteristics, task demands, and conditions under which the task is carried out (e.g., environment) (see Schack and Hackfort, 2007 for review). According to this concept, every action consists of some kind of mental representation network, which not only consists of action representation, but also relies on a mental schema with neural pathways to information labeled as emotions, actions, and coping plans. Even when motor movement becomes automated, intentions still guide actions through mental representation, which in this case are not extensively activated, allowing the perceptual-attention–cognitive system to work in parallel. Thus, one should consider the person-task-environment within a system framework where social and physical systems must be taken into consideration for capturing the perceptual-cognitive–motor linkage (Schack and Hackfort, 2007). Figure 1 captures the basic framework of action theory (Nitsch and Hackfort, 1981; Newell, 1986; Nitsch, 2004). One should consider all components of the system as dynamic components, which alter in time and space. However, changes in restructuring the neural network defined as a schema take time, and depend largely on the amount and frequency of exposure to similar tasks and environmental conditions.

An important tenet of intentional (goal)-directed motor action assumes that voluntary movements are stored in memory and contain anticipated consequences in the form of neural schema, and thus are of extreme relevance to the control of movement production (Schack and Tenenbaum, 2004a, b). Thus, when an action must be initiated, planned, and executed, the sensory system, mainly the visual system, is guided by the neural schema through long-term working memory (Ericsson and Kintsch, 1995). Access to long-term working memory consists of two routes: a knowledge-based route and a retrieval-based route. These allow the neural system to control and guide the visual system efficiently under conditions that vary in mental, temporal, and environmental pressure. The control of the visual system allows anticipatory decisions to be made, and consequently actions to be retrieved from long-term memory (LTM). Furthermore, knowledge structures in the form of mental representations not only allow efficient control over decision making and action, but allow the perceptual-cognitive–motor linkage for anticipated changes to occur once fast changes in the environment necessitate such alteration. When such representations are nonexistent, alternative decision plans and actions are likely to result in failure of the motor system to respond appropriately (see Tenenbaum, 2003 for details). Furthermore, mental representations consist of action plans associated with affective and motivational components, such as self-efficacy. If an appropriate action is retrieved, it may still

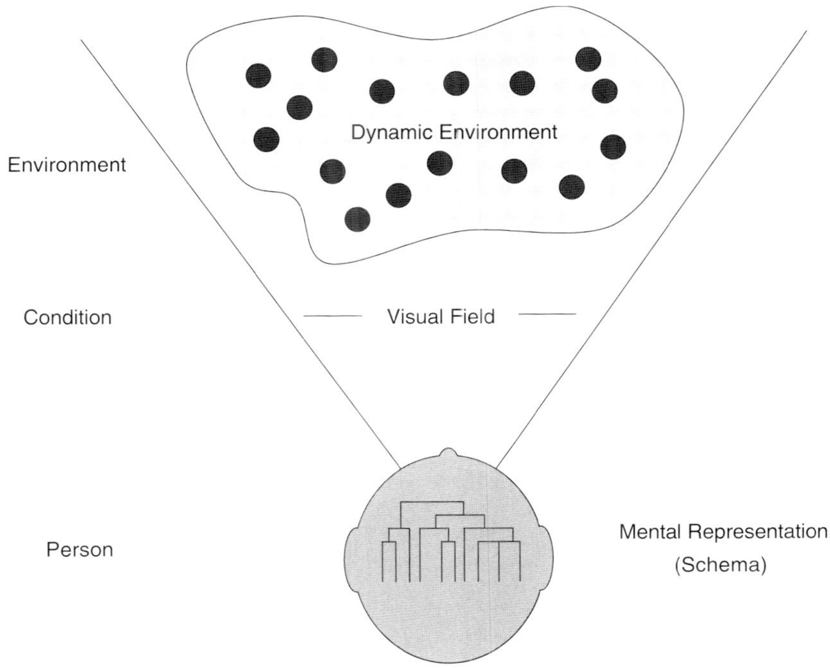

Fig. 1. Mental representation network guides the motor system in a given task within the environment.

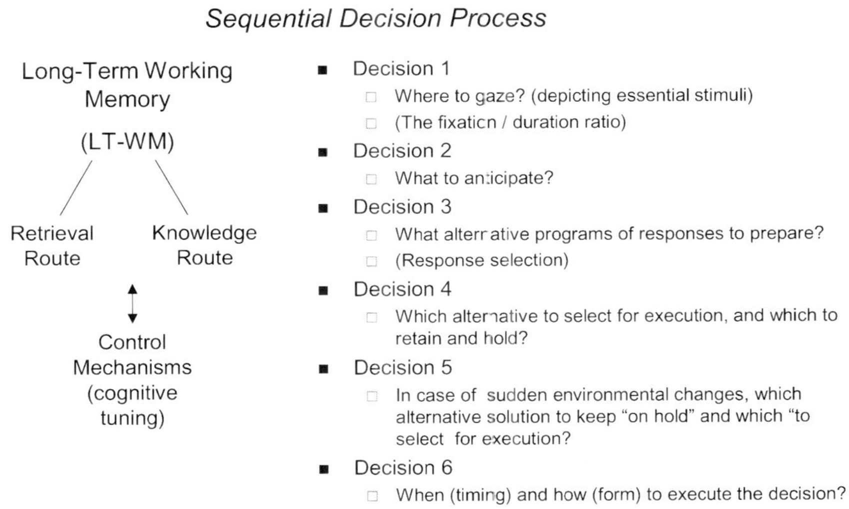

Fig. 2. A sequence of decision making controlled by long-term working memory and knowledge base and structure.

be interfered with once self-efficacy and emotions are deemed nonoptimal. Assuming that a knowledge base in the form of mental representations guides the motor system, and at the same time is fed by it, several decisions must be made under varying environmental conditions. These perceptual-cognitive decisions are presented as a logical sequence in Fig. 2.

Each decision in the sequence is related to a general plan stored in LTM, but may have different consequences if inappropriately solved. The first decision is related to the visual system; its goal is to direct the visual attention to the cues deemed essential for response selection. Thus, the first decision to make is "where to gaze, and to what cues to attend?" When the environmental cues are fed forward to the neural system, the mental representation network (i.e., schema) initiates some elaboration in the form of neural activity, which results in anticipating upcoming events with certain probability (Tenenbaum and Lidor, 2005). The main question here is, "what should be anticipated given the situation?" Anticipation is the most crucial component in the sequence, because it activates alternate and competing solutions from LTM in the form of response selection. Thus, the decision here is "what alternative solution should be sent to activate the motor system, and which one to leave as an alternative?" At this stage, the motor system is initiated, while in some cases this system remains "on alert" anticipating possible environmental alterations, and ready to respond to the question, "what alternative response must be selected to replace the previous response?" The last decision is related to the timing of the response, which is crucial in dynamic environments. Once a correct decision has been selected, activation of the motor system with improper timing will result in system failure. The decision here is, "when to activate the response selection in the motor system?"

The five-step sequential decision making is classified into two main components. The sequence offered here however, allows for an easy detection of action breakdown, but does not infer ultimate dependence of each preceding stage on the proceeding one. One component is related to decision making regarding environmental decisions. These components consist of the visual system and the visual strategy used to gather information, thus consists mainly of the perceptual system. The second system consists of mental and cognitive operations, which process the information gathered by the perceptual system, and fed forward via working memory for response selection (see Fig. 3). The neural schema controls the performance of the two systems.

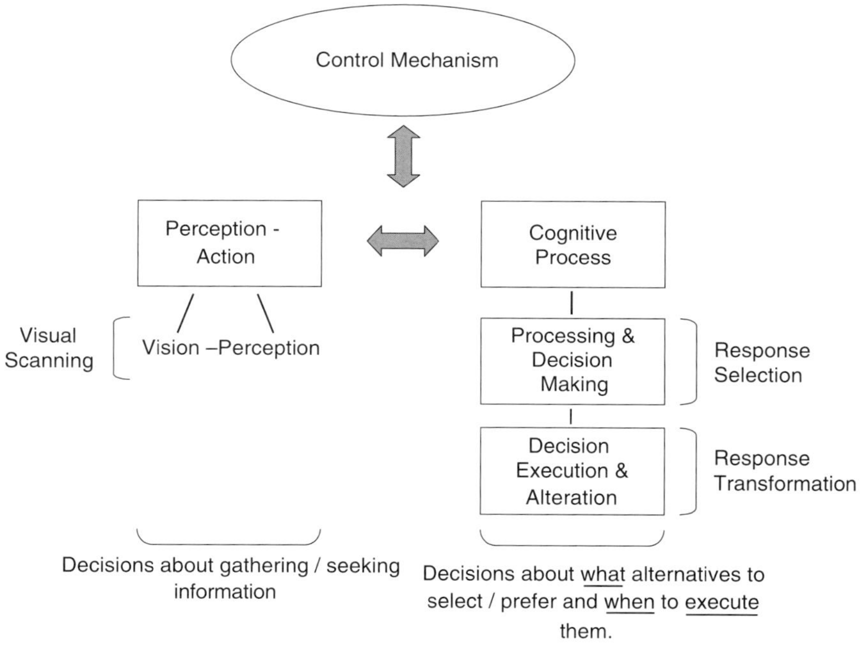

Fig. 3. Perception-action and cognitive components related to action guided by mental representation network (schema).

Linking perceptual-cognitive systems to the motor system

Under pressure, which may be emotionally induced, or result from demanding temporal conditions (e.g., when a batter must respond to an incoming baseball pitch), the perceptual-cognitive systems (A and B in Fig. 4) and the control system (E in Fig. 4) change their operational mode. More specifically, attention narrows (Landers, 1980), limiting the cognitive processing system to operate under goal-directed orientation while interfering thoughts compete for attention (Abernethy, 1993). The resulting collapse of the control system is unable to keep the perceptual-cognitive–motor flow intact. Consequences of the collapse are seen in a delayed response, increased response-selection errors, and diffuse and non-proficient motor responses. Once pressure is perceived as debilitating, usually by telic-oriented people (e.g., goal-directed and avoiding risk-taking) facing pressure, which alters their hedonic tone into experiencing anxiety (see Apter, 1982), the perceptual schema is no longer valuable resulting in the well-established LTM schema to operate under nonoptimal conditions. As a result, the control schema suffers collapse. However, under pressure conditions, which are typically perceived as challenging by paratelic-oriented people (e.g., taking risks and challenges; Apter, 1982), there is no collapse of the perceptual, cognitive, and control systems schema. Thus, the systems function efficiently by providing the necessary environmental information needed for response selection and coordinated motor action. Figure 4 portrays these linkages considering the perceptual, cognitive, and motor operating systems separately, but as depending on each other. It should be noted that the control system (Fig. 4(E) is viewed as a system that controls all

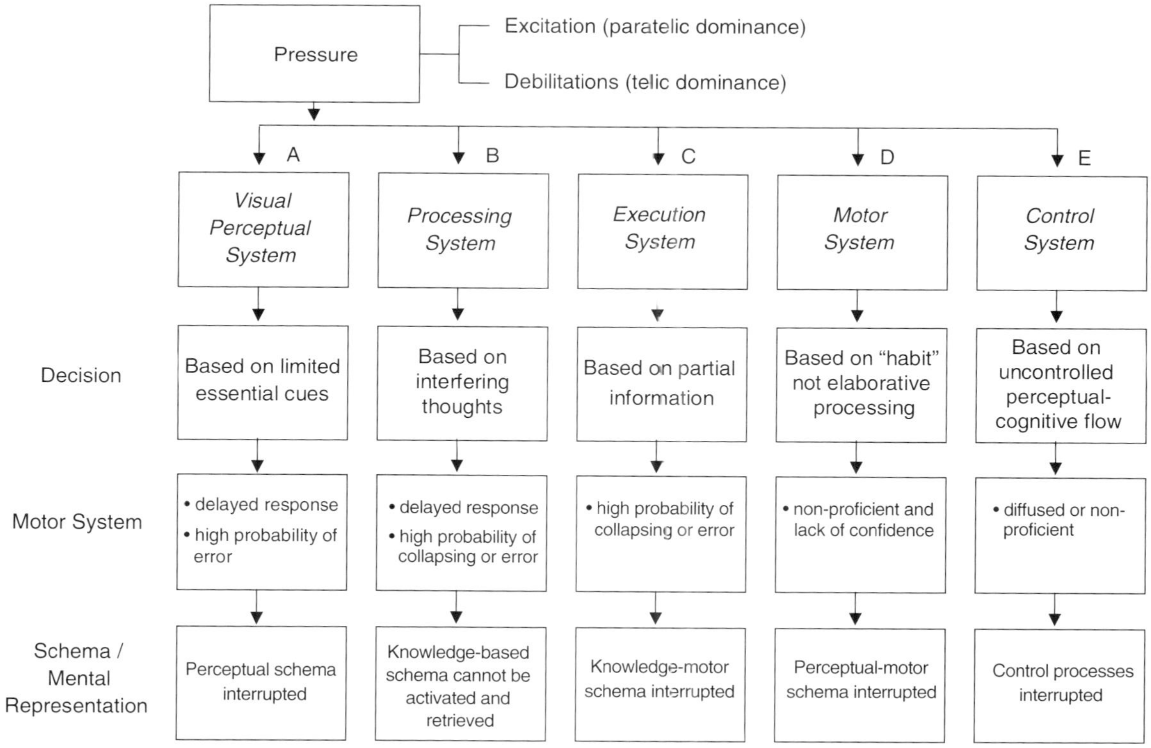

Fig. 4. A proposed scheme linking perceptual-cognitive components to the motor system using mental representations as underlying this linkage.

other systems, though graphically shown as one component in a sequence. It should also be noted that the systems illustrated here function differently under automated mode of operation, but they still are activated and functional (i.e., operate under unconscious control), especially when the system shifts from unintentional to intentional mode of operation. We also claim that impaired perceptual, cognitive, and motor processes and responses can be observed and detected through using common research paradigms (Abernethy, 1993; Tenenbaum and Lidor, 2005), which can be matched with their underlying mechanisms in the form of impaired schemata (Schack and Mechsner, 2006). These schemata are linked to the emotional centers in the brain (Hatfield and Kerick, 2007), and when primed they activate the associated neural network, which holds particular information labeled "emotion" (see Bower, 1981). This conceptual framework constitutes Fig. 4.

Changes in visual attention and information processing under pressure

To verify how the system works interactively, we must study how emotion and affective states, which vary in valence and intensity, provoke neural changes in the brain pathways and centers, while simultaneously observing behavioral changes. For example, retrieval of information is facilitated when context during retrieval extract the same emotional states experienced when encoding (*affect-state dependence*) (Bower, 1981). This linkage, and its relation to "knowledge structure," has not been studied in performing perceptual-motor tasks, particularly under pressure. Emotions are linked directly to memory and mental representation, thus directly activating mood consistent event files in memory and event-related parts of knowledge structures. When heeding information, subjects selectively pay attention to cues that are congruent with current emotional states (*affect-congruent attention*). In other words, direction of attention is influenced by the emotional content of the encountered stimuli. Positive or negative moods increase attention and the amount of rehearsal of mood-congruent facts, which results in stronger associations with positive or negative information, respectively. This notion must be further studied with respect to pressure conditions and skill level of the performer. In addition, how emotions and concentration vary when performers of different expertise levels perform a task is of interest.

Emotions can be viewed as memory units (Bower, 1981). They are components linked to the memory system that facilitate access to mental representations associated with targets of judgment (Forgas, 1991). Due to prior associations, innate and learned environmental situations activate particular emotion nodes stored in the memory. This activation spreads throughout neuronal circuits to mental representations of events associated with that emotion, influencing encoding and retrieval of material, as well as the valence of judgments of people, events, objects, and behaviors (Bower, 1991). Emotions are activated by re-experiencing the emotion, or by activation of any of their links (Barry et al., 2004). The stronger the activation of particular emotional nodes, the greater the mood-congruent effect. When emotions are strongly activated, emotion-congruent constructs (e.g., concepts, words, themes, and rules of inference) become primed and available for use, which brings into readiness certain perceptual categories, themes, or ways of interpreting the world congruent with current emotional states. In affect-priming terms, an emotion node spreads activation throughout the memories to which it is connected, increasing the chance that those memories will be retrieved (Bower, 1981).

This is important for capturing human performance because perception and action are based on the same representation structures (see coding theory; Prinz, 2005). Such an emotion, based activation of representation, affects the perception of the actual situation, and the focus of attention. Therefore, there is the risk that a "vicious circle" (a self-perpetuating process that returns to its starting point with no improvement from when it was begun) between threatening stimuli and negative emotional states activates event files, representation structures, and focuses attention on threatening stimuli, which most probably results in performance decrement.

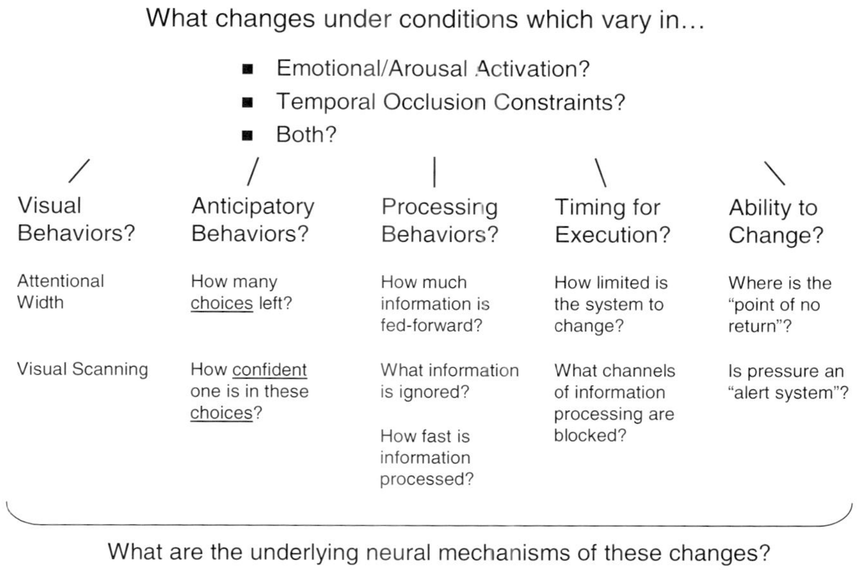

Fig. 5. Consequences of pressure on visual attention, processing information, and decision alteration.

Figure 5 illustrates possible behavioral outcomes that may occur under pressure conditions associated with emotions. We note that the emotion-associative schema is not presented in this figure, because of the limited evidence we present next. However, pressure affects visual scanning, usually in the form of narrowing the attention span, resulting in errors, which are attributed to information omission. When pressure and/or emotions affect the visual attention system, a collapse of all other systems is evident (see Fig. 6; components A, B, C, D, and E are derived from Fig. 4). This collapse results from a limitation in the information fed-forward, which is not comprehensive enough to make a reliable response selection. Furthermore, this is linked directly to alterations in the affective state, which takes the form of an increase in anxiety and a decrease in self-efficacy (Tenenbaum, 2003). Thus, the motor system is affected directly by nonoptimal emotional states, and indirectly through the effect of pressure and emotional states on the perceptual-cognitive system, resulting in a delayed response, inappropriate response, and uncoordinated action. We claim that these can be detected while observing respective changes in the mental representation schema

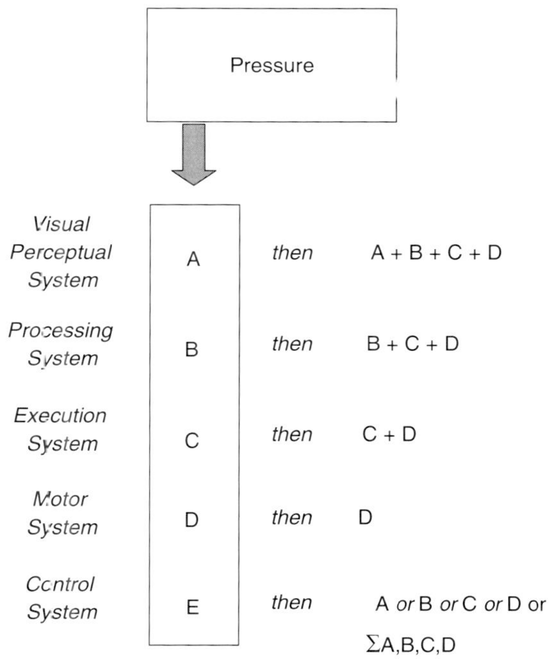

Fig. 6. A "domino effect" of the systems linking perceptual-cognitive information to the motor system under pressure and/or non optimal emotional state.

(see Figs. 7 and 8). In this line of conceptualization, anticipatory failure under high-pressure and nonoptimal emotional states result in limited choices for response selection, and decreases in confidence of these choices resulting in poor decision making and collapse of components C, D, and E (see Figs. 4–8).

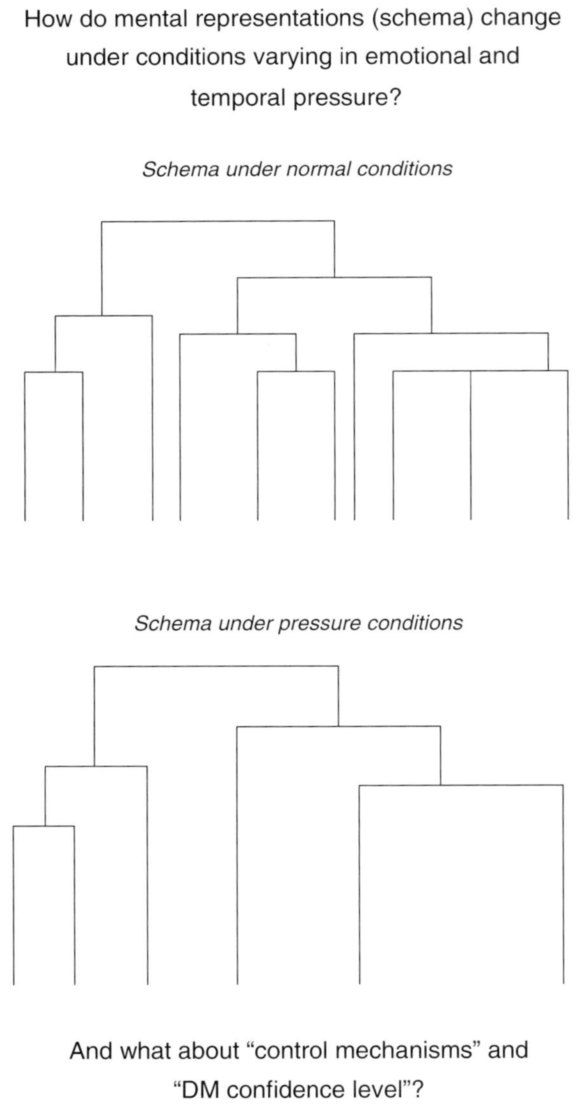

Fig. 7. A theoretical schema under normal and pressure conditions — some neural linkages, which represent motor elements, are nonactivated under pressure. DM, decision making.

Assuming that both the visual and anticipatory systems remain intact under nonoptimal conditions, the processing of information, however, may experience functional difficulties. Vital to the framework of expert performance is the cognitive control structure that governs information processing and task execution. Central to the optimal functioning of the cognitive control system is the proper allocation of attentional resources (Wulf, 2007). The fundamental assumption underlying the allocation of attention is that successful task performance is dependent upon attending to certain relevant information, while ignoring others (Lewis and Linder, 1997; Beilock and Gray, 2007). As such, an individual's focus of attention can have important consequences for the management and allocation of information-processing resources.

Under pressure, performance breakdown can result from compromises and disruptions to the attentional mechanisms that underlie skill execution. Understanding the impact of pressure on performance requires consideration of the attentional demands imposed by the task as well as consideration of skill level (Abernethy et al., 2007). Recent literature on "choking under pressure" has proposed two main attentional theories that seek to describe the mechanisms that account for pressure-induced performance failure (Wine, 1971; Baumeister, 1984; Lewis and Linder, 1997; Beilock and Carr, 2001). Distraction theories give an account of performance breakdown on tasks that rely heavily on working memory and fact retrieval (i.e., cognitive tasks) (Beilock and Carr, 2001). Explicit monitoring theories, in contrast, account for disruptions in tasks that are largely automated (i.e., proceduralized) and motoric in nature (Beilock and Carr, 2001).

Based upon resource allocation models of information processing, distraction theories propose that performance failures result from attention being diverted away from task-relevant information (Wine, 1971). As a consequence, increased focus on task-irrelevant cues such as worries and consequences consume vital working memory and attentional resources. As specified by resource allocation models, attention and

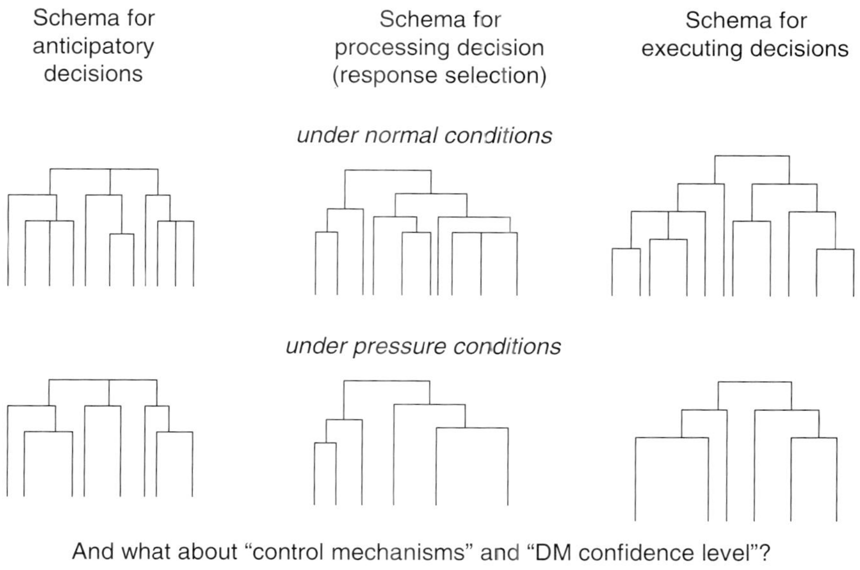

Fig. 8. Theoretical schemas for anticipatory, processing, and execution processes under normal and pressure conditions — some neural linkages, which represent motor elements, are nonactivated under pressure. DM, decision making.

working memory are both limited in capacity (Schmidt and Lee, 1999). This limited capacity impacts the ability to handle information from the task and environment efficiently. Under pressure, attentional resources become divided between task-relevant and task-irrelevant cues leaving inadequate working memory and attentional capacity to attend fully to the primary task to the extent required for successful performance (Hardy et al., 1996). As a result, compromises to attentional resources by task-irrelevant information lead to a disruption in performance.

The distraction theory best accounts for performance disruptions in tasks that place heavy demands on attention and working memory capacity. As a result, academic- or cognitive-based tasks have typically been the primary source of investigation supporting a distraction account of pressure-induced performance failure (e.g., Wine, 1971; Nottelman and Hill, 1977; Deffenbacher, 1978; Beilock et al., 2004b). However, any task that utilizes online attentional resources is potentially susceptible to pressure-induced compromises to attentional capacity. Such tasks may include decision-making, problem-solving, or reaction-time tests (Lewis and Linder, 1997; Beilock and Gray, 2007). Sports that rely heavily on such processes are likely to be most susceptible to a distraction account of choking.

Not all skills, however, place large demands on attention and working memory capacity. Specifically, high-level motor skills comprising of proceduralized knowledge do not require constant online attention, and run largely outside of working memory (Fitts and Posner, 1967; Beilock et al., 2002). Therefore, high-level skills that are assumed to be automated, requiring little online attention, should be relatively robust against conditions that draw attention away from task execution (Beilock and Carr, 2001). As a result, explicit monitoring theories propose an alternate attentional mechanism to account for pressure-induced performance failure for heavily proceduralized skills.

According to the explicit monitoring theory, performance pressure increases self-focused attention to the step-by-step processes of skill execution (Lewis and Linder, 1997). For highly proceduralized motor tasks, this shift in attention to the details of movement disrupts the automaticity of well-learned skills, resulting in performance breakdown (i.e., choking) (Fitts and Posner, 1967; Baumeister, 1984; Beilock and Gray, 2007). The tenants of explicit monitoring theory are anchored in both self-awareness and skill acquisition theories (Hardy et al., 1996; Lewis and Linder, 1997). The assumption that pressure increases self-focused attention draws on the idea that task importance becomes more salient under pressure. In an attempt to ensure correctness of movement, attention is turned inward toward monitoring the processes of skill execution (Baumeister, 1984; Lewis and Linder, 1997). Research in the self-awareness literature offers support of this assumption as situational factors, such as audience pressure, competition, ego-relevance, and reward and punishment contingency, have been shown to increase pressure and self-focus (Baumeister and Showers, 1986; Lewis and Linder, 1997).

The disruptive nature of skill-focused attention on performance is based on the tenants of skill acquisition theory (Lewis and Linder, 1997). As athletic skill develops, athletes progress through different phases of skill acquisition (Fitts and Posner, 1967). Each phase is characterized by differences in both attentional requirements and cognitive control structures (Beilock and Carr, 2001). At the beginning of skill acquisition, skill execution is governed by slow, declarative, and attention-demanding cognitive processes. However, as skill develops through practice, skill execution becomes governed by fast, automated, procedural knowledge that runs largely outside of attention (Fitts and Posner, 1967). It is these highly automated skills that suffer from increases in self-focused attention brought about by performance pressure. Baumeister (1984) suggested that the increased attention to the details of movements disrupts the automatic nature of skill execution through a return to more controlled processing, which is characteristic of earlier stages of skill acquisition. Numerous studies have found support for the contention that increased attention to skill execution harms performance in well-learned skills (e.g., Kimble and Perlmuter, 1970; Langer and Imber, 1979; Wulf and Weigelt, 1997; Beilock and Carr, 2001; Beilock et al., 2002, 2004a).

While pressure can lead to disruptions in information processing caused by improper attentional focus, pressure and nonoptimal emotional states may also result in the slowing down of decision-making processes, response selection, and neurological signals to the motor system, resulting in a collapse of the motor system (D) and the control system (E). It is very rare that the motor system collapses under pressure without any precedent changes in the perceptual-cognitive systems. However, behaviors such as this are common when the motor skill is not well learned and acquired, and the motor schema, which supports the motor action, is not well inherited within the mental representation scheme. When the control system is impaired, all the systems suffer, and the mental representation schema is no longer securing smooth and efficient communications among the five systems (see Fig. 6).

Changes in mental representations under pressure

A method eliciting the cognitive (i.e., knowledge) structure in the form of mental representations was recently introduced by Schack and Mechsner (2006). They showed that performers who differ in skill level differ significantly in the structure and complexity of their mental representations' structure when asked to reflect on their actions. Once a performance collapsed under pressure (e.g., spectators, media, importance of the event, life threat, etc.), the mental structure of the performance remained stable, but that the level of mental control collapsed as it could not meet the environmental or inner requirements of the cognitive system. Mental control (e.g., self-regulation) broke down because the performers lacked the sufficient strategies required for coping under external stressful conditions.

In a dual-task auto-racing simulation, drivers who were highly anxious experienced an altered ability to acquire peripheral information at the perceptual level. At higher levels of anxiety, the identification of peripheral lights became slower and less accurate, and significant performance decrements occurred in central and peripheral tasks (Janelle et al., 1999). A variety of negative and positive emotions are also associated with increased activation making it unclear whether it is the arousal or the valence of the emotions responsible for alterations on information processing.

Variation of psychological and physiological activation due to the stress response has an effect on the width of attention field, level of distractibility, amount of investment in controlled processing, and efficiency of attention processing (Janelle, 2002). These processes depend largely on the emotional experience (Mellalieu, 2003). In line with the research on specific psychological and physiological appraisal responses (e.g., Tomaka et al., 1997), it would be important to determine how the cognitive changes are influenced by different appraisals of stressful situations (i.e., threatening vs. challenging situation), and how these changes affect the motor system. Figures 7 and 8 assume that pressure and/or nonoptimal emotions result is collapse of the mental representation, which supports the motor program under normal operating conditions, and this mental collapse can be detected using behavioral and neural measures. One should note that in our conceptualized framework, appraisal and coping strategies in the form of self-regulation are part of the mental representation schema. When these schema are not well established, collapse of the perceptual-cognitive system and its linkage to the motor system is highly probable under conditions that evoke perceived pressure and elevated emotions.

Along with mental representations, the underlying knowledge structure that supports skilled performance also plays a role in resiliency to disruptions under pressure. Recent evidence suggests that learning environments can have a profound impact on how skill execution is controlled under pressure (e.g., Masters, 1992; Hardy et al., 1996). Specifically, the way in which knowledge is acquired may dictate the extent to which performance pressure leads to disruptions in skill execution. Traditionally, training systems are associated with deliberate attempts to learn, and are comprised of explicit instructions about the rules governing how a task is to be performed. The explicit instructions associated with conventional modes of learning lead to a large verbalizable knowledge base that is easily accessible and available for articulation. Such explicit knowledge about the complex chain of rules and techniques guiding performance, however, may be more susceptible to the debilitative effects of pressure as opposed to other types of knowledge structure.

According to explicit monitoring theories of choking, performance pressure turns attention inward toward the step-by-step processes of skill execution. This shift in attention toward the details of movement results in the disruption of automaticity resulting in performance breakdown (Baumeister, 1984; Lewis and Linder, 1997; Beilock and Carr, 2001). Masters' (1992) *Reinvestment hypothesis* provides an account of the specific mechanism underlying this breakdown. According to Masters, the increased self-focus resulting from pressure causes performers to reinvest declarative/explicit knowledge acquired during early learning (Gray, 2004; Mullen et al., 2005). This "reinvestment" of explicit knowledge results in the "dechunking" of automatic control structures that normally run uninterrupted. Once the control structure has been dechunked into smaller sequences of independent units, each unit must be separately activated and run, increasing the likelihood of errors at each transition (Beilock and Carr, 2001). As a result, execution becomes slow and error prone; which is characteristic of novice performance.

The accumulation of explicit knowledge through conventional modes of learning may facilitate the likelihood of performance failure under pressure. In contrast, knowledge acquired through more implicit means has been shown to be more robust under situations of increased psychological stress (MacMahon and Masters, 2002). Implicit learning is characterized by the acquisition of knowledge without deliberate attempts to learn. In addition, implicit knowledge is largely unavailable to verbal report. The inability to access verbalizable knowledge associated with implicit learning has led some to suggest that implicitly acquired skills are less likely to be disrupted under pressure compared to more explicitly acquired skills (e.g., Masters, 1992). Under pressure, knowledge acquired through implicit means would be inaccessible to be "reinvested" into skill execution, thus automaticity and performance would remain intact. Knowledge structure, therefore, can have important implications for the resiliency of the perceptual-cognitive–motor system to perform optimally under conditions that vary in situational demands.

Changes in the motor system under pressure

Under the framework proposed here, any collapse of the previous components underlying skilled performance will have a direct impact on the functioning of the motor system. However, to date, little research has examined the inter-relationship between these mediating components and the resulting motor response. While less is known about this inter-relationship, the direct impact of pressure and its associated emotions on movement behavior has been more thoroughly investigated. Research in this area has identified various biomechanical changes associated with stress and anxiety (e.g., Weinberg and Hunt, 1976; Beuter and Duda, 1985; Collins et al., 2001; Pijpers et al., 2003).

Findings revealed that feelings of anxiety and pressure can alter the characteristics of movement (Pijpers et al., 2005). More specifically, anxiety has been linked to movements that are less smooth, less efficient in terms of time and energy, and less variable (e.g., Weinberg and Hunt, 1976; Beuter and Duda, 1985; Pijpers et al., 2003). In a study examining British soldiers on a stepping task, Collins et al. (2001) found alterations in movement patterns across high- and low-anxiety conditions. Under high anxiety, reductions in movement variability were seen in the rigid coupling of the hip, knee, and ankle joints. Alternatively, performance under low anxiety revealed movements to be less limited and reflecting greater degrees of variability. Pijpers et al. (2003) found similar findings while investigating novice rock climbers. To manipulate anxiety, participants were required to climb at two differing heights (i.e., high and low) on a climbing wall. While climbing high on the wall, participants experienced increased levels of anxiety accompanied by changes in movement behavior. Specifically, movements were longer and less fluent resulting in less efficiency overall. Finally, Beuter and Duda (1985) reported anxiety-related changes in the coordination patterns of children on a stepping task. Under high-anxiety conditions, children produced movements that were less efficient and less smooth when compared to performance under low anxiety. Findings such as these provide a glimpse of the impact performance pressure can have on biomechanical processes.

According to Bernstein (1967) the central task involved with motor learning requires the solving of the *degrees of freedom (DOF) problem*. The DOF problem is concerned with how the body controls the numerous separate and independent elements (e.g., wrist, elbow, joints, etc.) in the production of coordinated motion. According to Bernstein, and later Vereijken et al. (1992), early learning is typified by an attempt to "freeze" extraneous DOF in order to reduce task complexity. As such, novice motor movements appear rigid, uncoordinated, and stiff. As skill develops through practice, novices begin to "thaw" or release previously frozen DOF resulting in greater independent motion, efficiency, and accuracy (Vickers, 2007). As high-level skill continues to develop, the motor system becomes able to exploit the release of additional DOF making use of the built in mechanical-inertial properties of the limbs (Schmidt and Lee, 1999). As a result, motor performance becomes smoother, faster, and more fluent. However, under pressure, Berstein (1967) postulated that skilled performers may attempt to "refreeze" DOF in an attempt to reduce task complexity and simplify the problem of movement control. As a result, movement execution regresses and parallels earlier stages of skill acquisition (Pijpers et al., 2003). The return to novice "freezing" strategies produces motor movements typified by motions that are rigid and jerky.

Examination of the biomechanical changes associated with anxiety and pressure mirror to a great extent the alterations of the cognitive processes associated with pressure-induced performance failure. In each case, pressure and anxiety appears to return the individual to a previous lower level of skill functioning (Beilock and Carr, 2001). Cognitive theories of choking under pressure state that performance pressure disrupts automaticity through a return to lower level conscious processing. Biomechanical theories, similarly, propose that pressure and anxiety revert experts back to previous novice "freezing" strategies. These similarities bear witness to the parallels and interactions that each system has with each other.

Cerebral activity related to perceived pressure and the motor system

As with the perceptual-cognitive systems, cerebral activity can play a fundamental role in performance under pressure. More specifically, depending on the valence of a stimuli, the lateral nuclei can communicate with the central nucleus in each amygdale, and subsequent connections can travel to critical forebrain, brainstem, autonomic, and endocrine structures that mediate the expression of emotion. Specifically, there are interconnections from the central nuclei to the (1) hypothalamus, which results in sympathetic arousal and stimulation of stress hormones via the hypothalamic-pituitary-adrenal (HPA) axis, (2) the periaqueductal gray, which results in motor responses, and (3) the cingulate cortex, which results in additional cortico-cortical communication with neocortical association regions such as the temporo-parietal regions. Additionally, interconnections to pontine nuclei in the reticular formation result in an increase in overall arousal. In this manner orchestrated sequelae occur in response to a stressful environment, which collectively, can change the performer's mental and physical state in a profound manner. For example, heart rate and cortisol levels rise, as does muscle tension, and the performer may concomitantly experience excessive self-talk and "too much thinking." As a result, their attention may become compromised and become explicitly managed, resulting in timing and coordination that is altered and likely reduced in quality.

In support of the "overthinking" hypothesis, Hung et al. (2005) recently provided psychophysiological evidence of increased neural activity between the left temporal region and the motor planning regions of the brain, by assessment of T3-Fz alpha electroencephalogram (EEG) coherence levels, when participants were asked to perform a dart-throwing task under pressure of social evaluation. Relative to a nonstress control condition, the increased neural activity was accompanied by heightened reports of state anxiety and reductions in self-reported confidence levels. As expected, accuracy of performance (i.e., visuomotor coordination) was reduced. However, how these changes were associated with mental representational changes was not studied, but now may contribute to bridging the understanding of covert and overt behaviors.

In light of the mental and physical change alterations that accrue, the activation of the amygdalae serves as a pivotal event in the manifestation of stress. The control of activity in the amygdalae can exert a powerful influence on the performer's mental and physical state. Beyond the structures and processes outlined by Bear et al. (2001), a critical component of the neurobiology of fear is the executive control over limbic function and subcortical emotional circuits, which is housed anatomically in the frontal regions of the forebrain. Importantly, the anterior cortical regions have extensive anatomical connections with several subcortical limbic structures implicated in emotional behavior, particularly the amygdala (Davidson, 2002, 2004). Davidson and colleagues have generated a significant body of literature that clearly shows a positive association between left frontal activation and positive affect while relative right activation is associated with negative affect (Davidson, 1998; Tomarken et al., 1992). Although the lateralization of frontal activation is robustly related to the valence of emotion, recent evidence points to a more fundamental association such that left frontal activation mediates approach-oriented behavior while right frontal activation is associated with avoidance or withdrawal-oriented behavior (Davidson, 2004). For example, left frontal activation is manifest during hostile behavior, which is certainly not a positive affective state, but most definitely involves approach toward an intended target. Whether positive in nature, approach oriented, or a combination of the two dimensions, it would appear that such a neurobiological state would be highly adaptive for the performer who must control his/her emotional level while actively engaged with challenging tasks while under great pressure. How these neural activities are related to perceived pressure, emotions, perceptual, cognitive, and motor systems' functioning interactively, must be studied within the framework we have outlined.

Conclusion

This chapter introduced a conceptual framework, which links the emotion, cognitive, and motor systems via a mental representation network. This symbolic network is believed to initiate, execute, and control motor actions while interacting with the environment. More specifically, mental representations allow the system to perceive, anticipate, make, and alter decision, and execute actions in the form of neural activity and muscle–joint activation. Under pressure, or any other mental state, impairment in the mental representation network changes either the perceptual, cognitive, or motor systems. These changes must be studied using an integrated method where perceptual, cognitive, and neural activities are studied simultaneously, and both overt and covert behaviors are linked to mental representations, which govern all actions.

References

Abernethy, B. (1993). Attention. In R. N. Singer, M. Murphey, & L. K. Tennant (Eds.), *Handbook of research on sport psychology* (1st ed., pp. 127–170). New York: Macmillan.

Abernethy, B., Masters, R. S. W., Maxwell, J. P., Kamp, J., & Jackson, R. C. (2007). Attentional processes in skill learning and expert performance. In G. Tenenbaum & R. C. Eklund (Eds.), *Handbook of sport psychology* (3rd ed., pp. 245–263). New York: Wiley.

Apter, M. J. (1982). *The experience of motivation: the theory of psychological reversals*. London: Academic Press.

Barry, E. S., Naus, M. J., & Rehm, L. P. (2004). Depression and implicit memory: understanding mood congruent bias. *Cognitive Therapy and Research*, *28*, 387–414.

Baumeister, R. F. (1984). Choking under pressure: self-consciousness and paradoxical effects of incentives on skillful performance. *Journal of Personality and Social Psychology*, *46*, 610–620.

Baumeister, R. F., & Showers, C. J. (1986). A review of paradoxical performance effects: choking under pressure in sports and mental tests. *European Journal of Social Psychology*, *16*, 361–383.

Bear, M. F., Connors, B., & Paradiso, M. A. (2001). *Neuroscience: exploring the brain*. Philadelphia, PA: Lippincott Williams & Wilkins.

Beilock, S. L., Bertenthal, B. I., McCoy, A. M., & Carr, T. H. (2004a). Haste does not always make waste: expertise, direction of attention, and speed versus accuracy in performing sensorimotor skills. *Psychonomic Bulletin & Review*, *11*, 373–379.

Beilock, S. L., & Carr, T. H. (2001). On the fragility of skilled performance: what governs choking under pressure? *Journal of Experimental Psychology — General*, *130*, 701–725.

Beilock, S. L., & Gray, R. (2007). Why do athletes choke under pressure? In G. Tenenbaum & R. C. Eklund (Eds.), *Handbook of sport psychology* (3rd ed., pp. 425–444). New York: Wiley.

Beilock, S. L., Holt, L. E., Kulp, C. A., & Carr, T. H. (2004b). More on the fragility of performance: choking under pressure in mathematical problem solving. *Journal of Experimental Psychology — General*, *133*, 584–600.

Beilock, S. L., Wierenga, S. A., & Carr, T. H. (2002). Expertise, attention, and memory in sensorimotor skill execution: impact of novel task constraints on dual-task performance and episodic memory. *Quarterly Journal of Experimental Psychology*, *55*, 1211–1240.

Bernstein, N. A. (1967). *The coordination and regulation of movements*. Oxford: Permagon Press.

Beuter, A., & Duda, J. L. (1985). Analysis of the arousal/motor performance relationship in children using movement kinematics. *Journal of Sport Psychology*, *7*, 229–243.

Bower, G. H. (1981). Mood and memory. *American Psychologist*, *36*, 129–148.

Bower, G. H. (1991). Mood congruity of social judgments. In J. P. Forgas (Ed.), *Emotion and social judgments* (pp. 31–53). New York: Pergamon Press.

Collins, D., Jones, B., Fairweather, M., Doolan, S., & Priestley, N. (2001). Examining anxiety associated changes in movement patterns. *International Journal of Sport Psychology*, *31*, 223–242.

Davidson, R. J. (1998). Affective style and affective disorders: perspectives from affective neuroscience. *Cognition & Emotion*, *12*, 307–320.

Davidson, R. J. (2002). The neural circuitry of emotion and affective style: prefrontal cortex and amygdale contributions. *Social Science Information*, *40*, 11–37.

Davidson, R. J. (2004). Affective style: causes and consequences. In J. T. Cacioppo & G. G. Berntson (Eds.), *Essays in social neuroscience* (pp. 77–91). Cambridge: MIT Press.

Deffenbacher, J. L. (1978). Worry, emotionality, and task-generated interference in test anxiety: an empirical test of attentional theory. *Journal of Educational Psychology*, *70*, 248–254.

Ericsson, K. A., & Kintsch, W. (1995). Long-term working memory. *Psychological Review*, *102*, 211–245.

Fitts, P. M., & Posner, M. I. (1967). *Human performance*. Belmont, CA: Brooks/Cole.

Forgas, J. P. (1991). Affect and social judgments: an introductory review. In J. P. Forgas (Ed.), *Emotion and social judgments* (pp. 3–30). New York, NY: Pergamon Press.

Gray, R. (2004). Attending to the execution of a complex sensorimotor skill: expertise differences, choking and slumps. *Journal of Experimental Psychology — Applied*, *10*, 42–54.

Hardy, L., Mullen, R., & Jones, G. (1996). Knowledge and conscious control of motor actions under stress. *British Journal of Psychology*, *87*, 621–636.

Hatfield, B. D., & Kerick, S. E. (2007). The psychology of superior sport performance: a cognitive and affective neuroscience perspective. In G. Tenenbaum & R. C. Eklund (Eds.), *Handbook of sport psychology* (3rd ed., pp. 84–109). New York: Wiley.

Hung, T. M., Lin, J. H., Lo, L. C., Kao, J. F., Hung, C. L., Chen, Y. J., & Lai, Z. (2005). *Effects of anxiety on EEG coherence during dart throw*. Paper presented at the 2005 World Congress, International Society for Sport Psychology, Sydney, Australia.

Janelle, C. M. (2002). Modification of visual attention parameters under conditions of heightened anxiety and arousal. *Journal of Sport and Exercise Psychology*, *21*, S61.

Janelle, C. M., Singer, R. N., & Williams, M. A. (1999). External distraction and attentional narrowing: visual search evidence. *Journal of Sport & Exercise Psychology*, *21*, 70–91.

Kimble, G. A., & Perlmuter, L. C. (1970). The problem of volition. *Psychological Review*, *77*, 361–384.

Landers, D. M. (1980). The arousal-performance relationship revisited. *Research Quarterly for Exercise and Sport*, *51*, 77–90.

Langer, E., & Imber, G. (1979). When practice makes imperfect: debilitating effects of overlearning. *Journal of Personality and Social Psychology*, *37*, 2014–2024.

Lewis, B. P., & Linder, D. E. (1997). Thinking about choking? Attentional processes and paradoxical performance. *Personality and Social Psychology Bulletin*, *23*, 937–944.

MacMahon, K. M. A., & Masters, R. S. W. (2002). The effects of secondary tasks on implicit motor skill performance. *International Journal of Sport Psychology*, *33*, 307–324.

Masters, R. S. W. (1992). Knowledge, knerves and know-how: the role of explicit versus implicit knowledge in the breakdown of a complex motor skill under pressure. *British Journal of Psychology*, *83*, 343–358.

Mellalieu, S. D. (2003). Mood matters: but how much? A comment on Lane and Terry (2000). *Journal of Applied Sport Psychology*, *15*, 99–114.

Mullen, R., Hardy, L., & Tattersall, A. (2005). The effect of anxiety on motor performance: a test of the conscious processing hypothesis. *Journal of Sport & Exercise Psychology*, *27*, 212–225.

Newell, K. M. (1986). Constraints on the development of coordination. In M. G. Wade & H. T. A. Whiting (Eds.), *Motor development in children: aspects of coordination and control* (pp. 341–360). Dordrecht: Martinus Nijhoff.

Nitsch, J. R. (2004). The action-theoretical perspective: a frame of reference for research and application in sport psychology. *Zeitschrift für Sportpsychologie*, *11*, 10–23.

Nitsch, J. R., & Hackfort, D. (1981). Stress in schule und hochschule: Eine handlungspsychologische funktionsanalyse [Stress in school and university: an action psychology functional analysis]. In J. R. Nitsch (Ed.), *Stress* (pp. 263–311). Bern, Switzerland: Huber.

Nottelman, E., & Hill, K. (1977). Test anxiety and off-task behavior in evaluative situations. *Child Development*, *48*, 225–231.

Pijpers, J. R., Oudejans, R. R. D., & Bakker, F. C. (2005). Anxiety-induced changes in movement behavior during the execution of a complex whole-body task. *Quarterly Journal of Experimental Psychology*, *58A*, 421–445.

Pijpers, J. R., Oudejans, R. R. D., Holsheimer, F., & Bakker, F. C. (2003). Anxiety-performance relationships in climbing: a process-oriented approach. *Psychology of Sport and Exercise*, *4*, 283–304.

Prinz, J. J. (2005). Emotions, embodiment, and awareness. In L. F. Barrett, P. M. Niedenthal, & P. Winkeilman (Eds.), *Emotion and consciousness* (pp. 363–383). New York: Guilford Press.

Schack, T., & Hackfort, D. (2007). An action theory approach to applied sport psychology. In G. Tenenbaum & R. C. Eklund (Eds.), *Handbook of sport psychology* (3rd ed., pp. 332–351). New York: Wiley.

Schack, T., & Mechsner, F. (2006). Representation of motor skills in human long-term memory. *Neuroscience Letters*, *391*, 77–81.

Schack, T., & Tenenbaum, G. (Eds.), (2004a). The construction of action — new perspectives in science: Pt. I. perceptual and cognitive control [special issue]. *International Journal of Sport Psychology*, *2*.

Schack, T., & Tenenbaum, G. (Eds.). (2004b). The construction of action — new perspectives in science: Pt. II. representation and planning [special issue]. *International Journal of Sport Psychology*, *2*.

Schmidt, R. A., & Lee, T. D. (1999). Attention and performance. In R. Schmidt & T. Lee (Eds.), *Motor control and learning: a behavioral emphasis* (pp. 61–91). Champaign, IL: Human Kinetics.

Tenenbaum, G. (2003). Expert athletes: an integrated approach to decision making. In J. L. Starkes & K. A. Ericsson (Eds.), *Expert performance in sports* (pp. 191–218). Champaign, IL: Human Kinetics.

Tenenbaum, G., & Lidor, R. (2005). Research on decision-making and the use of cognitive strategies in sport settings. In D. Hackfort, J. Duda, & R. Lidor (Eds.), *Handbook of research on applied sport psychology* (pp. 75–91). Morgantown, WV: Fitness Information Technology.

Tomaka, J., Blascovich, J., Kibler, J., & Ernst, J. M. (1997). Cognitive and physiological antecedents of threat and challenge appraisals. *Journal of Personality and Social Psychology*, *73*, 63–72.

Tomarken, A. J., Davidson, R. J., Wheeler, R. E., & Doss, R. C. (1992). Individual differences in anterior brain asymmetry and fundamental dimensions of emotion. *Journal of Personality and Social Psychology*, *62*, 676–687.

Vereijken, B., van Emmerik, R. E. A., Whiting, H. T. A., & Newell, K. M. (1992). Free(z)ing degrees of freedom in skill acquisition. *Journal of Motor Behavior*, *24*, 133–142.

Vickers, J. N. (2007). *Perception, cognition, and decision training: the quiet eye in action*. Champaign, IL: Human Kinetics.

Weinberg, R. S., & Hunt, V. V. (1976). The interrrelationships between anxiety, motor performance, and electromyography. *Journal of Motor Behavior*, *8*, 219–224.

Wine, J. (1971). Test anxiety and direction of attention. *Psychological Bulletin*, *76*, 92–104.

Wulf, G. (2007). *Attention and motor skill learning*. Champaign, IL: Human Kinetics.

Wulf, G., & Weigelt, C. (1997). Instructions about physical principles in learning a complex motor skill: to tell or not to tell.... *Research Quarterly for Exercise and Sport*, *68*, 362–367.

M. Raab et al. (Eds.)
Progress in Brain Research, Vol. 174
ISSN 0079-6123

CHAPTER 21

Biases and optimality of sensory-motor and cognitive decisions

Julia Trommershäuser*

Department of Psychology, Giessen University, Giessen, Germany

Abstract: I review a variety of behavioral studies directed at understanding how probability and value information is represented in motor and cognitive tasks. Subjects in (cognitive) decision-making tasks often misrepresent the frequency of rare events and typically fail to maximize expected gain. In contrast, subjects in mathematically equivalent movement tasks are generally found to be very good at choosing motor strategies that come close to maximizing expected gain. I discuss the differences between the sources of uncertainty for decisions in the motor and cognitive domain and summarize experimental evidence about how information about uncertainty is acquired in motor and cognitive decision tasks. Finally, I briefly review the evidence concerning the neural coding of probability, expected gain, and other information in decision-making tasks.

Keywords: Bayes rule; decision making; motor decision; cognitive decision; optimization

Introduction

Stochastic uncertainty plays a fundamental role in perception, cognition, and motor control and a wide variety of biological tasks can be formulated in statistical terms, drawing on the language of statistical decision theory. In the following, I provide an overview of recent evidence from a variety of behavioral studies directed at understanding how probability information is represented in motor and cognitive tasks. I will argue that framing behavioral tasks in the language of Bayesian and statistical decision theory enables a comparison of performance in perceptual, motor, and cognitive decision tasks.

Research on decision making seeks to understand how subjects choose between plans of action (lotteries, gambles, prospects) that differ in their consequences (Kahneman and Tversky, 2000). A subject in a typical decision task might be given a choice between a 10% chance of \$5000 (and otherwise nothing) and a 95% percent chance of \$300 (and otherwise nothing). If subjects choose among options with explicitly stated probabilities, they make "decisions under risk" and otherwise they make "decisions under uncertainty" (Knight, 1921). In the task described above, most subjects would prefer to receive \$5000 rather than \$300, or \$300 rather than \$0. The key difficulty in making such decisions is that typically no plan of action available to the subject guarantees a specific outcome. In the example above, there are possible outcomes of \$0, \$300, and \$5000 and it is the subject's task to choose between the different combinations of outcomes and probabilities.

*Corresponding author.
Tel.: +49-641-992-6108; Fax: +49-641-992-6119;
E-mail: Julia.Trommershaeuser@psychol.uni-giessen.de

DOI: 10.1016/S0079-6123(09)01321-1

The subject's choice is between one lottery

$$L_1 = [0.9, \$0;\ 0.0, \$300;\ 0.1, \$5000]$$

and a second lottery

$$L_2 = [0.05, \$0;\ 0.95, \$300;\ 0.0, \$5000].$$

Recent work in motor control formulates the problem of movement planning in terms of statistical and Bayesian decision theory, which effectively converts the problem of movement planning to the mathematically equivalent problem of decision making under risk (Trommershäuser et al., 2006). Generally, statistical decision theory including Bayesian decision theory serves as an appropriate common model for how organisms represent and use uncertainty in a wide range of tasks (Landy et al., 1995; Knill and Richards, 1996; Maloney, 2002).

Reformulating the problem of movement planning in terms of statistical and Bayesian decision theory leads to new and interesting ways to model the types of decisions that underlie movement and to examine the interplay between sensory uncertainty and movement planning, and movement execution in tasks in which the outcome of an action has consequences for the organism (Trommershäuser et al., 2003a, b; Körding and Wolpert, 2004).

Statistical decision theory and sensory-motor control

More than a half century ago, researchers developed and tested signal-detection models of simple perceptual decisions in the recently developed framework of statistical decision theory (Blackwell and Girshick, 1954; Green and Swets, 1966/1974). The framework of statistical decision theory is remarkably general and appropriate for modeling many realistic tasks in cognition, perception, and planning of movement (e.g., see Maloney, 2002).

In statistical decision theory, the organism's choice is limited to a set of possible plans of actions. Any motor plan specifies the intended movement of each part of the arm and body, in particular the velocity of the movement and the pattern of neural signals (in time) needed to execute the movement. In this particular case, the velocity of the movement, fast or slow, will affect how accurately the subject can control his/her movement. The success or failure of the movement depends on the actual outcome and how it interacts with the state of the environment, specifically the objects present within the arm's reach.

In the following, I briefly summarize how statistical decision theory can be applied to define the best possible movement strategy by taking into account the uncertainty associated with the sensory estimate, the error in executing a motor response and the possible gains and losses associated with each possible motor outcome.

Evidently, movement planning should take into account how accurate the *sensory information* is and how much *prior* information the subject has about the object location. If his/her sensory and prior information are poor, he/she might do well to choose a path that avoids any possible objects in the scene and proceed very slowly. Movement planning corresponds to a choice of a movement plan or strategy s that specifies the aim point, velocity, and other properties of movement execution. The velocity of the movement determines its accuracy (Fitts and Petersen, 1964; Schmidt et al., 1979; Meyer et al., 1988).

In detail, a movement strategy is a mapping from sensory input[1] V to a movement plan $s(V)$ (Fig. 1). The expected gain associated with the choice of strategy $s(V)$ is given by:

$$EG(s) = \iiint g(t, w)\, p_T(t|s(v)) p_v(v|w) p_w(w) dv\, dt\, dw,$$

where W is the random state of the world (i.e., positions of arm, object, and possible obstacles in the scene, etc.) with prior distribution $p_w(w)$, V is sensory information about the state of the world with likelihood distribution $p_v(v|w)$ and T is the stochastic movement trajectory resulting from the executed movement plan $s_T(V)$. The term $g(t,w)$ specifies the gain resulting from an actual trajectory t in the actual state of the world w. The optimal movement strategy is one that maximizes the expected gain $EG(s)$.

[1]Here, I follow the convention that random variables are in uppercase (e.g., X) while the corresponding specific values that those variables can take on are in lowercase, e.g., $p(x)$.

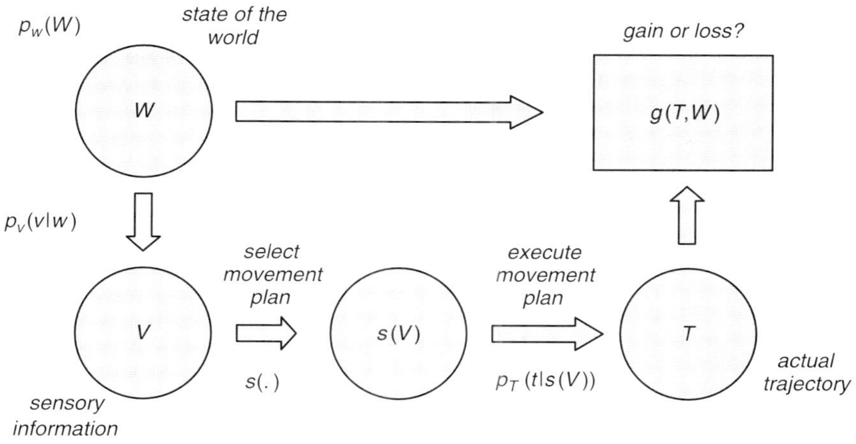

Fig. 1. Application of statistical decision theory in the context of visuc-motor tasks to define the best possible movement plan. A movement strategy is a mapping from sensory input V to a movement plan $s(V)$. Gains and losses $g(T,W)$ are determined by the actual trajectory T executed in the actual state of the world W (redrawn from Trommershäuser et al., 2008).

Once we realize that we can frame a wide span of cognitive, perceptual, and motor tasks in the common framework of statistical decision theory, several questions arise: how is probability represented and used? Is there a common mechanism for representing and processing uncertainty across all of these domains? Alternatively, are there fundamental differences in how probability is represented and processed in different domains: cognition, perception, and action? One benefit of a common framework is that we can compare performance across mathematically equivalent tasks in cognition, visuo-motor control, and sensory processing to give us a better understanding of how the brain represents and processes stochastic uncertainty.

Optimality and movement planning

In planning a movement, the brain has to select one of many possible movement plans or strategies. It is evidently desirable that the choice of strategy serves the demands of the particular task and also minimizes the "wear and tear" on the organism (see also Chapter 9: (Ir)rationality in action: do soccer players and goalkeepers fail to learn how to best perform during a penalty kick?; Chapter 10: Getting around: making fast and frugal navigation decisions; Chapter 23: Juggling with the brain — thought and action in the human motor system).

Typical research in the field of human motor control combines theoretical and experimental approaches. For example, a subject in a motor control experiment might perform simple reaching movements to a target, often constrained to two dimensions, that is, along the top of a table. The recorded movement trajectories are then compared to the predictions of a computational model, mimicking the conditions of the experiment. Early approaches to modeling movement planning take the form of an optimization problem in which the cost function to be minimized is biomechanical and the optimization goal is to minimize some measure of stress on the muscles and joints (e.g., Kaminsky and Gentile, 1986; Uno et al., 1989). These models differ primarily in the choice of the cost function.

Possible biomechanical cost functions include measures of joint mobility (Soechting and Lacquaniti, 1981; Kaminsky and Gentile, 1986), muscle tension changes (Dornay et al., 1996), mean squared rate of change of acceleration (Flash and Hogan, 1985), mean torque change (Uno et al., 1989), total energy expenditure (Alexander, 1997), and peak work (Soechting et al., 1995). The outcome of applying these models is typically a single, deterministic trajectory that optimizes the tradeoff between the goal of the movement and the biomechanical costs for the organism. These models are successful in explaining the human ability to adapt to forces applied during movement execution (Burdet et al., 2001; Franklin et al., 2007). A combination of biomechanical constraints was demonstrated by Cuijpers et al. (2004) who showed that humans will grasp an elliptical cylinder either along its major or minor axis, resulting in a stable grip, but will tend to choose the axis closer to that used for circular cylinders (i.e., the grasp that is more comfortable).

However, the models just described yield single deterministic trajectories that do not take into account the possible consequences of motor errors due to noise in the motor system. Uncertainty in the motor system originates from noisy neural control signals that lead to variable motor output. The noise is signal dependent; larger control signals lead to larger variability in motor outcome (Harris and Wolpert, 1998). Thus, faster movements, which require larger control signals, are more variable, resulting in the well-known speed-accuracy tradeoff (Fitts and Petersen, 1964; Schmidt et al., 1979; Meyer et al., 1988).

Experimental work concerned with the planning and execution of speeded eye and arm movements indicates that the complex sequences of neural events that underlie voluntary movements are selected so as to minimize movement error (Sabes and Jordan, 1997; Harris and Wolpert, 1998; Todorov, 2004). Note that this approach is based on the notion that the endpoint variability is a consequence of the "biological noise" in the motor control system and therefore unavoidable.

Following the observation that movements are corrupted by considerable motor noise and do not always follow the same deterministic trajectory, Harris and Wolpert (1998) suggested that movement trajectories are selected to minimize the variance of the final eye or arm position. They proposed that the underlying determinant of trajectory planning is the minimization of the noise in the neural control signal that activates the muscles during the execution of a motor command and in the post-movement period. In their model, the final arm or eye position is computed as a function of a (deterministic) biomechanical expression and a noisy neural signal, whereas the noise increases with the magnitude of the neural signal. According to the model of Harris and Wolpert (1998), the planned trajectory of the eye and arm is chosen to minimize variance in the endpoint of the movement. The idea behind this approach is that the variability in the final position of a saccade or pointing movement is the result of the accumulated deviations of the executed trajectory from the planned trajectory over the duration of the movement. The minimum variance model by Harris and Wolpert (1998) managed to successfully describe horizontal saccadic eye movements, hand paths for a set of point-to-point movements (Harris and Wolpert, 1998), as well as the movement trajectories measured in an obstacle avoidance task (Hamilton and Wolpert, 2002).

In a similar approach, Sabes and Jordan (1997) studied the contribution of kinematic and dynamic properties of the arm to movement planning. Subjects moved their hands between pairs of targets, avoiding an obstacle along the path. To carry out this task rapidly while avoiding the obstacle, subjects should choose a trajectory so that the direction from the obstacle to the trajectory at its nearest approach was one for which movement variability was minimal. Sabes and Jordan (1997) compared three sensitivity measures (kinematic, inertial, and elastic), each of which provided predictions of the point of closest approach to the obstacle that was most safe. The data were in qualitative agreement with the predictions of all three sensitivity measures, and best predicted by their inertial sensitivity model (*mobility*).

Movement under risk is decision making under risk

I next describe a movement task used in one of our previous studies that is equivalent to decision making under risk (Fig. 2; see also Trommershäuser et al., 2003a, b). On each trial, subjects are asked to reach out and touch a computer screen within

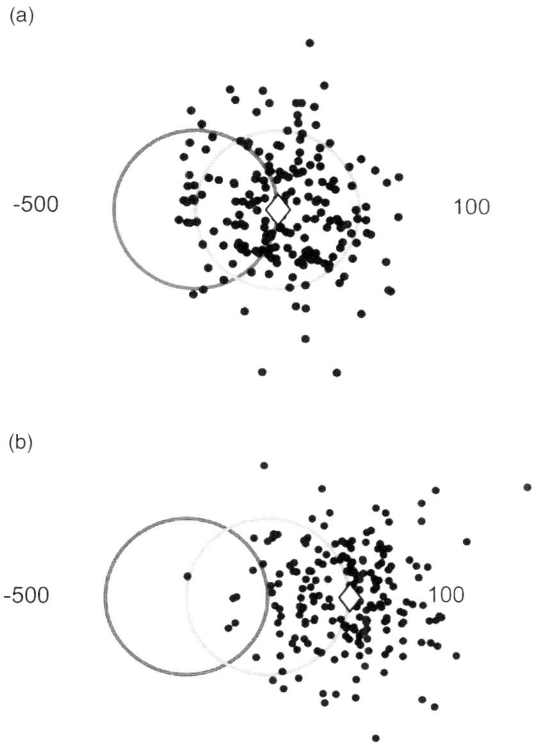

Fig. 2. Equivalence of a movement task and decision making under risk. Subjects touch a computer screen within a short period of time (e.g., 700 ms). Subjects can win 100 points by hitting inside the target circle or lose 500 points by hitting inside the penalty circle. (a) Expected gain for a subject aiming at the center of the target (aim point indicated by the white diamond). This subject has a motor uncertainty of 5.6 mm (standard deviation); target and penalty circle have radii of 9 mm. This motor strategy yields an expected gain of −111.39 points/trial. (b) Expected gain for a subject with the same motor uncertainty as in (a). The subject is aiming towards the right of the target center to avoid accidental hits inside the penalty circle. This strategy results in an expected gain of 20.68 points/trial and corresponds to the optimal strategy maximizing expected gain.

a short period of time (e.g., 700 ms). There are colored circular regions on the screen and subjects can win or lose points by touching them. Hits inside a target region yield a gain of 100 points; accidental hits inside a nearby penalty region incur losses of 500 points. Movements that do not reach the screen within the time limit are heavily penalized and following training they almost never occur.

Under these conditions, a movement that reaches the screen within the time limit can end in one of four possible regions: penalty only (region R_1, gain[2] $G_1 = -500$), target/penalty overlap (region R_2, gain $G_2 = -400$), target only (region R_3, gain $G_3 = 100$), or neither/background (region R_4, gain $G_4 = 0$). Subjects know that, at the end of the experiment, they will be paid a small bonus of typically 2.5 cents per 100 points, proportional to the overall sum of points they have won. To maximize the score, the subject should touch the target region as often as possible and avoid accidentally hitting the overlapping penalty region. However, under the tight time constraint of our task, motor responses are variable (Fitts and Petersen, 1964; Schmidt et al., 1979; Meyer et al., 1988), and thus subjects cannot completely control their movements.

A visuo-motor strategy s is a movement plan that the subject could adopt in this task (see also Trommershäuser et al., 2003a). The choice of visuo-motor plan depends on stochastic sensory input V and we can also write it as $s(V)$ to emphasize this dependence. The plan controls the timing and force prior to movement initiation and during movement execution. It incorporates the possibility of using visual and proprioceptive feedback to guide the movement. There are many possible visuo-motor plans. Ignoring possible biological costs to the organism, visuo-motor plans that lead to a touch on the screen within the time limit differ to the extent that they affect the probability $P_s(R_i)$ of hitting each of the four regions R_i ($i = 1,\ldots,4$).

The combination of event probabilities $P_s(R_i)$ resulting from a particular visuo-motor plan and associated gains G_i form a lottery,

$$L(s) = [P_s(R_1), G_1;\ P_s(R_2), G_2;\ P_s(R_3), G_3;\ P_s(R_4), G_4].$$

[2]Here, as well as in our previous work, we refer to rewards and penalties associated with outcomes as "gains" denoted G_i with losses coded as negative gains. In some fields of research (e.g., economics) the term "loss" is preferred with gains coded as negative losses. A third alternative is the term "value."

An alternative movement strategy s' corresponds to a second lottery,

$$L(s') = [P_{s'}(R_1), G_1;\ P_{s'}(R_2), G_2;\ P_{s'}(R_3), G_3;\ P_{s'}(R_4), G_4].$$

In planning movement in this task, subjects are effectively choosing among infinitely many lotteries of the form above — each lottery being associated with a particular movement strategy. They are engaged in decision making but in a decision-making task of remarkable complexity.

Nearly optimal strategies during movement under risk

Our previous research indicates that participants in our experiments are highly efficient at choosing visuo-motor plans that come close to maximizing expected gain. In the simple task of Trommershäuser et al. (2003a, b) a movement plan corresponds to the mean endpoint of touches resulting from repeated execution of that plan. Figure 2 shows a movement strategy whose mean endpoint (marked as a white diamond) falls on the center of the target circle. The black points represent the typical scatter that would result if this subject executed this strategy many times. This particular choice of strategy optimizes the chances of touching within the target circle but, if touches within the penalty circle carry an appreciable penalty, this choice of strategy is unlikely to win much money. A large proportion of the black dots lie in the penalty region and, over the course of many trials, a subject executing this strategy has an expected *loss* of 111.39 points per trial.[3]

In the experiment described above, subjects know that they will be paid a bonus proportional to the sum of the points they have won. To maximize the score, the subject should touch the target region as often as possible and avoid the overlapping penalty region. The gray diamond in Fig. 2(b) is the mean endpoint corresponding to the optimal strategy maximizing expected gain. It is displaced away from the penalty circle and the center of the target region. The subject's expected winnings with this choice of movement strategy are 20.68 points per trial. Our previous results show that subjects' mean endpoints are very close to the endpoint that would maximize expected gain, that is, the endpoint that corresponds to the best tradeoff between the risk of hitting the penalty region and missing the reward region (e.g., see Trommershäuser et al., 2003a, b, 2005).

Subjects in these movement-planning experiments are compensating for their own motor uncertainty by effectively solving a difficult decision-making problem. Generally, human subjects have been observed to choose nearly optimal strategies in motor tasks with changing stochastic variability (Baddeley et al., 2003; Trommershäuser et al., 2005). Additional research has extended this conclusion to tasks that involve precise timing and optimal use of available time (Dean et al., 2007) and tasks involving rapid choice between possible movement targets (Trommershäuser et al., 2006). Performance is typically less efficient for pointing movements towards noisy and blurry targets (Körding and Wolpert, 2004; Ma-Wyatt et al., 2006; Tassinari et al., 2006; Vaziri et al., 2006; Schlicht and Schrater, 2007).

Suboptimal cognitive decisions

In contrast, subjects in cognitive decision-making tasks, like the ones described above, typically fail to maximize expected gain (e.g., see Kahneman and Tversky, 2000). Human performance in decision making under risk is markedly suboptimal by any relevant criterion, fraught with cognitive biases (Kahneman and Tversky, 2000) that result in serious deficits in performance. These deviations from optimality include a tendency to frame positive and negative outcomes differentially with an exaggerated aversion to losses (Kahneman and Tversky, 1979) and to misrepresent and exaggerate small probabilities (Allais, 1953; Attneave, 1953; Lichtenstein et al., 1978; Tversky and Kahneman, 1992). This exaggeration of the frequency of low-frequency events is observed in many, but not all decision-making studies (Sedlmeier et al., 1998).

[3]See Trommershäuser et al. (2003a) for details of these computations and characterization of subjects' motor uncertainty.

Subjects do not always come close to maximizing expected gain in movement planning. Performance has been found to be suboptimal when the number of penalty/reward regions is increased (Wu et al., 2006), when the reward or penalty received is stochastic rather than determined by outcome of the subject's movement (Maloney et al., 2007), or when the uncertainty associated with the mean spatial orientation of the target region is varied trial-by-trial in an analogous unspeeded perceptual orientation discrimination task (Landy et al., 2007). Thus, while there is a collection of motor tasks where performance is remarkably good, we cannot simply claim that performance in any task with a speeded motor response is "near optimal."

Learning probabilities versus practicing the task

Of course, there are several factors that may have contributed to the near-optimal performance of subjects in the study of Trommershäuser et al. (2003a, b). In these experiments, the subject makes a long series of choices and over the course of the experiment his/her accumulated winnings increase. This is typically very different in cognitive decision-making experiments, in which the subject typically makes a single "one shot" choice. Indeed, when decision makers are faced with a series of repeated decisions they tend to move closer to maximum expected gain (Redelmeier and Tversky, 1992; Wakker et al., 1997; "the house money effect:" Thaler and Johnson, 1990). Further, the gain or loss associated with each trial is small in the movement tasks described above. Studies of risky choice find that subjects are closer to maximizing expected gain for small stakes (Camerer, 1992; Holt and Laury, 2002) and when subjects receive feedback over the course of the experiment or have prior experience with the task (Hertwig et al., 2004).

It is surprising, but subjects do not show such a trend of gradually approaching maximum expected gain during these movement-planning tasks. Before the "decision-making" phase of the experiment, subjects practiced the speeded motor task extensively by simply touching targets. During this initial training period, the experimenter monitored their motor performance until it stabilized and the experimenter could measure each subject's residual motor variability. Following training, subjects learned about the gains and losses assigned to each region and were asked to try to earn as much money as they could. They were not explicitly instructed to work out a motor strategy that took into account the spatial locations of reward and penalty regions and the magnitude of penalty and reward, but their highly efficient performance indicates that they did so from the first trial in which rewards and penalties were specified.

To summarize, in the design of the studies of Trommershäuser et al. (2003a, b) and later work (Dean et al., 2007), subjects were first trained to be "motor experts" in a simple task where they were instructed to touch targets on the screen. Only then were they confronted with a task involving tradeoffs between possible rewards and penalties. As Trommershäuser et al. (2003a, b) reported, there were no obvious trends in subjects' aim points that would suggest that subjects were modifying their decision-making strategy in response to their experience with the decision-making task. Examination of the initial trials of the decision phase of the experiment suggests that subjects immediately changed their movement strategy from that used in training to that required to optimally tradeoff the probabilities of hitting the reward and penalty regions.

This apparent lack of learning is of great interest in that it suggests that, while subjects certainly learned to carry out the motor task in the training phases of these experiments, and learned their own motor uncertainty, they seemed not to need further experience once rewards and losses were introduced into the motor task to perform as well as they did. The trends in performance found with repetition of cognitive-decision tasks seem absent in equivalent movement-planning tasks.

The surprising contrast between success in "movement planning under risk" and performance during cognitive decision making under risk is heightened by the realization that, in cognitive decision making under risk, subjects are told the exact probabilities of outcomes and thus have perfect knowledge of how their choice of strategy changes the probability of attaining each outcome. The knowledge of probabilities in equivalent motor tasks is never stated explicitly and thus can equal but never exceed the knowledge available under cognitive decision making under risk. These results suggest that humans are able to estimate the uncertainties associated with sensory and motor noise and make use of this knowledge to improve their performance.

Neural coding of uncertain decisions

Lastly, I summarize recent experimental work directed at understanding the neural coding of decisions. There is considerable experimental work along two lines. First, electrophysiological recordings have been used in monkeys to measure single-cell activity during binary-decision tasks in response to manipulations of reward and reward uncertainty (e.g., see Sugrue et al., 2005 for a review). Second, in a typical functional magnetic resonance imaging (fMRI) decision-making experiment, subjects perform a series of risky or ambiguous decisions, that is, they choose between pairs of lotteries that differ with respect to the subject's knowledge of outcomes and probabilities of each outcome.

Following Herrnstein's (1961) pioneering behavioral work, electrophysiological studies typically employ a paradigm in which a monkey chooses between two alternative responses that may differ with respect to the sensory information available on each trial, the prior odds, and the outcome assigned to each response alternative (e.g., see Gold & Shadlen, 2002, 2007; Heekeren et al., 2008). These experiments yield insight into how sensory information is integrated during so-called perceptual decisions. Electrophysiological evidence from recordings from monkeys indicates that optimal integration of the noisy neural responses to motion of a small set of neurons in areas MT (Britten et al., 1992) or lateral intraparietal area (LIP) (Shadlen and Newsome, 2001) can achieve performance at levels as good as the monkey, and may be used to predict the monkey's response reliably on a trial-by-trial basis.

Value-based decisions, as opposed to perceptual decisions, are decisions in which different values are associated with a response alternative, so that the sensory input interacts with the physiological needs of the animal in the presence of reward. In these experiments, reward is typically manipulated by assigning variable amounts of juice to different response alternatives (Platt and Glimcher, 1999; Sugrue et al., 2004). When rewards were assigned stochastically, the monkey's choices appeared to be based on an estimate of the probability of reward gained by sampling over the last few trials (Sugrue et al., 2004). Single-cell activity in response to stochastic variations of reward has been found in ventral midbrain areas (Fiorillo et al., 2003). This phasic activity of the dopamine neurons correlated with the so-called prediction error, that is, with the difference between actual and expected reward (Schultz et al., 1997; Morris et al., 2004). These same neurons also produced a tonic response that was highest in conditions of highest risk, that is, in trials in which the probability of receiving the reward was 0.5. However, the behavioral relevance of midbrain dopaminergic single-cell activity recorded in response to changes in reward probability remains controversial (Bayer and Glimcher, 2005).

Similarly, using fMRI techniques in humans, dopamine-dependent modulation of striatal and brainstem activity has been identified as reward prediction error (Pessiglione et al., 2006; D'Ardenne et al., 2008). A variety of subcortical and cortical areas have been implicated in the coding of various other decision variables such as gain, probability of reward, risk, and ambiguity (for more extensive reviews, see Glimcher and Rustichini, 2004; O'Doherty, 2004; Trepel et al., 2005; Daw and Doya, 2006; Montague et al., 2006). Most paradigms employ a visual representation of a gambling task and, after a delay of several seconds, subjects are instructed to choose between pairs of options by key-press. Brain activity is monitored during the delay period and correlated with various decision variables.

If the probability of reward is known to the subject, the task is decision making under risk. Neural activity may be correlated with gain or loss of a potential reward, the probability of reward, their product (expected gain) or risk (the variance of gain). A number of studies suggest that reward value is encoded in the striatum and portions of prefrontal cortex (PFC) and orbitofrontal cortex (OFC) (O'Doherty, 2004; Knutson et al., 2005; Tanaka et al., 2005; Daw et al., 2006; Tom et al., 2007; D'Ardenne et al., 2008). Learning models are often based on a prediction of reward, with learning driven by reward-prediction errors. Reward-prediction error signals are generally localized to the striatum, although also seen in the striatum, OFC, and amygdala (O'Doherty, 2004; Daw et al., 2006; Pessiglione et al., 2006; Yacubian et al., 2006). It has been difficult to disentangle probability of reward from expected gain, and most studies see responses correlated with expected gain in the striatum, OFC, and medial PFC (Delgado et al., 2005; Hsu et al., 2005; Knutson et al., 2005; Daw et al., 2006; Preuschoff et al., 2006). Several of these studies see an increase in activity with increasing expected value whether or not the outcome is an expected loss or gain. In contrast, Yacubian et al. (2006) suggest that while expected gains are encoded in the striatum, expected losses result in responses in the amygdala, perhaps associated with negative emotion. This is supported by the finding that decisions consistent with framing effects (e.g., preferring the certain option when a gamble is framed in terms of gains, but taking the risky option when framed in terms of losses), are correlated with increased response in the amygdala. PFC response is higher in subjects that have less of a framing effect (De Martino et al., 2006) suggesting a requirement for cognitive control to suppress this cognitive bias.

An important distinction in these gambling tasks is between risk (in which the probabilities of the outcomes are known precisely) and ambiguity (when they are not).[4] Responses correlated with risk have been found in anterior insula, OFC, and striatum (Preuschoff et al., 2006) as well as dorsolateral PFC and posterior parietal cortex (Huettel et al., 2005). It has also been found that risky choices (choice of a "stock") lead to responses in ventral striatum, while risk-avoiding choices (choice of a "bond") lead to responses in anterior insula (Kuhnen and Knutson, 2005).

In behavioral experiments, subjects typically exhibit a preference for options in which the probabilities for incurring rewards are clearly stated, a phenomenon called ambiguity aversion (Ellsberg, 1961; Fox and Tversky, 1995). Responses correlated with the ambiguity of a decision have been found in the posterior part of the inferior frontal sulcus (Huettel et al., 2006), OFC, amygdala, and dorsomedial PFC, along with a negative correlation with responses in the striatum (Hsu et al., 2005). The correlation with amygdala activity suggests a role for emotions in ambiguous decisions. Subjects with a preference for ambiguity over risk show stronger responses in lateral PFC, while subjects with a preference for risk over ambiguity show stronger responses in posterior parietal cortex (Huettel et al., 2006).

A reward received immediately is generally valued more than one that will be delayed, a phenomenon known as temporal discounting. If a reward will be received immediately, a variety of brain areas respond including striatum and OFC, but the inclusion of a potentially delayed reward recruits other areas including PFC (McClure et al., 2004; Tanaka et al., 2005) suggesting the need for cognitive control for choices involving delayed gratification. Prefrontal activity has also been found to correlate with correct performance in a noisy perceptual-discrimination task (Heekeren et al., 2004), suggesting that the human prefrontal cortex serves as an integrator of noisy sensory evidence (see also Rorie and Newsome, 2005; Heekeren et al., 2008).

To summarize, the results presented here indicate that movement planning under risk shares the same formal structure as cognitive decision making. Subjects in movement tasks are generally found to be very good at choosing motor strategies that come close to maximizing expected gain. In contrast, subjects in cognitive decision-making tasks typically fail to maximize expected

[4]In the movement planning experiments described above, subjects know their motor uncertainty very well, and, thus, have a near precise knowledge of the probability to touch reward or loss region; we therefore refer to the motor tasks described above as movement planning under risk (e.g., see Trommershäuser et al., 2003a, 2005).

gain. Moreover, the sources of uncertainty in motor tasks are endogenous: they reflect the organism's own uncertainty in planning movement while, in contrast, uncertainty in cognitive tasks is typically imposed by the experimenter. Thus, probabilistic information from cognition, perception, and movement has different origins and it would be of interest to compare the neural circuits underlying the representation of probability in movement and cognitive decision making.

In the latter, feedback about outcomes typically reduces biases and misperceptions in the representation of probability estimates, moving behavior closer to strategies maximizing expected gain. On the contrary, in movement planning under risk subjects' performance is initially already close to optimal and does not appear to change with feedback. Reward clearly affects and changes cognitive decisions, but this does not seem to be the case for movement under risk — which suggests that feedback about errors and reward is integrated differently for motor and cognitive decisions. A comparison of cognitive decision-making tasks and equivalent movement tasks can provide insight how the brain makes two seemingly very different types of decisions where the formal structures of the decisions are identical.

Acknowledgments

I thank Hauke Heekeren and Markus Raab for helpful comments on an earlier version of the manuscript. This work was supported by the Deutsche Forschungsgemeinschaft (Emmy-Noether-Programme; grant TR 528/1-3; 1-4).

References

Alexander, R. M. (1997). A minimum energy cost hypothesis for human arm trajectories. *Biological Cybernetics*, *76*, 97–105.

Allais, M. (1953). Le comportement de l'homme rationnel devant la risque: critique des postulats et axiomes de l'école Américaine. *Econometrica*, *21*, 503–546.

Attneave, F. (1953). Psychological probability as a function of experienced frequency. *Journal of Experimental Psychology*, *46*, 81–86.

Baddeley, R. J., Ingram, H. A., & Miall, R. C. (2003). System identification applied to a visuomotor task: near-optimal human performance in a noisy changing task. *Journal of Neuroscience*, *7*, 3066–3075.

Bayer, H. M., & Glimcher, P. W. (2005). Midbrain dopamine neurons encode a quantitative reward prediction error signal. *Neuron*, *47*, 129–141.

Blackwell, D., & Girshick, M. A. (1954). *Theory of games and statistical decisions*. New York: Wiley.

Britten, K. H., Shadlen, M. N., Newsome, W. T., & Movshon, J. A. (1992). The analysis of visual motion: a comparison of neuronal and psychophysical performance. *Journal of Neuroscience*, *12*, 4745–4765.

Burdet, E., Osu, R., Franklin, D. W., Milner, T. E., & Kawato, M. (2001). The central nervous system stabilizes unstable dynamics by learning optimal impedance. *Nature*, *414*, 446–449.

Camerer, C. (1992). The rationality of prices and volume in experimental market. *Organizational Behavior and Human Decision Processes*, *51*, 237–272.

Cuijpers, R. H., Smeets, J. B. J., & Brenner, E. (2004). On the relation between object shape and grasping kinematics. *Journal of Neurophysiology*, *91*, 2598–2606.

Daw, N. D., & Doya, K. (2006). The computational neurobiology of learning and reward. *Current Opinions in Neurobiology*, *16*, 199–204.

Daw, N. D., O'Doherty, J. P., Dayan, P., Seymour, B., & Dolan, R. J. (2006). Cortical substrates for exploratory decisions in humans. *Nature*, *441*, 876–879.

D'Ardenne, K., McClure, S. M., Nystrom, L. E., & Cohen, J. D. (2008). BOLD responses reflecting dopaminergic signals in the human ventral tegmental area. *Science*, *319*, 1264–1267.

Dean, M., Wu., S.-W., & Maloney, L. T. (2007). Trading off speed and accuracy in rapid, goal-directed movements. *Journal of Vision*, *7*, 10.1–10.12.

Delgado, M. R., Miller, M. M., Inati, S., & Phelps, E. A. (2005). An fMRI study of reward-related probability learning. *Neuroimage*, *24*, 862–873.

De Martino, B., Kumaran, D., Seymour, B., & Dolan, R. J. (2006). Frames, biases, and rational decision-making in the human brain. *Science*, *313*, 684–687.

Dornay, M., Uno, Y., Kawato, M., & Suzuki, R. (1996). Minimum muscle-tension change trajectories predicted by using a 17-muscle model of the monkey's arm. *Journal of Motor Behavior*, *2*, 83–100.

Ellsberg, D. (1961). Risk, ambiguity, and the Savage axioms. *Quarterly Journal of Economics*, *75*, 643–669.

Fiorillo, C. D., Tobler, P. N., & Schultz, W. (2003). Discrete coding of reward probability and uncertainty by dopamine neurons. *Science*, *299*, 1898–1902.

Fitts, P. M., & Petersen, J. R. (1964). Information capacity of discrete motor responses. *Journal of Experimental Psychology*, *67*, 103–112.

Flash, T., & Hogan, N. (1985). The coordination of arm movements: an experimentally confirmed mathematical model. *Journal of Neuroscience*, *5*, 1688–1703.

Fox, C. R., & Tversky, A. (1995). Ambiguity aversion and comparative ignorance. *Quarterly Journal of Economics*, *110*, 585–603.

Franklin, D. W., Liaw, G., Milner, T. E., Osu, R., Burdet, E., & Kawato, M. (2007). Endpoint stiffness of the arm is directionally tuned to instability in the environment. *Journal of Neuroscience*, *27*, 7705–7716.

Glimcher, P. W., & Rustichini, A. (2004). Neuroeconomics: the consilience of brain and decision. *Science*, *306*, 447–452.

Gold, J. I., & Shadlen, M. N. (2002). Banburismus and the brain: decoding the relationship between sensory stimuli, decisions, and reward. *Neuron*, *36*, 299–308.

Gold, J. I., & Shadlen, M. N. (2007). The neural basis of decision making. *Annual Review of Neuroscience*, *30*, 535–574.

Green, D. M., & Swets, J. A. (1966/1974). *Signal detection theory and psychophysics*. New York: Wiley, Reprinted 1974, Huntington, NY: Krieger.

Hamilton, A. F. C., & Wolpert, D. M. (2002). Controlling the statistics of action: obstacle avoidance. *Journal of Neurophysiology*, *87*, 2434–2440.

Harris, C. M., & Wolpert, D. M. (1998). Signal-dependent noise determines motor planning. *Nature*, *394*, 780–784.

Heekeren, H. R., Marrett, S., Bandettini, P. A., & Ungerleider, L. G. (2004). A general mechanism for perceptual decision-making in the human brain. *Nature*, *431*, 859–862.

Heekeren, H. R., Marrett, S., & Ungerleider, L. G. (2008). The neural systems that mediate human perceptual decision making. *Nature Reviews. Neuroscience*, *9*, 467–479.

Herrnstein, R. J. (1961). Relative and absolute strength of response as a function of frequency of reinforcement. *Journal of the Experimental Analysis of Behavior*, *4*, 267–272.

Hertwig, R., Barron, G., Weber, E. U., & Erev, I. (2004). Decisions from experience and the effect of rare events in risky choice. *Psychological Science*, *15*, 534–539.

Holt, C., & Laury, S. (2002). Risk aversion and incentive effects. *American Economic Review*, *92*, 1644–1655.

Hsu, M., Bhatt, M., Adoplhs, R., Tranel, D., & Camerer, C. F. (2005). Neural systems responding to degrees of uncertainty in human decision-making. *Science*, *310*, 1680–1683.

Huettel, S. A., Song, A. W., & McCarthy, G. (2005). Decisions under uncertainty: probabilistic context influences activation of prefrontal and parietal cortices. *Journal of Neuroscience*, *25*, 3304–3311.

Huettel, S. A., Stowe, C. J., Gordon, E. M., Warner, B. T., & Platt, M. L. (2006). Neural signatures of economic preferences for risk and ambiguity. *Neuron*, *49*, 765–775.

Kahneman, D., & Tversky, A. (1979). Prospect theory: an analysis of decision under risk. *Econometrica*, *47*, 263–292.

Kahneman, D., & Tversky, A. (Eds.). (2000). *Choices, values and frames*. Cambridge: Cambridge University Press.

Kaminsky, T., & Gentile, A. M. (1986). Joint control strategies and hand trajectories in multijoint pointing movements. *Journal of Motor Behavior*, *18*, 261–278.

Knight, F. H. (1921). *Risk, uncertainty and profit*. Boston: Houghton Mifflin.

Knill, D., & Richards, W. (1996). *Perception as Bayesian inference*. New York: Cambridge University Press.

Knutson, B., Taylor, J., Kaufman, M., Peterson, R., & Glover, G. (2005). Distributed neural representation of expected value. *Journal of Neuroscience*, *25*, 4806–4812.

Körding, K. P., & Wolpert, D. M. (2004). Bayesian integration in sensorimotor learning. *Nature*, *427*, 244–247.

Kuhnen, C. M., & Knutson, B. (2005). The neural basis of financial risk taking. *Neuron*, *47*, 763–770.

Landy, M. S., Goutcher, R., Trommershäuser, J., & Mamassian, P. (2007). Visual estimation under risk. *Journal of Vision*, *7*, 1–15.

Landy, M. S., Maloney, L. T., Johnston, E. B., & Young, M. (1995). Measurement and modeling of depth cue combination: in defense of weak fusion. *Vision Research*, *35*, 389–412.

Lichtenstein, S., Slovic, P., Fischhoff, B., Laymann, M., & Coombs, B. (1978). Judged frequency of lethal events. *Journal of Experimental Psychology — Learning Memory and Cognition*, *4*, 551–578.

Maloney, L. T. (2002). Statistical decision theory and biological vision. In D. Heyer & R. Mausfeld (Eds.), *Perception and the physical world: psychological and philosophical issues in perception* (pp. 145–189). Chichester: Wiley.

Maloney, L. T., Trommershäuser, J., & Landy, M. S. (2007). Questions without words: a comparison between decision making under risk and movement planning under risk. In W. Gray (Ed.), *Integrated models of cognitive systems* (pp. 297–315). New York: Oxford University Press.

Ma-Wyatt, A., Stritzke, M., & Trommershäuser, J. (2006). Eye-hand coordination can be altered for rapid pointing under risk. *Journal of Vision*, *6*, 920a.

McClure, S. M., Laibson, D. I., Loewenstein, G., & Cohen, J. D. (2004). Separate neural systems value immediate and delayed monetary rewards. *Science*, *306*, 503–507.

Meyer, D. E., Abrahams, R. A., Kornblum, S., Wright, C. E., & Smith, J. E. (1988). Optimality in human motor performance: ideal control of rapid aimed movements. *Psychological Review*, *95*, 340–370.

Montague, P. R., King-Casas, B., & Cohen, J. D. (2006). Imaging valuation models in human choice. *Annual Review of Neuroscience*, *29*, 417–448.

Morris, G., Arkadir, D., Nevet, A., Vaadia, E., & Bergman, H. (2004). Coincident but distinct messages of midbrain dopamine and striatal tonically active neurons. *Neuron*, *43*, 133–143.

O'Doherty, J. P. (2004). Reward representations and reward-related learning in the human brain: insights from neuroimaging. *Current Opinion in Neurobiology*, *14*, 769–776.

Pessiglione, M., Seymour, B., Flandin, G., Dolan, R. J., & Frith, C. D. (2006). Dopamine-dependent prediction errors underpin reward-seeking behaviour in humans. *Nature*, *442*, 1042–1045.

Platt, M. L., & Glimcher, P. W. (1999). Neural correlates of decision variables in parietal cortex. *Nature*, *400*, 233.

Preuschoff, K., Bossaerts, P., & Quartz, S. R. (2006). Neural differentiation of expected reward and risk in human subcortical structures. *Neuron*, *51*, 381–390.

Redelmeier, D. A., & Tversky, A. (1992). On the framing of multiple prospects. *Psychological Science*, *3*, 191–193.

Rorie, A. E., & Newsome, W. T. (2005). A general mechanism for decision-making in the human brain? *Trends in Cognitive Science*, *9*, 41–43.

Sabes, P. N., & Jordan, M. I. (1997). Obstacle avoidance and a perturbation sensitivity model for motor planning. *Journal of Neuroscience*, *17*, 7119–7128.

Schlicht, E. J., & Schrater, P. R. (2007). Impact of coordinate transformation uncertainty on human sensorimotor control. *Journal of Neurophysiology*, *97*, 4203–4214.

Schmidt, R. A., Zelaznik, H., Hawkins, B., Frank, J. S., & Quinn, J. T. (1979). Motor output variance: a theory for the accuracy of rapid motor acts. *Psychological Review*, *86*, 415–451.

Schultz, W., Dayan, P., & Montague, P. (1997). A neural substrate of prediction and reward. *Science*, *275*, 1593–1599.

Sedlmeier, P., Hertwig, R., & Gigerenzer, G. (1998). Are judgments of the positional frequencies of letters systematically biased due to availability? *Journal of Experimental Psychology — Learning Memory and Cognition*, *24*, 754–770.

Shadlen, M. N., & Newsome, W. T. (2001). Neural basis of a perceptual decision in the parietal cortex (area LIP) of the rhesus monkey. *Journal of Neurophysiology*, *86*, 1916–1936.

Soechting, J. F., Buneo, C. A., Herrmann, U., & Flanders, M. (1995). Moving effortlessly in three dimensions: does Donders' law apply to arm movement? *Journal of Neuroscience*, *15*, 6271–6280.

Soechting, J. F., & Lacquaniti, F. (1981). Invariant characteristics of a pointing movement in man. *Journal of Neuroscience*, *1*, 710–720.

Sugrue, L. P., Corrado, G. S., & Newsome, W. T. (2004). Matching behavior and the representation of value in the parietal cortex. *Science*, *304*, 1782–1787.

Sugrue, L. P., Corrado, G. S., & Newsome, W. T. (2005). Choosing the greater of two goods: neural currencies for valuation and decision making. *Nature Reviews. Neuroscience,*, *6*, 363–375.

Tanaka, S. C., Doya, K., Okada, G., Ueda, K., Okamoto, Y., & Yamawaki, S. (2005). Prediction of immediate and future rewards differentially recruits cortico-basal ganglia loops. *Nature Neuroscience*, *7*, 887–893.

Tassinari, H., Hudson, T. E., & Landy, M. S. (2006). Combining priors and noisy visual cues in a rapid pointing task. *Journal of Neuroscience*, *26*, 10154–10163.

Thaler, R., & Johnson, E. J. (1990). Gambling with the house money and trying to break even: the effects of prior outcomes on risky choice. *Management Science*, *36*, 643–660.

Todorov, E. (2004). Optimality principles in sensorimotor control. *Nature Neuroscience*, *9*, 907–915.

Tom, S. M., Fox, C. R., Trepel, C., & Poldrack, R. A. (2007). The neural basis of loss aversion in decision-making under risk. *Science*, *315*, 515–518.

Trepel, C., Fox, C. R., & Poldrack, R. A. (2005). Prospect theory on the brain? Toward a cognitive neuroscience of decision under risk. *Cognitive Brain Research*, *23*, 34–50.

Trommershäuser, J., Gepshtein, S., Maloney, L. T., Landy, M. S., & Banks, M. S. (2005). Optimal compensation for changes in task-relevant movement variability. *Journal of Neuroscience*, *25*, 7169–7178.

Trommershäuser, J., Landy, M. S., & Maloney, L. T. (2006). Humans rapidly estimate expected gain in movement planning. *Psychological Science*, *17*, 981–988.

Trommershäuser, J., Maloney, L. T., & Landy, M. S. (2003a). Statistical decision theory and rapid, goal-directed movements. *Journal of the Optical Society of America A*, *20*, 1419–1433.

Trommershäuser, J., Maloney, L. T., & Landy, M. S. (2003b). Statistical decision theory and trade-offs in the control of motor response. *Spatial Vision*, *16*, 255–275.

Trommershäuser, J., Maloney, L. T., & Landy, M. S. (2008). Decision making, movement planning, and statistical decision theory. *Trends in Cognitive Science*, *12*, 291–297.

Tversky, A., & Kahneman, D. (1992). Advances in prospect theory: cumulative representation of uncertainty. *Journal of Risk and Uncertainty*, *5*, 297–323.

Uno, Y., Kawato, M., & Suzuki, R. (1989). Formation and control of optimal trajectory in human multijoint arm movement: minimum torque-change model. *Biological Cybernetics*, *61*, 89–101.

Vaziri, S., Diedrichsen, J., & Shadmehr, R. (2006). Why does the brain predict sensory consequences of oculomotor commands? Optimal integration of the predicted and the actual sensory feedback. *Journal of Neuroscience*, *26*, 4188–4197.

Wakker, P. P., Thaler, R. H., & Tversky, A. (1997). Probabilistic insurance. *Journal of Risk and Uncertainty*, *15*, 7–28.

Wu, S.-W., Trommershäuser, J., Maloney, L. T., & Landy, M. S. (2006). Limits to human movement planning in tasks with asymmetric gain landscapes. *Journal of Vision*, *6*, 53–63.

Yacubian, J., Glascher, J., Schroeder, K., Sommer, T., Braus, D. F., & Büchel, C. (2006). Dissociable systems for gain- and loss-related value predictions and errors of prediction in the human brain. *Journal of Neuroscience*, *26*, 9530–9537.

M. Raab et al. (Eds.)
Progress in Brain Research, Vol. 174
ISSN 0079-6123

CHAPTER 22

Advances in coupling perception and action: the quiet eye as a bidirectional link between gaze, attention, and action

Joan N. Vickers*

Faculty of Kinesiology, University of Calgary, Calgary, AB, Canada

Abstract: One of the most elusive mysteries in psychology is perception–action coupling and the extent vision for perception is distinct from vision for action. In this chapter, I explore research on the control of the gaze during well-known sport tasks (vision for action) and the bidirectional link between perceptual and cognitive processes and optimal/nonoptimal motor performance. Considerable evidence now exists showing that specific gaze characteristics underlie higher levels of sport performance. The quiet eye has emerged as a characteristic of higher levels of performance and is the final fixation or tracking gaze that occurs prior to the final movement. Cognitive and ecological accounts of the quiet eye are presented and current controversies and future directions explored.

Keywords: quiet eye; decision making; attention; perception

Introduction

All actions have the three qualities that make up the themes of this workshop (entitled *Mind and motion: the bidirectional link between thought and action*; 29–31 May 2008; Bielefeld, Germany) — the individual must first be able to perceive what needs to be done and represent it within neural, perceptual, and/or cognitive structures (group 1); they must be able to select the best course of action from the many options that may be present efficiently (group 2); and they must be able to implement a cognitively planned course of action so that an intended outcome occurs (group 3). My goal in this chapter is to summarize my research in terms of the last theme — how is a cognitively intended course of action physically implemented. To answer this question I review the eye movements and gaze research in which my students, colleagues, and I have coupled perception and action in experiments using well-known sport and other motor tasks. Following this, I explain how the results we have obtained, in particular around the quiet eye, have led to successful training interventions. Finally, I cover the theoretical bases for the research and the two competing explanations currently in the literature, ending in the final section with a discussion on some current controversies and future directions.

My own work has concentrated on using eye movements, and more specifically measures of gaze, as a way of looking into the mind of motor performers. What is it that they see that contributes,

*Corresponding author
Tel.: +1-403-220-3420; Fax: +1-403-284-3553;
E-mail: vickers@ucalgary.ca

DOI: 10.1016/S0079-6123(09)01322-3

either positively or negatively, to their motor success or failure? By concentrating on performance in well-known sport tasks it is possible to access the minds of individuals who have trained at least 10,000 hours in specific tasks and reached the highest levels of achievement as confirmed by independent statistics, or alternatively, have trained for 10,000 hours and not reached those heights (Ericsson, 1996; Helsen et al., 1998). Why is it that some athletes achieve the ultimate in performance statistics while many others, who are as gifted physically and have trained as hard, do not achieve these levels? Wrapped up within elite and near-elite athletes are living examples of intended actions going well and not so well.

In a typical research study carried out in my laboratory, elite and near-elite athletes perform experimental tasks in conditions that are very similar to those found within their sport. Their gaze is recorded by a mobile eye tracker, which is integrated with a motion analysis system that records their movements at the same time. The athletes perform until an equal number of successful and nonsuccessful trials are achieved under various experimental conditions in which we may vary task complexity, competitive pressure, athlete anxiety, and/or physiological arousal. Our goal is to determine the types of gaze behaviors that underlie both successful and nonsuccessful performances. From these gaze behaviors we then attempt to explain the perceptual and cognitive processes that define optimal and nonoptimal motor performance.

Linking gaze to attention and other cognitive processes has only been possible with important changes in the literature linking gaze and attention. For many years, it was difficult to link shifts in gaze with shifts in attention (e.g., Posner, 1980) but more recent studies show that under certain conditions a shift in the gaze is invariably preceded by a shift in attention (Shepherd et al., 1986; Kowler et al., 1995; Deubel and Schneider, 1996; Corbetta, 1998; Henderson, 2003). There is now strong evidence that when a saccade is made to a new location there is a corresponding shift in attention in the direction of the saccade. This means that when athletes shift their gaze to a new location, they also shift their attention to that location at least for a brief period. However, it is important to stress that once the gaze and attention have arrived at a new location, the duration of the gaze may not always be an indicator of attention. Athletes may still covertly divert their attention elsewhere even as fixation remains on a location.

In a typical study all the gazes are coded and analyzed coupled with the distinct motor phases that make up the movement. Over many studies, one gaze has emerged as being a characteristic of elite performance in a wide range of targeting tasks. For a given motor task, the quiet eye is defined as the final fixation or tracking gaze that is located on a specific location or object in the visuomotor workspace within 3° of visual angle for a minimum of 100 ms. The onset of the quiet eye occurs prior to the final movement in the task and the offset occurs when the gaze deviates off the object or location by more than 3° of visual angle for a minimum of 100 ms, therefore the quiet eye can carry through and beyond the final movement of the task. The quiet eye of elite performers is significantly longer than that of near-elite, or lower-skilled performers, meaning those who consistently achieve high levels of performance have learned to fixate or track critical objects or locations for longer durations irrespective of the conditions encountered. The quiet eye onset of elite performers is invariably earlier; elite performers have found a way to see critical information earlier than near-elite and lower-skilled performers and to process this information longer prior to making the final movement. Finally, the quiet eye of elite performers is of an optimal duration, being neither too long nor too short, but ideal given the constraints of the task being performed. What is striking about the quiet eye is the robustness of the finding that has been found in targeting tasks (Vickers, 1992, 1996; Janelle et al., 2000; Williams et al., 2002; Vickers, 2004; Oudejans et al., 2005; Oliveira, 2007; Vickers and Williams, 2007), interceptive timing tasks (Adolphe et al., 1997; Vickers and Adolphe, 1997; Rodrigues et al., 2002; McPherson and Vickers, 2004; Panchuk and Vickers, 2006), as well as tactical tasks (Martell and Vickers, 2004; Vickers, 2006, 2007). The quiet eye has also been identified as a gaze affected by high levels of performance pressure and anxiety (Vickers and Williams, 2007; Behan and Wilson, 2008). In

Fig. 1. A frame of vision-in-action data showing the quiet eye gaze of the player on the front hoop as indicated by the black cursor.

addition, a recent meta-analysis by Mann et al. (2007) has identified the quiet eye as one of only three predictors of perceptual-motor expertise (along with fixation location and a low frequency of fixation). Figure 1 presents a typical frame of the vision-in-action data showing the quiet-eye location of an elite basketball player. Each frame of vision-in-action data is comprised of four parts: an eye image, a gaze image that below is shown on the right with the gaze indicated by the black gaze cursor on the hoop, a motor image shown below on the left, and a time code that records time synchronized in all three images.

Quiet-eye training

In addition to carrying out research to determine the characteristics of the quiet eye in various motor tasks, research has also determined the effectiveness of quiet-eye training to improve sports performance. In order for a perceptual and attention training program to be successful in any domain, four conditions must be met (Gopher, 1993; Williams and Grant, 1999). First, it must be shown that control over the gaze and attention leads to higher levels of motor performance. Second, individuals must be identified who have difficulty controlling their gaze and attention in a task and classified according to the depth and quality of training needed. Third, there must be evidence that these individuals have the ability to improve their gaze and attention with proper training. Finally, it must be shown that the training of a more optimal control of the gaze and attention contributes to improvements in motor performance in both the research and competitive setting. The first two conditions have been met by the many studies cited above showing that elite sports performers possess unique gaze and quiet-eye characteristics compared to those who are nonelite or near-elite. The third and fourth requirements have been met in sport studies where improvements in gaze and motor performance have been found under both experimental and field conditions (Adolphe et al., 1997; Harle and Vickers, 2001; Oudejans et al., 2005; Vickers, 2007).

Quiet-eye training involves using both video modeling and video feedback of the gaze coupled with the motor behaviour followed by training in drills that help athletes develop the same quiet-eye focus as found in elite performers. Seven steps are involved.

1. It is first critical to carry out research in the task and identify the quiet-eye characteristics of elite performers during successful trials. Elite athletes are tested and their quiet-eye location, onset, offset, and duration identified during successful and unsuccessful trials identified. Once this information is obtained then task-specific norms and standards for quiet-eye training can be applied.
2. The athlete is tested *in situ* using a mobile eye tracker and motion analysis system while performing the task in conditions similar to those found in competition. The duration of the quiet-eye location, onset, offset, and duration is determined during successful and unsuccessful shots.
3. Using an elite prototype (derived from step 1), the athlete is taught the gaze characteristics identified in the literature and reflected through the elite prototype in terms of a low frequency of gaze, the final fixation allocated to a specific location, and an early onset and duration of the quiet eye coupled with the final movement. The athlete is taught to reduce the frequency of fixation or tracking gaze to fewer locations with the final fixation, or quiet eye, being located on a critical location.
4. The athlete is then shown his/her own coupled gaze and motor data and given feedback about their quiet eye. An important part of this step is to ask the athlete questions about how their gaze and attention differs from, or is similar to, that of the elite prototype using frame-by-frame video comparison. The key is to probe cognitively how much the athlete understands about the control of their attention as they perform. Most athletes are adept at identifying the differences between their gaze and that of expert performers; many are surprised at how erratic their gaze and attention is compared to that of elite performers.
5. The athlete is asked to select an aspect of their quiet eye they want to change during subsequent attempts. During this decision-training phase it is important to encourage the athlete to concentrate only on adopting one or more of the quiet-eye attributes (location, onset, offset, duration) with limited or no coaching of changes in their technique. Preliminary studies have shown that when the gaze control improves athletes change their technique to accommodate the more effective control of their gaze and attention (Harle and Vickers, 2001; Oudejans et al., 2005).
6. The athlete should then practice drills designed to promote the desired quiet-eye focus. The goal is to have the athlete practice in drills designed to promote the attention, quiet eye, and other gaze characteristics of elite performers. A variety of drills should be designed using a number of decision-training tools that have strong support in the motor learning literature including variable and random practice, bandwidth feedback, questioning, and an external focus of attention rather than an internal focus. Using this approach, unusually high increases in performance have been found in a number of sport areas (Adolphe et al., 1997; Vickers et al., 1999; Harle and Vickers, 2001; Vickers, 2003; Vickers et al., 2004; Oudejans et al., 2005; Raab et al., 2005; Vickers, 2007).
7. The athletes' performance in competition should be assessed and follow-up quiet-eye tests carried out as is needed to improve the athlete's performance permanently.

We used this training process with elite and near-elite varsity basketball players (Harle and Vickers, 2001). We found a significant increase in quiet-eye duration and free-throw accuracy in the experimental setting in year 1 and in the second year, the team improved their free-throw shooting accuracy in games over a full season from 54 to 76% (an increase of 22%, which was significantly higher than two control teams who did not receive similar training). Oudejans et al. (2005) investigated the effects of "visual control training" on basketball jump shooting by expert male players. The goal of the training was to improve the athletes' pickup of information during the final period before ball release. Two methods were used over a training period of 8 weeks. The players wore liquid-crystal goggles that occluded the hoop so they could only see the hoop during the final 350 ms. The goggles forced the players to attend maximally to the hoop during the short amount of time it was visible. In addition, the players were required to shoot from behind a screen set up at the free-throw line and placed at a height that blocked their view of the hoop. The players increased the duration of time the goggles were open from a mean of 353 ms before training to 386 ms following training. Field-shooting accuracy improved in games from a mean of 46% before training to 61% following training, for a mean increase of 15%. The control group, who did not receive similar training, did not improve but maintained the same shooting percentage of 42%. The amount of improvement in these studies is considerable and shows that athletes who are trained to control their gaze, attention and decision making while performing in drills that simulate events within the game are much greater than when physical and/or psychological training are used alone.

Theoretical accounts of the quiet eye

Although the quiet eye has been found to underlie higher levels of skill and performance in a wide variety of sport tasks, two different theoretical accounts have been put forward to explain why this gaze is important in motor performance. Below these two theoretical perspectives are reviewed, with the first from cognitive psychology/neuroscience and the second from an ecological psychology/dynamic systems perspective.

Cognitive psychology/neuroscience

Because the quiet-eye onset occurs prior to the final movement in a task and is of longer duration when performance is higher, the quiet-eye period represents the period of time when the neural networks are organized to control the movement (Vickers, 1996; Williams et al., 2002; Behans and Wilson, 2008). The quiet-eye period represents the time needed to process cognitively the information that is being fixated or tracked and to focus attention on the demands of the task. In this view the neural networks underlying higher levels of performance must be "fed" very precise external visual information in order for the complex neural systems underlying control of the limbs to be assembled and activated. When the location, onset, offset, and duration of the quiet eye are all optimal then the resultant performance is superior; when any one of these dimensions is nonoptimal then performance is inferior.

Results in support of this view were found by Williams et al. (2002) who recorded the gaze of highly skilled and novice billiard players as they performed shots of varying complexity. In two experiments, they manipulated the quiet-eye duration during easy, intermediate, and difficult shots. Since more complex motor responses require longer pre-programming times (e.g., Henry, 1953) it was expected that if the quiet-eye duration was related to cognitive programming the more complex shots would require a longer quiet-eye period. The results showed that the quiet-eye period was significantly longer for the highly skilled players than for the novices in all levels of shots, and it was also longer on hits compared with misses. When the preparation time was reduced by 25% and 50% of what each player normally used, shorter quiet-eye periods were a characteristic of poorer performance, irrespective of skill level. Williams et al. (2002) interpreted the quiet-eye duration as the critical period when cognitive processing was carried out.

It is clear from this and other studies that the neural, perceptual, and cognitive systems need an optimal amount of time to process critical visual information prior to an action being carried out. Posner and Raichle (1994) have identified three attention neural networks that may be central to this process. The posterior orienting network is responsible for controlling the gaze and attention in space. This network, which is located in the parietal region, directs the gaze to specific locations of importance in a task. It is also responsible for preventing the disengagement of the gaze to other locations. Free-throw shooters, golfers, rifle shooters, and cricket players may use the posterior network to align their gaze to specific locations in space and maintain the gaze at a single location. The anterior executive network is responsible for bringing into consciousness critical aspects of what is being fixated. This network interprets what is being viewed and imposes a higher-order understanding on the task based on past experience and knowledge. Skilled players bring a richer knowledge base and more refined rules than less-skilled performers, who are often unsure of what they need to see as they perform. Finally, the vigilance network is responsible for coordinating the posterior and anterior networks and preventing unwanted or distracting information from gaining access to the other networks during periods of sustained focus. The vigilance network is responsible for the sustained concentration seen in elite players, especially during pressure-filled games of long duration.

More recently, neurophysiological studies in monkeys have suggested that the brain regions that are involved in selecting and planning a certain action have an important role in forming decisions that lead to that action. Heekeren et al. (2003) asked human observers to make direction-of-motion judgments about dynamic random-dot-motion stimuli and indicate their judgments with an eye movement to one of two visual targets. The authors localized regions that are part of the oculomotor network. Importantly, during the period of decision formation between the onset of visual motion and the cue to respond, the percent change in the blood-oxygen-level-dependent (BOLD) signal in the oculomotor network was highly correlated with the strength of the motion signal in the stimuli. These data are thus consistent with the single-unit studies in monkeys that identified similar regions in the process of forming a perceptual decision.

The results are also similar to those of Heinen et al. (2006) who had participants play "ocular baseball" while undergoing functional magnetic resonance imaging (fMRI). In this game, the subjects had to decide whether or not the trajectory of a dot moving across a computer screen was likely to cross into a visible "strike" zone. If the participants decided that the dot was likely to enter the strike zone, they had to make an eye movement; in the other case, their eyes had to remain fixed on a point in the centre of the screen. The results showed that when a decision was associated with a specific movement, the formation of the decision and the preparation of the behavioral response had a common neural substrate. Put more generally, the findings support the view that the human oculomotor system also has an important role in perceptual decision making.

In addition to the quiet eye being identified as an indicator of optimal focus and attention during low-pressure situations, the influence of anxiety and physiological arousal on the quiet-eye period has been examined (Vickers and Williams, 2007; Behan and Wilson, 2008). Behan and Wilson, in a simulated archery task, found that under conditions of elevated cognitive anxiety, optimal visual orientation, as indexed by quiet-eye duration, was altered. Participants generally showed reductions in the duration of quiet eye, as they took more fixations around the vicinity of the target than they did in the low-pressure condition. These results show that the quiet-eye period is sensitive to increases in anxiety and may be a useful index of the efficiency of visual orientation in aiming tasks. Vickers and Williams (2007) found that elite biathletes who increased their quiet-eye duration during high-pressure competition, as opposed to low-pressure practice, were able to overcome the normally debilitating effects of the high physiological workload, high competitive pressure, and anxiety.

These results raise the question of why a long-duration quiet eye should improve motor performance under conditions of high pressure and/or very high physiological arousal? A possible reason may lie in the work of Setchenov (1903/1935), a Russian physiologist who showed that when individuals were fatigued to exhaustion they could do more physiological work when a "diverting" activity was used to direct their attention to an external target. Assmussen and Mazin (1978a, b) subsequently found the phenomenon applied in a wide variety of tasks (both mental and physical) and that the amount of work that could be performed was greater with eyes open compared to eyes closed. Even when complete exhaustion was reached, opening the eyes led to a 15–30%

increase in the amount of exercise that could be performed. They reasoned that the input of afferent information acted to redirect attention away from the physiological demands of the task resulting in an ability to perform at a higher level. The Setchenov phenomenon is, therefore, not related to an internal focus of attention, but instead to an external focus of attention mediated by vision. These results further suggest that the ability to overcome the normally debilitating effects of maximum exercise can be aided through the use of an appropriate external focus of attention, as highlighted by the changes in the quiet eye found in the Behans and Wilson (2008) and Vickers and Williams (2007) studies.

Ecological psychology and dynamic systems

Alternate theoretical explanations for the quiet eye have also been proposed from an ecological psychology and dynamic systems perspective. Researchers from these approaches state that people perceive environments directly unaided by inference, memories, or other neural representations as suggested by cognitive psychologists (Gibson, 1979/1986; Michaels and Carello, 1981). Skilled movement depends on the establishment of direct optical relationships that develop without any apparent need for the processing stages that define cognitive psychology. Over time, and with experience and training, some of these relationships become invariant, which Michaels and Carello (1981) describe as "those high-order patterns of stimulation that underlie perceptual constancies, or more generally, the persistent properties of the environment that an animal is said to know. Invariant structures in light and sound not only specify objects, places and events in the environment, but also the activities of the organism… Thus invariants are, by virtue of the laws that support them, information about the environment and the animal's relation to it" (p. 40).

Researchers from an ecological or dynamic systems approach have argued that the quiet-eye period facilitates the orientation of the body in space and allows for the execution of movements that are more attuned to the affordances and other constraints that are present (Oudejans et al., 2002, 2005; Oliveira et al., 2007). Since the quiet eye has invariant characteristics of location, onset, offset, and duration relative to the final movement in a specific task, it optimizes optic flow and permits a better orientation of the performer relative to critical environmental constraints. For example, Oudejans et al. (2005) explains that the quiet eye is a factor in basketball shooting because it permits "a continuous updating of the relation between shooter and rim, up until ball release, as this relation at ball release provides the best determination of force, direction, and velocity needed to make a successful shot." This updating is not carried out by an internal feedback system but through the generation of dynamical relationships between the position of the gaze in space and gaze relative to the target that are subconscious and require no cognitive processing.

Controversies and future directions

It is clear that cognitive, neuroscience, and ecological theories provide distinct, and in some ways competing, ways of understanding the role of a quiet eye in visuomotor coordination. To be successful, a theoretical framework must account for how human are able to perform both rapid dynamic tasks (such as ice hockey or soccer goaltending, cricket batting) as well as those that are slower (as found when walking, or shooting a free throw). Slow and fast movements are normally defined by the duration of their movement times since this dictates the extent to which feedback and additional cognitive processing can be used to modify or change the movement. Generally, cognitive theories are good at explaining how actions with movement times in excess of 200 ms are controlled, while ecological and dynamic systems accounts are best at accounting for rapid movements under 200 ms. Each theory is characterized by a number of emerging dichotomies, the main ones being focal and ambient vision; top-down and bottom-up processing; ventral and dorsal processing; and closed- and open-loop motor control. The focal, top-down, ventral, closed system is tailored for situations where movement times are more than 200 ms and there is adequate time for cognitive processing to occur, while the ambient, bottom-up, dorsal, and open-loop control systems are specialized for tasks performed when movement times are less than 200 ms therefore affected by time constraints. In all motor tasks the two systems work together and permit the great range of actions that humans perform. There is considerable empirical evidence showing that a long-duration quiet eye is a characteristic of higher levels of motor performance is tasks where movement times are both below and above the 200 ms threshold. Elite athletes find a way to get the information they need earlier and they process this information longer irrespective of task constraints than do nonelite athletes.

However, ecological psychologists have argued that it is the final information within the quiet-eye period that is most important and not the earlier information that is fixated (Oudejans et al., 2005; Oliveira et al., 2007). Cognitive psychologists argue just the opposite. Personally I do not know 100%, but to date, the weight of evidence is in support of early rather than late quiet-eye information. I believe the final answer will depend on how gaze data are ultimately coded and analyzed in perception–action studies. Currently this is being done in two ways, with each approach consistent with either a cognitive or ecological/dynamic systems view. Cognitive psychologists code gaze data using precedents arising from the eye-movements literature (Bridgeman et al., 1975; Optican, 1985; Carl and Gellman, 1987; Carpenter, 1988; Coren et al., 2004). Cognitive psychologists identify at least three types of gaze behaviors in the data stream — fixations, pursuit tracking, and saccades. Each gaze is defined according to rules that have been established over decades of eye-tracking research. Briefly, a fixation occurs when the gaze is held on an object or location within 3° of visual angle for 100 ms or longer (Optican, 1985; Carl and Gellman, 1987; Carpenter, 1988). The 100-ms threshold is the minimum amount of time needed to recognize or become aware of stimuli. Additional time is required to make a movement, with about 180 ms needed to actually see an object and initiate a simple movement, such as pressing a key. Pursuit tracking occurs when the gaze follows a moving object, such as a ball or a person. The 100-ms threshold is used for pursuit tracking for the same reason it is used for fixations; it is only when the gaze is stabilized on the moving object or person that the individual is able to process the information provided by that object or person. During both fixations and pursuit tracking information can be processed. Saccades occur when the eyes move quickly from one fixated or tracked location to another. Saccades are rapid eye movements that bring the point of maximal visual acuity onto to the fovea so that it can be seen with clarity. We average about three saccades each second when viewing a normal scene, and these range in duration from 60 to 100 ms. In order to see and comprehend a scene, we must move our eyes rapidly from one fixated location or object to another using saccades. During saccades, information is suppressed (Bridgeman et al., 1975). Information gained during fixation or tracking is maintained across saccades so that a stable, coherent scene is viewed (Irwin, 1996). We do not perceive the blur as our eyes move, neither are we able to see a new object that appears during a saccade. However, we do possess an object-file transsaccadic memory (Irwin, 1996) that allows us to perceive scenes that are cohesive and meaningful.

In contrast, ecological psychologists hold true to the teachings of Gibson (1979/1986) and treat every gaze as being equal to all others. Ecological psychologists do not use any of the definitions for the gaze arising from the eye-movements literature, but instead assume that each gaze detects critical affordances, invariants, or elements of optic flow in the dynamic environment. They are not concerned about whether visual information is processed by the brain, only that it is detected by the visual system. Since most eye trackers have rates of 30 or 60 Hz, then ecologists recognize that visual information detected in as little as 16.67 ms may be valuable in terms of affecting a movement (Oliveira, 2007). While the coding rules used by cognitive psychologists have a long history in eye tracking and are good at explaining contributions made by the focal system, these rules often do not recognize potential contributions made by the ambient system. Indeed, gaze data that do not meet the rules for fixations, pursuit tracking, or saccades are usually identified as "other" and are not reported in a major way. Conversely, the coding grids used by ecological psychologists do not recognize that humans possess both a focal and an ambient system, and that humans use eye movements during which information is both processed (fixations, saccades) and suppressed (saccades). One way to resolve this issue is to code and analyze the same data set using both methods. We are in the process of doing this within my laboratory in a study using both cognitive and ecological coding approaches.

As is evident, there are many issues still to be resolved in terms of understanding the contributions of the gaze to performance in motor skills. Regardless of the theoretical perspective taken, there is considerable research evidence showing that the quiet-eye period is a perception–action variable that defines higher levels of skill and performance. Gaze-training studies show that training the gaze improves performance and early evidence indicates that research insights into how the gaze functions in various motor tasks have a profound effect on the training of athletes and education of coaches. Despite considerable advances by a growing number of research teams around the world, there is still more research that needs to be done before we have a complete understanding of this intriguing phenomenon. In many ways the goal of gaze and quiet-eye research is similar to that carried out in the past that looked at the merits of open- versus closed-loop control, or dorsal versus ventral processing. The goal is to understand how humans perform so well under impossible time constraints. In many respects the questions and challenges are still the same; the theatre of investigation has just changed to investigating the role of the gaze in action.

References

Adolphe, R., Vickers, J. N., & LaPlante, G. (1997). The effects of training visual attention on gaze behaviour and accuracy: a pilot study. *International Journal of Sports Vision*, *4*(1), 28–33.

Assmussen, E., & Mazin, B. (1978a). A central nervous component in local muscular fatigue. *European Journal of Applied Physiology*, *38*, 9–15.

Assmussen, E., & Mazin, B. (1978b). Recuperation after muscular fatigue by "diverting activities". *European Journal of Applied Physiology*, *38*, 1–8.

Behan, M., & Wilson, M. (2008). State anxiety and visual attention: the role of the quiet eye period in aiming to a far target. *Journal of Sports Sciences*, *26*(2), 207–215.

Bridgeman, G., Hendry, D., & Start, L. (1975). Failure to detect displacement of visual world during saccadic eye movements. *Vision Research*, *15*, 719–722.

Carl, J., & Gellman, R. (1987). Human smooth pursuit: stimulus-dependent responses. *Journal of Neurophysiology*, *57*, 1446–1463.

Carpenter, R. H. S. (1988). *Movements of the eyes* (2nd ed.). London: Pion.

Corbetta, M. (1998). Frontoparietal cortical networks for directing attention and the eye to visual locations: identical, independent, or overlapping. *Proceedings of the National Academy of Sciences of the United States of America*, *95*, 831–838.

Coren, S., Ward, L., & Enns, J. T. (2004). *Sensation and perception* (6th ed.). New York: Wiley.

Deubel, H., & Schneider, W. X. (1996). Saccade target selection and object recognition: evidence of a common attentional mechanism. *Vision Research*, *36*, 1827–1837.

Ericsson, K. A. (1996). The acquisition of expert performance: an introduction to some of the issues. In K. A. Ericsson (Ed.), *The road to excellence: the acquisition of expert performance in the arts and sciences, sports, and games* (pp. 1–50). Mahwah, NJ: Lawrence Erlbaum Associates.

Gibson, J. J. (1979/1986). *The ecological approach to visual perception*. Hillsdale, NJ: Lawrence Erlbaum Associates.

Gopher, D. (1993). The skill of attention control: acquisition and execution of attention strategies. In D. Meyer & S. Kornblum (Eds.), *Attention and performance XIV* (pp. 299–322). Cambridge, MA: MIT Press.

Harle, S., & Vickers, J. N. (2001). Training quiet eye (QE) improves accuracy in the basketball free throw. *The Sport Psychologist*, *15*, 289–305.

Heekeren, H. R., Marrett, S., Bandettini, P. A., & Ungerleider, L. G. (2003). Human fMRI evidence for representation of a perceptual decision in oculomotor areas. *Proceedings of the Society for Neuroscience*, (Abstr. 228.8). Washington, DC.

Heinen, S. J., Rowland, J., Lee, B. T., & Wade, A. R. (2006). An oculomotor decision process revealed by functional magnetic resonance imaging. *Journal of Neuroscience*, *26*, 13515–13522.

Helsen, W. F., Starkes, J. L., & Hodges, N. J. (1998). Team sports and the theory of deliberate practice. *Journal of Sport & Exercise Psychology*, *20*, 12–34.

Henderson, J. M. (2003). Human gaze control during real-world scene perception. *Trends in Cognitive Science*, *7*, 498–504.

Henry, F. M. (1953). Dynamic kinaesthetic perception and adjustment. *Research Quarterly*, *24*, 176–187.

Irwin, D. (1996). Integrating information across saccadic eye movements. *Current Directions in Psychological Science*, *5*, 94–100.

Janelle, C. M., Hillman, C. H., Apparies, R. J., Murray, N. P., Meili, L., Fallon, E. A., & Hatfield, D. B. (2000). Expertise differences in cortical activation and gaze behavior during rifle shooting. *Journal of Sport & Exercise Psychology*, *22*, 167–182.

Kowler, E., Anderson, E., Dosher, B., & Blaser, E. (1995). The role of attention in the programming of saccades. *Vision Research*, *35*, 1897–1916.

Mann, D. T. Y., Williams, A. M., Ward, P., & Janelle, C. M. (2007). Perceptual-cognitive expertise in sport: a meta-analysis. *Journal of Sport & Exercise Psychology*, *29*, 457–478.

Martell, S. G., & Vickers, J. N. (2004). Gaze characteristics of elite and near-elite ice hockey players. *Human Movement Science*, *22*, 689–712.

McPherson, S. L., & Vickers, J. N. (2004). Cognitive control in motor expertise. *International Journal of Sport & Exercise Psychology*, *2*, 274–300.

Michaels, C. F., & Carello, C. (1981). *Direct perception*. Englewood Cliffs, NJ: Prentice-Hall.

Oliveira, R., Huys, R., Oudejans, R. R. D., van de Langenberg, R., & Beek, P. J. (2007). Basketball jump shooting is controlled online by vision. *Experimental Psychology*, *54*, 180–186.

Oliveira, R. (2007). *Visual perception for basketball shooting*. Unpublished Dissertation, The Vrieje University, Amsterdam.

Optican, L. M. (1985). Adaptive properties of the saccadic system. In A. Berthoz & M. Melvile-Jones (Eds.), *Adaptative mechanisms in gaze control: facts and theories* (pp. 71–79). New York: Elsevier.

Oudejans, R. R. D., Koedijker, J. M., Bleijendaal, I., & Bakker, F. C. (2005). The education of attention in aiming at a far target: training visual control in basketball jump shooting. *International Journal of Sport Psychology*, *3*, 197–221.

Oudejans, R. R. D., van de Langenberg, R. W., & Hutter, R. I. (2002). Aiming at a far target under different viewing conditions: visual control in basketball jump shooting. *Human Movement Science*, *21*, 457–480.

Panchuk, D., & Vickers, J. N. (2006). Gaze behaviours of goaltenders under spatial temporal constraints. *Human Movement Science*, *25*, 733–752.

Posner, M. I. (1980). Orienting of attention. *Quarterly Journal of Experimental Psychology*, *32*, 3–25.

Posner, M. I., & Raichle, M. E. (1994). *Images of mind*. New York: Scientific American Library.

Raab, M., Masters, R. S. W., & Maxwell, J. P. (2005). Improving the 'how' and 'what' decisions of elite table tennis players. *Human Movement Science*, *24*, 326–344.

Rodrigues, S. T., Vickers, J. N., & Williams, A. M. (2002). Head, eye and arm co-ordination in table tennis: an exploratory study. *Journal of Sport Sciences*, *20*, 171–186.

Setchenov, I. M. (1903/1935). In *Selected works* (pp. 246–240). Moscow, Russia: Hause Academy of Sciences of the USSR.

Shepherd, M., Findlay, J. M., & Hockey, R. J. (1986). The relationship between eye movements and spatial attention. *Quarterly Journal of Experimental Psychology*, *38A*, 475–491.

Vickers, J. N. (1992). Gaze control in putting. *Perception*, *21*, 117–132.

Vickers, J. N. (1996). Visual control when aiming at a far target. *Journal of Experimental Psychology – Human Perception and Performance*, *22*, 342–354.

Vickers, J. N. (2003). Decision training: an innovative approach to coaching. *Canadian Journal for Women Coaches Online*. <http://www.coach.ca/WOMEN/e/journal/> Accessed 30.01.09.

Vickers, J. N. (2004). The quiet eye: it's the difference between a good putter and a poor one, here's proof. *Golf Digest* (January), 96–101.

Vickers, J. N. (2006). Gaze of Olympic speed skaters while skating at full speed on a regulation oval. Perception–action coupling in a dynamic performance environment. *Cognitive Processing*, *7*, 102–105.

Vickers, J. N. (2007). *Perception, cognition, and decision training: the quiet eye in action*. Champaign, IL: Human Kinetics.

Vickers, J. N., & Adolphe, R. A. (1997). Gaze behaviour during a ball tracking and aiming skill. *International Journal of Sports Vision*, *4*, 18–27.

Vickers, J. N., Livingston, L., Umeris, S., & Holden, D. (1999). Decision training: the effects of complex instruction, variable practice and reduced delayed feedback on the acquisition and transfer of a complex motor skill. *Journal of Sport Sciences*, *17*, 357–367.

Vickers, J. N., Reeves, M. A., Chambers, K. L., & Martell, S. T. (2004). Decision training: cognitive strategies for enhancing performance. In A. M. Williams & N. J. Hodges (Eds.), *Skill acquisition in sport: research, theory and practice* (pp. 103–120). London: Routledge.

Vickers, J. N., & Williams, A. M. (2007). Performing under pressure: the interactive effects of physiological arousal, cognitive anxiety and gaze control in elite biathlon shooters. *Journal of Motor Behavior*, *39*, 381–394.

Williams, A. M., & Grant, A. (1999). Training perceptual skill in sport. *International Journal of Sports Psychology*, *30*, 194–220.

Williams, A. M., Singer, R. A., & Frehlich, S. (2002). Quiet eye duration, expertise, and task complexity in a near and far aiming task. *Journal of Motor Behavior*, *34*, 197–207.

M. Raab et al. (Eds.)
Progress in Brain Research, Vol. 174
ISSN 0079-6123

CHAPTER 23

Juggling with the brain — thought and action in the human motor system

Uta Wolfensteller[1,2,*]

[1]*Max Planck Institute for Human Cognitive and Brain Sciences, Leipzig, Germany*
[2]*Section of Systems Neuroscience, Department of Psychiatry and Psychotherapy, Medical Faculty Carl Gustav Carus, Technical University of Dresden, Dresden, Germany*

Abstract: Empirical findings from various research fields indicate that cognitive and motor processes are far less dissimilar than previously thought. The present chapter takes a neuroscientific perspective and offers evidence for similarities between cognition and action focusing on three key players of the classical motor system: the primary motor cortex, the cerebellum, and the premotor cortex. Briefly, although movement execution is apparently supported in part by the same cerebral resources engaged in cognitive processes, the three brain regions reviewed here are differentially engaged in more or less action-bound cognitive processes.

Keywords: motor system; brain areas; cognition; premotor cortex

Introduction

Perceiving and *representing* options, *selecting* between them, and *implementing* a particular action cannot be considered mutually exclusive processes. Rather, they necessarily inform and influence each other, not only in the serial manner just outlined (perception–selection–implementation), but also in a more intertwined manner (as suggested by the facilitating or disturbing impact action can have on perception, see for instance Müsseler and Wühr, 2002; Danielmeier et al., 2004). Ultimately, this interaction enables us to perform as true actors rather than mere reactors. Let us take the instance of juggling. This is a motor skill that does not come easily to most people but needs extensive practice. It constitutes a movement pattern involving the coordination of multiple limbs as well as at least two objects. Experienced jugglers are able to turn, run, jump, and quickly change the overall pattern of the flying objects or duet with a partner, while still managing to keep the juggled objects in the air. Importantly, the joy and beauty of juggling are less defined by the coordinated movement pattern itself, but rather by the effect of these movements, that is, the flying pattern of the objects. To achieve this, the juggler must adjust his/her body movements intentionally and flexibly based on (a) the juggling goal and (b) the estimated effects of a performed movement on the juggled items. In that sense, this form of behavior necessarily requires a combination of thought and action, where both mutually inform each other. The present chapter presents a neuroscientific perspective

*Corresponding author.
Tel.: +49-351-4634-2212; Fax: +49-351-463-42202;
E-mail: uta.wolfensteller@tu-dresden.de

DOI: 10.1016/S0079-6123(09)01323-5

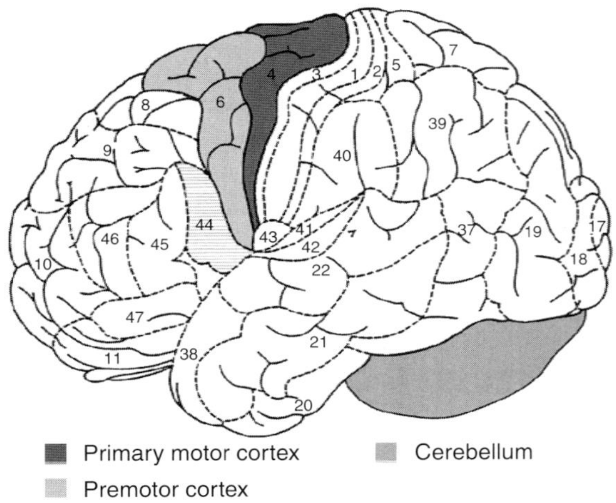

Fig. 1. Lateral view on the left human brain showing three areas of the classical motor system, which are discussed in the present paper with respect to their role in thought and action. Modified from Nieuwenhuys et al. (1991), p. 10, Fig. 5a, © Springer Verlag Berlin Heidelberg, with kind permission of Springer Science and Business Media. (See Color Plate 23.1 in color plate section.)

on the similarities between thought and action exemplifying three parts of the classical cerebral motor system, namely the primary motor cortex, the cerebellum, and the premotor cortex (see Fig. 1). To this end, the separate sections will start with a short macro-anatomical description of each particular area, followed by an elaboration on their involvement in action and thought, that is, their motor and cognitive functions.

The primary motor cortex

The human primary motor cortex (M1) covers the central sulcus and the posterior stripe of the precentral gyrus (Preuss et al., 1996; Geyer et al., 2000). Based on cytoarchitectonic and receptor architectonic measures, M1 can be subdivided into an anterior and a posterior portion (area 4a and 4p, Geyer et al., 1996). Its direct connections to the spinal cord have long been considered to be a property that differentiated the primary motor cortex from so-called "nonprimary motor areas," such as the premotor cortex, which were thought to have only indirect connections to the spinal cord (Levin, 1944). Partly based on these differences in connectivity, the primary motor cortex has classically been implicated in low-level motor functions, whereas nonprimary motor areas were considered to subserve higher-level motor functions. However, in the light of findings showing that some nonprimary motor areas apparently have direct access to the spinal cord as well (Dum and Strick, 1991), this functional distinction appears to be less clear-cut than initially thought.

Within the primary motor cortex (M1), body parts are represented in a somatotopic manner less fine-grained, but comparable to the somatotopic organization of the primary somatosensory cortex (Penfield and Rasmussen, 1950). These original findings obtained using electric stimulation on the open scalp, have since been replicated multiple times in a less invasive manner by using functional magnetic resonance imaging (fMRI) in healthy human volunteers (e.g., Ehrsson et al., 2003; Hanakawa et al., 2005). The representations of leg, arm and facial movements are distributed along the central sulcus and precentral gyrus in superior-to-inferior direction, either in the hemisphere contralateral to the moving limb, such as in the case of arm or leg movements, or in both hemispheres, such as in the case of mouth movements. However, notwithstanding this broad somatotopy of body parts, substantial overlap has been observed within each of these body part representations (Schieber, 2001; Gentner and Classen, 2006). This cortical overlap has been argued to reflect the fact that motor effectors are represented in terms of their biomechanical interdependency. According to this principle, effectors that are highly dependent on one another are represented in close spatial neighborhood (Schieber, 2001). To understand the functional implication of these M1 activations further, one might ask what the primary motor neurons are coding. Unfortunately, the answer to this question is rather complicated given that "neural correlates have been found for virtually every movement parameter examined such as spatial target location, hand position or velocity, joint motion and muscle activation" (Scott, 2008, p. 1219). In order to understand the role of M1 in cognitive and motor aspects of voluntary behavior, it might therefore be worthwhile to examine different situations engaging this region.

One important function of M1 pertains to the acquisition of motor skills, especially those involving the coordination of different body parts, as in the above-mentioned juggling example. Consequently, skill-learning paradigms have been used to investigate the plasticity of M1. In nonhuman primates, functional plasticity of M1 was observed during motor sequence learning, where a substantial proportion of neurons changed their firing pattern after extensive training (Matsuzaka et al., 2007). Similar results have been obtained in humans showing, for instance, that M1 activation decreases during the course of skill acquisition (Debaere et al., 2004; Floyer-Lea and Matthews, 2005). However, the work of M1 does not end once the initial learning success has been achieved. Instead, early motor consolidation of a new skill seems to depend crucially on M1 function (Muellbacher et al., 2002). When M1 was disrupted by means of rapid transcranial magnetic stimulation (TMS) immediately after an initial motor training, the beneficial effects of this training session were diminished whereas the ability to learn in a new training session was not affected. A relative increase of M1 activation has been shown after prolonged skill practice (Floyer-Lea and Matthews, 2005; for a review see Ungerleider et al., 2002). As the activation of M1 is not tied to the movement execution phase alone, but seems to span a broad temporal range from preparation to (early) consolidation following execution, it becomes apparent that there is more to M1 function than low-level motor processes.

Following up on that notion, there is now ample evidence for higher-level motor and cognitive functions of the primary motor cortex. Several recent functional magnetic resonance imaging (fMRI) and TMS studies found this area to be engaged during the mere imagination of body movements without actual execution (Facchini et al., 2002; Ehrsson et al., 2003; Wolfensteller et al., 2007; Munzert et al., 2008). In one study, it has been shown that execution and imagery of hand movements activate both the posterior and the anterior portion of M1, though the anterior portion was not fully covered during motor imagery (Sharma et al., 2008). The activation of M1 is typically found to be weaker during imagined than actually performed movements (Ehrsson et al., 2003; Hanakawa et al., 2008; Sharma et al., 2008). Furthermore, in some motor imagery studies, the primary motor cortex was not found to be activated at all (e.g., see Dechent et al., 2004; de Lange et al., 2006). The discrepancy between studies reporting M1 activation and those who do not report M1 activation is considered to reflect methodological differences (Sharma et al., 2008), such as movement complexity, the involved effectors, and the amount of training administered. The role of M1, however, clearly extends beyond its involvement in pure movement execution, as for instance shown by its involvement in processing sensory information in the absence of a motor command, such as in the case of kinesthetic illusions (Naito et al., 2002). Furthermore, movement-related activation in M1 is modulated by attention, such that directing attention away from a to-be-performed motor task results in activation decreases in a number of brain regions, including M1 (Johansen-Berg and Matthews, 2002). In a series of TMS studies on inhibitory control it has been shown that, apart from movement initiation, the stopping of a prepared movement in the last minute also involves the contralateral M1 (Coxon et al., 2006; for a review see Band and van Boxtel, 1999). Together, these findings support the view that M1 is engaged in cognitive processes that are extremely close to motor actions (Sanes and Donoghue, 2000).

Recently, the engagement of the primary motor cortex in cognitive processes that are less directly linked to motor actions has been investigated using an experimental procedure that does not even require explicitly imagining or preparing a movement, but only the simple reading of, or listening to, words. These words could either be related or unrelated to motor actions, such as "grasp" and "cheat", respectively. Some of these studies revealed specific activation patterns for M1 following a rough somatotopy for read action-related words (Hauk et al., 2004; Ruschemeyer et al., 2007). The findings on that topic, however, are not unequivocal. Hearing action-related words for instance led to a somatotopic activation pattern in premotor and parietal cortex rather than in M1 (Tettamanti et al., 2005). It has also been shown that transcranial magnetic stimulation of M1 during reading action-related words facilitated responses in the context of motor imagery only, but not during silent reading or judging the word frequency (Tomasino et al., 2008). Furthermore, the study by Ruschemeyer and colleagues (2007) suggests that there is a limit to the looseness of the action-relation of a word triggering activation in M1. While M1 was activated during reading the German word equivalent to "physically grasping something" (greifen), reading the German word equivalent to "grasping something in the sense of understanding it" (begreifen), which includes the same word stem, did not lead to activation in M1 (Ruschemeyer et al., 2007). Taken together, in close keeping with Sanes and Donoghue (2000), M1 can not be denied to have cognitive functions, and these appear to be linked to actions in a rather direct, although not necessarily extremely close manner.

The cerebellum

Another part of the central motor system that has long been considered to subserve classical motor functions exclusively is the cerebellum. This structure, which is located posterior to the brainstem and inferior to the occipital lobe, can be roughly divided into a midline structure (the vermis), and two laterally adjacent hemispheres. By means of fMRI it has been shown that, just as in the primary motor cortex, the representation of movements within the human cerebellum roughly follows a somatotopic pattern (Rijntjes et al., 1999; Grodd et al., 2001; for a review see Manni and Petrosini, 2004). In the cerebellum, movements are represented in the lobe ipsilateral to the moving limb and there is evidence for two separate body maps, one in the anterior portion of the cerebellum, and another within the posterior portion of the cerebellum (Rijntjes et al., 1999; Grodd et al., 2001; Manni and Petrosini, 2004).

The major role in motor control ascribed to the cerebellum, especially during the acquisition of motor skills, is supported by its projections to the primary motor cortex and the premotor cortex (for an overview see Middleton and Strick, 2000; Kelly and Strick, 2003). By influencing these cortical motor structures, the cerebellum is known to control movement rate and regularity, that is, the temporal organization of motor behavior (Ramnani, 2006). Clinically, lesions within the cerebellum can lead to the so-called cerebellar motor syndrome, which encompasses, among others, impairments of gait (ataxia) and extremity coordination (dysmetria) as well as disordered eye movements (for a review see Schmahmann, 2004), which is supposed to reflect a disturbed timing of muscle agonists and antagonists (Salman, 2002). Related to this timing function, it has been argued that the cerebellum is required in situations calling for high temporal acuity in general (Ivry, 1996; Tesche and Karhu, 2000).

During the past decades, the cerebellum has been primarily discussed within the framework of internal models (Miall et al., 1993; Wolpert and Miall, 1996; Miall, 1998). These internal models refer to simulated natural processes, such as movements, and can be subdivided into forward and inverse models. Forward models establish representations of input–output mappings (Ramnani, 2006) estimating the to-be-expected sensory feedback given a certain input. In the case of motor control, this input would be an efference copy of a motor command. If the actual outcome of a given input does not match the expected one, an error signal will be fed back. Learning based on error feedback is defined as "supervised learning" (Doya, 1999). In motor learning, an error would be, for instance, a body posture different from the intended or expected one. Importantly, to enable error detection and feedback, a given motor commands' expected effect has to be estimated in the first place. This supervised motor learning based on forward modeling is thought to involve the cerebellum (Doya, 1999, 2000), which is accordingly considered to be a sensorimotor predictor (Bastian, 2006).

In nonhuman primates, lesions in the cerebellum have been shown to impair late stages of visuomotor sequence learning (Nixon and Passingham, 2001). These findings support a role of the cerebellum in optimizing movements on the basis of predictable events. Comparable results have been obtained in humans, showing an involvement of the cerebellum in predictive grip force modulation (for a review see Blakemore and Sirigu, 2003). Consistent with a predictor role, the cerebellum has been implicated in the central cancellation of self-produced tactile stimuli (Blakemore et al., 1998). Activation in the cerebellum during self-applied tactile stimulation modulated activity within the somatosensory cortex, thereby diminishing the tickling sensation. Ultimately, the exact estimation of an upcoming sensory event provided by the cerebellum is at the core of our inability to tickle ourselves. In line with the notion that temporal acuity depends on the cerebellum, it has been shown that sensory cancellation crucially depends on timing. By artificially introducing a temporal delay between the self-produced movement and the tactile event it causes, the perceived ticklishness increases (Blakemore et al., 2000). Similarly, a self-initiated touch of one's own finger is associated with a lowered tactile sensitivity. Importantly, this is still the case if the fingers unintentionally miss each other and no touch was applied, which suggests that the state estimation process is anticipatory (Bays et al., 2006).

In computational terms, the predictive function of the cerebellum is thought to rely on Bayesian state estimation (Paulin, 2005), that is, on an estimation of a future state (e.g., the future position of a limb) based on the current state (e.g., the current position of the limb), the efference copy of a motor command (e.g., a reaching movement), and prior knowledge about these contingencies (e.g., the probability that a particular limb position results from a particular reaching movement that started from a particular limb position). Direct evidence for a cerebellar role in state estimation stems from recent fMRI and TMS studies (Diedrichsen et al., 2007; Miall et al., 2007). When simultaneously performing arm and hand movements in a motor learning paradigm, participants appear to rely on state estimation to coordinate the movement sequence, whereas when performing the movements successively, that is, without temporal overlap, they rely on estimates of absolute time. Estimating states of the arm compared to estimating the absolute time of the movement onsets was associated with enhanced activation in the anterior cerebellum (Diedrichsen et al., 2007). Another study revealed specific impairments of reaching movements following TMS applied to the cerebellum, which can be taken to reflect a disruption of state estimation (Miall et al., 2007). Taken together, these studies further support the notion that the cerebellum serves as a predictor in the motor system.

An fMRI study extended this notion by showing that the cerebellum is also involved in forward modeling of purely perceptual events (O'Reilly et al., 2008). After temporarily occluding the moving stimulus, participants were asked to either judge the deviation in terms of spatial location or underlying movement velocity, once it reappeared. Activation in the cerebellum was restricted to predicting temporal-spatial patterns, that is, when an estimation of velocity was required. Conceptually, this lends further support to the notion that forward models might be exploited for many other purposes besides motor control (Wolpert et al., 2003; Ito, 2008). The notion that the cerebellum subserves higher-level functions as well is further supported by the fact that besides its projections to the primary motor cortex and premotor cortex, it is also connected to the prefrontal cortex (Middleton and Strick, 2000). Translating this into the functionality of the cerebellum, some researchers have claimed "that the cerebellum is not only a motor control device, but it is also an essential component of the brain mechanisms for personality, mood, and intellect" (Schmahmann, 2004) or that "interactions between the lateral cerebellum and neocortical association areas might underlie the automation of cognitive processes such as thought, planning, decision making, and even some forms of attention" (Nixon, 2003).

The premotor cortex

The lateral premotor cortex, which will be the focus of the following section, extends over the anterior portion of the precentral gyrus and the posterior part of the superior frontal gyrus (Preuss et al., 1996; Rizzolatti et al., 1998). When following the assumed homology between the ventral premotor field F5 in nonhuman primates and parts of

Broca's area in humans (BA 44), the posterior portion of the inferior frontal gyrus might also be termed premotor cortex (Rizzolatti et al., 1998; Nelissen et al., 2005).

Originally, Penfield and Rasmussen (1950) suggested one body map that encompassed both the lateral premotor and primary motor cortex. In line with this, fMRI studies have also revealed a rough somatotopic representation of executed movements in the premotor cortex (Ehrsson et al., 2003). The premotor cortex seems to be engaged during far more complex movements than M1 as indicated by several studies using different methods. For instance, electric stimulation of the precentral gyrus (including premotor and primary motor cortex) in nonhuman primates elicits complex movement patterns, such as bringing the arm in front of the face in a defensive gesture (Graziano et al., 2002). Motor difficulties of human patients with lesions in the premotor cortex are most pronounced for complex movements involving the coordination of more than one limb (Freund and Hummelsheim, 1985). Finally, neuroimaging studies in humans have revealed enhanced premotor activation for complex movement sequences engaging fingers (e.g., Boecker et al., 2002; Grafton et al., 2002; Haslinger et al., 2002) or eyes (Heide et al., 2001).

In line with labeling the premotor cortex a nonprimary motor area, there is converging evidence that it is involved in higher-level motor functions such as motor imagery (e.g., Ehrsson et al., 2003; Hanakawa et al., 2005; Wolfensteller et al., 2007; Munzert et al., 2008). Notably, activation of the premotor cortex during imagined movements has been reported more consistently than activation of M1. The activation pattern observed for different imagined movements follows a rough somatotopy (Ehrsson et al., 2003; Wolfensteller et al., 2007), thereby confirming premotor involvement in cognitive functions that are rather close to motor action. Another cognitive process that is related to motor action, but in a more prospective manner, is the process of action selection. In most situations, acting in a meaningful manner requires selecting the currently appropriate option out of several possible actions. If we, for the sake of the argument, neglect motivational and volitional factors, the implication of "appropriate" can be reduced to "correct according to a valid stimulus–response (S-R) rule." Converging evidence suggests that this form of rule-based motor behavior crucially relies on the premotor cortex, more specifically the dorsolateral premotor cortex (e.g., see Germain and Lamarre, 1993; Wise et al., 1996; Cavina-Pratesi et al., 2006; Crone et al., 2006) and the pre-supplementary motor area (Sakai et al., 1998; Toni et al., 1999; Cavina-Pratesi et al., 2006; Crone et al., 2006). In this context, the firing of premotor neurons has been suggested to reflect the association between the stimulus and the response, rather than either of the two alone (di Pellegrino and Wise, 1993).

Interestingly, the premotor cortex is apparently involved in processes that are even less directly related to motor actions in the sense that they are distinguishable from selecting or planning an action. Besides motor properties, a large proportion of premotor neurons additionally have perceptual properties, which is reflected by the fact that they fire during visual stimulation (e.g., Gentilucci et al., 1988; Rizzolatti et al., 1988). A particular class of these sensorimotor neurons, the mirror neurons, discharge during the observation of another individual's actions in a wide range of situations: when they are presented visually (di Pellegrino et al., 1992; Rizzolatti et al., 1996) or auditorily (Kohler et al., 2002; Keysers et al., 2003), when they are partly occluded (Umiltà et al., 2001), and even when they involve tools (Ferrari et al., 2005). In humans, there is, due to ethical and methodological reasons, only indirect evidence for the existence of such neurons. In fMRI studies, for example, activation within the premotor cortex has been consistently reported when participants observed actions presented visually (e.g., Buccino et al., 2001; Manthey et al., 2003; Schubotz and von Cramon, 2004; Sakreida et al., 2005; Zentgraf et al., 2005; Munzert et al., 2008) or auditorily (Gazzola et al., 2006). Similar to executed and imagined movements, the premotor representations of observed movements were distributed following a broad somatotopy, irrespective of whether the presented movement involved an object or not (Buccino et al., 2001; Sakreida et al., 2005). These findings have been taken to support the existence of a "mirror neuron system" in humans, which in turn has been functionally linked to inferring the goal of an observed action (Rizzolatti and Fadiga, 1998; Umiltà et al., 2001; Buccino et al., 2004). Going one step further, as inferring someone else's goal might be extremely relevant for one's own behavior, the mirror neuron system including the premotor cortex has been suggested to be important in social cognition (Grèzes and Decety, 2001; Decety and Grezes, 2006; but see Jacob and Jeannerod, 2005). At this juncture, it has to be pointed out again that Broca's area, covering the posterior part of the inferior frontal gyrus, is the supposed human homolog of the premotor area F5 in macaques where mirror neurons have been found (e.g., Rizzolatti et al., 2002; Rizzolatti and Craighero, 2004). However, as outlined above, action observation engages the entire lateral premotor cortex covering the precentral gyrus. This in turn suggests that in order to understand the goal underlying another person's action it might be necessary and relevant to predict and integrate a variety of environmental dynamics. For instance, when trying to understand a fellow juggler's goal, that is, whether and when he/she wants me to catch something or whether he/she rather wants to catch it him/herself, it is relevant to anticipate his/her movements and the juggled objects' directions.

Various findings emphasize a role of the premotor cortex in more attention-related processes beyond action observation (Boussaoud, 2001; Simon et al., 2002). Based on clinical studies and neurophysiological research on nonhuman primates, Rizzolatti and colleagues (1987) have proposed a premotor theory of attention, claiming that spatial attention and the control of arm, head, and eye movements rely on the same neural network, which includes the dorsal premotor cortex. The framework was later extended to object-related attention and the programming of hand movements (Craighero et al., 1999, 2002). Using an experimental paradigm that required prospective attention to one out of several stimulus properties, fMRI evidence for an attentional modulation of the premotor cortex has been obtained for the visual domain (Schubotz and von Cramon, 2001; Wolfensteller et al., 2004, 2007) and for the auditory domain (Schubotz et al., 2003). Specifically, attending to spatial properties engaged the dorsal premotor cortex, while attending to object-specific properties engaged the superior part of the ventral premotor cortex (see also Schubotz and von Cramon, 2002) and attending to rhythm or pitch properties engaged the inferior part of the ventral premotor cortex. These findings show that the premotor cortex is engaged in processes that clearly extend beyond motor behavior. More specifically, the engagement of the premotor cortex in anticipatory processes, such as the prediction of stimulus sequences employed in these studies, suggests that the premotor activation reflects a facet of forward modeling (e.g., Schubotz, 2007).

Final considerations

The evidence reviewed here clearly shows that implementing movements relies, at least partly, on the same cerebral sources engaged in selecting and planning movements. Conceptually, it has been argued that motor systems are "anticipation engines" whose purpose is to predict action consequences (Knoblich and Prinz, 2005). When considering the initial juggling example, the advantages of anticipation are evident. Applying a certain force when throwing one ball will, for instance, result in the ball reaching a certain peak height with a certain velocity. Estimates of these two among many other parameters must be taken into account when planning the next movement in order to catch the ball and throw it up again without dropping the others. Notably, it has been shown that ball players trying to hit a ball, instead of actually predicting the final position of the ball and moving towards this spot, apply a simple heuristic, the so-called gaze heuristic, which basically comprises fixating on the ball and adjusting one's running speed in order to keep the angle between ball and ground constant (McLeod and Dienes, 1996; for a recent overview see Hutchinson and Gigerenzer, 2005). Taking this into account, it is conceivable that the anticipatory processes involved in the complex skill of juggling might be supported by a combination of simple heuristics, resulting in a reduced number of to-be-predicted parameters. Given the issue of parsimony, it is not surprising that these highly interdependent processes are implemented in the same or in densely interconnected brain structures. Apparently, the three motor areas reviewed here are differentially engaged in cognitive processes that are more or less action-bound. The cognitive functions of the primary motor cortex, for instance, appear to be the most directly related to actions. They encompass action preparation and changing a prepared action, which could still be argued to reflect the transmission and adaptation of motor commands to the periphery. Moreover, some aspects of motor imagery and action observation have been shown to draw also on the primary motor cortex. This putatively reflects a combination of simultaneously transmitting the efference copy of a motor command to cerebral areas engaged in forward modeling, namely the premotor cortex (Miall, 2003; Schubotz, 2007) and the cerebellum (Miall, 2003; Ito, 2006), while inhibiting the transmission of the actual motor command. In that sense, the primary motor cortex is more than a simple relay-station delivering the output of motor commands computed elsewhere. Rather it has to be acknowledged for its role in providing the input to (predictive) computational processes, thereby enabling prospective behavior. This input can then be used in the cerebellar and premotor parts of the "anticipation machine."

Initially, mainly the cerebellum was discussed in the framework of forward modeling, but in the past years, this was extended to the premotor cortex (e.g., Miall, 2003; Schubotz, 2007). An important, but yet unresolved question is how the contributions of the cerebellum and the premotor cortex to anticipatory processes in movement and cognition can be differentiated. As the premotor cortex is considered to be a nonprimary motor area, the anticipatory function of the premotor cortex might be more biased towards external nonmotor events, such as the course of a flying object, whereas the anticipatory function of the cerebellum might be more biased towards internal motor events, such as the configuration of one's joints and muscles while moving. However, these different biases do not preclude shared functions such as predicting the effects of one's own actions on the external environment. Related to that, one account holds that the cerebellum and premotor cortex support supervised and unsupervised learning, respectively (e.g., Doya, 1999, 2000; Schubotz, 2007). In contrast to supervised learning, which is based on error feedback, in unsupervised learning associations are formed without such feedback. Accordingly, anticipation would incorporate prediction errors as feedback teaching signals (which are more readily available for motor events) in the case of supervised, but not in unsupervised learning.

Effective anticipation is necessarily based on detecting regularities, that is, on learning contingencies between motor actions and their consequences (in terms of resulting body posture, physical effects in outside world, as well as the altered perception of the outside world), and between events in the outside world and their consequences (in terms of events in the outside world and on oneself). Once the basic action-outcome contingencies are learned and under automatic control, the very same cerebral resources might be used to process other, putatively less directly action-related contingencies. Furthermore, learning these contingencies enables the inverse computational process, that is, the selection and planning of an action based on a desired outcome, as can be illustrated by taking the juggling example. Here, it would amount to estimating the required movement pattern resulting in one flaming torch being high in the air, while the juggler is temporarily holding the other two without burning his/her hands.

As a final note, it is important to stress that the brain, including the central motor system, has to be considered as a functionally and structurally plastic organ: "Behavior will lead to changes in brain circuitry, hence establishing organic underpinnings of learned attitudes, dispositions, or thinking styles, as much as faulty brain circuits will lead to specific behavioral patterns" (Robertson et al., 2003, p. 108). Motor training is, for instance, assumed to cause plasticity at several levels (Adkins et al., 2006). In the first place it is associated with specific plastic changes in particular brain areas, such as an enlargement of the hand representation within primary motor cortex after years of extensive piano playing (Amunts et al., 1997; Gaser and Schlaug, 2003), or an increase in gray matter in motion-sensitive visual areas after juggling training (Draganski et al., 2004; Driemeyer et al., 2008). This might be accompanied by, or cause plastic changes in connectivity within the cerebral network engaged in that particular task, such as between the lateral and mesial premotor cortex after extensive piano practice (Bengtsson et al., 2005), or between left and right sensorimotor cortex during early motor skill learning (Sun et al., 2007). Finally, it has been argued that training-induced plastic changes in one area might cause plasticity in another area that is connected to the first one, but functionally unrelated to the trained motor skill. This form of meta-plasticity (Adkins et al., 2006) describes a global change in the state of the system as a whole, in addition to plastic changes in directly skill-related areas. Ultimately, this might be at the core of the interdependence of thought and action. All sorts of expertise, including motor expertise, change the way we perceive and weigh new inputs (see Chapter 1: Grounding cognition in action: expertise, comprehension, and judgment), including our choice options and actions, which is accompanied, followed, or caused by changes in brain circuits subserving these processes. In that sense, the brain areas underlying the processes of *representing* options and *selecting* and *implementing* actions are neither exclusive nor independent from one another.

Acknowledgments

I cordially thank Anna Abraham, Andreja Bubic, and Birte U. Forstmann for their helpful suggestions and Heike Schmidt for helping with the figure.

References

Adkins, D. L., Boychuk, J., Remple, M. S., & Kleim, J. A. (2006). Motor training induces experience-specific patterns of plasticity across motor cortex and spinal cord. *Journal of Applied Physiology*, *101*(6), 1776–1782.

Amunts, K., Schlaug, G., Jaencke, L., Steinmetz, H., Schleicher, A., Dabringhaus, A., & Zilles, K. (1997). Motor cortex and hand motor skills: Structural compliance in the human brain. *Human Brain Mapping*, *5*, 206–215.

Band, G. P., & van Boxtel, G. J. (1999). Inhibitory motor control in stop paradigms: review and reinterpretation of neural mechanisms. *Acta Psychologica*, *101*(2–3), 179–211.

Bastian, A. J. (2006). Learning to predict the future: the cerebellum adapts feedforward movement control. *Current Opinions in Neurobiology*, *16*(6), 645–649.

Bays, P. M., Flanagan, J. R., & Wolpert, D. M. (2006). Attenuation of self-generated tactile sensations is predictive, not postdictive. *PloS Biology*, *4*(2), e28.

Bengtsson, S. L., Nagy, Z., Skare, S., Forsman, L., Forssberg, H., & Ullén, F. (2005). Extensive piano practicing has regionally specific effects on white matter development. *Nature Neuroscience*, *8*(9), 1148–1150.

Blakemore, S. J., & Sirigu, A. (2003). Action prediction in the cerebellum and in the parietal lobe. *Experimental Brain Research*, *153*(2), 239–245.

Blakemore, S. J., Wolpert, D. M., & Frith, C. D. (1998). Central cancellation of self-produced tickle sensation. *Nature Neuroscience*, *1*(7), 635–640.

Blakemore, S. J., Wolpert, D., & Frith, C. D. (2000). Why can't you tickle yourself? *Neuroreport*, *11*(11), R11–R16.

Boecker, H., Ceballos-Baumann, A. O., Bartenstein, P., Dagher, A., Forster, K., Haslinger, B., Brooks, D. J., Schwaiger, M., & Conrad, B. (2002). A H(2)(15)O positron emission tomography study on mental imagery of movement sequences — the effect of modulating sequence length and direction. *Neuroimage*, *17*(2), 999–1009.

Boussaoud, D. (2001). Attention versus intention in the primate premotor cortex. *Neuroimage*, *14*(1, Pt.2), S40–S45.

Buccino, G., Binkofski, F., Fink, G. R., Fadiga, L., Fogassi, L., Gallese, V., Seitz, R. J., Zilles, K., Rizzolatti, G., & Freund, H. J. (2001). Action observation activates premotor and parietal areas in a somatotopic manner: an fMRI study. *European Journal of Neuroscience*, *13*(2), 400–404.

Buccino, G., Lui, F., Canessa, N., Patteri, I., Lagravinese, G., Benuzzi, F., Porro, C. A., & Rizzolatti, G. (2004). Neural circuits involved in the recognition of actions performed by nonconspecifics: an FMRI study. *Journal of Cognitive Neuroscience*, *16*(1), 114–126.

Cavina-Pratesi, C., Valyear, K. F., Culham, J. C., Kohler, S., Obhi, S. S., Marzi, C. A., & Goodale, M. A. (2006). Dissociating arbitrary stimulus–response mapping from movement planning during preparatory period: evidence from event-related functional magnetic resonance imaging. *Journal of Neuroscience*, *26*(10), 2704–2713.

Coxon, J. P., Stinear, C. M., & Byblow, W. D. (2006). Intracortical inhibition during volitional inhibition of prepared action. *Journal of Neurophysiology*, *95*(6), 3371–3383.

Craighero, L., Bello, A., Fadiga, L., & Rizzolatti, G. (2002). Hand action preparation influences the responses to hand pictures. *Neuropsychologia*, *40*(5), 492–502.

Craighero, L., Fadiga, L., Rizzolatti, G., & Umiltà, C. (1999). Action for perception: a motor-visual attentional effect.

Journal of Experimental Psychology — Human Perception and Performance, *25*(6), 1673–1692.

Crone, E. A., Wendelken, C., Donohue, S. E., & Bunge, S. A. (2006). Neural evidence for dissociable components of task-switching. *Cerebral Cortex*, *16*(4), 475–486.

Danielmeier, C., Zysset, S., Müsseler, J., & von Cramon, D. Y. (2004). Where action impairs visual encoding: an event-related fMRI study. *Cognitive Brain Research*, *21*(1), 39–48.

de Lange, F. P., Helmich, R. C., & Toni, I. (2006). Posture influences motor imagery: an fMRI study. *Neuroimage*, *33*(2), 609–617.

Debaere, F., Wenderoth, N., Sunaert, S., Van Hecke, P., & Swinnen, S. P. (2004). Changes in brain activation during the acquisition of a new bimanual coordination task. *Neuropsychologia*, *42*(7), 855–867.

Decety, J., & Grezes, J. (2006). The power of simulation: imagining one's own and other's behavior. *Brain Research*, *1079*(1), 4–14.

Dechent, P., Merboldt, K. D., & Frahm, J. (2004). Is the human primary motor cortex involved in motor imagery? *Cognitive Brain Research*, *19*(2), 138–144.

di Pellegrino, G., Fadiga, L., Fogassi, L., Gallese, V., & Rizzolatti, G. (1992). Understanding motor events: a neurophysiological study. *Experimental Brain Research*, *91*(1), 176–180.

di Pellegrino, G., & Wise, S. P. (1993). Visuospatial versus visuomotor activity in the premotor and prefrontal cortex of a primate. *Journal of Neuroscience*, *13*(3), 1227–1243.

Diedrichsen, J., Criscimagna-Hemminger, S. E., & Shadmehr, R. (2007). Dissociating timing and coordination as functions of the cerebellum. *Journal of Neuroscience*, *27*(23), 6291–6301.

Doya, K. (1999). What are the computations of the cerebellum, the basal ganglia and the cerebral cortex? *Neural Networks*, *12*(7–8), 961–974.

Doya, K. (2000). Complementary roles of basal ganglia and cerebellum in learning and motor control. *Current Opinions in Neurobiology*, *10*(6), 732–739.

Draganski, B., Gaser, C., Busch, V., Schuierer, G., Bogdahn, U., & May, A. (2004). Neuroplasticity: Changes in grey matter induced by training. *Nature*, *427*(6972), 311–312.

Driemeyer, J., Boyke, J., Gaser, C., Büchel, C., & May, A. (2008). Changes in gray matter induced by learning — revisited. *PLoS ONE*, *23*(7), e2669.

Dum, R. P., & Strick, P. L. (1991). The origin of corticospinal projections from the premotor areas in the frontal lobe. *Journal of Neuroscience*, *11*(3), 667–689.

Ehrsson, H. H., Geyer, S., & Naito, E. (2003). Imagery of voluntary movement of fingers toes and tongue activates corresponding body-part-specific motor representations. *Journal of Neurophysiology*, *90*(5), 3304–3316.

Facchini, S., Muellbacher, W., Battaglia, F., Boroojerdi, B., & Hallett, M. (2002). Focal enhancement of motor cortex excitability during motor imagery: a transcranial magnetic stimulation study. *Acta Neurologica Scandinavica*, *105*(3), 146–151.

Ferrari, P. F., Rozzi, S., & Fogassi, L. (2005). Mirror neurons responding to observation of actions made with tools in monkey ventral premotor cortex. *Journal of Cognitive Neuroscience*, *17*(2), 212–226.

Floyer-Lea, A., & Matthews, P. M. (2005). Distinguishable brain activation networks for short- and long-term motor skill learning. *Journal of Neurophysiology*, *94*(1), 512–518.

Freund, H. J., & Hummelsheim, H. (1985). Lesions of premotor cortex in man. *Brain*, *108*(Pt.3), 697–733.

Gaser, C., & Schlaug, G. (2003). Gray matter differences between musicians and nonmusicians. *Annals of the New York Academy of Society*, *999*, 514–517.

Gazzola, V., Aziz-Zadeh, L., & Keysers, C. (2006). Empathy and the somatotopic auditory mirror system in humans. *Current Biology*, *16*(18), 1824–1829.

Gentilucci, M., Fogassi, L., Luppino, G., Matelli, M., Camarda, R., & Rizzolatti, G. (1988). Functional organization of inferior area 6 in the macaque monkey. I. Somatotopy and the control of proximal movements. *Experimental Brain Research*, *71*(3), 475–490.

Gentner, R., & Classen, J. (2006). Modular organization of finger movements by the human central nervous system. *Neuron*, *52*(4), 731–742.

Germain, L., & Lamarre, Y. (1993). Neuronal activity in the motor and premotor cortices before and after learning the associations between auditory stimuli and motor responses. *Brain Research*, *611*(1), 175–179.

Geyer, S., Ledberg, A., Schleicher, A., Kinomura, S., Schormann, T., Bürgel, U., Klingberg, T., Larsson, J., Zilles, K., & Roland, P. E. (1996). Two different areas within the primary motor cortex of man. *Nature*, *382*(6594), 805–807.

Geyer, S., Matelli, M., Luppino, G., & Zilles, K. (2000). Functional neuroanatomy of the primate isocortical motor system. *Anatomy and Embryology*, *202*(6), 443–474.

Grafton, S. T., Hazeltine, E., & Ivry, R. B. (2002). Motor sequence learning with the nondominant left hand. A PET functional imaging study. *Experimental Brain Research*, *146*(3), 369–378.

Graziano, M. S., Taylor, C. S., & Moore, T. (2002). Complex movements evoked by microstimulation of precentral cortex. *Neuron*, *34*(5), 841–851.

Grèzes, J., & Decety, J. (2001). Functional anatomy of execution, mental simulation, observation, and verb generation of actions: a meta-analysis. *Human Brain Mapping*, *12*(1), 1–19.

Grodd, W., Hulsmann, E., Lotze, M., Wildgruber, D., & Erb, M. (2001). Sensorimotor mapping of the human cerebellum: fMRI evidence of somatotopic organization. *Human Brain Mapping*, *13*(2), 55–73.

Hanakawa, T., Dimyan, M. A., & Hallett, M. (2008). Motor planning, imagery, and execution in the distributed motor network: a time-course study with functional MRI. *Cerebral Cortex*, *18*, 2775–2788.

Hanakawa, T., Parikh, S., Bruno, M. K., & Hallett, M. (2005). Finger and face representations in the ipsilateral precentral motor areas in humans. *Journal of Neurophysiology*, *93*(5), 2950–2958.

Haslinger, B., Erhard, P., Weilke, F., Ceballos-Baumann, A. O., Bartenstein, P., Einsiedel, H. G. V., Schwaiger, M.,

Conrad, B., & Boecker, H. (2002). The role of lateral premotor-cerebellar-parietal circuits in motor sequence control: a parametric fMRI study. *Cognitive Brain Research, 13*(2), 159–168.

Hauk, O., Johnsrude, I., & Pulvermüller, F. (2004). Somatotopic representation of action words in human motor and premotor cortex. *Neuron, 41*(2), 301–307.

Heide, W., Binkofski, F., Seitz, R. J., Posse, S., Nitschke, M. F., Freund, H. J., & Kömpf, D. (2001). Activation of frontoparietal cortices during memorized triple-step sequences of saccadic eye movements: an fMRI study. *European Journal of Neuroscience, 13*(6), 1177–1189.

Hutchinson, J. M., & Gigerenzer, G. (2005). Simple heuristics and rules of thumb: where psychologists and behavioural biologists might meet. *Behavioural Processes, 69*(2), 97–124.

Ito, M. (2006). Cerebellar circuitry as a neuronal machine. *Progress in Neurobiology, 78*(3–5), 272–303.

Ito, M. (2008). Control of mental activities by internal models in the cerebellum. *Nature Reviews. Neuroscience,, 9*(4), 304–313.

Ivry, R. B. (1996). The representation of temporal information in perception and motor control. *Current Opinions in Neurobiology, 6*(6), 851–857.

Jacob, P., & Jeannerod, M. (2005). The motor theory of social cognition: a critique. *Trends in Cognitive Science, 9*(1), 21–25.

Johansen-Berg, H., & Matthews, P. M. (2002). Attention to movement modulates activity in sensori-motor areas, including primary motor cortex. *Experimental Brain Research, 142*(1), 13–24.

Kelly, R. M., & Strick, P. L. (2003). Cerebellar loops with motor cortex and prefrontal cortex of a nonhuman primate. *Journal of Neuroscience, 23*(23), 8432–8444.

Keysers, C., Kohler, E., Umiltà, M. A., Nanetti, L., Fogassi, L., & Gallese, V. (2003). Audiovisual mirror neurons and action recognition. *Experimental Brain Research, 153*(4), 628–636.

Knoblich, G., & Prinz, W. (2005). Linking perception and action: an ideomotor approach. In R. Leiguarda (Ed.), *Higher-order motor disorders: from neuroanatomy and neurobiology to clinical neurology* (pp. 79–104). Oxford: Oxford University Press.

Kohler, E., Keysers, C., Umiltà, M. A., Fogassi, L., Gallese, V., & Rizzolatti, G. (2002). Hearing sounds, understanding actions: action representation in mirror neurons. *Science, 297*(5582), 846–848.

Levin, P. M. (1944). Afferent connections. In P. C. Bucy (Ed.), *The precentral motor cortex* (pp. 133–148). Urbana, IL: University of Illinois Press.

Manni, E., & Petrosini, L. (2004). A century of cerebellar somatotopy: a debated representation. *Nature Reviews. Neuroscience, 5*(3), 241–249.

Manthey, S., Schubotz, R. I., & von Cramon, D. Y. (2003). Premotor cortex in observing erroneous action: an fMRI study. *Cognitive Brain Research, 15*(3), 296–307.

Matsuzaka, Y., Picard, N., & Strick, P. L. (2007). Skill representation in the primary motor cortex after long-term practice. *Journal of Neurophysiology, 97*(2), 1819–1832.

McLeod, P., & Dienes, Z. (1996). Do fielders know where to go to catch the ball or only how to get there? *Journal of Experimental Psychology — Human Perception and Performance, 22*, 531–543.

Miall, R. C. (1998). The cerebellum, predictive control and motor coordination. *Novartis Foundation Symposium, 218*, 272–284. Discussion 284–290.

Miall, R. C. (2003). Connecting mirror neurons and forward models. *Neuroreport, 14*(17), 2135–2137.

Miall, R. C., Christensen, L. O., Cain, O., & Stanley, J. (2007). Disruption of state estimation in the human lateral cerebellum. *PloS Biology, 5*(11), e316.

Miall, R. C., Weir, D. J., Wolpert, D. M., & Stein, J. F. (1993). Is the cerebellum a smith predictor? *Journal of Motor Behavior, 25*(3), 203–216.

Middleton, F. A., & Strick, P. L. (2000). Basal ganglia and cerebellar loops: motor and cognitive circuits. *Brain Research. Brain Research Review, 31*(2–3), 236–250.

Muellbacher, W., Ziemann, U., Wissel, J., Dang, N., Kofler, M., Facchini, S., Boroojerdi, B., Poewe, W., & Hallett, M. (2002). Early consolidation in human primary motor cortex. *Nature, 415*(6872), 640–644.

Munzert, J., Zentgraf, K., Stark, R., & Vaitl, D. (2008). Neural activation in cognitive motor processes: comparing motor imagery and observation of gymnastic movements. *Experimental Brain Research, 188*(3), 437–444.

Müsseler, J., & Wühr, P. (2002). Response-evoked interference in visual encoding. In W. Prinz & B. Hommel (Eds.), *Attention and performance XIX: common mechanisms in perception and action* (pp. 520–537). Oxford: Oxford University Press.

Naito, E., Roland, P. E., & Ehrsson, H. H. (2002). I feel my hand moving: a new role of the primary motor cortex in somatic perception of limb movement. *Neuron, 36*(5), 979–988.

Nelissen, K., Luppino, G., Vanduffel, W., Rizzolatti, G., & Orban, G. A. (2005). Observing others: multiple action representation in the frontal lobe. *Science, 310*(5746), 332–336.

Nieuwenhuys, R., Voogd, J., & van Huijzen, C. (1991). *Das Zentralnervensystem des Menschen: Ein Atlas mit Begleittext [The central nervous system of humans: an atlas with accompanying text]* (2nd ed.). Berlin: Springer-Verlag.

Nixon, P. D. (2003). The role of the cerebellum in preparing responses to predictable sensory events. *Cerebellum, 2*(2), 114–122.

Nixon, P. D., & Passingham, R. E. (2001). Predicting sensory events. The role of the cerebellum in motor learning. *Experimental Brain Research, 138*(2), 251–257.

O'Reilly, J. X., Mesulam, M. M., & Nobre, A. C. (2008). The cerebellum predicts the timing of perceptual events. *Journal of Neuroscience, 28*(9), 2252–2260.

Paulin, M. G. (2005). Evolution of the cerebellum as a neuronal machine for Bayesian state estimation. *Journal of Neural Engineering, 2*(3), S219–S234.

Penfield, W., & Rasmussen, T. (1950). *The cerebral cortex of man. A clinical study of localization of function*. New York: Macmillan.

Preuss, T. M., Stepniewska, I., & Kaas, J. H. (1996). Movement representation in the dorsal and ventral premotor areas of owl monkeys: a microstimulation study. *Journal of Comparative Neurology*, *371*(4), 649–676.

Ramnani, N. (2006). The primate cortico-cerebellar system: anatomy and function. *Nature Reviews. Neuroscience*, *7*(7), 511–522.

Rijntjes, M., Buechel, C., Kiebel, S., & Weiller, C. (1999). Multiple somatotopic representations in the human cerebellum. *Neuroreport*, *10*(17), 3653–3658.

Rizzolatti, G., Camarda, R., Fogassi, L., Gentilucci, M., Luppino, G., & Matelli, M. (1988). Functional organization of inferior area 6 in the macaque monkey. II. Area F5 and the control of distal movements. *Experimental Brain Research*, *71*(3), 491–507.

Rizzolatti, G., & Craighero, L. (2004). The mirror-neuron system. *Annual Review of Neuroscience*, *27*, 169–192.

Rizzolatti, G., & Fadiga, L. (1998). Grasping objects and grasping action meanings: the dual role of monkey rostroventral premotor cortex (area F5). *Novartis Foundation Symposium*, *218*, 81–95. Discussion 95–103.

Rizzolatti, G., Fadiga, L., Gallese, V., & Fogassi, L. (1996). Premotor cortex and the recognition of motor actions. *Brain Research. Brain Research Review*, *3*(2), 131–141.

Rizzolatti, G., Fogassi, L., & Gallese, V. (2002). Motor and cognitive functions of the ventral premotor cortex. *Current Opinions in Neurobiology*, *12*(2), 149–154.

Rizzolatti, G., Luppino, G., & Matelli, M. (1998). The organization of the cortical motor system: new concepts. *Electroencephalography and Clinical Neurophysiology*, *106*(4), 283–296.

Rizzolatti, G., Riggio, L., Dascola, I., & Umiltà, C. (1987). Reorienting attention across the horizontal and vertical meridians: evidence in favor of a premotor theory of attention. *Neuropsychologia*, *25*(1A), 31–40.

Robertson, E., Theoret, H., & Pascual-Leone, A. (2003). Skill learning. In U. Ziemann (Ed.), *Plasticity in the human nervous system: investigations with transcranial magnetic stimulation* (pp. 107–134). Cambridge: Cambridge University Press.

Ruschemeyer, S. A., Brass, M., & Friederici, A. D. (2007). Comprehending prehending: neural correlates of processing verbs with motor stems. *Journal of Cognitive Neuroscience*, *19*(5), 855–865.

Sakai, K., Takino, R., Hikosaka, O., Miyauchi, S., Sasaki, Y., Pütz, B., & Fujimaki, N. (1998). Separate cerebellar areas for motor control. *Neuroreport*, *9*(10), 2359–2363.

Sakreida, K., Schubotz, R. I., Wolfensteller, U., & von Cramon, D. Y. (2005). Motion class dependency in observers' motor areas revealed by functional magnetic resonance imaging. *Journal of Neuroscience*, *25*(6), 1335–1342.

Salman, M. S. (2002). The cerebellum: it's about time! But timing is not everything — new insights into the role of the cerebellum in timing motor and cognitive tasks. *Journal of Child Neurology*, *17*(1), 1–9.

Sanes, J. N., & Donoghue, J. P. (2000). Plasticity and primary motor cortex. *Annual Review of Neuroscience*, *23*, 393–415.

Schieber, M. H. (2001). Constraints on somatotopic organization in the primary motor cortex. *Journal of Neurophysiology*, *86*(5), 2125–2143.

Schmahmann, J. D. (2004). Disorders of the cerebellum: ataxia, dysmetria of thought, and the cerebellar cognitive affective syndrome. *Journal of Neuropsychiatry and Clinical Neurosciences*, *16*(3), 367–378.

Schubotz, R. I. (2007). Prediction of external events with our motor system: towards a new framework. *Trends in Cognitive Science*, *11*(5), 211–218.

Schubotz, R. I., & von Cramon, D. Y. (2001). Functional organization of the lateral premotor cortex: fMRI reveals different regions activated by anticipation of object properties, location and speed. *Brain Research Cognitive Brain Research*, *11*(1), 97–112.

Schubotz, R. I., & von Cramon, D. Y. (2002). Dynamic patterns make the premotor cortex interested in objects: influence of stimulus and task revealed by fMRI. *Brain Research Cognitive Brain Research*, *14*(3), 357–369.

Schubotz, R. I., & von Cramon, D. Y. (2004). Sequences of abstract nonbiological stimuli share ventral premotor cortex with action observation and imagery. *Journal of Neuroscience*, *24*(24), 5467–5474.

Schubotz, R. I., von Cramon, D. Y., & Lohmann, G. (2003). Auditory what, where, and when: a sensory somatotopy in lateral premotor cortex. *Neuroimage*, *20*(1), 173–185.

Scott, S. H. (2008). Inconvenient truths about neural processing in primary motor cortex. *Journal of Physiology*, *586*(5), 1217–1224.

Sharma, N., Jones, P. S., Carpenter, T. A., & Baron, J.-C. (2008). Mapping the involvement of BA 4a and 4p during Motor Imagery. *Neuroimage*, *41*(1), 92–99.

Simon, S. R., Meunier, M., Piettre, L., Berardi, A. M., Segebarth, C. M., & Boussaoud, D. (2002). Spatial attention and memory versus motor preparation: premotor cortex involvement as revealed by fMRI. *Journal of Neurophysiology*, *88*(4), 2047–2057.

Sun, F. T., Miller, L. M., Rao, A. A., & D'Esposito, M. (2007). Functional connectivity of cortical networks involved in bimanual motor sequence learning. *Cerebral Cortex*, *17*(5), 1227–1234.

Tesche, C. D., & Karhu, J. J. (2000). Anticipatory cerebellar responses during somatosensory omission in man. *Human Brain Mapping*, *9*(3), 119–142.

Tettamanti, M., Buccino, G., Saccuman, M. C., Gallese, V., Danna, M., Scifo, P., Fazio, F., Rizzolatti, G., Cappa, S. F., & Perani, D. (2005). Listening to action-related sentences activates fronto-parietal motor circuits. *Journal of Cognitive Neuroscience*, *17*(2), 273–281.

Tomasino, B., Fink, G. R., Sparing, R., Dafotakis, M., & Weiss, P. H. (2008). Action verbs and the primary motor cortex: a comparative TMS study of silent reading, frequency judgments, and motor imagery. *Neuropsychologia*, *46*, 1915–1926.

Toni, I., Schluter, N. D., Josephs, O., Friston, K., & Passingham, R. E. (1999). Signal-, set- and movement-related activity in the human brain: an event-related fMRI study. *Cerebral Cortex*, *9*(1), 35–49.

Umiltà, M. A., Kohler, E., Gallese, V., Fogassi, L., Fadiga, L., Keysers, C., & Rizzolatti, G. (2001). I know what you are doing. A neurophysiological study. *Neuron*, *31*(1), 155–165.

Ungerleider, L. G., Doyon, J., & Karni, A. (2002). Imaging brain plasticity during motor skill learning. *Neurobiology of Learning and Memory*, *78*(3), 553–564.

Wise, S. P., di Pellegrino, G., & Boussaoud, D. (1996). The premotor cortex and nonstandard sensorimotor mapping. *Canadian Journal of Physiology and Pharmacology*, *74*(4), 469–482.

Wolfensteller, U., Schubotz, R. I., & von Cramon, D. Y. (2004). "What" becoming "where": functional magnetic resonance imaging evidence for pragmatic relevance driving premotor cortex. *Journal of Neuroscience*, *24*(46), 10431–10439.

Wolfensteller, U., Schubotz, R. I., & von Cramon, D. Y. (2007). Understanding non-biological dynamics with your own premotor system. *Neuroimage*, *36*(Suppl. 2), T33–T43.

Wolpert, D. M., Doya, K., & Kawato, M. (2003). A unifying computational framework for motor control and social interaction. *Philosophical Transactions of the Royal Society of London Series B — Biological Sciences*, *358*(1431), 593–602.

Wolpert, D. M., & Miall, R. C. (1996). Forward models for physiological motor control. *Neural Networks*, *9*(8), 1265–1279.

Zentgraf, K., Stark, R., Reiser, M., Künzell, S., Schienle, A., Kirsch, P., Walter, B., Vaitl, D., & Munzert, J. (2005). Differential activation of pre-SMA and SMA proper during action observation: effects of instructions. *Neuroimage*, *26*(3), 662–672.

M. Raab et al. (Eds.)
Progress in Brain Research, Vol. 174
ISSN 0079-6123

CHAPTER 24

How are actions physically implemented?

Karen Zentgraf[1,*], Nikos Green[2], Jörn Munzert[1], Thomas Schack[3,7], Gershon Tenenbaum[4], Joan N. Vickers[5], Matthias Weigelt[3], Uta Wolfensteller[6] and Hauke R. Heekeren[2]

[1]*Institute for Sports Science, Justus-Liebig University, Giessen, Germany*
[2]*Max Planck Institute for Human Development, Berlin, Germany*
[3]*Neurocognition and Action Research Group and Center of Excellence "Cognitive Interaction Technology" (CITEC), Bielefeld University, Bielefeld, Germany*
[4]*Department of Educational Psychology and Learning Systems, College of Education, Florida State University, Tallahassee, FL, USA*
[5]*Faculty of Kinesiology, University of Calgary, Calgary, AB, Canada*
[6]*Max Planck Institute for Human Cognitive and Brain Sciences, Leipzig, Germany*
[7]*Research Institute for Cognition and Robotics (CoR-Lab), Bielefeld University, Bielefeld, Germany*

Abstract: This chapter focuses on the interdisciplinary discussion between cognitive psychologists and neuroscientists on how actions, the results of decision processes, are implemented. After surveying the approaches used in action implementation research, we analyze the contributions of these different approaches in more detail. Topics covered include expertise research in sports science, knowledge structures, neuroscientific research on motor imagery and decision making, computational models in motor control, robotics, and brain–machine interfaces. This forms the basis for discussing central issues for interdisciplinary research on action implementation from different viewpoints. In essence, most findings show the need to abandon serial frameworks of information processing suggesting a step-by-step pattern from perception, evaluation, and selection to execution. Instead, an outlook on new approaches is given, opening a route for future research in this field.

Keywords: action implementation; sports science; neuroscience; cognitive psychology; decision making; embodiment

Introduction

One set of questions that has experienced a major renaissance in science during the last two decades is: how can thought and action be linked, what is the neural basis for such processes, and how might this knowledge be used in applied settings, for example, to enable patients with motor disorders to regain control over their limbs? One study by Velliste et al. (2008) links thought and action well: two monkeys were trained to control a robotic arm by pure thought. The robotic arm could bring food to their mouth and was controlled by the neural activity in the monkeys' own primary motor cortex as recorded with implanted intracortical microelectrode arrays. This is an impressive example of how a brain–machine interface (BMI)

*Corresponding author.
Tel.: +49-641-992-5223; Fax: +49-641-991-9861;
E-mail: karen.zentgraf@sport.uni-giessen.de

DOI: 10.1016/S0079-6123(09)01324-7

can implement a real physical interaction with the world: an arm movement to perform everyday actions is coded in motor neurons, and these learned activity patterns can be used by monkeys to move a device other than their own arm. In the development of prosthetic devices for patients, this could prove to be a major step forward.

What does this tell us about the present question asking how actions are implemented from an interdisciplinary perspective covering both cognitive psychology and neuroscience? Various lines of research have focused on this issue. Traditional serial concepts view action implementation as the final stage after options have been perceived, evaluated, and one option has been selected. It has been argued, however, that "*perceiving* and *representing* options, *selecting* between them, and *implementing* a particular action, cannot be considered to be mutually exclusive processes. Rather, they necessarily inform and influence each other, not only in the serial manner (perception-selection-implementation), but also in a more intertwined manner" (Chapter 23: Juggling with the brain—thought and action in the human motor system; see also Heekeren et al., 2008). This chapter will consider insights that are relevant for the implementation stage.

In a first step, we shall clarify what we mean by "action implementation." As our objective is also to link various ideas on the underlying processes of action implementation taken from different fields, our second step will be to introduce four research perspectives and further elaborate these in terms of how they address the phenomenon of action implementation. It should be pointed out here that some of the ideas outlined in this chapter do not necessarily accord with the idea that pre-action cognitive stages clearly precede action implementation. Finally, we shall present controversial issues and open questions that arise when the aforementioned research perspectives are integrated.

What do we mean by "action implementation"?

Following the logic of the previous chapters on option representation and evaluation, as well as action selection based on option evaluation, we use the term "action implementation" to refer to the enactment of voluntary movements. Different views have to be distinguished: a neuroscientist, for instance, will ask how and where action implementation is reflected in brain activity, whereas a cognitive psychologist will ask how action implementation can be conceptualized in terms of information processing, for example, how building blocks of action implementation should be defined and measured. In philosophical action theory, the crucial question is which processes cause intentional human movements to happen and how much so-called "wants" are causal to motor behavior (e.g., Habermas, 1989). From a systems-theoretical view, it has been suggested that human movements are dynamically constrained by the *performer*, the *task* itself, and the *context* (Newell, 1986; Newell and McDonald, 1994). Consequently, when dealing with the question of how actions are implemented, all three aspects should be considered.

Looking at *performer-related factors*, especially motor expertise has been shown to influence pre-enactment and enactment stages of actions greatly. In this category of performer-related factors, emotional aspects during execution also need to be considered (Beilock and McConnell, 2004; Chapter 20: Mental representations as an underlying mechanism for human performance). It has been demonstrated that many performers exhibit high-level skills in a practice setup, but sometimes struggle under stressful conditions (Chapter 20: Mental representations as an underlying mechanism for human performance). How is it that the linkage between the emotional-cognitive-motor systems changes under pressure? What are the underlying mechanisms that permit or prevent an effective implementation of action?

With respect to Newell's framework, we should also clearly define the *task at hand* when suggesting mechanisms at work. As an example, research questions need to be as clear and precise as the following: how does eye-movement control for information pickup highlight mechanisms of whole-body motor control in an interceptive sport task, or how do task-specific representations mediate between perception and action in athletes or humanoid robots? Or, what does the consideration of different task–environment–performer interactions tell us about the functional mechanism at work in the human motor system?

It has been shown that *context-related conditions* such as uncertainty influence the action implementation system (Trommershäuser et al., 2006). In cognitive decision-making tasks such as choosing between two wagers, participants typically fail to maximize expected gain. In contrast, in mathematically equivalent movement tasks, participants are typically very good at choosing motor strategies that come close to maximizing expected gain (Chapter 21: Biases and optimality of sensory-motor and cognitive decisions). Trommershäuser concludes that the probabilistic information in motor tasks is endogenous and is therefore represented in a different way than the more exogenous probabilistic information in cognitive tasks. Another example for a context-related condition that clearly influences motor behavior is the instructional setting (Gray, 2004; Wulf, 2007 for a review). In these studies, the attentional focus is changed by instructions given to participants prior to practicing a motor task. For instance, participants are asked to focus either on their own body parts during movement execution or on external events that might be movement related in some conditions but not in others. In general, these studies show that contextual variables need to be taken into account if we want to understand the full picture of action implementation.

To date, different research fields have addressed these questions mostly independently from each another. Cognitive research has provided conceptual models describing how movements might be internally controlled in experts or in specific settings, whereas neuroscientists provide neurobiological models of the mechanisms through which movements and actions are implemented in the brain. Although cognitive psychologists agree on the idea that cognitive functions are implemented in the brain, cognitive models do not necessarily refer to biological aspects or principles of brain functioning, but instead concentrate on how information is processed. In turn, neuroscientists investigating cognitive or motor functions of the brain do not necessarily take cognitive models into account or feed their findings back into these models. The next section will give an overview of these different perspectives as well as examples for approaches used in different fields to tackle the question of how actions are implemented.

Overview: how are actions physically implemented?

Perceptual-cognitive perspective

Cognitive psychologists argue that actions are performed to achieve certain goals, that is, to produce changes in the environment. The term "effect" refers to these intended consequences of motor actions. One dominant idea in the perceptual-cognitive perspective is that action implementation is functionally related to perceptual action effects, accounting for the perception–action interface missing in traditional approaches. Hommel et al. (2001) proposed an influential theoretical framework in which perceptual contents and action plans are coded in a common representational medium (i.e., event codes). Hence, intended action effects are an essential control variable for action and for assessing action effects at each stage of action implementation (see also Nattkemper and Ziessler, 2004; Rosenbaum et al., 2007).

The perceptual-cognitive perspective suggests that movements are organized and stored in memory as perceivable events through a representation of anticipated (e.g., perceptual) effects, with the corresponding motor activity automatically and flexibly tuned to serve these effects (e.g., Mechsner et al., 2001). Several approaches have emerged in cognitive psychology in recent years that systematically study the function of effect anticipation and effect representations in action implementation and action control. These include the ideomotor approach (Knuf et al., 2001) and the cognitive-architecture-of-action approach (Schack and Mechsner, 2006; Chapter 19: The cognitive nature of action — functional links between cognitive psychology, movement science and robotics).

Neuroscience approaches

As illustrated in the introduction, neuroscience has studied the neurophysiological level of actions with a wide range of methods ranging from single-unit recordings and transcranial magnetic stimulation to neuroimaging methods such as functional magnetic resonance imaging (fMRI) and electroencephalography (EEG). Recent research indicates that action execution does not progress serially from preparatory activation in the premotor cortex to primary motor cortex (M1) as the final central output machinery, but that other areas also connect to the subcentral system and work in parallel to implement actions (Dum and Strick, 1996; Akkal et al., 2007). Another influential aspect in neuroscience refers to the fact that M1 can be activated without peripheral muscle contractions, as revealed in recent imagery or action observation studies (see Rizzolatti and Craighero, 2004 for a review on the mirror-neuron system; Lotze and Halsband, 2006 for a review on motor imagery; cf. Chapter 18: Motor imagery and its implications for understanding the motor system). This implies that just assigning the function of motor output to the motor system provides a poor description of it, and that it also subserves more cognitive functions (Sanes and Donoghue, 2000; Lu and Ashe, 2005). Consequently, the classical distinction between "cognitive" and "motor" areas (and functions) needs to be revised.

Computational models of motor control

Computational models developed in movement sciences provide frameworks explaining how the central nervous system relates sensory signals to motor commands. An essential aspect of most computational models is the predictive nature of motor control (Wolpert et al., 1995). In brief, the inverse model is taken to generate an appropriate motor command, and the forward model is taken to map this motor command (or its efference copy) onto the predicted outcome of the action, thereby building a template against which the incoming information (reafferences) can be compared. There is usually little discrepancy between the anticipated outcome and the real sensory feedback during moving. Greater discrepancies result in rapid adjustments of the motor command and, on this basis, in modified anticipated consequences of actions. Recently, computational models from the motor control domain have also been linked to the domain of social interaction, for example, when observing the actions of another person, changes in the internal mental states of this observer may lead to specific actions that will also be perceived in turn by the previously acting person as well (see Wolpert et al., 2003 for details).

Ecological views on action research

Ecological approaches were put forward by James Gibson (1979) and these oppose cognitive approaches in some ways. The ecological idea of *direct* perception suggests that people perceive the environments in which they perform unaided by inference, memories, or other neural representations (Michaels and Carello, 1981). In this framework, movements are dependent on the establishment of direct relationships that develop without any apparent need for the many processing stages described in cognitive psychology. The ecological approach stresses the concept of "perception for action" by identifying two main problems: that of affordances and that of control parameters that guide movements (Turvey and Kugler, 1984). Affordances refer to opportunities for action that are offered by the actor's environment and that are relative to the actor's motor competencies (Van der Kamp et al., 2001). It is stressed as a main aspect of this approach that sources of information are selected in a task-specific manner (Van der Kamp et al., 1997).

Specific approaches to the study of action implementation

This section will describe specific empirical approaches used within the disciplines outlined above.

Expertise research in sports sciences

Expertise research is one way to elucidate how actions are put into effect. In visual search studies, participants are requested to make decisions on the basis of specific visual input, and their gaze behavior is studied during this process. The visual-search paradigm has now been applied in research on reading, art, mathematics, and chess for more than 20 years. It has also been used extensively in sports, especially in temporal and spatial occlusion paradigms in which aspects of a video display are masked. Experts typically exhibit a lower frequency of gaze shifts and keep their eyes longer on critical aspects of the display (Williams et al., 1999; Williams and Ward, 2003). The visual search paradigm has provided valuable insights into how novice and expert athletes differ, but the whole skill is rarely performed physically in these studies. Thus, only limited insight is gained into the gaze, attention, and decision-making characteristics during successful and unsuccessful motor performance. Hence, researchers started to study gaze behavior while experts were actually doing sports, that is, *during* motor performance. In the typical study by Vickers (see Vickers, 2007 for an overview), high- and low-skilled individuals perform motor tasks under conditions similar to those found within their sport. Their gaze is recorded by a mobile eye tracker coupled with a motion analysis system that records their movements synchronously. An equal number of successful and nonsuccessful trials are analyzed under various experimental conditions in which task complexity, competitive pressure, athlete anxiety, and/or physiological arousal are manipulated. The goal is to determine the bidirectional linkages that may exist between gaze, attention, and decision-making processes underlying both successful and nonsuccessful performances. Such studies then use these gaze behaviors to try to explain the perceptual and cognitive processes that define optimal and nonoptimal motor performance. For many years, it proved difficult to link shifts in gaze with shifts in attention, but more recent studies have shown that, under certain conditions, a shift in gaze is invariably preceded by a shift in attention (Kowler et al., 1995; Deubel and Schneider, 1996; Corbetta, 1998; Henderson, 2003). There is now strong evidence that when a saccade is made to a new location, there is a corresponding attention shift in the direction of the saccade. When athletes shift their gaze to a new location, it means that they have also shifted their attention to that location for at least a brief period of time. Over many studies, one gaze called the "quiet eye" (Vickers, 1996; Vickers, 2006; Chapter 22: Advances in coupling perception and action: the quiet eye as a bidirectional link between gaze, attention, and action) has emerged as being a characteristic of successful performance on a wide range of motor tasks (Janelle et al., 2000; Rodrigues et al., 2002; Williams et al., 2002; Panchuk and Vickers, 2006). For a given motor task, the quiet eye is the final fixation or tracking gaze that is located on a specific location or object in the visuomotor workspace within 3° of visual angle (or less) for a minimum of 100 ms. The onset of the quiet eye occurs prior to the final movement on the task and the offset when the gaze deviates off the object or location by more than 3° of visual angle (or less) for a minimum of 100 ms. Therefore, the quiet eye can carry through and beyond the final movement of the task. The quiet eye of elite performers is significantly longer than that of near-elite or lower-skilled performers; that is, those who consistently achieve high levels have learned to fixate or track critical objects or locations for longer durations prior to final critical movement irrespective of the conditions encountered. Additionally, elite performers have found a way to see critical visual information earlier than near-elite and lower-skilled performers and to process this information longer prior to making the final movement, because the quiet eye onset of elite performers is invariably earlier. Finally, the quiet eye of elite performers is of optimal duration, being neither too long nor too short, but ideal given the constraints of the task being performed. A meta-analysis by Mann et al. (2007) has identified the quiet eye as being one of three predictors of perceptual-motor expertise (along with fixation location and a low frequency of gaze), and it is also affected by high levels of pressure and high anxiety

(Vickers and Williams, 2007; Behan and Wilson, 2008).

Knowledge structures

Cognitive structures related to the task itself, the environment, and the anticipated steps of action are fundamental key elements of action implementation. In different fields such as bimanual coordination (Weigelt et al., 2006), manual action (Rosenbaum et al., 2007), complex sport movements (Schack and Mechsner, 2006), robotics (Chapter 19: The cognitive nature of action — functional links between cognitive psychology, movement science and robotics), and decision making (Chapter 14: A conceptual framework for studying emotions–cognitions–performance linkage under conditions that vary in perceived pressure), it has been shown that central costs and interference in actions depend greatly on how these movements are represented on a cognitive level. Perceptual-cognitive representations in action implementation might involve different formats such as propositions, relational structures of many kinds, and concepts. Some studies (see Schack, 2004a, b; Schack and Mechsner, 2006; Chapter 14: A conceptual framework for studying emotions–cognitions–performance linkage under conditions that vary in perceived pressure Chapter 19: The cognitive nature of action — functional links between cognitive psychology, movement science and robotics) have provided evidence for so-called basic action concepts (BACs) in analogy to the well-established notion of basic concepts in object representation (Rosch, 1978). BACs can be viewed as the mental counterparts of functionally relevant elementary components or transitional states (body postures) of complex movements.

The integration of representation units (BACs) into structures of representation has been studied with a wide range of methods (e.g., Hodges et al., 2007). But, in principle, two methodological approaches are used when studying representation structures in complex actions. These derive them from either response behavior or reaction times. Whereas the first approach has been used to study order formation in long-term memory (LTM), the second approach has been used to ascertain chunk structures in working memory (Schack, 2004b). Schack and Mechsner (2006) studied the tennis serve to investigate the nature and role of LTM in skilled athletic performance. In high-level experts, these representational frameworks were organized in a distinctive hierarchical tree-like structure, remarkably similar between individuals, and well matched to the functional and biomechanical demands of the task. In comparison, action representations in low-level players and nonplayers were organized less hierarchically and more variably between individuals.

The results of studies in golf, soccer, windsurfing, volleyball, gymnastics, or dancing (Schack, 2003, 2004b; Schack and Bar-Eli, 2007; Schack and Hackfort, 2007) show that the mental representation structures relate to performance. These representation structures are the outcome of an increasing and effort-reducing formation of order in LTM. This order formation reveals a clear relation to the structure of the movement. With increasing expertise, the representation of the movement corresponds more and more exactly to its spatiotemporal structure. The representation structures are built up from sensory movement effects of distinctive nodal points (e.g., body postures) of the movement. Hence, the representation structure itself possesses spatiotemporal properties and corresponds well with the movement structure. Accordingly, movement control becomes possible by representing the anticipated intermediate effects of the movement and comparing them with incoming effects. It also means that no special translation mechanism is required between perception, representation, and movement.

Results from another line of experimental research showed that not only the structure formation of mental representations in LTM but also chunk formation in working memory are built upon BACs and relate systematically to movement structures. These studies suggest a *movement-based chunking*, implying a relation between chunking processes in working memory and the movement structure; in other words, structures in movement and memory mutually overlap.

Based on these findings, research lines were created to improve action implementation in high-performance sport, rehabilitation, and robotics. One issue in traditional procedures for mental training is the attempt to optimize performance by means of repeated movement imagination without taking the athlete's mental technique representation into account (i.e., they are representation-blind). However, if the movement's cognitive reference structure has structural gaps or errors, these will tend to be stabilized rather than overcome by repeated practice. The alternative is to measure the mental representation of the movement before mental training and then integrate these results into training. Mental training procedures based on mental representations have been applied in professional sports (Schack and Bar-Eli, 2007 for an overview; Schack and Hackfort, 2007) as well as recently in the rehabilitation of hand functions in patients after stroke (Braun et al., 2007, 2008).

Neuroscience research on motor imagery and the motor system

Recently, the concept of how the motor system works has changed fundamentally, not only through the discovery of supposedly nonmotor functions within the motor system (Rizzolatti and Craighero, 2004 for a review on the mirror-neuron system) but also through a re-evaluation of motor output during stimulation of the motor cortex (Graziano et al., 2002). One approach for gaining insight into the functioning of the motor system is to study motor-related "cognitive" states without moving as in action observation, silent action verbalization, or motor imagery. In his simulation concept, Jeannerod (2001) called these states "S-states" (S for simulation). Within this framework, similarities on the behavioral side and neural overlap between S-states and execution are interpreted as providing the common basis for the motor representations implemented within the motor system. As a result, motor imagery as a state without moving but with activation of the motor system provides insight into the functions implemented in the motor system. What can be concluded from motor imagery studies is that "cognitive" or "sensory" functions are also implemented in the motor system, and this clearly delivers a new view on motor system functioning. Jeannerod (2001) provocatively formulated that covert actions include everything that is involved in overt actions except for the muscular contractions, joint rotations, and reafferent signals generated by real movement. Therefore, Jeannerod's theory predicts neural identity of motor imagery and the point of time just prior to execution as well as neural similarity between covert and overt actions. In general, brain-imaging studies have supported these claims (see Grèzes and Decety, 2001; Jeannerod, 2001 for meta-analyses on this topic), because the activated network during motor imagery involves dorsal and ventral motor cortex, cerebellar and subcortical structures, and sometimes even the primary motor cortex (M1; Munzert et al., 2008). M1 has traditionally been viewed as the definite executive structure, implying that it serves as a "gate" for signals to enter the spinal chord. This raises the question whether motor imagery and the activation of motor representations might even play a causal role in motor behavior. Another issue that might be disentangled by the usage of covert action paradigms relates to the implied continuum (Heekeren et al., 2008) between action selection, generating a motor intention, preparing for action, and motor execution.

fMRI studies suggest that cognitive, but motor-related processes are in fact integrated within the motor system. Neuroimaging findings of activation in the ventral premotor cortex reflecting mirror-neuron activity (Rizzolatti and Craighero, 2004), as well as findings on attention-related activation of the premotor cortex (Johansen-Berg and Matthews, 2002; Rowe et al., 2002) and M1 (Binkofski et al., 2002), strongly suggest that motor functions need to be reconceptualized. Additional empirical evidence for a new view on the motor system has come from studies by Schubotz and Von Cramon (2001). Schubotz et al. (2003) and Wolfensteller et al. (2007) focused on the predictive nature of premotor cortex activation (for an overview see Schubotz, 2007). These studies established that the premotor cortex is not only engaged in the prediction of biological dynamics (e.g., as in action observation and movement imagination) but also in the prediction of nonbiological dynamics (e.g., spatial, object, or rhythmic dynamics of abstract visual or auditory stimulus sequences). But why should the motor system perform prediction in general? Prinz (2006, p. 516) argues that there is a "crucial advantage of leaving the job of anticipation to the motor system" because it is implemented as an anticipating device (Wolpert and Flanagan, 2001).

Neuroscience research on decision making

Looking at the monkey brain, it is worth noting that regions in the monkey brain that have been implicated both in representing decision variables and in performing the comparator operation are the very same areas that select, plan, and execute motor responses (Gold and Shadlen, 2007). In other words, the boundaries between sensory, decision-related, and motor-related processing are not as distinct as often thought. For example, when monkeys have to decide in which direction a random-dot motion stimulus moves and indicate their decision with an eye movement, decision-related as well as saccade-related activity can be recorded from the frontal eye fields (FEF) (Gold and Shadlen, 2003). Similarly, when monkeys perform a vibrotactile discrimination task, activity in medial and ventrolateral premotor cortex reflects the temporal evolution of the decision-making process leading to action selection (Hernandez et al., 2002; Romo et al., 2004). Other neurophysiological studies have revealed that decision variables are represented in the superior colliculus, a midbrain region involved in the generation of saccadic eye movements (Gold and Shadlen, 2000; Horwitz et al., 2004). These studies thus support theoretical and modeling studies (Verschure and Althaus, 2003; Wyss et al., 2004) suggesting that the brain regions involved in selecting and planning a certain action play an important role in forming decisions that lead to that action.

It should be noted, however, that in most of the monkey studies, the monkeys were trained to indicate their perceptual decision with a *particular* action. In other words, the monkeys could treat the perceptual decision as a problem of movement selection. Seen in this light, it is not surprising that motor structures appear to play a role in decision formation. Nonetheless, it is not yet clear how these structures contribute to decisions that are not linked to particular actions.

So far, neurophysiological studies in monkeys, as well as modeling studies, suggest that the brain regions involved in selecting and planning a certain action play an important role in forming decisions that lead to that action. To test whether this result also holds for the human brain, Heinen et al. (2006) had participants play "ocular baseball" while undergoing fMRI in a study of oculomotor decision making. In this game, participants had to indicate by eye movements whether they thought a dot moving across a computer screen would cross into a visible strike zone or not. They scored a point when they correctly predicted a "strike," so that their eye movements pursued a dot that eventually crossed into the strike zone. They also scored a point on trials when they correctly predicted a "ball" and withheld an eye movement (e.g., remained fixated) when the dot missed the strike zone. When the results of a task with identical oculomotor behavior were compared to the "baseball" trials, decision-related signals were found in the superior parietal lobule, the FEF, and the ventrolateral prefrontal cortex. In line with the monkey data, these results suggest that when a decision is associated with a specific movement, formation of the decision and preparation of the behavioral response share a common neural substrate. Put more generally, the finding supports the view that the human motor system also plays an important role in perceptual decision making.

More recently, Heekeren et al. (2006) have investigated whether decisions may be transformed into motor actions in the human brain independent of motor planning and execution, that is, at an abstract level. Human participants performed the direction-of-motion discrimination task also used by Gold and colleagues (see above) and responded with either button presses or saccadic eye movements. They predicted that areas representing decision variables at a more abstract level would show a greater response to high-coherence (easy) relative to low-coherence (difficult) trials — independent of the motor system used to express the decision. Four areas displayed this pattern of activity: left posterior dorsolateral prefrontal cortex (DLPFC), left posterior cingulate cortex, left intraparietal sulcus (IPS), and left fusiform/parahippocampal gyrus. Most importantly, the blood-oxygen-level-dependent (BOLD) signal increase in these regions was independent of the motor system the participants used to express their decision. The results from this fMRI study are in line with findings reported by Kim and Shadlen (1999). They showed that the neural activity in the DLPFC of monkeys increases proportionally to the strength of the motion signal in the stimulus. The findings in humans, however, suggest that the posterior DLPFC is an important component of a network that not only accumulates sensory evidence to compute a decision but also translates this evidence into an action independent of response modality.

Notably, to date, neurophysiological studies in monkeys have not found neurons whose activity reflects decisions independently of response modality. In fact, one could conclude from the neurophysiological studies in monkeys that "to see and decide is, in effect, to plan a motor response" (Rorie and Newsome, 2005, p. 43). In humans, in contrast, Heekeren et al. (2006) found regions of the cortex that responded independently from the motor effectors used. Based on these findings, one could speculate that a more abstract decision-making network has evolved in humans providing a more flexible link between decision and action (Heekeren et al., 2004).

Computational models of motor control

These models provide frameworks on how the central nervous system relates sensory signals and motor commands. An essential aspect of most computational models is their predictive nature (e.g., Wolpert et al., 1995; Dean and Cruse, 1998; Kalveram, 2004), because the inverse model generates an appropriate motor command and the forward model maps this motor command (or the efference copy) with the anticipated outcome of the action. Whereas past computational models were mainly used as simulator tools to investigate small-range motor actions (such as reaching and grasping movements), they have recently been linked to the domain of complex social interactions.

In a general perspective, computational models are important tools for simulating different stages of action implementation and action control (Dean and Cruse, 1998). It is important to distinguish the mechanism and possibilities of computation (e.g., the simulation of information processing via neuronal networks) from the simulated model. Computational models (e.g., artificial neural networks) are now used jointly by researchers in movement science, biocybernetics, and in robotics (Ritter et al., 2003; Blaesing, and Cruse, 2004; Pfeifer and Bongard, 2006) to investigate the functioning of motor control and to implement the results of motor control research for technical platforms. This joint effort is a new way to investigate action implementation and to build such functions as action implementation into simulation models and apply them to different types of robots and technical platforms.

Robotics

Cognitive robotics is another important field of research that has contributed to the experimental study of action implementation. Experimental approaches to study the action implementation in human actions benefit from insights gained from artificial control architectures for robot systems and vice versa (Pfeifer and Bongard, 2006). The research lines and areas created in the field of robotics aim to investigate systematically the principles needed to build artificial cognitive systems that can interact with a human in an intuitive way — including the acquisition of new skills by learning.

From explorations of the possibilities and limits of artificial control architectures for robot systems, it is well known that feedback connections based on technical systems can determine dynamic systems behavior almost entirely — effectively covering almost all dynamic details (and imperfections) that would be exhibited if the modules were to operate in open-loop mode. Therefore, even rather crude approximations to the different cognitive module functions in human action implementation grant us the opportunity to gain significant insights into the functioning and cognitive background of robots and human–robot interactions. Therefore, a major part of the challenge facing cognitive robotics will be to explore architectural patterns in action implementation that organize the computational contributions of a collection of sensory, motor, and memory modules into a coherent sensorimotor processing activity (Ritter et al., 2003).

Current robot technology has matured to the point at which it can approximate a reasonable spectrum of isolated perceptual, cognitive, and motor capabilities, allowing the exploration of architectures for the integration of these functions into robot action control. This gives us the opportunity to fit models dealing with action implementation in human action together with implementation architectures generated for robot actions. Among the key issues to be addressed is the question how structured representations can arise during skill acquisition, and how the underlying processes can be understood in order to replicate them on robot platforms. Researchers translate their findings in studies of human movement into models that can guide the implementation of cognitive robot architectures.

BMIs

The introduction of this chapter started with an impressive example of what can be learned about action implementation by means of BMIs. In the study by Velliste et al. (2008), researchers used an invasive technology to link the motor cortex of monkeys with a robotic arm. The monkeys were trained to control this so that it would bring food to their mouth. This example shows that BMIs are an interesting tool to study action implementation by linking brain activity with external technical devices that enable the whole system to attain anticipated effects. BMI technology is defined by the decoding of brain activity with the help of invasive, partially invasive, or noninvasive devices and the use of the extracted data for a near real-time operation of a machine (e.g., a robot or a computer). Such brain-adequate interfaces should diminish the barrier and functional differences between the human brain and the technical system as much as possible. BMI research started in the 1980s, but Kennedy and Bakay (1998) were the first researchers to install a brain implant in a human that produced signals of sufficient quality to allow the BMIs to simulate movements on a computer screen. Their patient suffered from locked-in syndrome; his implant was installed in 1998 and enabled him to control a computer cursor. In noninvasive human–brain interfaces (HBI), researchers often use EEG or fMRI methods to extract brain activity and to link it to robot hands or to virtual-reality setups (Ritter et al., 2007). Based on such research, it is possible to isolate important modules of action implementation in humans. This approach makes it possible to assess what kind of activation pattern is responsible for implementing the anticipated action pattern and how these brain activation patterns are translated into motor commands for a robotic hand.

Controversies/open questions

Is there clear evidence for a functional role of mental representations?

One major issue is related to the functional role of cognitive representations. Is it conclusive that mental representations are the *cause* of better motor performance? To put it in other words, what evidence clearly shows that these representations are not just an epiphenomenon? Is there a way of thinking about expertise behavior in other terms?

Traditionally, the primary problem of motor control has been taken to be organizing the correct pattern of muscular activation (e.g., Schmidt and Lee, 1998). The perceptual-cognitive approach, in contrast, proposes that the crucial step is constructing the appropriate mental representations, because these representations primarily govern the tuning of motor commands and muscular activity patterns (e.g., Rosenbaum et al., 2007). From this point of view, the intended action effect is an essential control variable for implementing action, for assessing action effects, and for organizing action control. As noted above, representation structures are built up from the sensory movement effects of distinctive nodal points (e.g., body postures) of the movement. Accordingly, movement control becomes possible by representing the anticipated intermediate effects of the movement and comparing them with incoming effects. It also means that no special translation mechanism is required between perception, representation, and action. It is assumed that actions are implemented, organized, and stored in memory as perceivable events through a representation of anticipated characteristic (e.g., perceptual) effects, with the corresponding motor activity flexibly tuned to serve these effects (Mechsner, 2004; but see also Carson, 2004).

The other perspective on the controversy addresses the point that knowledge structures must, at least to some degree, allow verbalization and conscious processing. Reflexive processes might be helpful for option generation and for decision making, but they may have no functional relevance for *action implementation*. At this stage, a direct link between perceptual representations and motor processes may not be influenced by knowledge structures as suggested by ecological approaches, but rather reflect the ongoing process of moving as can be found in skilled action.

Altogether, two conclusions can be derived from the controversy. The first is how knowledge and mental representations are defined. In the discussion, different kinds of concept labels are used such as motor representation, perceptual representation, perceptual-cognitive representation, mental representation, or knowledge. In general, there is a consensus that representation plays a functional role in action implementation, but this is not the case for all parts of knowledge. Humans have different kinds and different levels of knowledge. And, indeed, not all parts of knowledge structures, especially the more reflexive parts of them, have a functional meaning in action implementation. But in every case, the functionally important representation structures are a central part of knowledge. So, the second conclusion is that we have to develop valid experimental methods to measure representation structures without asking participants explicitly about their knowledge or representation structures. High performers may well not *report* the same knowledge they *use* for performance, or patients may well *report* knowledge they do not *use* for performance.

Does the perception-action link need cognitive processes?

One question that arises is whether linking an intended action to perceptual input relies on cognitive processes. From an ecological perspective, optical flow patterns potentiate motor activity directly, enabling organisms to interact with objects. The notion of object affordances (e.g., Ellis and Tucker, 2000; Tipper et al., 2006) suggests that objects are graspable, sit-uponable, climbable, stand-upon-able, and so on. This, however, already assumes that the organism *is* moving, and be it only through eye movements. Among others, *attention* and *memory* are basic components of cognition (Eysenck and Keane, 2000). It is difficult to understand how moving through the environment and interacting with objects (let alone other people as interaction partners) would work without directing one's attention. Selecting between two alternatives (*decision making* being a third component of cognition) from the same perceptual input (such as when passing the ball to player A, but not player B in a soccer game) requires shifts in attention, but also memory. Memory structures store, among others, movement knowledge and tactical knowledge. Furthermore, movement knowledge includes representations of the motor repertoire. Humans may, for instance, pass the ball to player A using an inside kick, or may score a goal with a header. Tactical knowledge decides about one's own action space. While much tactical knowledge may enlarge action space (i.e., raising the number of action alternatives in a particular situation), the absence of tactical knowledge inevitably limits a human's activities (i.e., decreasing the number of action alternatives). From these examples pertaining to *attention*, *memory*, and *decision making* as three basic components of cognition, it can be concluded that the perception-action link cannot be considered without including cognitive processes. In contrast, the concept of affordances assumes that a direct pickup of task-relevant cues provides an actor with information for skill selection that does not necessarily rely on knowledge structures. Experimental designs must be developed to reveal to what degree, if at all, knowledge structures are obligatory for action implementation.

Can there be a bidirectional link between cognitive psychology and neuroscience?

Up until now, research on action implementation has developed rather independently in cognitive psychology and neuroscience. A synthesis — a bidirectional link — of these two approaches offers the potential to further our understanding of action implementation.

So how can neuroscientific findings inform cognitive theories on action implementation and formation? Taking this path would entail a model-based approach in which researchers try to match distinct predictive behavioral models (derived from cognitive theories) with brain data. Using model selection, the "best" model (given the task context) could be derived to inform cognitive theories about structural as well as functional assumptions and parameters. This approach might lead to a better understanding of both human brain function and the cognitive models underlying action implementation and formation.

How do emotions and/or physiological arousal influence action implementation?

One area that has received less attention is the direct emotion-action link or the mediating role cognition plays within this linkage. The "hedonic tone" associated with the emotional state has been found to affect the behavioral and cognitive "repertoire" required to accomplish a given task (Fredrickson, 1998). Thus, the hedonic tone associated with emotions allows the generation of solutions and it impacts on attention focus, cognitive processes, action possibilities, and use of intellectual resources, as well as the use of social resources for action implementation. The facilitating and debilitating impact of emotions and their associated hedonic tone on action production, as well as the sequential emotion-action-cognition-emotion chain need to be explored along with both their behavioral consequences and their underlying neurobiological mechanisms.

Recent research has shown that the emotional state affects the active defensive circuitry resulting in faster and more variable voluntary movements (Coombes et al., 2006). More specifically, unpleasant emotions were subsequently reflected in more force production and acceleration of the central processing than pleasant emotions. Although the effect of emotions on attention and cognitive processes has received much attention (Chapter 14: A conceptual framework for studying emotions–cognitions–performance linkage under conditions that vary in perceived pressure), the underlying mechanisms of the emotion-cognition-action-emotion link have yet to be explored in relation to the specific cognitive and physical demands of a task.

For most individuals, high levels of physiological arousal and cognitive anxiety interact to affect the implementation of action negatively. However, this is not the case for all people (Hanin, 2000). Some athletes prevail despite insurmountable odds, and soldiers, medical personnel, and emergency workers perform superbly under extreme pressure. Mounting evidence shows that these individuals are able to overcome the debilitating effects of high levels of physiological arousal, anxiety, and pressure by adopting an external focus of attention rather than an internal one (directed toward internal physiological, technical, or emotional processes). Other empirical data supporting this view (see also Chapter 4: Perceiving and moving in sports and other high-pressure contexts) has been presented recently in studies on attention (Beilock et al., 2004) and gaze control (Vickers and Williams, 2007).

Future perspectives

The scientific disciplines of cognitive psychology and neuroscience have greatly advanced our knowledge on action control and implementation, as briefly reviewed above. Several promising fields have emerged that establish the organization of behavior from the bidirectional link between mind and motion. First, there has been a paradigm shift in understanding the brain as embedded in a body, and there have been many recent discoveries concerning body maps and brain plasticity. When modalities of the interactions with our surroundings are parts of human thinking, traditional disciplines merge into one field; this paradigmatic change has been referred to as the embodiment approach (e.g., Wilson, 2002). This knowledge has just begun to be viewed in a more complex way, namely, as an example of how these body representations interact and overlap with processes relevant for action control. From this perspective, research on action implementation will consolidate as a field in which interactions between perception, motor planning, kinesthetic sensations, and psychological processes such as social cognition or attentional focusing become fundamental for understanding systemic functioning.

Parallel to this, experts in the field of multisensory processing are starting to develop an interest in the perception-action link. There is some reappraisal of the rather unspectacular but essential insight that perception also "serves" action, and does not work in isolation (Stein and Stanford, 2008 for a review on multisensory integration). Additionally, we can see a growing interest in the field of how actions are controlled when they are performed together with others (Knoblich and Jordan, 2003; Sebanz et al., 2006; Newman-Norlund et al., 2007, 2008). New data and theories on the cognitive components that are taken to be critical for the implementation of cooperative and joint actions have been published, and it seems as if this topic has only just begun to further our understanding of action implementation in general (Sebanz et al., 2006). It will be interesting to see how future research linking thought and action will change our understanding of people acting in the world.

Abbreviations

BACs	basic action concepts
BMI	brain-machine interface
BOLD	blood-oxygenation-level-dependent
DLPFC	dorsolateral prefrontal cortex
EEG	electroencephalography
FEF	frontal eye field
fMRI	functional magnetic resonance imaging
HBI	human-brain interface
IPS	intraparietal sulcus
LTM	long-term memory
M1	primary motor cortex
ms	milliseconds
S-states	simulation states

References

Akkal, D., Dum, R. P., & Strick, P. L. (2007). Supplementary motor area and presupplementary motor area: targets of basal ganglia and cerebellar output. *Journal of Neuroscience*, *7*(40), 10659–10673.

Behan, M., & Wilson, M. (2008). State anxiety and visual attention: the role of the quiet eye period in aiming to a far target. *Journal of Sports Sciences*, *26*(2), 207–215.

Beilock, S., Bertenthal, B. I., McCoy, A. M., & Carr, T. H. (2004). Haste does not always make waste: expertise, direction of attention, and speed versus accuracy in performing sensorimotor skills. *Psychonomic Bulletin & Review*, *11*(2), 373–379.

Beilock, S., & McConnell, A. R. (2004). Stereotype threat and sport: can athletic performance be threatened? *Journal of Sport & Exercise Psychology*, *26*(4), 597–609.

Binkofski, F., Fink, G. R., Geyer, S., Buccino, G., Gruber, O., Shah, N. J., Taylor, J. G., Seitz, R. J., Zilles, K., & Freund, H.-J. (2002). Neural activity in human primary motor cortex areas 4a and 4p is modulated differentially by attention to action. *Journal of Neuroscience*, *88*(1), 514–519.

Blaesing, B., & Cruse, H. (2004). Stick insect locomotion in a complex environment: climbing over large gaps. *Journal of Experimental Biology*, *207*(8), 1273–1286.

Braun, S., Kleynen, M., Schols, J., Schack, T., Beursken, A., & Wade, D. (2008). Using mental practice in stroke rehabilitation: a framework. *Clinical Rehabilitation*, *22*(7), 579–591.

Braun, S. M., Beurskens, A. J., Schack, T., Marcellis, R. G., Oti, K. C., Schols, J. M., & Wade, D. T. (2007). Is it possible to use the SDA-M to investigate representations of motor actions in stroke patients? *Clinical Rehabilitation*, *21*(9), 822–832.

Carson, R. G. (2004). A simple and unified approach to human voluntary movement. *Journal of Motor Behavior*, *36*(4), 378–380.

Coombes, S. A., Cauraugh, J. H., & Janelle, C. M. (2006). Emotions and movement: activation of defensive circuitry alters the magnitude of sustained contraction. *Neuroscience Letters*, *396*(3), 192–196.

Corbetta, M. (1998). Frontoparietal cortical networks for directing attention and the eye to visual locations: identical, independent, or overlapping. *Proceedings of the National Academy of Sciences of the United States of America*, *95*(3), 831–838.

Dean, J., & Cruse, H. (1998). Motor pattern generation. In M. Arbib (Ed.), *Handbook for brain theory and neural networks* (pp. 600–605). Cambridge, MA: MIT Press.

Deubel, H., & Schneider, W. X. (1996). Saccade target selection and object recognition: evidence of a common attentional mechanism. *Vision Research*, *36*(12), 1827–1837.

Dum, R. P., & Strick, P. L. (1996). Spinal cord terminations of the medial wall motor areas in macaque monkeys. *Journal of Neurosciences*, *16*(20), 6513–6525.

Ellis, R., & Tucker, M. (2000). Micro-affordance: the potentiation of components of action by seen objects. *British Journal of Psychology*, *91*(4), 451–471.

Eysenck, M. W., & Keane, M. T. (2000). *Cognitive psychology — a student's handbook* (4th ed.). Philadelphia, PA: Psychology Press.

Fredrickson, B. (1998). What good are positive emotions? *Review of General Psychology*, *2*(3), 300–319.

Gibson, J. J. (1979). *The ecological approach to visual perception*. Boston, MA: Houghton Mifflin.

Gold, J. I., & Shadlen, M. N. (2000). Representation of a perceptual decision in developing oculomotor commands. *Nature*, *404*(6776), 390–394.

Gold, J. I., & Shadlen, M. N. (2003). The influence of behavioral context on the representation of a perceptual decision in developing oculomotor commands. *Journal of Neuroscience*, *23*(2), 632–651.

Gold, J. I., & Shadlen, M. N. (2007). The neural basis of decision making. *Annual Review of Neuroscience*, *30*(1), 535–574.

Gray, R. (2004). Attending to the execution of a complex sensorimotor skill: expertise differences, choking, and slumps. *Journal of Experimental Psychology — Applied*, *10*(1), 42–54.

Graziano, M. S. A., Taylor, C. S. R., Moore, T., & Cooke, D. F. (2002). The cortical control of movement revisited. *Neuron*, *36*(3), 349–362.

Grèzes, J., & Decety, J. (2001). Functional anatomy of execution, mental simulation, observation, and verb generation of actions: a meta-analysis. *Human Brain Mapping*, *12*(1), 1–19.

Habermas, J. (1989). *Theory of communicative action: the critique of functionalist reason*. Oxford: Blackwell Publishers.

Hanin, Y. L. (2000). *Emotions in sport*. Champaign, IL: Human Kinetics.

Heekeren, H. R., Marrett, S., Bandettini, P. A., & Ungerleider, L. G. (2004). A general mechanism for perceptual decision-making in the human brain. *Nature*, *431*(7010), 859–862.

Heekeren, H. R., Marrett, S., Ruff, D. A., Bandettini, P. A., & Ungerleider, L. G. (2006). Involvement of human left dorsolateral prefrontal cortex in perceptual decision making is independent of response modality. *Proceedings of the National Academy of Sciences of the United States of America*, *103*(26), 10023–10028.

Heekeren, H. R., Marrett, S., & Ungerleider, L. G. (2008). The neural systems that mediate human perceptual decision making. *Nature Reviews Neuroscience*, *9*(6), 467–479.

Heinen, S. J., Rowland, J., Lee, B. T., & Wade, A. R. (2006). An oculomotor decision process revealed by functional magnetic resonance imaging. *Journal of Neuroscience*, *26*(52), 13515–13522.

Henderson, J. M. (2003). Human gaze control during real-world scene perception. *Trends in Cognitive Sciences*, *7*(11), 498–504.

Hernandez, A., Zainos, A., & Romo, R. (2002). Temporal evolution of a decision-making process in medial premotor cortex. *Neuron*, *33*(6), 959–972.

Hodges, N., Huys, R., & Starkes, J. (2007). Methodological review and evaluation of research in expert performance in sport. In G. Tenenbaum & R. C. Eklund (Eds.), *Handbook of sport psychology* (3rd ed, pp. 161–183). New York: Wiley.

Hommel, B., Müsseler, J., Aschersleben, G., & Prinz, W. (2001). The theory of event coding (TEC): a framework for perception and action planning. *Behavioral and Brain Sciences*, *24*(5), 849–878.

Horwitz, G. D., Batista, A. P., & Newsome, W. T. (2004). Representation of an abstract perceptual decision in macaque superior colliculus. *Journal of Neurophysiology*, *91*(5), 2281–2296.

Janelle, C. M., Hillman, C. H., Apparies, R. J., Murray, N. P., Meili, L., Fallon, E. A., & Hatfield, D. B. (2000). Expertise differences in cortical activation and gaze behavior during rifle shooting. *Journal of Sport & Exercise Psychology*, *22*(2), 167–182.

Jeannerod, M. (2001). Neural simulation of action: a unifying mechanism for motor cognition. *Neuroimage*, *14*(1), S103–S109.

Johansen-Berg, H., & Matthews, P. M. (2002). Attention to movement modulates activity in sensori-motor areas, including primary motor cortex. *Experimental Brain Research*, *142*(1), 13–24.

Kalveram, T. (2004). The inverse problem in cognitive, perceptual, and proprioceptive control of sensorimotor behaviour: towards a biologically plausible model of the control of aiming movements. Special issue: the construction of action — new perspectives in movement science. Part I: Perceptual and cognitive control in action. *Journal of Sport & Exercise Psychology*, *2*(3), 255–273.

Kennedy, P. R., & Bakay, R. A. (1998). Restoration of neural output from a paralysed patient by a direct brain connection. *Neuroreport*, *9*(8), 1707–1711.

Kim, J. N., & Shadlen, M. N. (1999). Neural correlates of a decision in the dorsolateral prefrontal cortex of the macaque. *Nature Neuroscience*, *2*(2), 176–185.

Knoblich, G., & Jordan, S. (2003). Action coordination in individuals and groups: learning anticipatory control. *Journal of Experimental Psychology — Learning Memory and Cognition*, *29*(5), 1006–1016.

Knuf, L., Aschersleben, G., & Prinz, W. (2001). An analysis of ideomotor action. *Journal of Experimental Psychology — General*, *130*(4), 779–798.

Kowler, E., Anderson, E., Dosher, B., & Blaser, E. (1995). The role of attention in the programming of saccades. *Vision Research*, *35*(13), 1897–1916.

Lotze, M., & Halsband, U. (2006). Motor imagery. *Journal of Neurophysiology*, *99*(4–6), 386–395.

Lu, X. F., & Ashe, J. (2005). Anticipatory activity in primary motor cortex codes memorized movement sequences. *Neuron*, *45*(6), 967–973.

Mann, D., Williams, A., & Ward, P. (2007). Perceptual-cognitive expertise in sport: a meta-analysis. *Journal of Sport & Exercise Psychology*, *29*(4), 457–478.

Mechsner, F. (2004). A psychological approach to human voluntary movement. *Journal of Motor Behavior*, *36*(4), 355–370.

Mechsner, F., Kerzel, D., Knoblich, G., & Prinz, W. (2001). Perceptual basis of bimanual coordination. *Nature*, *414* (6859), 69–73.

Michaels, C. F., & Carello, C. (1981). *Direct perception*. Englewood Cliffs, NJ: Prentice-Hall.

Munzert, J., Zentgraf, K., Stark, R., & Vaitl, D. (2008). Neural activation in cognitive motor processes: comparing motor imagery and observation of gymnastic movements. *Experimental Brain Research*, *188*(3), 437–444.

Nattkemper, D., & Ziessler, M. (2004). Cognitive control of action: the role of action effects. *Psychological Research*, *68*(2–3), 71–73.

Newell, K. M. (1986). Constraints of the development of coordination. In M. G. Wade & H. T. A. Whiting (Eds.), *Motor development in children: aspects of coordination and control* (pp. 341–360). Dordrecht, NL: Martinus Nijhoff.

Newell, K. M., & McDonald, P. V. (1994). Learning to coordinate redundant biomechanical degrees of freedom. In S. Swinnen, H. Heuer, J. Massion, & P. Casaer (Eds.), *Interlimb coordination: neural, dynamical, and cognitive constraints* (pp. 517–531). New York: Academic Press.

Newman-Norlund, R. D., Bosga, J., Meulenbroek, J. G. R., & Bekkering, H. (2008). Anatomical substrates of cooperative joint-action in a continuous motor task: virtual lifting and balancing. *Neuroimage*, *41*(1), 169–177.

Newman-Norlund, R. D., Noordzij, M. L., Meulenbroek, R. G. J., & Bekkering, H. (2007). Exploring the brain basis of joint action: coordination of actions, goals and intentions. *Society Neuroscience*, *2*(1), 48–65.

Panchuk, D., & Vickers, J. N. (2006). Gaze behaviours of goaltenders under spatial temporal constraints. *Human Movement Science*, *25*(6), 733–752.

Pfeifer, R., & Bongard, J. (2006). *How the body shapes the way we think — a new view of intelligence*. Cambridge, MA: MIT Press.

Prinz, W. (2006). What re-enactment earns us. *Cortex*, *42*(4), 515–517.

Ritter, H., Kaper, M., Lenhardt, A., & Ontrup, J. (2007). Making human-machine-interfaces more brain-adequate. *International Congress Series*, *1301*, 15–21.

Ritter, H., Steil, J. J., Noelker, C., Roethling, F., & McGuire, P. (2003). Neural architectures for robotic intelligence. *Reviews in the Neurosciences*, *14*(1–2), 121–143.

Rizzolatti, G., & Craighero, L. (2004). The mirror-neuron system. *Annual Review of Neuroscience*, *27*(1), 169–192.

Rodrigues, S. T., Vickers, J. N., & Williams, A. M. (2002). Head, eye and arm co-ordination in table tennis: an exploratory study. *Journal of Sports Sciences*, *20*(3), 171–186.

Romo, R., Hernandez, A., & Zainos, A. (2004). Neuronal correlates of a perceptual decision in ventral premotor cortex. *Neuron*, *41*(1), 165–173.

Rorie, A. E., & Newsome, W. T. A. (2005). A general mechanism for decision-making in the human brain? *Trends in Cognitive Sciences*, *9*(2), 41–43.

Rosch, E. (1978). Principles of categorization. In E. Rosch & B. B. Lloyd (Eds.), *Cognition and categorization* (pp. 27–48). Hillsdale, NJ: Lawrence Erlbaum Associates.

Rosenbaum, D. A., Cohen, R. G., Jax, S. A., Van Der Wel, R., & Weiss, D. J. (2007). The problem of serial order in behavior: Lashley's legacy. *Human Movement Science*, *26*(4), 525–554.

Rowe, J., Friston, K., Frackowiak, R., & Passingham, R. (2002). Attention to action: specific modulation of cortico-cortical interactions in humans. *Neuroimage*, *17*(2), 988–998.

Sanes, J. N., & Donoghue, J. P. (2000). Plasticity and primary motor cortex. *Annual Review of Neuroscience*, *23*(1), 393–415.

Schack, T. (2003). The relationship between motor representation and biomechanical parameters in complex movements — towards an integrative perspective of movement science. *European Journal of Sport Science*, *3*(2), 1–13.

Schack, T. (2004)a. Knowledge and performance in action. *Journal of Knowledge Management*, *8*(4), 38–53.

Schack, T. (2004)b. The cognitive architecture of complex movement. Special issue: the construction of action — new perspectives in movement science. Part II: representation and planning. *Journal of Sport & Exercise Psychology*, *2*(4), 403–438.

Schack, T., & Bar-Eli, M. (2007). Psychological factors in technical preparation. In B. Blumenstein, R. Lidor, & G. Tenenbaum (Eds.), *Psychology of sport training* (pp. 62–103). Oxford: Meyer & Meyer Sport.

Schack, T., & Hackfort, D. (2007). An action theory approach to applied sport psychology. In G. Tenenbaum & R. C. Eklund (Eds.), *Handbook of sport psychology* (3rd ed., pp. 332–351). New York: Wiley.

Schack, T., & Mechsner, F. (2006). Representation of motor skills in human long-term memory. *Neuroscience Letters*, *391*(3), 77–81.

Schmidt, R., & Lee, T. (1998). *Motor control and learning — a behavioral emphasis* (3rd ed.). Champaign, IL: Human Kinetics.

Schubotz, R. I. (2007). Prediction of external events with our motor system: towards a new framework. *Trends in Cognitive Sciences*, *11*(5), 211–218.

Schubotz, R. I., & Von Cramon, D. Y. (2001). Functional organization of the lateral premotor cortex: fMRI reveals different regions activated by anticipation of object properties, location and speed. *Brain Research. Cognitive Brain Research*, *11*(1), 97–112.

Schubotz, R. I., Von Cramon, D. Y., & Lohmann, G. (2003). Auditory what, where, and when: a sensory somatotopy in lateral premotor cortex. *Neuroimage*, *20*(1), 173–185.

Sebanz, N., Bekkering, H., & Knoblich, G. (2006). Joint action: bodies and minds moving together. *Trends in Cognitive Sciences*, *10*(2), 70–76.

Stein, B., & Stanford, T. R. (2008). Multisensory integration: current issue from the perspective of the single neuron. *Nature Reviews Neuroscience*, *9*(4), 255–266.

Tipper, S., Paul, M. A., & Hayes, A. E. (2006). Vision-for-action: the effects of object property discrimination and action state on affordance compatibility effects. *Psychonomic Bulletin & Review*, *13*(3), 493–498.

Trommershäuser, J., Landy, M. S., & Maloney, L. T. (2006). Humans rapidly estimate expected gain in movement planning. *Psychological Science*, *17*(11), 981–988.

Turvey, M. T., & Kugler, P. N. (1984). An ecological approach to perception and action. In H. T. A. Whiting (Ed.), *Human motor actions — Bernstein reassessed* (pp. 373–412). Amsterdam: North-Holland.

Van der Kamp, J., Savelsbergh, G. J. P., & Rosengren, K. S. (2001). The separation of action and perception and the issue of affordances. *Ecological Psychology*, *13*(2), 167–172.

Van der Kamp, J., Savelsbergh, G. J. P., & Smeets, J. (1997). Multiple information sources in interceptive timing. *Human Movement Science*, *16*(6), 787–821.

Velliste, M., Perel, S., Spalding, C. M., Whitford, S. A., & Schwartz, B. A. (2008). Cortical control of a prosthetic arm for self-feeding. *Nature*, *453*(7198), 1098–1101.

Verschure, P. M. J. F., & Althaus, P. (2003). A real-world rational agent: unifying old and new AI. *Cognitive Science*, *27*(4), 561–590.

Vickers, J. N. (1996). Visual control when aiming at a far target. *Journal of Experimental Psychology — Human Perception and Performance*, *2*(22), 342–354.

Vickers, J. N. (2006). Gaze of Olympic speed skaters while skating at full speed on a regulation oval. Perception-action coupling in a dynamic performance environment. *Cognitive Processing*, *7*(1), 102–105.

Vickers, J. N. (2007). *Perception, cognition, and decision training: the quiet eye in action*. Champaign, IL: Human Kinetics.

Vickers, J. N., & Williams, A. M. (2007). Performing under pressure: the interactive effects of physiological arousal, cognitive anxiety and gaze control in elite biathlon shooters. *Journal of Motor Behavior*, *39*(5), 381–394.

Weigelt, M., Kunde, W., & Prinz, W. (2006). End-state comfort in bimanual object manipulation. *Experimental Psychology*, *53*(2), 143–148.

Williams, A. M., Davids, K., & Williams, J. G. (1999). *Visual perception and action in sports*. London: E & F N Spon.

Williams, A. M., Singer, R. A., & Frehlich, S. (2002). Quiet eye duration, expertise, and task complexity in a near and far aiming task. *Journal of Motor Behavior*, *34*(2), 197–207.

Williams, A. M., & Ward, P. (2003). Perceptual expertise: development in sport. In J. L. Starkes & K. A. Ericsson (Eds.), *Expert performance in sport* (pp. 219–251). Champaign, IL: Human Kinetics.

Wilson, M. (2002). Six views of embodied cognition. *Psychonomic Bulletin & Review*, *9*(4), 625–636.

Wolfensteller, U., Schubotz, R. I., & von Cramon, Y. (2007). Understanding non-biological dynamics with your own premotor system. *Neuroimage*, *36*(Suppl.), T33–T43.

Wolpert, D. M., Doya, K., & Kawato, M. (2003). A unifying computational framework for motor control and social interaction. *Philosophical Transactions of the Royal Society of London Series B — Biological Sciences*, *358*(1431), 593–602.

Wolpert, D. M., & Flanagan, J. R. (2001). Motor prediction. *Current Biology*, *11*(18), R729–R732.

Wolpert, D. M., Ghahramani, Z., & Jordan, M. I. (1995). An internal model for sensorimotor integration. *Science*, *269*(5232), 1880–1882.

Wulf, G. (2007). Attentional focus and motor learning: a review of 10 years of research. *E-Journal Bewegung und Training*, *1*(2–3), 1–11.

Wyss, R., Konig, P., & Verschure, P. F. (2004). Involving the motor system in decision making. *Proceedings. Biological sciences/The Royal Society*, *271*(3), 50–52.

M. Raab et al. (Eds.)
Progress in Brain Research, Vol. 174
ISSN 0079-6123

CHAPTER 25

Mind and motion: surveying successes and stumbles in looking ahead

Joseph G. Johnson[1], Markus Raab[2,*] and Hauke R. Heekeren[3,4]

[1]Department of Psychology, Miami University, Oxford, OH, USA
[2]Institute of Psychology, German Sport University, Cologne, Germany
[3]Max Planck Institute for Human Development, Berlin, Germany
[4]Max Planck Institute for Human Cognitive and Brain Sciences, Leipzig, Germany

Abstract: This concluding chapter is written as an exercise in evaluative and formative outcome assessment for the workshop that produced the individual content chapters in this edited volume (entitled *Mind and motion: the bidirectional link between thought and action*; 29–31 May 2008; Bielefeld, Germany). We go beyond simply summarizing and reiterating the academic progress that was made in each group that is presented in the group reports. Rather, we focus on the obstacles that we encountered along the way in hopes of recognizing what may impede future progress in the study of motor-cognitive links. In particular, we identified three key challenges that arose during our workshop discussions: the limitations of a reductionist tendency, the myopia associated with disciplinary biases, and the threats to inferential validity when dealing with mental phenomena. We offer prescriptive advice about how to overcome these limitations and provide a template for future interaction by interdisciplinary researchers who want to continue scientific research in this domain.

Keywords: perception; action; cognition; decision making; interdisciplinary research

Introduction

The behavioral sciences have a long tradition of pioneers who appreciate a systems approach to the study of behavior. For example, Egon Brunswik stressed the need for considering the environment in which an organism is embedded, noting that individual behavior did not occur — and thus could not be studied — in isolation. This connection between the organism and the environment may be apparent, and has been incorporated into psychological research in many domains including perception, cognition, and movement science. However, there still exists a relative disregard of the system existing *within* the individual; that is, the relationship between the body and the mind.

Admittedly, the relationship between the body and mind has received considerable attention in some academic disciplines — most notably in the dualist tradition of philosophy. However, in this context, the body and mind were viewed as disjoint, rather than as united components of a system of thought and action (although Kant advocated a more integrated view). Yet, as a result of the contributions of philosophy to

*Corresponding author.
Tel.: +49-221-4982-5491; Fax: +49-331-4982-8320;
E-mail: raab@dshs-koeln.de

DOI: 10.1016/S0079-6123(09)01325-9

cognitive science, this blossoming, multidisciplinary field recognizes the importance of the body and mind connection. For example, recent work on the hot topics of "embodied" and "situated" cognition suggests that our cognitive, conceptual frameworks are driven by metaphorical relations to our perceptual structures, both of which evolve in a context-dependent manner.

Researchers in cognitive psychology, neuroscience, psychophysics, movement science, robotics, social psychology, and other disciplines represented in this volume are uniquely positioned to take advantage of their collective expertise in dealing with the study of behavior as a tripartite system involving body, mind, and surroundings. This was the goal of the workshop that produced the chapters in this edited volume. The challenge of the workshop lay in integrating the vast body of knowledge in each domain to develop a systems approach. As representatives of different disciplines, we may possess differences in training or terminology, but as a group, through common interests, we should be concerned with the same issues to understand the same phenomena. However, we can only capitalize on the unique advantages of different disciplines through interaction and collaboration, often with excellent results that are mutually beneficial to all fields involved. The ZiF workshop that motivated the current volume was one example of such an endeavor.

The purpose of this concluding chapter is to summarize the progress that we see as emanating from the days spent together at the Center for Interdisciplinary Research at the University of Bielefeld, Germany. Specifically, we will first briefly summarize the consensus that each group and the workshop as a whole achieved. Second, we will discuss the obstacles that we discovered along the way, and how they allowed us to recognize our individual and collective shortcomings in preparedness to tackle such a broad and foreboding topic. Third, we offer prescriptive advice on how to overcome these complications and hindrances in future work that truly integrates various perspectives and human faculties (cognition and action). Finally, we provide an organizational roadmap that we hope is useful as guidance for future work in this arena. Throughout, our motivation is to reflect upon our experiences to inform those that would follow in our footsteps in pioneering research into the bidirectional links between the cognitive and motor systems.

Advances in knowledge gained from the workshop

The workshop format posed specific questions to each working group and assembled interdisciplinary groups of experts to construct theoretical and practical answers to these questions. These were framed in the context of a decision facing an agent immersed in a situation that required appraisal and action.

The first group was charged with identifying fundamental principles that could describe how alternative courses of action might be perceived and represented by the agent. For example, maybe these potential actions are represented as cognitive structures, motor programs, or diffuse patterns of activation. Are they represented as a collection of relative or comparative units, or as invariant absolutes? Are they comprised of "basic units" clustered into hierarchies, or as an immense assortment of individual exemplars?

The second group was charged with illuminating how an agent might recruit, manipulate, and evaluate these representations to derive a preferred course of action. Would such a transformation of input to intended output necessarily be conscious, accessible, and describable? Could it instead be more implicit and associative? Given the apparent complexity of many human actions and movements, would the mechanism(s) that produce a single action in a given situation, and perhaps inhibit others, necessarily then be complex?

Finally, the third group's task was to focus on the translation of intention into action. Whatever the nature of the representation of potential actions, and however one of these representations or actions was privileged in any sort of mental appraisal algorithm, how does the human body actually enact the physical and mental apparatus necessary to bring the physical world in line with the preferred representation? Is any sort of cognitive intervention necessary to achieve this result, or would direct perception–action links suffice in producing the same end result (and with more parsimony)?

Admittedly, these are rather lofty questions and given a few days of interaction it is understandable that conclusive, unanimous answers were not readily apparent in any of these groups. The group reports in the current volume provide a glimpse at the intellectual currency that was generated within each group; but none of these provides a definitive answer to their respective question as initially cast. Rather than view this outcome as an underachievement, we herald the progress made within each group at understanding the difficulties inherent in tackling such questions and the unique contributions and perspectives that were brought to bear. To frame the remainder of this concluding chapter, we begin with a brief review of the highlights from the workshop as a whole and the individual groups, in particular.

The first group studying how options are perceived and represented decided to open a broader perspective how the topic of mind and motion should be investigated from an interdisciplinary background. The discussion resulted in a framework that describes the dynamic bidirectional links between perception, decision making, and action. Furthermore it illustrates how personal, task, and environmental constraints may influence these links. Changes of these bidirectional links between perception, decision making, and action were described on a time axis representing short- and long-term adaptations. Finally the framework was applied to current research to convey new ideas for experimental paradigms, data interpretation, and applications (see Chapter 8: The bidirectional links between decision making, perception, and action).

The second group hoped to determine how the motor and cognitive systems necessarily worked together to arrive at decisions that might be cognitively assessed but are motorically enacted. Interestingly, this group realized that there were a number of qualifications that needed to be addressed before the question could be suitably phrased, let alone answered. As a result, the primary output of this group was an appreciation of the various levels by which decisions could be defined (see Chapter 16: How do motoric realities shape, and become shaped by, the way people evaluate and select potential courses of action? Towards a unitary framework of embodied decision making). With this understanding, one could view the answer to the original question as: “it depends.” Fortunately, this answer is not at all trivial, as the group report outlines a coherent approach to applying several successful frameworks once the appropriate level of analysis is determined.

Finally, the third group tackled the question of how actions are implemented from an interdisciplinary perspective covering both cognitive psychology and neuroscience. Similar to the other two groups, these authors felt the need to first create a common ground by sharpening the questions and by defining central concepts such as what is meant by “action implementation.” While cognitive psychologists agree on the idea that cognitive functions are implemented in the brain, cognitive models do not necessarily refer to biological aspects or principles of brain functioning, but instead concentrate on how information is processed. With the ultimate goal to link ideas on the underlying processes of action implementation taken from different fields, this group introduced four research perspectives and further elaborated these in terms of how they address the phenomenon of action implementation. An important part of this group’s answer to the question at hand was that findings from various fields show the need to abandon serial frameworks of information processing suggesting a step-by-step pattern from perception, evaluation, and selection to execution (see Chapter 24: How are actions physically implemented?).

A common theme that permeated each of these working groups was the challenge of studying a single process — mental or otherwise, observable or unobservable — in isolation. That is, each group found itself necessarily “trespassing” on the domain of the other groups. How can one describe representations in the human mind, without knowing what end these representations serve? How can one determine how a decision is made, without knowing what preceded the decision and how it will be realized? How can one specify how actions are implemented, without knowing the impetuses and goals of said action? As a result, the group outputs possess great variability, especially in the degree of relatedness to the originally posed questions. Of course, one can consult the individual chapters in this volume for details. The remainder of this chapter, then, will be dedicated to tracking and assessing the procedural quality of our workshop to provide formative input to future similar endeavors.

Speed bumps and road blocks

We are each proud of the production of each of our respective groups, and view the workshop as a success, even if we candidly must admit that the means to the end was sometimes turbulent. Granted, our approach to the workshop was risky in its ambition and format, and the payoff could have been inconsequential. Beyond the content of the group output, we believe that the most useful product of the workshop was in identifying the challenges that faced us — we now realize the speed bumps that may impede progress in this field, and the road blocks that may halt potential theoretical advances altogether. Discovering these obstacles is in itself a very productive result. Now we know the difficulties we must confront if we want to forge ahead in this area, and armed with this awareness we can avoid "treading water" or becoming bogged down with petty misunderstandings or semantic differences, for example. Encouragingly, no impasse surfaced in any of the groups. Therefore, we are confident that ultimately a true understanding of the scope and implications of bidirectional links between cognition and action can be achieved. Our path may be littered with speed bumps and road blocks, but at least we learned enough to realize that it is not a dead end. In this section, we explicitly enumerate the obstacles we encountered along the way.

Reductionism

As recognized in other chapters in this volume, perhaps the single largest threat to true and comprehensive understanding of the bidirectional links between cognition and action is the reductionist tendency. In the current workshop, this principle was manifest in many forms, none of which contributed very positively to our ultimate goals. As scientists we are often myopic and focused solely on the problems that face us directly. Even within a discipline, researchers are content, if not encouraged, to study problems limited in scope for the sake of manageability and the development of expertise. Cognitive psychology, for example, has such a broad expanse of topical interests that any single researcher can rarely possess even conversational, let alone expert, knowledge in more than a few constituent fields. To truly understand language, for example, is to understand speech production, speech comprehension, syntax, lexicon, categorization, and many other topics that individually can (and have) fueled entire careers. Directly studying the entire system is the exception rather than the rule, yet phenomena as complex as language or decision making — and indeed beings as complex as humans — are almost certainly greater than the sum of their parts. However, despite our best intentions, as disciplinarily trained scientists the default is to concern ourselves with only our chosen domain of study, and to impose our own unique perspectives on any research question.

The first, and most concrete, problematic instantiation of the reductionist philosophy was our organizing framework and the questions that it posed. In particular, this problem interacted with the disciplinary biases of us as organizers. It was pointed out explicitly in at least one of the working groups that the organizing questions suffered critically by adopting the information-processing assumptions of modern cognitive psychology. That is, by organizing the groups around themes of representation, decision, and implementation, we were *a priori* supposing the ability to reduce behavior into these constituent processes, and thereby assumed that motor-cognitive interdependencies in behavior could still be accurately studied in this manner. Entire approaches represented at the workshop, such as dynamic systems theory, strongly reject such artificial distinctions, which creates obvious tension. As mentioned earlier, this premise met with disdain in each of the groups. Furthermore, it turned out to be severely limiting and impaired the flexibility of the responses that seemed available to each group. This structure implicitly surfaced other problems as well, such as the assumption of a serial feed-forward progression across these processes. In the end, each group rejected this framework to some degree. In fact, one group report focuses solely on presenting an alternative conceptualization and thereby dismissing their "assigned" question almost entirely. Another group report goes to great length to qualify its response as taking this (disputed) position as the basis for inevitably defining a multilevel response.

The second challenge caused by reductionism is the obstacle to fruitful interaction among interdisciplinary groups. That is, because of our individual reductionist tendencies, there was fertile ground for disagreement concerning concepts that are seemingly straightforward when viewed only through the lens of a single discipline. For example, none of us probably would have predicted an inability to form a consensus on what even characterizes a "decision," or the target behavior of interest. Is a saccade a decision? In what sense is a reflex a decision? What level of consciousness, or complexity, or import is required to classify a "true decision" as having taken place? When all one studies is the saccades of primates, or a "go/no-go" task, or consumer choice, or jury deliberations, then certainly one naturally defines a "decision" only in the context of one's own relevant task. Lack of consensus in this sense also raises the issue of disciplinary differences addressed next.

Disciplinary biases

Few would dispute that a topic such as motor-cognitive interactions not only benefits from, but in fact *requires* perspectives from multiple disciplines. (Of course, if it were not for reductionism, this would not necessarily be the case: a single perspective would suffice if it alone were all-encompassing.) The second major obstacle we encountered was the disciplinary biases that each of us brought to the meeting. This presents difficulties for a number of reasons, but two of which we see as primarily manifest in the workshop.

First, each discipline demands certain ways of tackling problems, and has different criteria for "good" solutions. A persuasive argument to an economic historian might be considered "hand waving" to a biopsychologist. In contrast, compelling evidence to the neuroscientist could be disregarded as "mere correlations" with little theoretical significance to the political scientist. Consider one specific exchange that occurred in one of our groups, concerning topics such as degrees of freedom and model parameterization. To some researchers, a model's success in explaining, replicating, and predicting phenomena is sufficient grounds for endorsement. Others, however, are always concerned with criticisms regarding model flexibility and lack of parsimony. If a model's parameters are not uniquely identifiable, such as if the model contains too many parameters relative to the degrees of freedom in the data used to verify the model, it is discarded almost out of hand in some fields. A major challenge lay in reconciling these very disparate viewpoints about the quality of, and constraints on, our explanatory tools. In short, if a group cannot begin by agreeing on what constitutes a good answer, then how would they be expected to go about seeking exactly that?

Second, a discipline's methods can color how theories are formed as well as what variables and concepts are considered interesting or "worthy." Again we illustrate the point by criticizing our own development of the workshop's organizing framework. The information-processing framework of the human mind so popular in modern cognitive psychology is based on the metaphor of the mind as a computer. This stems in no small part from the introduction, evolution, and explosion of computing machines and power since World War II. Furthermore, beyond theory development, this can also play a role in determining which constructs or variables are important and worth studying and accounting for. Economists are engaged in applied settings that deal with markets and institutions; even microeconomics is concerned with behaviors on the order of the firm or household. Is it any wonder many economists have been reluctant to embrace the importance of the psychology of the individual in their theories? In this context, are human emotion and individual differences not just nuisance variables like the weather and natural disasters that are difficult to predict and beyond the explanatory goals of the models? It is not hard to imagine that the role of, for example, emotions would be of differential importance to cognitive psychology, social psychology, robotics, and neuroscience! Because representatives from each of these particular fields sat around a single table at the workshop, one can imagine the potential barriers to true collaborative and interdisciplinary discussion.

Inferential validity

The third and final general impediment we saw to achieving immediate and overwhelming progress in detailing the bidirectional links between thought and action is the lack of an ability to validate clearly claims we would make, or solutions that we would endorse. In other words, even if the groups were to have found seemingly "correct" answers to our questions, how would we ever know that we had in fact found them? As largely unobservable phenomena, many of the fundamental, decisive issues with which the groups struggled are difficult if not impossible to study empirically. Furthermore, even in instances where contrasting assertions lent themselves to empirical study, such work had not been undertaken, had not been entertained (to our knowledge), and perhaps had not even been conceived, for reasons raised earlier. Specifically, researchers in any single field might take a certain phenomenon for granted to the point that it is not even considered necessary to verify in the laboratory or the field. Perhaps a specific qualification to a phenomenon or crucial follow-up study was never identified as critically lacking. Only through the intermingling of disciplines were we able to recognize, by the virtue of another discipline's perspective, the need for addressing these situations.

Another incarnation of the threat to inferential validity arises in the context of generalizing across the specific conditions that might be used to provide evidence for any claim. Many individual chapters in the current volume (as well as many authors outside this collection) recognize and voice the need to consider the domain-specific nature of many phenomena, such as decision making. Granted, it is difficult to generalize from sterile laboratory conditions to many real-world environments, and it is easy to argue why ideally one would like to study the latter to make valid claims about true human behavior. In the limit, however, this dooms the assumption that there are any broader regularities in human behavior, that there are mental and physical mechanisms that can be applied across many situations. Essentially, the question becomes to what degree can we draw valid inferences across various domains, people, environments, time scales, etc.? Are there indeed behavioral invariants that we are trying to discover, or very specific trends?

How to overcome these obstacles?

A quick qualification is probably in order regarding the previous section before we continue. Without a doubt, we think the content outcome of the working groups was meaningful and time well spent. So, following a section that cataloged our most serious shortcomings we should at this point reiterate the purposes of such an exercise. By enumerating our challenges and critically assessing the workshop we hope to play a constructive, functional role in shaping subsequent work. In particular, in this section, we offer prescriptive advice for how to remedy the issues raised above and thereby steer clear of the most frustrating obstacles.

Beyond reductionism

We propose that to combat reductionistic tendencies and the trappings they bring to the scientific study of an issue, we need to change both the way we see ourselves as scientists as well as the way we tackle problems. First, we advocate for a style of "Renaissance researcher" that is versed in multiple disciplines and domains, as is suitable to study particular problems. Much like the drive in secondary education pedagogy to focus on problem-based learning, this emphasis should be beneficial for researchers of both the cognitive and motor systems. We envision a community where scientists are not identified by the label that is attached to their degrees, but rather by the problems with which they are engaged. Thereby researchers formerly known as sport scientists, cognitive psychologists, and neuroscientists could all instead be known as judgment bias researchers, or social-facilitation-in-sports researchers. This focus on distinct problems blurs disciplinary lines in what we believe to be a fruitful manner, and encourages individuals to obtain whatever knowledge is requisite for understanding phenomena, rather than focusing within-discipline. This community should work together daily on the same floor rather than meeting occasionally in workshops. The Center for Interdisciplinary Research in Bielefeld and other centers of interdisciplinary advanced studies should be the rule not the exception.

Second, we propose that the unit of analysis be expanded from a single hypothetical mental process to the broader scope of the behavior under examination; in particular, the largest identifiable chunk of behavior. We must resist the urge to isolate single events demarcated by convenience or academic specialization. In the current context, we must not presuppose discrete, identifiable "decision events" as the proper units of analysis (cf. Chapter 16: How do motoric realities shape, and become shaped by, the way people evaluate and select potential courses of action? Towards a unitary framework of embodied decision making). A decision event may not always have a clear onset or a clear resolution, but represent merely a temporary pattern of mental and physical behavior. We should study an entire stream of behavior to avoid excluding potentially important inputs and outputs of the mental and motor processes under study. In natural settings, we are not typically presented with an explicit set of options in tightly controlled circumstances, as some of our empirical work does. Rather, life presents itself in all of its splendid, untidy glory for us to parse (or not) and interact upon. Instead of observing, for example, the athlete's verbalizable decision when faced with a freeze-frame video of an offensive formation at a critical point, we should examine the athlete's behavior (perceptions, cognitions, and actions) during the entire development of the situation at hand. In doing so, we can follow a behavior from beginning to end to appreciate the interplay of constituent processes — if individually identifiable component processes even exist as such at all.

Appreciating disciplinary contributions

As we mentioned in the introduction of this chapter, the only way to truly foster interdisciplinary perspectives is to continue to interact and include multiple approaches. Some behavioral researchers promote the use of "adversarial collaboration," where the most enlightening flaws in one's line of reasoning are raised by one's fiercest competitors. Of course, only by first identifying shortcomings in our disciplinary modes of thought can we then correct them. We would continue to endorse the interdisciplinary nature of gatherings such as ours, and if anything would expand on this notion to include formal training in one another's means and methods. By "playing physicist for a day," we allow ourselves to be informed about the unique challenges and contributions the field of physics has to offer. Then we understand how to incorporate concepts such as dynamics, force, energy, entropy, and inertia into our own work — what are the behavioral and motoric analogs of these concepts? How might a field as seemingly esoteric (to a behaviorist) as quantum mechanics enlighten us to new avenues of thought? Could we apply Heisenberg's uncertainty principle to the study of, for example, actions and intentions rather than location and momentum? Certainly it is asking quite a lot to assume we will all earn multiple advanced degrees, but the more modest goal of short primers in one another's fields does not seem prohibitive.

A new empiricism

The toolbox of methods and metrics across disciplines exhibits considerable variability. We support a multimethod protocol for studying complex behaviors such as motor-cognitive interactions. The experimental method may be easily championed by some, but there is much to be gained from other methods such as computer simulation, neuroimaging, and self-report (not to mention introspection), to name only a few of those represented in the current volume. By combining multiple methods in the study of behavior, we can capitalize on the advantages of each method while allowing others to compensate for their shortcomings. Even within a specific field, such as neuroscience, one can use multiple methods to great benefit (e.g., combining high temporal resolution from one method with the greater spatial resolution of another).

The problem of domain specificity versus generality is a more challenging and complicated one. This is not simply a matter of basic versus applied research, or laboratory versus field research. The easiest recommendation would be to suggest including all of these types of research in our investigation of motor-cognitive links. We can study the athlete in the field to identify important environmental variables, and then target these variables in laboratory work. We can understand basic underpinnings of behavior that can then be used in worldly applications. These propositions are not novel, and are practiced everyday in many of the parent disciplines represented at the workshop.

Beyond this suggestion, however, we propose that considerable effort should be invested in a taxonomy of environmental characteristics or decision types where motor and cognitive systems are involved. Imagine if we could specify six or seven environmental primitives that could serve as orthogonal bases for constructing the majority of everyday situations. If we can accomplish this, then there is no pressing need to study a behavior in every possible application imaginable. We could then focus on these primitives and their interactions, and use interpolation and extrapolation methods to derive testable hypotheses for environments of direct interest to our individual research programs (or funding agencies!). In the interplay of thought and action, however, we are only at the initial steps of achieving something similar to the table of elements in chemistry or the laws in physics.

The road ahead

We would like to conclude this chapter by formalizing the preceding discussion into a concrete vision to organize future work in this area. We begin by summarizing from the group reports the open questions that face research on bidirectional communication and dependence between motion and cognition. Then, we offer an organizational framework for continued study in this area.

Recognizing what we have not learned

Each of the working groups contributed to the identification of outstanding issues and open questions that could not be resolved in the limited opportunity of the current workshop. The first group provided a global framework, however more specific assumptions were not yet achieved such as when do we assume sequential versus parallel processing? When are deterministic versus probabilistic models sufficient? How static or dynamic are interactions between components of the framework? The second group proposed different levels of decision, but could only speculate as to how the mind–body link influenced each absolutely and/or relatively (see Chapter 16: How do motoric realities shape, and become shaped by, the way people evaluate and select potential courses of action? Towards a unitary framework of embodied decision making). For example, in what sense do motoric realities influence very high-order, deliberate, preferential choice tasks, and how is this influence different than the influence on transient, intuitive tasks? The third group did an excellent job of formally phrasing open issues that arose in their discussions (see Chapter 24: How are actions physically implemented?): how can we define and measure knowledge structures? What role does cognition play, *per se*, beyond the explanatory power of direct perception-action links? And what about the role of emotions?

Practicing what we have learned

In this final section, we describe a potential framework that would incorporate the recommendations from the preceding discussion to "correct flaws" in the way we approach the mind–body linkage, and "break down barriers" to allow for its interdisciplinary investigation. We draw heavily on the principles described in the first group report (Chapter 8: The bidirectional links between decision making, perception, and action) and the task conceptualization of the second group report (Chapter 16: How do motoric realities shape, and become shaped by, the way people evaluate and select potential courses of action? Towards a unitary framework of embodied decision making).

We argue for a stance that is described by the four principles introduced in the first group report. First, it is necessary to truly consider the *bi*directional nature of motor-cognitive influences. In particular, the majority of progress made up to this point has focused on how the motor system may influence the performance of mental (cognitive) processes. Much less attention has been paid to the other important arrowhead — how higher-order processes influence motoric processes in a top-down fashion. By analogy, consider that the role of top-down influences has been very productive in the study of perception and classification, but has not been extended further down to the level of sensation. More ambitiously, rather than remembering to study both mind–body and body–mind links, we should realize the constant interplay of the two in a single system. Second, we believe that the joint motor-cognitive system architecture that we are studying is necessarily parallel or nonconsecutive. We learned that it is not profitable to assume a progression of activity that places differential emphasis on thought and action. Rather, at any moment in time, the whole of human activity (mental and motor) is engaged. Third, and related to the nonconsecutive tenet, is the fundamentally dynamic nature of behavior. We need to study streams, not segments, to determine how the mind and body interact over time. Finally, we must appreciate the constraints that exist on both the mind and the body. Only by understanding the realities of the limits on human mental function and physically conceivable action will we understand more than just *how* human behavior arises, but *why* it occurs the way it does.

Future study in the realm of motor-cognitive interactions will best be served, we believe, by a task-based division different than the one attempted in the current workshop. In particular, we propose that the functional division introduced by DeCaro et al. (Chapter 16: How do motoric realities shape, and become shaped by, the way people evaluate and select potential courses of action? Towards a unitary framework of embodied decision making) is a far more appropriate division of labor. Rather than have research teams study information-processing hallmarks such as representation, decision, and implementation, these processes should be jointly studied across three different decision styles. First are lower-order decisions that are almost exclusively sensory driven, where the decision representation is sensorimotor and/or proprioceptive, the decision (as such) is almost reflexive, and the enactment is served by stable motor programs. Second are associative decisions driven by representation as recognition and decision by classification, such as "if–then" rule-based learning. Finally, third are the higher-order, abstracted decisions such as preferential choice (consider voting in a presidential election), where the decision may be very deliberate, conscious, and algorithmic.

An important contribution will be to map these different types of decisions to corresponding neural systems. It is tempting to speculate that these may be ordered phylogenetically. Decisions on the lower levels may rely more on phylogenetically ancient structures such as the striatum, while higher-order, abstracted decisions will involve more recently evolved neocortical structures, such as the prefrontal cortex (PFC), to a greater degree.

What might these visions look like for a reprisal of the workshop that spawned this volume? In short, we would propose research teams guided by principles outlined by de Oliveira et al. (Chapter 8: The bidirectional links between decision making, perception, and action), consisting of disciplinary traditions reviewed by Zentgraf et al. (Chapter 24: How are actions physically implemented?) assigned to each of the three decision levels proposed by DeCaro et al. (Chapter 16: How do motoric realities shape, and become shaped by, the way people evaluate and select potential courses of action? Towards a unitary framework of embodied decision making). Specifically, we would organize a follow-up workshop by first recruiting from the same fields that contributed to the current workshop, thereby acknowledging the disciplinary diversity that produced the excellent results collected in this volume. As Zentgraf et al. (Chapter 24: How are actions physically implemented?) point out, the combination of these approaches is pleasingly complementary. Additionally, we would expand the concept of the current workshop by proposing a few days of tutorials and short courses in an attempt to better educate extradisciplinary researchers about the benefits and unique insights to be gained from each discipline. Finally, we would foster a sense of community by engaging in discussions for some length of time prior to a face-to-face workshop. In this, we envision something like a biweekly, global, technologically mediated reading group, where individual researchers would suggest papers that exemplify the relevant readings from their own field to the topic of motor-cognitive bidirectional links. This would greatly expand the preparation for the next workshop and provide a solid, common knowledge base across research teams.

With the proper mix of interdisciplinary scientists, and the foundation of a broad comprehension of the interdisciplinary literature, these teams would be prepared to work together in the antireductionist and crossdisciplinary spirit that we advocate. Of course, these teams need to work on a proper set of problems; we have learned the perils of mis-specifying the target questions. Rather than organize around an information-processing view of artificially segmented processes, we propose that each of three teams focus on one of the decision levels put forth by DeCaro et al. (Chapter 16: How do motoric realities shape, and become shaped by, the way people evaluate and select potential courses of action? Towards a unitary framework of embodied decision making). Specifically, we would propose that each group selects two specific tasks or situations that represent their particular decision level, and study these two tasks in their entirety, including the context of the task initiation, the consequences of the actions deployed to bring about resolution, and the full range of processes in between. This appreciation of the entire stream of behavior stands in stark contrast to the functional group divisions undertaken in the current workshop. Furthermore, by selecting two distinct tasks, each group could also then perform a comparative analysis to begin to understand the prospects for a domain-general versus domain-specific approach at each level.

Ultimately, we may find that disjoint, even if annual, workshops may be too transient to make the kind of consistent progress that is required to study motor-cognitive links. Even with extensive preparation, there is only so much that can be accomplished in a few days' or weeks' time. Ideally, we would propose a long-term, fully integrated and immersed working group to spend a considerable amount of consecutive time on these issues. Such a group would epitomize the ethos that we put forth in the discussion of this chapter by crossing disciplinary lines and resist carving a problem up into pieces that, while maybe convenient, preclude a true understanding of the relevant behaviors. Whichever route is taken, we highly anticipate continued interaction and persistence in pushing the bounds of our limited understanding. The journey on which we have embarked has only just begun, but we hope to here have helped provide a road map by which the bidirectional links between the mind and body, the motor and cognitive aspects of behavior, can be discovered.

Subject Index

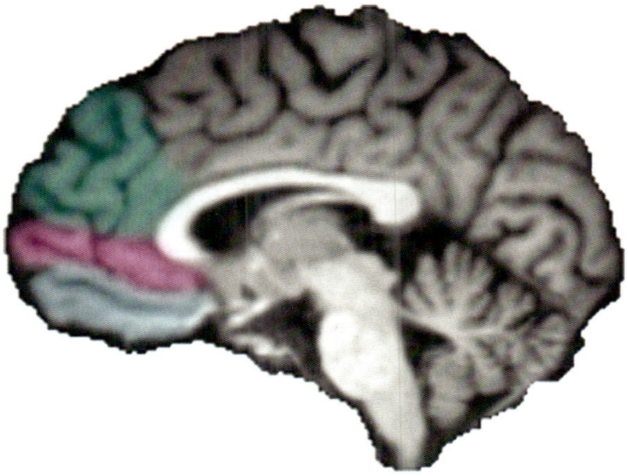

Plate 6.1. Basic prefrontal regions shown on a sagittal view of an individual brain normalized and aligned to the Talairach stereotactic space. Area shown in light blue illustrates orbitofrontal cortex (OFC), purple ventromedial prefrontal cortex (VMPFC), and green medial prefrontal cortex.

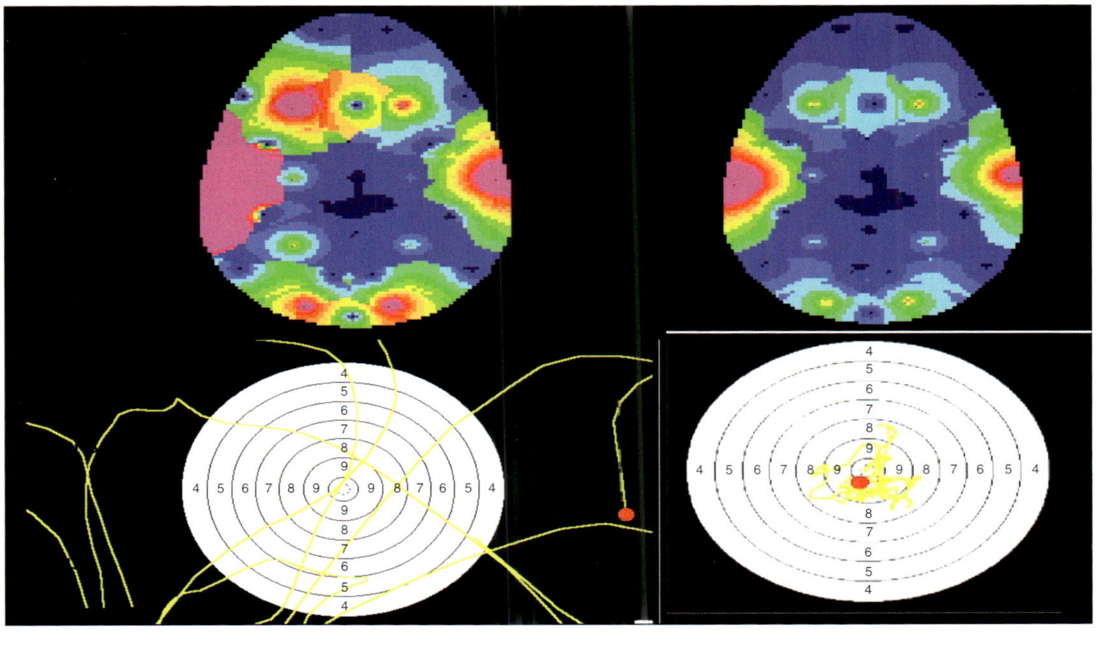

Plate 14.2. Cortical activity in novice and expert marksman.

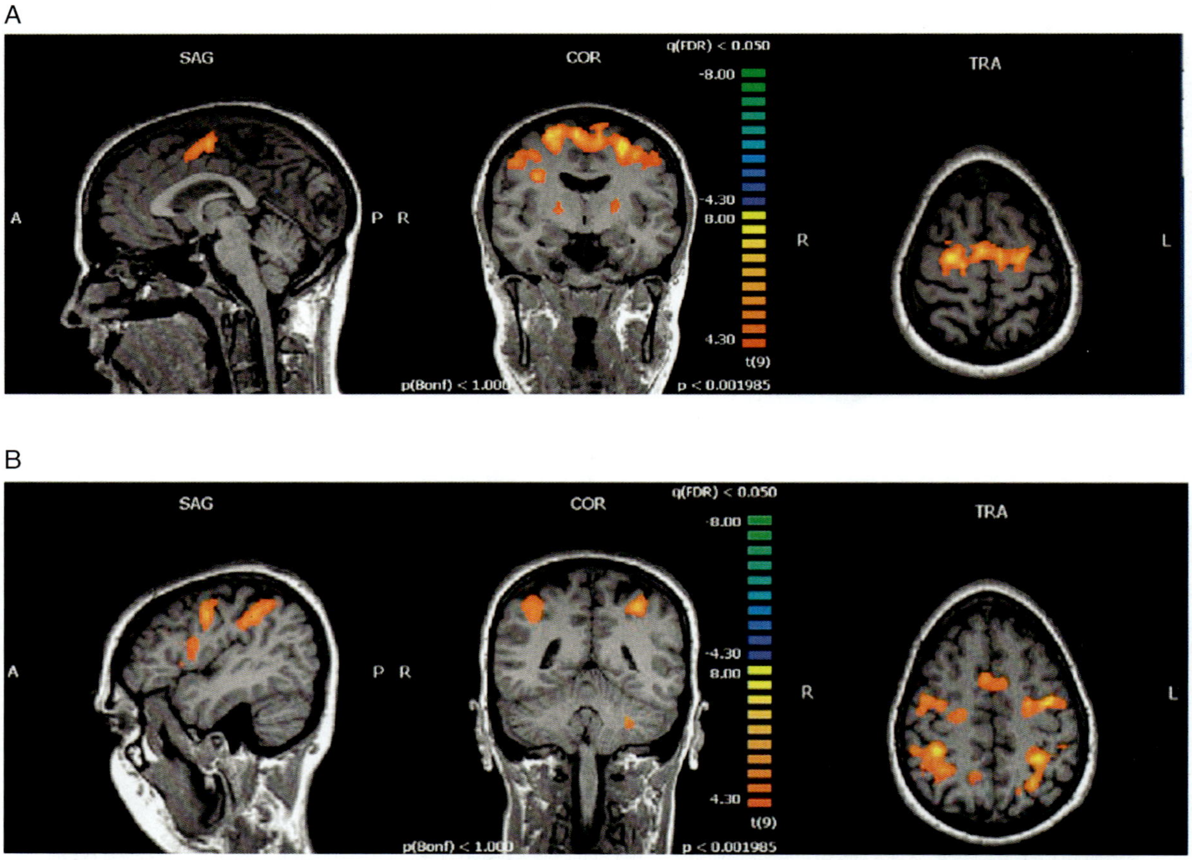

Plate 18.2. Neural activations of cortical areas in movement observation and motor imagery (MIm). SAG, sagittal; COR, coronal; TRA, transversal. Reprinted from Munzert et al. (2008) with permission.

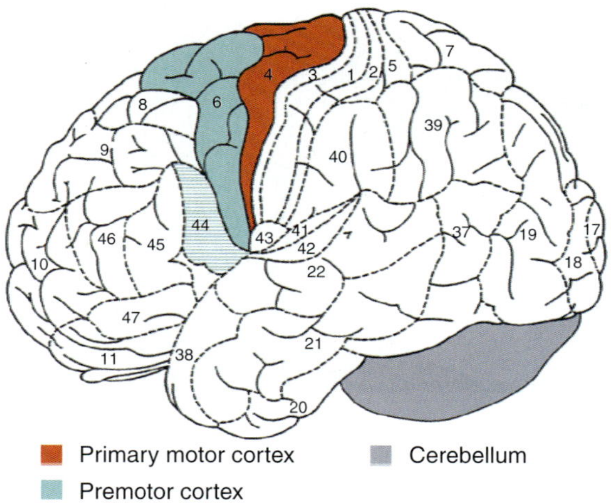

Plate 23.1. Lateral view on the left human brain showing three areas of the classical motor system, which are discussed in the present paper with respect to their role in thought and action. Modified from Nieuwenhuys et al. (1991), p. 10, Fig. 5a, © Springer Verlag Berlin Heidelberg, with kind permission of Springer Science and Business Media.